ANATOMY

OF

HATHA YOGA

ANATOMY
OF
HATHA YOGA

A Manual for Students, Teachers, and Practitioners

by

H. David Coulter

Foreword

by

Timothy McCall, M. D.

Body and Breath

Honesdale, PA, USA

Text and illustrations ©2001 by H. David Coulter
Foreword ©2001 by Timothy McCall

Body and Breath, Inc., 2114 Ames Hill Rd., Marlboro, VT 05344 USA

2017 2016 2015 2014 2013 2012 2011 2010   17 16 15 14 13 12 11 10 9 8

PRECAUTIONARY NOTE: *This is not a medical text, but a compendium of remarks concerning how anatomy and physiology relate to hatha yoga. Any medical questions regarding contraindications and cautions or any questions regarding whether or not to proceed with particular practices or postures should be referred either to health professionals who have an interest in medical problems associated with exercise, stretching, and breathing, or to hatha yoga teachers who have had experience working with medical problems in a therapeutic setting supervised by health professionals.*

Library of Congress Cataloging-in-Publication Data

Coulter, H. David (Herbert David), 1939–
    Anatomy of Hatha Yoga : a manual for students, teachers, and practitioners / by
    H. David Coulter ; foreword by Timothy McCall.
        p. cm.
    Includes bibilographical references and index.
    ISBN 978-0-9707006-1-2 (alk. paper)
        1. Yoga, Hatha—Physiological aspects. 2. Human mechanics. 3. Human anatomy.
    I. Title.

RA781.7.C685 2001
613.7′046—dc21

2001025691

To my parents,
who guided me lovingly,
watched my life with joy and enthusiasm,
supported my academic and personal interests,
and always thought the best of me.

"*Why do you never find anything written about that idiosyncratic thought you advert to, about your fascination with something no one else understands? Because it is up to you. There is something you find interesting, for a reason hard to explain. It is hard to explain because you have never read it on any page; there you begin. You were made and set here to give voice to this, your own astonishment.*"

— Annie Dillard, in *The Writing Life*

CONTENTS

FOREWORD

Hatha yoga. Its teachers and serious students are convinced of its power to build strength and confidence, to improve flexibility and balance, and to foster spiritual peace and contentment. And beyond its attributes as preventive medicine, many of us also believe in the power of yoga to heal, to aid in recovering from everything from low back strain to carpal tunnel syndrome and to help cope with chronic problems like arthritis, multiple sclerosis and infection with the human immunodeficiency virus (HIV).

But despite the recent boom in yoga's popularity, most scientists and physicians have been slow to embrace this discipline. To many of them perhaps, it seems like a mystical pursuit, a quasi-religion with little basis in the modern world of science. In a medical profession now itself dominated by a near religious reverence for the randomized, controlled study, knowledge acquired through thousands of years of direct observation, introspection, and trial and error may seem quaint.

But as the West has slowly opened in the past decades to Eastern, experientially based fields like acupuncture—as part of a greater acceptance of alternative medicine in general—yoga has begun to stake its claim. Concepts like prana or chi, however, are not warmly received by skeptical scientists. To win them over you need to provide the kind of evidence they buy. Studies. Preferably published in peer-review journals. And you need to propose mechanisms of action that conform with science as they understand it.

A significant breakthrough was provided by Dr. Dean Ornish, a California-based cardiologist who interrupted his college years to study with Sri Swami Satchidananda. His work, published in 1990 in the prestigious British medical journal the *Lancet*, showed that a program that combines hatha yoga with dietary changes, exercise, and group therapy can actually reverse blockages in the heart's main arteries—which doctors used to think wasn't possible.

In 1998, research led by Marian Garfinkel of the Medical College of Pennsylvania and published in the *Journal of the American Medical Association* found that Iyengar yoga could effectively reduce the symptoms of carpal tunnel syndrome, a malady of near epidemic proportions in this computer age. Of note, Garfinkel's study lasted only eight weeks, and yet the intervention proved efficacious. Serious yoga practitioners realize of course that although some benefit may be noticed after even a single class, yoga's most profound effects accrue over years—even decades—not weeks. Yoga is indeed powerful medicine but it is slow medicine.

More studies will be needed to convince the medical establishment, but that research could also be slow in coming. Funding is a perennial problem. Unlike the situation with, say, pharmaceuticals, there is no private industry to bankroll the scientific investigation of hatha yoga. Given the incredible cost of long-range studies—which are more likely to demonstrate effectiveness— I suspect that we're unlikely to see any time soon the kind of overwhelming proof that skeptical scientists want. This presents a philosophical question:

When you have an intervention which appears safe and effective—and when its side effects are almost entirely positive—should one wait for proof before trying it? This value judgment lies at the heart of the recent debate over many traditional healing methods.

Ironically, though, even within the world of alternative medicine yoga seems under-appreciated. Two years ago, I attended a four-day conference on alternative medicine sponsored by Harvard Medical School. A wide range of topics from herbs to prayer to homeopathy were covered in detail. Yet in the dozens of presentations I attended, yoga was mentioned just once: In a slide that accompanied the lecture on cardiovascular disease, yoga was one of several modalities listed under "Other Stress Reduction Techniques." Yoga is certainly a stress reduction device but to reduce it to just that misses so much.

Given the situation, how welcome then is David Coulter's *Anatomy of Hatha Yoga*. David combines the perspectives of a dedicated yogi with that of a former anatomy professor and research associate at two major American medical schools. He has set himself the ambitious goal of combining the modern scientific understanding of anatomy and physiology with the ancient practice of hatha yoga.

The result of an obvious labor of love, the book explains hatha yoga in demystified, scientific terms while at the same time honoring its traditions. It should go a long way to helping yoga achieve the scientific recognition it deserves. Useful as both a textbook and as a reference, *Anatomy of Hatha Yoga* is a book that all serious yoga teachers and practitioners will want on their shelves. It will also be welcomed by sympathetic physicians—and there are more of us all the time—as well as physical therapists and other health professionals. Speaking as a doctor who had already studied anatomy in detail (though forgotten more than I'd care to admit) and as a dedicated student of yoga, I can happily report that this book heightened my understanding of both hatha yoga and anatomy and—as a nice bonus—improved my personal practice.

I realize, however, that to those who lack scientific training *Anatomy of Hatha Yoga* may seem daunting. Some sections use terminology and concepts that may be challenging on first reading. If you feel intimidated, my suggestion is to adopt the mentality many employ when reading the ancient and sometimes difficult texts of the yoga tradition. Read with an open heart and if you get frustrated, try another part or come back to it another day. As with yoga itself, diligent students will be rewarded with an ever-greater understanding.

Timothy McCall, MD
Boston, Massachusetts
January, 2001

Dr. Timothy McCall is a board-certified specialist in internal medicine and the author of *Examining Your Doctor: A Patient's Guide to Avoiding Harmful Medical Care*. His work has appeared in more than a dozen major publications including the *New England Journal of Medicine*, the *Nation* and the *Los Angeles Times*. He can be found on the web at www.drmccall.com

PREFACE

The origins of this book date from twenty-five years ago when I was teaching various neuroscience, microscopic anatomy, and elementary anatomy courses in the Department of Cell Biology and Neuroanatomy at the University of Minnesota. At the same time I was learning about yoga in classes at the Meditation Center in Minneapolis. During those years, Swami Rama, who founded the Himalayan Institute, often lectured in Minnesota, and one of his messages was that yoga was neither exercise nor religion, but a science, and he wanted modern biomedical science to examine it in that light. One of his purposes in coming to the West was to bring this about, a purpose which is reflected by the name he selected for the institute that he founded—The Himalayan International Institute of Yoga Science and Philosophy. The idea of connecting yoga with modern science resonated with me, and the conviction grew that I could be a part of such a quest. Soon after I communicated my interest, Swamiji called and suggested that I pay him a visit to talk about writing a book on anatomy and hatha yoga. And that is how this project began in 1976.

Apart from several false starts and near-fatal errors, I did little writing on this subject between 1976 and 1988, but still I benefited from students' questions in courses on anatomy and hatha yoga at the University of Minnesota (Extension Division), more comprehensive courses on yoga anatomy for graduate students at the Himalayan Institute in the late 1980s, anatomy and physiology courses in the mid-1990s for the Pacific Institute for Oriental Medicine (NYC), and from 1990 to the present, teaching anatomy for students of Ohashiatsu®, a method of Oriental bodywork. These courses brought me in touch with many telling questions from students interested in various aspects of holistic medicine; without them, the seed planted by Swamiji would never have matured.

And so it went, from a working draft in the summer of 1976 to 1995, when after many gentle and not-so-gentle nudges, Swamiji insisted that my time was up, I was to finish the book, finish it now, and not run away. If I tried to escape, he avowed, he would follow me to the ends of the earth; what he would do upon finding me is better left unsaid. Happily, he saw an early but complete draft of the text a year before his passing in November of 1996.

IMAGE COLORIZATION KEY

- Fat
- Skin
- Muscle
- Tendon, ligaments, & cartilage
- Bone
- Internal organ
- Oxygenated blood
- Deoxygenated blood
- Airway
- Nerve
- Central Nervous System

INTRODUCTION

A comprehensive statement on the anatomy and physiology of hatha yoga ought to have been written years ago. But it hasn't happened, and my aim is to remedy the deficiency. After considering the subject for twenty-five years, it's clear that such a work might well interweave two themes: for the benefit of completeness, a traditional treatment of how to do yoga postures (yoga asanas) using anatomically precise terminology, and, for correlations with medical science, an objective analysis of how those postures are realized in some of the great systems of the body. In that regard, special emphasis is placed here on the musculoskeletal, nervous, respiratory, and cardiovascular systems—the musculoskeletal system because that is where all our actions are expressed, the nervous system because that is the residence of all the managerial functions of the musculoskeletal system, the respiratory system because breathing is of such paramount importance in yoga, and the cardiovascular system because inverted postures cannot be fully comprehended without understanding the dynamics of the circulation. Most of the emphasis is practical—doing experiments, learning to observe the body, and further refining actions and observations.

The discussion is intended for an audience of yoga teachers, health professionals, and anyone else who is interested in exploring some of the structural and functional aspects of hatha yoga. The work can also serve as a guide for students of alternative medicine who would like to communicate with those who place their faith more strictly in contemporary science. To help everyone in that regard I've included only material that is generally accepted in modern biomedical sciences, avoiding comment on non-physical concepts such as prana, the nadis, and the chakras, none of which are presently testable in the scientific sense, and none of which have obvious parallels in turn-of-the-millennium biology.

The book begins with an introductory discussion of some basic premises that set a philosophical tone and suggest a consistent mental and physical approach to postures. Ten chapters follow, the first three fundamental to the last seven. Chapter 1 summarizes the basic principles of the anatomy and physiology of hatha yoga. Breathing is next in chapter 2 since the manner in which we breathe in hatha yoga is important for expediting movement and posture. Breathing is followed by pelvic and abdominal exercises in chapter 3 for three reasons: many of those exercises use specialized methods of breathing, they are excellent warm-ups for other postures, and the pelvis and abdomen form the foundation of the body. Standing postures will then be covered in chapter 4 because these poses are so important for beginning students, and because they provide a preview of backbending, forward bending, and twisting postures, which are covered in detail in chapters 5, 6, and 7. The headstand and shoulderstand, including a brief introduction to cardiovascular function, are included in chapters 8 and 9. Postures for relaxation and meditation are treated last in chapter 10.

It will be helpful to experiment with each posture, preferably in the order given. This approach will lead you logically through a wealth of musculoskeletal anatomy, bring the academic discourse to life, and permit you to understand the body's architecture and work with it safely. If some of the sections on anatomy and physiology seem formidable, there is an easy solution. Turn the page. Or turn several pages. Go directly to the next section on postures, in which most of the discussion can be understood in context. Just keep in mind, however, that knowledge is power, and that to communicate effectively with laypeople who have technical questions as well as with health professionals to whom you may go for advice, it may be desirable to refer back to the more challenging sections of this book as the need arises. And those who do not find these sections particularly demanding can look to Alter's definitive *Science of Flexibility*, as well as to other sources that are listed after the glossary, if they require more technical details than are provided here.

BASIC PREMISES

The last half of the twentieth century saw many schools of hatha yoga take root in the West. Some are based on authentic oral traditions passed down through many generations of teachers. Some are pitched to meet modern needs and expectations but are still consistent with the ancient art, science, and philosophy of yoga. Still others have developed New Age tangents that traditionalists view with suspicion. Picture this title placed near the exit of your local bookstore: *Get Rich, Young, and Beautiful with Hatha Yoga.* I've not seen it, but it would hardly be surprising, and I have to admit that I would look carefully before not buying it . . . .

Given human differences, the many schools of hatha yoga approach even the most basic postures with differing expectations, and yoga teachers find themselves facing a spectrum of students that ranges from accomplished dancers and gymnasts to nursing home residents who are afraid to lie down on the floor for fear they won't be able to get back up. That's fine; it's not a problem to transcend such differences, because for everyone, no matter what their age or level of expertise, the most important issue in hatha yoga is not flexibility and the ability to do difficult postures, but awareness—awareness of the body and the breath, and for those who read this book, awareness of the anatomical and physiological principles that underlie each posture. From this awareness comes control, and from control comes grace and beauty. Even postures approximated by beginning students can carry the germ of poise and elegance.

How to accomplish these goals is another matter, and we often see disagreement over how the poses should be approached and taught. Therefore, the guidelines that follow are not set in stone; their purpose is to provide a common point of reference from which we can discuss the anatomy and physiology of hatha yoga.

FOCUS YOUR ATTENTION

Lock your attention within the body. You can hold your concentration on breathing, on tissues that are being stretched, on joints that are being stressed, on the speed of your movements, or on the relationships between breathing and stretching. You can also concentrate on your options as you move in and out of postures. Practicing with total attention within the body is advanced yoga, no matter how easy the posture; practicing with your attention scattered is the practice of a beginner, no matter how difficult the posture. Hatha yoga trains the mind as well as the body, so focus your attention without lapse.

BE AWARE OF YOUR BREATH

We'll see in chapters 2–7 that inhalations lift you more fully into many postures and create a healthy internal tension and stability in the torso. You can test this by lying prone on the floor and noticing that lifting up higher in the cobra posture (fig. 2.10) is aided by inhalation. Paradoxically, however, exhalations rather than inhalations carry you further into many other postures. You can test this by settling into a sitting forward bend and noticing that exhalation allows you to draw your chest down closer to your thighs (fig. 6.13). But in either case you get two benefits: diaphragmatic breathing assists the work of stretching the tissues, and your awareness of those effects directs you to make subtle adjustments in the posture.

While doing postures, as a general rule keep the airway wide open, breathe only through the nose, and breathe smoothly, evenly, and quietly. Never hold the breath at the glottis or make noise as you breathe except as required or suggested by specific practices.

BUILD FOUNDATIONS

As you do each asana, analyze its foundation in the body and pinpoint the key muscles that assist in maintaining that foundation: the lower extremities and their extensor muscles in standing postures; the shoulders, neck, spine (vertebral column), and muscles of the torso in the shoulderstand; and the entirety of the musculoskeletal system, but especially the abdominal and deep back muscles, in the peacock. Focus your attention accordingly on the pertinent regional anatomy, both to prevent injury and to refine your understanding of the posture.

Then there is another kind of foundation, more general than what we appreciate from the point of view of regional anatomy—the foundation of connective tissues throughout the body, especially those that bind the musculoskeletal system together. The connective tissues are like steel reinforcing rods in concrete; they are hidden but intrinsic to the integrity of the whole. To strengthen these tissues in preparation for more demanding work with postures, concentrate at first on toughening up joint capsules, tendons, ligaments, and the fascial sheaths that envelop muscles. The practical method for accomplishing these aims is to build strength, and to do this from the inside out, starting with the central muscles of the torso and then moving from there to the extremities. Aches and pains frequently develop if you attempt extreme stretches before you have first developed the strength and skill to protect the all-important joints. Unless you are already a weightlifter or body builder, stretching and becoming flexible should be a secondary concern. Only as your practice matures should your emphasis be changed to cultivate a greater range of motion around the joints.

MOVING INTO AND OUT OF POSTURES

Being in a state of silence when you have come into a posture is soothing and even magical, but you cannot connect with that state except by knowing how you got there and knowing where you're going. If you jerk from posture to posture you cannot enjoy the journey, and the journey is just as important as the destination. So move into and out of postures slowly and consciously. As you move, survey the body from head to toe: hands, wrists, forearms, elbows, arms, and shoulders; feet, ankles, legs, knees, thighs, and hips; and pelvis, abdomen, chest, neck, and head. You will soon develop awareness of how the body functions as a unit and notice quirks and discontinuities in your practice which you can then smooth out. Finally, as you learn to move more gracefully, the final posture will seem less difficult.

HONOR THE SUGGESTIONS OF PAIN

Do you honor or ignore messages from aches and pains? If you have back pain, do you adjust your posture and activities to minimize it, or do you just tough it out? And do you keep a deferential eye on your body, or do you find that you get so wrapped up in some challenge that you forget about it? If you do not listen to messages from your body you will be a candidate for pulled muscles, tendinitis, pinched nerves, and ruptured intervertebral disks. To avoid injury in hatha yoga you have to develop a self-respecting awareness.

Begin your program of hatha yoga with a resolution to avoid pain. Unless you have had years of experience and know exactly what you are doing, pushing yourself into a painful stretch will not only court injury, it will also create a state of fear and anxiety, and your nervous system will store those memories and thwart your efforts to recreate the posture. Pain is a gift; it tells us that some problem has developed. Analyze the nature of the problem instead of pushing ahead mindlessly. With self-awareness and the guidance of a competent teacher, you can do other postures that circumvent the difficulty.

CULTIVATE REGULARITY, ENTHUSIASM, AND CAUTION

Try to practice at the same time and in the same place every day. Such habits will make it easier to analyze day-to-day changes. Mornings are best for improving health—stiffness in the early morning tells you where you need the most careful work and attention. Later in the day, you lose that sensitivity and incur the risk of injury. Cultivate a frolicsome enthusiasm in the morning to counter stiffness, and cautiousness in the evening to avoid hurting yourself. And at any time, if you start feeling uncommonly strong, flexible, and frisky, be careful. That's when it is easy to go too far.

TAKE PERSONAL RESPONSIBILITY

Study with knowledgeable teachers, but at the same time take responsibility for your own decisions and actions. Your instructor may be strong and vigorous, and may urge you on, but you have to be the final arbiter of what you are capable of doing. Because many hatha yoga postures make use of unnatural positions, they expose weaknesses in the body, and it is up to you to decide how and whether to proceed. One criterion is to make sure you not only feel fine an hour after your practice, but twenty-four hours later as well. Finally, honor the contraindications for each posture and each class of postures; if in doubt, consult with a medical practitioner who has had experience with hatha yoga.

CULTIVATE PATIENCE

Learn from the tortoise. Cultivate the patience to move forward steadily, no matter how slow your progress. Remember as well that the benefits of hatha yoga go beyond getting stronger and more flexible, and that if you are monitoring only that realm, you may be disappointed. For any kind of beneficial result you have to be patient. The main culprit is thinking that you should be able to accomplish something without making consistent effort. That attitude has two unfortunate side effects: first, it diverts your attention from the work before you to what you believe you are entitled to; and second, it makes it impossible to learn and appreciate what is taking place this minute. So resolve to practice being with your experience in the present moment, enjoy yourself no matter what, and let go of expectations.

CHAPTER ONE
MOVEMENT AND POSTURE

"Every year I tell my students in my first lecture that at least half of what I am about to teach them will eventually be shown to be wrong. The trouble is I do not know which half. The future is a rough taskmaster. Nevertheless, a herd instinct often grips the imaginations of scientists. Like lemmings, we are prone to charge over cliffs when a large enough pack of us moves in that direction."

— Michael Gershon, in *The Second Brain*, p. 34.

The first organizing principle underlying human movement and posture is our existence in a gravitational field. Imagine its absence in a spacecraft, where astronauts float unless they are strapped in place, and where outside the vessel little backpack rockets propel them from one work site to another. To get exercise, which is crucial for preventing loss of bone calcium on long voyages, they must work out on machines bolted to the floor. They can't do the three things that most of us depend on: walking, running, and lifting. If they tried to partner up for workouts, all they could do is jerk one another back and forth. And even hatha yoga postures would be valueless; they would involve little more than relaxing and squirming around.

Back on earth, it is helpful to keep recalling how the force of gravity dominates our practice of hatha yoga. We tend to overlook it, forgetting that it keeps us grounded in the most literal possible sense. When we lift up into the cobra, the locust, or the bow postures, we lift parts of the body away from the ground against the force of gravity. In the shoulderstand the force of gravity holds the shoulders against the floor. In a standing posture we would collapse if we did not either keep antigravity muscles active or lock joints to remain erect. And even lying supine, without the need either to balance or to activate the antigravity muscles, we make use of gravity in other creative ways, as when we grasp our knees, pull them toward the chest, roll from side to side, and allow our body weight to massage the back muscles against the floor.

Keeping in mind that the earth's gravitational field influences every movement we make, we'll turn our attention in the rest of this chapter to the mechanisms that make movement and posture possible. First we'll look at how the skeletal muscles move the body, then we'll discuss the way the nervous system controls the operation of the skeletal muscles, and then we'll examine how connective tissues restrict movement. If we understand how these three function together within the field of gravity, we can begin to understand some of the principles underlying hatha yoga. Finally, we'll put it all together in a discussion of three postures. We'll begin with the role of skeletal muscles.

THE NEURO-MUSCULOSKELETAL SYSTEM

To any informed observer, it is plain that the musculoskeletal system executes all our acts of will, expresses our conscious and unconscious habits, breathes air into the lungs, articulates our oral expression of words, and implements all generally recognized forms of nonverbal expression and communication. And in the practice of hatha yoga, it is plainly the musculo-skeletal system that enables us to achieve external balance, to twist, bend, turn upside down, to be still or active, and to accomplish all cleansing and breathing exercises. Nevertheless, we are subtly deceived if we think that is the end of the story. Just as we see munchkins sing and dance in *The Wizard of Oz* and do not learn that they are not autonomous until the end of the story, we'll find that muscles, like munchkins, do not operate in isolation. And just as Dorothy found that the wizard kept a tether on every-thing going on in his realm, so we'll see that the nervous system keeps an absolute rein on the musculoskeletal system. The two systems combined form a neuro-musculoskeletal system that unifies all aspects of our actions and activities.

To illustrate how the nervous system manages posture, let's say you are standing and decide to sit. First your nervous system commands the *flexor muscles* (muscles that fold the limbs and bend the spine forward) to pull the upper part of the trunk forward and to initiate bending at the hips, knees, and ankles. A bare moment after you initiate that movement, gravity takes center stage and starts to pull you toward the sitting position. And at the same time—accompanying the action of gravity—the nervous system commands the *extensor muscles* (those that resist folding the limbs) to counteract gravity and keep you from falling in a heap. Finally, as soon as you are settled in a secure seated position, the nervous system permits the extensor muscles and the body as a whole to relax.

The musculoskeletal system does more than move the body, it also serves as a movable container for the internal organs. Just as a robot hous-es and protects its hidden supporting elements (power plant, integrated circuits, programmable computers, self-repairing components, and enough fuel to function for a reasonable length of time), so does the musculoskeletal system house and protect the delicate internal organs. Hatha yoga postures teach us to control both the muscles that operate the extremities and the muscles that form the container.

SKELETAL MUSCLE

The term *"muscle"* technically includes both its central fleshy part, the *belly* of the muscle, and its tendons. The belly of a muscle is composed of individual *muscle fibers* (muscle cells) which are surrounded by *connective tissue fibers* that run into a tendon. The *tendon* in turn connects the belly of the muscle to a bone.

Under ordinary circumstances muscle cells *contract*, or shorten, only because *nerve impulses* signal them to do so. When many nerve impulses per second travel to most of the individual fibers in a muscle, it pulls strongly on the tendon; if only a few nerve impulses per second travel to a smaller population of fibers within the muscle, it pulls weakly on the tendon; and if nerve impulses are totally absent the muscle is totally relaxed.

[Technical note: One of the most persistent misconceptions doggedly surviving in the biomedical community is that all muscles, even those at rest, always keep receiving at least some nerve impulses. Fifty years of electromyography with fine-wire needle electrodes is at odds with this belief, documenting from the 1950s on that it's not necessarily true, and that with biofeedback training we can learn to relax most of our skeletal muscles completely.]

A muscle usually operates on a movable *joint* such as a hinge or a ball and socket, and when a muscle is stimulated to contract by the nervous system, the resulting tension is imparted to the bones on both sides of the fulcrum of the joint. In the case of a hinge such as the elbow that opens to about 180°, any muscle situated on the face of the hinge that can close will decrease the angle between the two bones, and any muscle situated on the back side of the hinge will open it up from a closed or partially closed position. For example, the *biceps brachii* muscle lies on the inside of the hinge, so it acts to flex the *forearm* (by definition, the segment of the upper extremity between the wrist and the elbow), pulling the hand toward the shoulder. The *triceps brachii* is situated on the back side of the *arm* (the segment of the upper extremity between the elbow and the shoulder) on the outside of the hinge, so it acts to extend the elbow, or unfold the hinge (fig. 1.1).

ORIGINS AND INSERTIONS

We use the words *"origin"* and *"insertion"* to indicate where muscles are attached to bones in relation to the most common movement at a joint. The origin of a muscle is on the bone that is relatively (or usually) stationary, and the insertion of a muscle is on the bone that is most generally moved. Flexion of the elbow is again a good example. Since ordinarily the arm is fixed and the forearm is moved, at least in relative terms, we say that the biceps brachii and triceps brachii take origin from the arm and shoulder, and that they insert on the forearm (fig. 1.1).

The origins and insertions of a muscle can be functionally reversed. When the *latissimus dorsi* muscle (figs. 8.9–10) pulls the arm down and back in a swimming stroke, its textbook origin is from the lower back and *pelvis*, and its insertion is on the *humerus* in the arm. But when we do a chin-up the arm is the relatively stable origin, and the lower back and pelvis become the insertion for lifting the body as a whole. In the coming chapters we'll see many examples of how working origins and insertions are reversed.

AGONIST AND ANTAGONIST MUSCLES

The muscles surrounding a joint act cooperatively, but one of them—the *agonist*—ordinarily serves as the *prime mover*, assisted in its role by functionally related muscles called *synergists*. While the agonist and its synergists are acting on one side of the joint, muscles on the opposite side act as *antagonists*. As suggested by the name, antagonists monitor, smooth, and even retard the movement in question. For example, when the biceps brachii and the *brachialis* in the arm (the agonist and one of its synergists) shorten to flex the elbow, the triceps brachii (on the opposite side of the arm) resists flexion antagonistically while incidentally holding the joint surfaces in correct apposition (fig. 1.1).

Muscles also act in relation to the force of gravity. In the lower extremities extensor muscles act as *antigravity muscles* to keep you upright and resist crumpling to the floor. Examples: the *quadriceps femoris* muscle (figs. 1.2, 3.9, and 8.11) on the front of the *thigh* (the segment of the lower extremity between the hip joint and the knee joint) extends the knee joint as you step onto a platform, and the calf muscles extend the ankles as you lift your heels to reach an object on a high shelf. Flexor muscles are antagonists to the extensors. They can act in two ways. They often aid gravity, as when you settle into a standing forward bend and then pull yourself down more insistently with your hip flexors—the *iliopsoas* muscles (figs. 2.8, 3.7, 3.9, and 8.13). But they also act to oppose gravity: if you want to run in place the iliopsoas muscle complex flexes the hip joint, lifting the thigh and drawing the knee toward the chest; and if you want to kick yourself in the buttocks the *hamstrings* (fig. 3.8, 3.10, 8.10, and 8.12) flex the knee, pulling the *leg* (the segment of the lower extremity between the knee and the ankle) toward the thigh. Even so, the flexor muscles in the lower extremities are not classified as antigravity muscles, because under ordinary postural circumstances they are antagonists to the muscles that are supporting the body weight as a whole.

For the upper extremities the situation is different, because unless you are doing something unusual like taking a walk in a handstand with slightly bent elbows (which necessitates a strong commitment from the triceps brachii muscles), the extensor muscles do not support the weight of the body. In most practical circumstances, it is likely to be the flexors rather than extensors that act as antigravity muscles in the upper extremities, as when you flex an elbow to lift a package or complete a chin-up.

[Technical note: Throughout this book, in order to keep terminology simple and yet precise, I'll stick with strict anatomical definitions of arm, forearm, thigh, and leg, which means never using ambiguous terms such as "upper arm," "lower arm," "upper leg," and "lower leg." The same goes for the careless use of the term "arm" to encompass an undetermined portion of the upper extremity and the careless use of the term "leg" to encompass an undetermined portion of the lower extremity.]

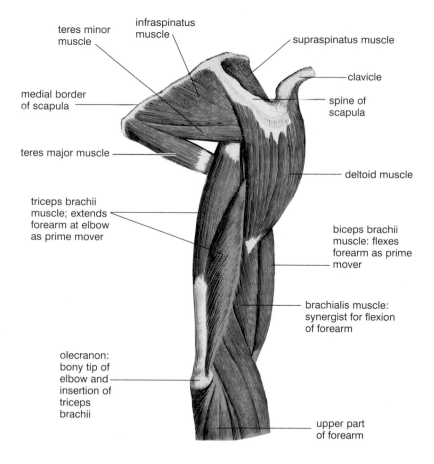

Figure 1.1. View of the right scapula, arm, and upper part of the forearm from behind and the side (from Sappey; see "Acknowledgements" for discussion of credits regarding drawings, illustrations, and other visual materials).

[And another technical note: Just to avoid confusion, I'll not use the word flex except in regard to the opposite of extend. Everyone knows what a first grader means by saying "look at me flex my muscles," but beyond this childhood expression, it can lead to ambiguity. For example, having someone "flex their biceps" results in flexion of the forearm, but "flexing" the gluteal muscles—the "gluts"—results not in flexion but in extension of the hips. For describing yoga postures it's better just to avoid the usage altogether.]

[And one more: Until getting used to terminology for movements of body parts, it is often a puzzle whether it's better to refer to moving a joint through some range of motion, or to moving the body part on the far side of the joint. For example, the choice might be between saying: extend the knee joint versus extend the leg, abduct the hip joint versus abduct the thigh, flex the ankle joint versus flex the foot, extend the elbow joint versus extend the forearm, or hyperextend the wrist joint versus hyperextend the hand. Even though the two usages are roughly equivalent, the context usually makes one or the other seem more sensible. For example, sometimes we refer specifically to the joint, as in "flex the wrist." In that case, saying "flex the hand" would be ambiguous because it could mean any one of three things: making a fist, flexing the wrist, or both in combination. On the other hand, referring to the body part is often more self-explanatory, as in "flex the arm forward 90°." Although the alternative—"flex the shoulder joint 90°"—isn't nonsensical, it's a little arcane for the non-professional.]

CONCENTRIC SHORTENING AND ECCENTRIC LENGTHENING

To understand how the musculoskeletal system operates in hatha yoga we must look at how individual muscles contribute to whole-body activity. The simplest situation, concentric contraction, or *"concentric shortening,"* is one in which muscle fibers are stimulated by nerve impulses and the entire muscle responds by shortening, as when the biceps brachii muscle in the forearm shortens concentrically to lift a book.

When we want to put the book down the picture is more complicated. We do not ordinarily drop an object we have just lifted—we set it down carefully by slowly extending the elbow, and we accomplish that by allowing the muscle as a whole to become longer while keeping some of its muscle fibers in a state of contraction. Whenever this happens—whenever a muscle increases in length under tension while resisting gravity—the movement is called *"eccentric lengthening."*

We see concentric shortening and eccentric lengthening in most natural activities. When you walk up a flight of stairs, the muscles that are lifting you up are shortening concentrically; when you walk back down the stairs, the same muscles are lengthening eccentrically to control your descent. And when you haul yourself up a climbing rope hand over hand, muscles of the upper extremities shorten concentrically every time you pull yourself up; as you come back down, the same muscles lengthen eccentrically.

In hatha yoga we see concentric shortening and eccentric lengthening in hundreds of situations. The simplest is when a single muscle or muscle group opposes gravity, as when the back muscles shorten concentrically to lift the torso up from a standing forward bend. Then as you slowly lower back down into the bend, the back muscles resist the force of gravity that is pulling you forward, lengthening eccentrically to smooth your descent.

ISOTONIC AND ISOMETRIC ACTIVITY

Most readers are already familiar with the terms "*isotonic*" and "*isometric*." Strictly speaking, the term isotonic refers solely to shortening of a muscle under a constant load, but this never happens in reality except in the case of vanishingly small ranges of motion. Over time, however, the term isotonic has become corrupted to apply generally to exercise that involves movement, usually under conditions of moderate or minimal resistance. Isometric exercise, on the other hand, refers to something more precise— holding still, often under conditions of substantial or maximum resistance. Raising and lowering a book repetitively is an isotonic exercise for the biceps brachii and its synergists, and holding it still, neither allowing it to fall nor raising it, is an isometric exercise for the same muscles. Most athletic endeavors involve isotonic exercise because they involve movement. Japanese sumo wrestling between equally matched, tightly gripped, and momentarily immobile opponents is one obvious exception. And isometric exercise is also exemplified by any and every hatha yoga posture which you are holding steadily with muscular effort.

RELAXATION, STRETCH, AND MOBILITY

If few or no nerve impulses are impinging on muscle fibers, the muscle tissue will be relaxed, as when you are in the corpse posture (fig. 1.14). But if a relaxed muscle is stretched, the situation becomes more complex. Working with a partner can make this plain. If you lie down and lift your hands straight overhead, and then ask someone to stretch you gently by pulling on your wrists, you will notice that you can easily go with the stretch provided you have good flexibility. But if your partner pulls too suddenly or if there is any appreciable pain, the nervous system will resist relaxation and keep the muscles tense; or at the least, you will sense them tightening up to resist the stretch. Finally, if you allow yourself to remain near your limit of passive but comfortable stretch for a while longer, you may feel the muscles relax again, allowing your partner to pull more insistently.

Many of these same responses are apparent if you set up similar conditions of stretching on your own, as when you place your hands overhead against a wall and stretch the underside of the arms. This is more demanding of your concentration than relaxing into someone else's work, however, because you are concentrating on two tasks at the same time: creating the necessary conditions for the stretch, and relaxing into that effort. But the same rules apply. If you go too far and too quickly, pain inhibits lengthening, prevents relaxation, and spoils the work.

MUSCULAR ACTIVITY IN A LUNGING POSTURE

To discover for yourself how skeletal muscles operate in hatha yoga, try a warrior posture (warrior I) with the feet spread wide apart, the hands stretched overhead, and the palms together (figs. 1.2 and 7.20). Feel what happens as you slowly pull the arms to the rear and lower your weight. To pull the arms up and back, the muscles facing the rear have to shorten concentrically, while antagonist muscles facing the front passively resist the stretch and possibly completion of the posture. As you lower your weight the quadriceps femoris muscle on the front of the flexed thigh resists gravity and lengthens eccentrically. Finally, as you hold still in the posture, muscles throughout the body will be in a state of isometric contraction.

Several important principles of musculoskeletal activity cannot be addressed until we have considered the nervous system and the connective tissues in detail. For now, it is enough to realize that all muscular activity, whether it be contraction of individual cells, isotonic or isometric exercise, agonist or antagonist activity, concentric shortening, or eccentric lengthening, takes place strictly under the guidance of the nervous system.

muscles facing the front resist pulling of the arms to the rear

muscles facing the rear shorten concentrically, thus pulling the arms backward

right quadriceps femoris muscle lengthens eccentrically

Figure 1.2. Warrior I pose

THE NERVOUS SYSTEM

We experience all—or at least everything pertaining to the material world—through the agency of specialized, irreplaceable cells called *neurons*, 100 billion of them in the brain alone, that channel information throughout the body and within the vast supporting cellular milieu of the *central nervous system* (the brain and spinal cord). This is all accomplished by only three kinds of neurons: *sensory neurons*, which carry the flow of sensation from the *peripheral nervous system* (by definition all parts of the nervous system excepting the brain and spinal cord) into the central nervous system and consciousness; *motor neurons*, which carry instructions from the brain and spinal cord into the peripheral nervous system, and from there to muscles and glands; and *interneurons*, or *association neurons*, which are interposed between the sensory neurons and the motor neurons, and which transmit our will and volition to the motor neurons. The sensory information is carried into the *dorsal horn* of the spinal cord by way of *dorsal roots*, and the motor information is carried out of the *ventral horn* of the spinal cord by way of *ventral roots*. The dorsal and ventral roots join to form mixed (motor and sensory) *spinal nerves* that in turn innervate structures throughout the body (figs. 1.3–9).

[Technical notes: Because this is a book correlating biomedical science with yoga, which many consider to be a science of mind, a few comments are required on a subject of perennial, although possibly overworked, philosophical interest—the nature of mind *vis-a-vis* the nervous system. Speaking for neuroscientists, I think I can say that most of us accept as axiomatic that neurons are collectively responsible for all of our thinking, cognition, emotions, and other activities of mind, and that the totality of mind is inherent in the nervous system. But I also have to say as a practicing yogi that according to that tradition, the principle of mind is separate from and more subtle than the nervous system, and is considered to be a life principle that extends even beyond the body.]

[How and whether these questions become resolved in the third millennium is anyone's guess. They are topics that are not usually taken seriously by working scientists, who usually consider it a waste of time to ponder non-testable propositions, which are by definition propositions that cannot possibly be proven wrong. Such statements abound in new age commentaries, and are a source of mild embarrassment to those of us who are trying to examine older traditions using techniques of modern science. This says nothing about the accuracy of such proposals. It may be true, for example, that "life cannot continue in the absence of prana." The problem is that short of developing a definition and assay for prana, such a statement can not be tested—it can only be accepted, denied, or argued *ad infinitum*.]

[This approach to experimentation and observation doesn't require a lot of brilliance. It simply stipulates that you must always ask yourself if the nature and content of a statement make it potentially refutable with an experimental approach. If it's not, you will be accurate 90% of the time if you conclude that the idea is spurious, even though it may sound inviting or may even appear self-evident, as did the chemical theory of phlogiston in the mid-18th century. To give the benefit of the doubt to the purveyors of such statements, it's rare that they are outright fabrications. On the other hand, one should always keep in mind that all of us (including scientists) have a huge capacity for deceiving ourselves when it comes to defending our ideas and innovations. The problem is that it's often impossible to distinguish fantasy, wishful thinking, mild exaggeration, and imprecise language from out-and-out fraud. What to do? In the end it's a waste of time to make a career of ferreting out errors—one can't get rid of bad ideas by pointing them out. On the other hand, if we turn our attention to propositions that can be tested, the creative attention this requires sometimes brings inspiration and better ideas, which in turn disposes of bad ideas by displacing them. Lavoisier discredited the theory of phlogiston by pointing to brilliant experiments (many of them carried out by others), not by crafting cunning arguments.]

[One last concern: if your complaint is that you can't understand a particular concept and do not feel competent to criticize it, don't assume that the problem is your own lack of intelligence or scientific background. More than likely, the idea wasn't presented in a straightforward manner, and it usually happens that this masks one

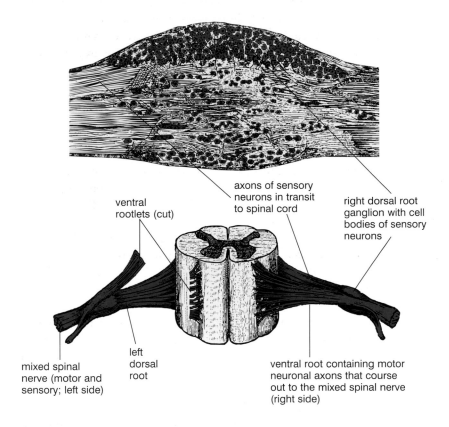

Figure 1.3. Microscopic section of dorsal root ganglion (above), and three-dimensional view of the first lumbar segment (L1) of the spinal cord, showing paired dorsal and ventral roots and mixed (motor and sensory) spinal nerves (from Quain).

or more fatal flaws in the putative reasoning. One dependable test of a concept is whether you can convincingly explain it, along with the mechanics of how it operates, to a third party. If you find yourself getting your explanation garbled, or if your listener does not comprehend your arguement or is unpersuaded, please examine and research the idea more critically, and if it still does not pass muster, either discard it or put it on the back burner. I invite the reader to hold me to these standards. To honor them, I'll limit inquiries to what we can appraise and discuss in the realm of modern biomedical science, and to refine and improve my presentation, I ask for your written input and cordial criticism.]

Returning to our immediate concerns, it is plain that neurons channel our mindful intentions to the muscles, but we still need working definitions for will and volition. In this book I'm arbitrarily defining will as the decision-making process associated with mind, and I'm defining volition as the actual initiation of the on and off commands from the cerebral cortex and other regions of the central nervous system that are responsible for commanding our actions. So "will" is a black box, the contents of which are still largely unknown and at best marginally accessible to experimentation. The nature and content of volition, by contrast, can be explored with established methods of neuroscience.

NEURONS

The neuron is the basic structural and functional unit of the nervous system. Although there are other cell types in the nervous system, namely the *neuroglia*, or "nerve glue cells," which outnumber neurons 10:1, these supporting cells do not appear, as do the neurons, to be in the business of transmitting information from place to place. So the neuron is our main interest. It has several components: a nucleated cell body that supports growth and development, and cellular extensions, or processes, some of them very long, that receive and transmit information. The cellular processes are of two types: *dendrites* and *axons*. Picture an octopus hooked on a fishing line. Its eight arms are the dendrites, and the fishing line is the axon. A typical motor neuron contains many dendrites that branch off the cell body. Its single axon—the fishing line—may extend anywhere from a fraction of an inch away from the cell body to four feet in the case of a motor neuron whose cell body is in the spinal cord and whose terminal ends in a muscle of the foot, or even fifteen feet long in the case of similar neurons in a giraffe. The axon may have branches that come off the main trunk of the axon near the cell body (*axon collaterals*), and all branches, including the main trunk, subdivide profusely as they near their targets.

Dendrites are specialized to receive information from the environment or from other neurons, and an axon transmits information in the form of nerve impulses to some other site in the body. Dendrites of sensory neurons are in the skin, joints, muscles, and internal organs; their cell bodies are in *dorsal root ganglia*, which are located alongside the spine, and their axons carry sensory information into the spinal cord (figs. 1.3–9). Dendrites of motor neurons are located in the central nervous system, and axons of motor neurons fan out from there (in peripheral nerves) to innervate muscle cells and glands throughout the body. Between the sensory and motor neurons are the association neurons, or interneurons, whose dendrites receive information from sensory neurons and whose axons contact other interneurons or motor neurons that innervate muscles (fig. 1.4). As a class, the interneurons comprise most of the neurons within the brain and spinal cord, including secondary and tertiary linking neurons that relay sensory signals to the cerebrum, projection neurons that relay motor signals from the cerebrum and cerebellum to intermediary neurons that eventually contact motor neurons of the spinal cord, and commissural neurons that connect the right and left cerebral hemispheres—that is, the "right brain" and the "left brain."

Interneurons put it all together. You sense and ultimately do, and between sensing and doing are the integrating activities of the interneurons. It's true, as the first-grade reader suggests, that you can think and do, but more often you sense, think, and do.

To operate the entire organism, neurons form networks and chains that contact and influence one another at sites called "*synapses*." Synaptic terminals of axons at such sites release chemical transmitter substances that affect the dendrite of the next neuron in the chain (fig. 1.4). The first neuron is the pre-synaptic neuron, and the neuron affected is the post-synaptic neuron. The pre-synaptic axon terminal transmits to the post-synaptic dendrite—not the other way around; it's a one-way street.

Two types of transmitter substances are released at the synapse: one *facilitates* (speeds up) the activity of the post-synaptic neuron; the other *inhibits* (slows down) the activity of the post-synaptic neuron. Thousands of axon terminals may synapse on the dendrites of one post-synaptic neuron, and the level of activity of the recipient neuron depends on its pre-synaptic input. More *facilitation* yields more activity in the post-synaptic neuron in the form of increasing numbers of nerve impulses that travel down its axon; more *inhibition* yields diminished activity. For example, the pre-synaptic input of association neurons to motor neurons either facili-tates the activity of motor neurons, causing them to fire more nerve impulses per second to skeletal muscles, or it inhibits their activity, causing them to fire fewer nerve impulses per second. The peacock posture (fig. 3.23d) requires maximum facilitation and diminished inhibition of the motor neurons that innervate the abdominal muscles, deep back muscles, muscles that stabilize the scapulae, and flexors of the forearms. On the other hand, muscular relaxation in the corpse posture (fig. 1.14) requires reduced facilitation and possibly increased inhibition of motor neurons

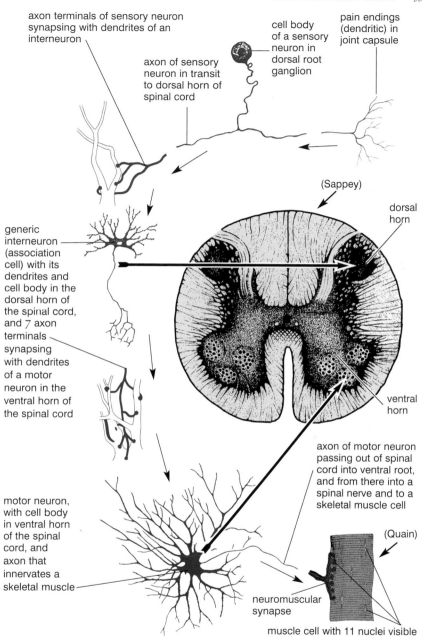

axon terminals of sensory neuron synapsing with dendrites of an interneuron

axon of sensory neuron in transit to dorsal horn of spinal cord

cell body of a sensory neuron in dorsal root ganglion

pain endings (dendritic) in joint capsule

(Sappey)

dorsal horn

generic interneuron (association cell) with its dendrites and cell body in the dorsal horn of the spinal cord, and 7 axon terminals synapsing with dendrites of a motor neuron in the ventral horn of the spinal cord

ventral horn

axon of motor neuron passing out of spinal cord into ventral root, and from there into a spinal nerve and to a skeletal muscle cell

(Quain)

motor neuron, with cell body in ventral horn of the spinal cord, and axon that innervates a skeletal muscle

neuromuscular synapse

muscle cell with 11 nuclei visible

Figure 1.4. Cross-section of the fifth lumbar segment (L5) of the spinal cord, with sensory input from a joint receptor, a generic interneuron, and motor output to a skeletal muscle cell. The small arrows indicate the direction of nerve impulses and pre- to post-synaptic interneuronal relationships. The long heavy arrows indicate the locations of the generic interneuron in the spinal cord dorsal horn and of the motor neuron in the spinal cord ventral horn.

throughout the central nervous system (see fig. 10.1 for a summary of possible mechanisms of muscular relaxation).

VOLITION: THE PATHWAYS TO ACTIVE VOLUNTARY MOVEMENT

Exercising our volition to create active voluntary movement involves dozens of well-known circuits of association neurons whose dendrites and cell bodies are in the cerebrum, cerebellum, and other portions of the brain, and whose axons terminate on motor neurons. A small but important subset of projection neurons, the subset whose cell bodies are located in the *cerebral cortex* and whose axons terminate on motor neurons in the spinal cord, are known as *"upper motor neurons"* because they are important in controlling willed activity. These are differentiated from the main class of motor neurons, the *"lower motor neurons,"* whose cell bodies are located in the spinal cord. Collectively, the lower motor neurons are called the *"final common pathway"* because it is their axons that directly innervate skeletal muscles. In common parlance, if someone refers simply to *"motor neurons,"* they are invariably thinking of lower motor neurons (fig. 1.5).

LOWER MOTOR NEURON PARALYSIS: FLACCID PARALYSIS

The best way to understand how the motor pathways of the nervous system operate is to examine the classic neurological syndromes that result from illnesses, or from injuries that have an impact on some aspect of motor function. We'll start with one of the most famous: poliomyelitis, commonly known as polio, which destroys lower motor neurons. Anyone who grew up in the 1940s and early 1950s will remember the dread of this disease. And then a miracle—the Salk vaccine—came in 1954, putting an end to the fear.

Poliomyelitis can be devastating because it destroys the lower motor neurons and deprives the muscles of nerve impulses from the spinal cord, and this results in muscular paralysis. Our power of volition in the cerebral cortex has been disconnected from the pathway of action out of the spinal cord because the final common pathway has been destroyed. In its most extreme form the resulting paralysis causes muscles to become completely flaccid, and this accounts for its medical name: *flaccid paralysis*. The same thing happens in a less global fashion when a peripheral nerve is severed or crushed. Destruction of the lower motor neurons or their axons at any site in the spinal cord or peripheral nerves causes paralysis of all their muscular targets. Will, volition, and active voluntary movement are totally frustrated.

UPPER MOTOR NEURON PARALYSIS: SPASTIC PARALYSIS

When the upper motor neurons or their axons are destroyed as in an injury or stroke (the interruption of blood supply to the brain) that destroys the motor region of the cerebral cortex, we lose much of our voluntary control

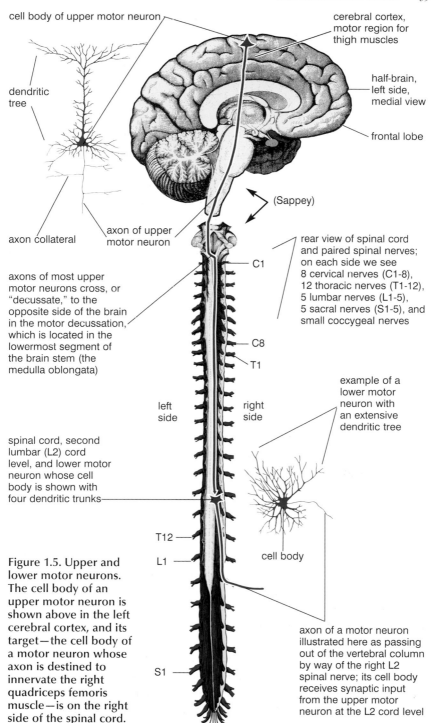

cell body of upper motor neuron

dendritic tree

axon collateral

axon of upper motor neuron

cerebral cortex, motor region for thigh muscles

half-brain, left side, medial view

frontal lobe

(Sappey)

C1

C8

T1

axons of most upper motor neurons cross, or "decussate," to the opposite side of the brain in the motor decussation, which is located in the lowermost segment of the brain stem (the medulla oblongata)

rear view of spinal cord and paired spinal nerves; on each side we see 8 cervical nerves (C1-8), 12 thoracic nerves (T1-12), 5 lumbar nerves (L1-5), 5 sacral nerves (S1-5), and small coccygeal nerves

example of a lower motor neuron with an extensive dendritic tree

left side

right side

spinal cord, second lumbar (L2) cord level, and lower motor neuron whose cell body is shown with four dendritic trunks

T12

L1

cell body

S1

axon of a motor neuron illustrated here as passing out of the vertebral column by way of the right L2 spinal nerve; its cell body receives synaptic input from the upper motor neuron at the L2 cord level

Figure 1.5. Upper and lower motor neurons. The cell body of an upper motor neuron is shown above in the left cerebral cortex, and its target—the cell body of a motor neuron whose axon is destined to innervate the right quadriceps femoris muscle—is on the right side of the spinal cord.

of the lower motor neurons, especially on the side opposite to the site of the injury. Our will can no longer be expressed actively and smoothly. The ultimate result of this, at least in severe cases in which a vascular mishap occurs at a site where the axons of other motor systems are interrupted along with those of the upper motor neurons, is not flaccid paralysis but *spastic paralysis*, in which the muscles are rigid and not easily controlled. A semblance of motor function remains because other parts of the nervous system, parts that have been spared injury, also send axon terminals to the lower motor neurons and affect motor function. The problem is that these supplemental sources of input cannot be controlled accurately, and some of them facilitate the lower motor neurons to such an extent that skeletal muscles are driven into strong and uncontrolled states of contraction. Although most of the time the condition does not result in total dysfunction, severe spastic paralysis is only mildly less devastating than flaccid paralysis; some active voluntary movements are possible, but they are poorly coordinated, especially those that make use of the distal muscles of the extremities (fig. 1.6).

SPINAL CORD INJURIES

If the entire spinal cord is severed or severely damaged at some specific level, there are two main problems. First, sensory information that comes into the spinal cord from below the level of the injury cannot get to the cerebral cortex and thereby to conscious awareness. The patient is not aware of touch, pressure, pain, or temperature from the affected region of the body. Second, motor commands from the brain cannot get to the lower motor neurons that are located below the injury. Spinal cord injuries at different levels illustrate these conditions: a spinal cord transection in the thoracic region would result in *paraplegia*—paralysis and loss of sensation in the lower extremities; and a spinal cord transection in the lower part of the neck would cause *quadriplegia*—paralysis and loss of sensation from the neck down, including all four extremities (fig. 2.12). Injuries such as these are usually the result of either automobile or sports accidents.

REFLEXES

So far our discussion has focused on neuronal connections from the top down—from our intention, to the cerebral cortex, to upper motor neurons, lower motor neurons, and skeletal muscles. But there is something else to consider, something much more primitive and elemental in the nervous system that bypasses our conscious choices: *reflexes*, or unconscious motor responses to sensory stimuli. In this context reflexes have nothing at all to do with the lightning-fast reactions ("fast reflexes") that are needed for expertise in video games or quick-draw artistry. These reactions refer to unconscious responses carried out at the spinal level.

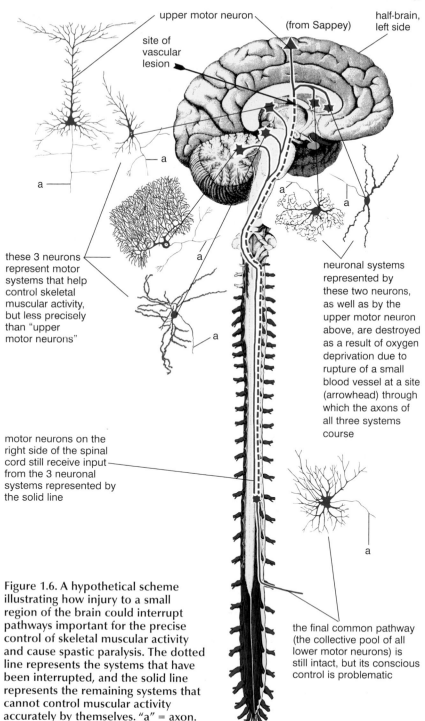

upper motor neuron

(from Sappey)

half-brain,
left side

site of
vascular
lesion

these 3 neurons
represent motor
systems that help
control skeletal
muscular activity,
but less precisely
than "upper
motor neurons"

neuronal systems
represented by
these two neurons,
as well as by the
upper motor neuron
above, are destroyed
as a result of oxygen
deprivation due to
rupture of a small
blood vessel at a site
(arrowhead) through
which the axons of
all three systems
course

motor neurons on the
right side of the spinal
cord still receive input
from the 3 neuronal
systems represented by
the solid line

Figure 1.6. A hypothetical scheme
illustrating how injury to a small
region of the brain could interrupt
pathways important for the precise
control of skeletal muscular activity
and cause spastic paralysis. The dotted
line represents the systems that have
been interrupted, and the solid line
represents the remaining systems that
cannot control muscular activity
accurately by themselves. "a" = axon.

the final common pathway
(the collective pool of all
lower motor neurons) is
still intact, but its conscious
control is problematic

Reflexes are simple. That is why they are called reflexes. They always include four elements: a sensory neuron that receives a stimulus and that carries nerve impulses into the spinal cord, an integrating center within the spinal cord, a motor neuron that relays nerve impulses back out to a muscle, and the muscular response that completes the action. More explicitly, the sensory neurons carry nerve impulses from a muscle, tendon, ligament, joint, or the skin to an integrating center in the spinal cord. This integrating center might be as simple as one synapse between the sensory and motor neuron, or it might involve one or more interneurons. The motor neuron, in its turn, innervates muscle cells that complete the action. By definition, the reflex bypasses higher centers of consciousness. Awareness of the accompanying sensation gets to the cerebral cortex after the fact and only because it is carried there independently by other circuits. There are dozens of well-known reflexes. We'll examine three, all of which are important in hatha yoga.

THE MYOTATIC STRETCH REFLEX

The *myotatic stretch reflex*, familiar to everyone as the *"kneejerk,"* is actually found throughout the body, but is especially active in antigravity muscles (fig. 1.7). You can test it in the thigh. Cross your knees so that one foot can bounce up and down freely, and then tap the *patellar tendon* just below the kneecap with the edge of your hand. Find just the right spot, and the big set of quadriceps femoris muscles on the front of the thigh will contract reflexly and cause the foot to fly up. You have to remain relaxed, however, because it is possible to override the reflex with a willed effort to hold the leg in place.

The receptors for the myotatic stretch reflex are located in the belly of the muscle, where the dendrites of sensory neurons are in contact with *muscle spindles*—specialized receptors barely large enough to be visible with the naked eye. Named for their shapes, each of these muscle spindles contains a spindle-shaped collection of specialized muscle fibers that are loaded with sensory receptors (fig. 1.7).

The reflex works this way: When you tap the patellar tendon to activate the reflex at the knee joint, the impact stretches muscle spindles in the quadriceps femoris muscle on the front of the thigh. This stretch is as fast as an eyeblink, but it nevertheless stimulates the specific sensory neurons whose dendrites end in the muscle spindles and whose axons terminate directly on motor neurons back in the spinal cord. Those axon terminals strongly facilitate the cell bodies of the motor neurons whose axon terminals stimulate the quadriceps femoris muscle, causing it to shorten and jerk the foot up. The myotatic stretch reflex is specific in that it feeds back only to the muscle in which the spindle is located.

As with all reflexes, this one takes place a fraction of a second before you

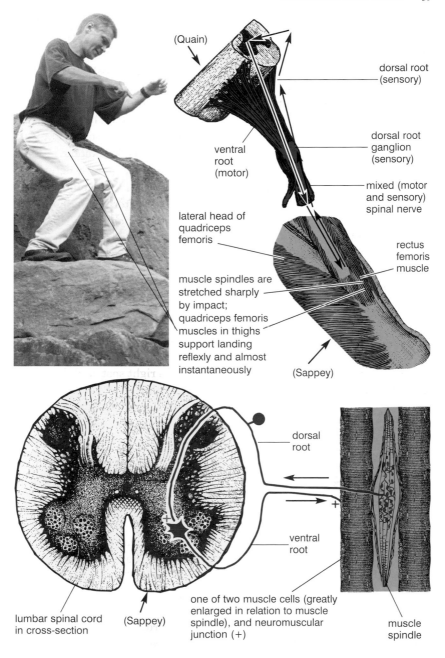

(Quain)

dorsal root (sensory)

dorsal root ganglion (sensory)

mixed (motor and sensory) spinal nerve

ventral root (motor)

lateral head of quadriceps femoris

rectus femoris muscle

muscle spindles are stretched sharply by impact; quadriceps femoris muscles in thighs support landing reflexly and almost instantaneously

(Sappey)

dorsal root

ventral root

lumbar spinal cord in cross-section

(Sappey)

one of two muscle cells (greatly enlarged in relation to muscle spindle), and neuromuscular junction (+)

muscle spindle

Figure 1.7. The myotatic stretch reflex. A 3-foot vertical jump momentarily stretches muscle spindles in all the extensor (anti-gravity) muscles of the lower extremities. The spindles then provide direct (monosynaptic) and almost immediate facilitatory input (+ in ventral horn of spinal cord) to extensor motor neurons, resulting in strong reflex contraction of the individual muscles.

are aware of it consciously. You feel it happen after the fact, after the reflex has already completed its circuit. And you notice the sensation consciously only because separate receptors for the modality of touch send messages to the cerebral cortex and thus into the conscious mind.

You can feel the myotatic stretch reflex in operation in many sports in which your muscles absorb dynamic shocks. For example, when you are water skiing on rough water outside the wake of a boat, the muscle spindles in the knee extensors of the thighs are stretched by the impact of hitting each wave, and absorbing one bump after another would quickly collapse your posture were it not for the myotatic stretch reflex. Instead, what happens is that each impact activates the reflex for the quadriceps femoris muscles in a few milliseconds, thus stabilizing the body in an upright position. You can also feel the reflex when you attack moguls aggressively on a ski slope, run down the boulder field of a mountain (fig. 1.7), or simply jump off a chair onto the floor—any activity in which an impact shocks the muscle spindles. The reflex is therefore a major contributor to what we interpret as "strength" in our dynamic interactions with gravity. Athletes depend on it far more than most of them realize.

Stimulating myotatic stretch reflexes repetitively has another important effect: it shortens muscles and diminishes flexibility. We can see this most obviously in jogging, which only mildly engages the reflexes each time your front foot hits the ground, but engages them thousands of times in a halfhour. This can cause problems if taken to an extreme, and if you tend to be tight you should always do prolonged slow stretching after a run. On the other hand, if the muscles, tendons, and ligaments are overly loose from too much stretching and too few repetitive movements, joints can become destabilized, and in such cases an activity that tightens everything down is one of the best things you can do.

In hatha yoga we usually want to minimize the effects of the myotatic stretch reflex because even moderately dynamic movements will fire the receptors, stimulate the motor neurons, shorten the muscles, and thereby limit stretch. Any dynamic movement in hatha yoga activates the myotatic stretch reflex—bouncy sun salutations, jumping in and out of standing postures, and joints and glands exercises carried off with flair and toss. These are all fine, especially as warm-ups, but if you wish to lengthen muscles and increase flexibility it is better to move into postures slowly.

THE CLASP KNIFE REFLEX

The *clasp knife reflex* acts like the blade of a pocket knife when it resists closure up to a certain point and then suddenly snaps into its folded position. It is another stretch reflex, but this one causes the targeted muscle to relax rather than contract. The stimulus for the reflex is not dynamic stretch of a muscle spindle, but contractile tension on a sensory receptor in a tendon. This tension reflexly causes the muscle attached to that tendon to relax and the joint to buckle (fig. 1.8).

The sensory receptor for the clasp knife reflex is the *Golgi tendon organ.* Most of the receptors are actually located near musculotendinous junctions, where they link small slips of connective tissue with their associated muscle fibers. The Golgi tendon organ is therefore activated by the contraction of muscles cells that are in line (in series) with the receptor. Recent studies have clarified that the Golgi tendon organ is relatively insensitive to passive stretch, but that it begins to fire nerve impulses back to the spinal cord as soon as muscle fibers start tugging on it.

And then what happens? This is the main idea: unlike the myotatic stretch reflex, here the incoming sensory axons do not terminate directly on motor neurons (which would increase their activity and stimulate a muscular contraction), but on *inhibitory interneurons* that diminish the activity of motor neurons and thereby cause the muscle to relax. If you stimulate the receptor, the reflex relaxes the muscle (fig. 1.8). It is a precise feedback loop in which the contraction of muscle fibers shuts down their own activity. This feedback loop works something like a thermostat that shuts off the heat when the temperature rises. Anecdotal reports of super-human strength in which a parent is able to lift an automobile off her child might be due to a massive central nervous system inhibition of this reflex, like a thermostat that stops working and overheats a house. In ordinary life we see the clasp knife reflex in action, at least in a gross form, when two unequally matched arm wrestlers hold their positions for a few seconds, and suddenly the weaker of the two gives way (fig. 1.8).

Whether intentional or not, we constantly make use of the clasp knife reflex while we are practicing hatha yoga. To see it most effectively and to begin to gain awareness of its utility, measure roughly how far you can come into a forward bend with your knees straight, preferably the first thing in the morning. Then bend the knees enough to flatten the torso against the thighs. Hold that position firmly, keeping the arms tightly wrapped around the thighs to stabilize the back in a comfortable position in relation to the pelvis. Then try to straighten the knees while keeping the chest tightly in place, and hold that position in an intense isometric pull for 30 seconds. This is the hamstrings-quadriceps thigh pull (fig. 1.16), and we'll examine it in more detail later in this chapter. Release the pose and then check to see how much further you can come into a forward bend with the knees straight. The difference will be a measure of how much the Golgi tendon organs "stimu-lated" the hamstring muscles to relax by way of the clasp knife reflex.

The Golgi tendon organs are sensitive to manual stimulation as well as to muscular tension. If you manipulate any musculotendinous junction in the body vigorously, its Golgi tendon organs will reflexly cause their associated muscle fibers to relax. This is one of the reasons why deep massage is relaxing. This is also why body therapists wanting to reduce tension in a specific muscle will work directly on its musculotendinous junctions. It's an old chiropractic trick—manual stimulation stimulates the clasp knife reflex almost as efficiently as contractile tension. Surprisingly, the results last for a day or two, during which time the recipient of the work has a chance to correct the offending musculoskeletal habit that gave rise to the excess tension in the first place.

Although you can test the effects of manual stimulation on tendons anywhere in the body, let's experiment with the *adductor muscles* on the inside of the thighs because tight adductors, more than any other muscles, limit your ability to sit straight and comfortably in the classic yoga sitting postures. First test your ability to sit in either the auspicious or accomplished posture (figs. 10.11 and 10.14). Then release the pose and lie with the hips butted up firmly against a wall with the knees extended and the thighs spread out as much as possible for an adductor stretch. With the help of a partner to hold your thighs abducted, try to pull the thighs together isometrically, engaging the adductors as much as possible, and at the same time stimulate the Golgi tendon organs in the adductor muscles with vigorous rubbing. Some of the adductor tendons are the cordlike structures in the inner thighs near the genitals. Others are more flattened and are located further to the rear. All of them take origin from a pair of bones, the *inferior pubic rami* (fig. 1.12), that together form the rear-facing V which accommodates the genitals.

As you massage the adductors for a minute or so while keeping them under tension, you will feel them gradually release, as evidenced by being able to abduct the thighs more completely. Then sit up and check for improvement in your sitting posture. The combination of massaging the adductor tendons plus making an isometric effort with stretched adductors powerfully inhibits the motor neurons that innervate these muscles, and this allows them to release and permits you to sit straighter and more comfortably.

The hamstrings-quadriceps thigh pull and the adductor massage give us obvious examples of how the clasp knife reflex operates. It is also invoked in a milder form any time you are able to stay comfortable in an active posture for more than 10–15 seconds, which is what we often do in hatha yoga. In this case don't bounce unless you want to induce the myotatic stretch reflex, and don't take a posture into the discomfort zone unless you are prepared to trigger flexion reflexes, which we'll discuss next.

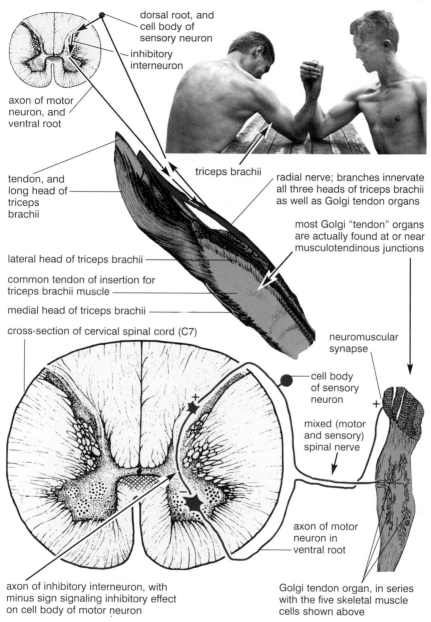

dorsal root, and cell body of sensory neuron

inhibitory interneuron

axon of motor neuron, and ventral root

tendon, and long head of triceps brachii

triceps brachii

radial nerve; branches innervate all three heads of triceps brachii as well as Golgi tendon organs

most Golgi "tendon" organs are actually found at or near musculotendinous junctions

lateral head of triceps brachii

common tendon of insertion for triceps brachii muscle

medial head of triceps brachii

cross-section of cervical spinal cord (C7)

neuromuscular synapse

cell body of sensory neuron

mixed (motor and sensory) spinal nerve

axon of motor neuron in ventral root

axon of inhibitory interneuron, with minus sign signaling inhibitory effect on cell body of motor neuron

Golgi tendon organ, in series with the five skeletal muscle cells shown above

Figure 1.8. The clasp knife reflex. Muscular effort stimulates Golgi tendon organs, whose sensory input to the spinal cord activates inhibitory interneurons (+ in dorsal horn); the inhibitory interneurons then inhibit motor neurons (– in ventral horn), resulting in fewer nerve impulses per second to the skeletal muscle cells (+ effects at the neuromuscular synapse are minimized). Final result is relaxation of the muscle, or in this case, loss of the armwrestling match (Sappey).

FLEXION REFLEXES

The flexion reflexes (fig. 1.9) are pain reflexes. If you inadvertently touch a hot skillet you jerk your hand back reflexly. You don't have to think about it, it just happens. As with the other reflexes, awareness comes a moment later. Flexion reflexes are more complex than stretch reflexes, but they are easier to comprehend because pain is such an obvious part of everyone's conscious experience. Even if it is no more than a feeling of stretch that went too far while you were gardening, a pain in the knee or hip that developed after a strenuous hike, or a neck problem you didn't notice until you started to turn too far in a certain direction, with rare exceptions your automatic response will be flexion. You may be only vaguely aware of the reflex itself, but you will certainly be aware of the fear and tension that accompanies it.

The sensory neurons (including their axons) that carry the modalities of pain and temperature conduct their nerve impulses more slowly than those that activate the myotatic stretch reflex. What is more, flexion reflexes are polysynaptic—that is, they involve one or more interneurons in addition to the sensory and motor neurons—and each synapse in the chain of neurons slows down the speed of the reaction. You can estimate the conduction time for temperature by licking your finger and touching a coffee pot that is hot enough to hurt but not hot enough to cause injury. It will take almost a second for the sensation to reach consciousness from a finger, well over a second from a big toe, and, for the adventuresome, about a tenth of a second from the tip of the nose. Such slow conduction times from the extremities would not serve the myotatic stretch reflex. If, for that reflex, it took a full second for nerve impulses to reach the spinal cord, you would be in serious trouble jumping off a platform onto the floor with bent knees—you would collapse and shatter your kneecaps before the extensor muscles could react enough to support your weight.

Like the two stretch reflexes we have just considered, the motor reflexes for flexion are spinal, not cerebral. So even if the spinal cord were cut off from the brain, the flexion reflex would still withdraw a foot from a toxic stimulus. That's why neurologists have little reason to be encouraged when the foot of a patient with a spinal cord injury responds to a pinch.

RECIPROCAL INHIBITION

Flexion reflexes not only activate flexor muscles to pull the hand or the foot toward the torso, they also relax the extensors, which then allows flexion to take place freely. This is done through the agency of inhibitory inter-neurons. While facilitatory interneurons impinge on motor neurons that innervate flexors, thus causing them to contract, inhibitory interneurons impinge on motor neurons that innervate extensors, causing them to relax.

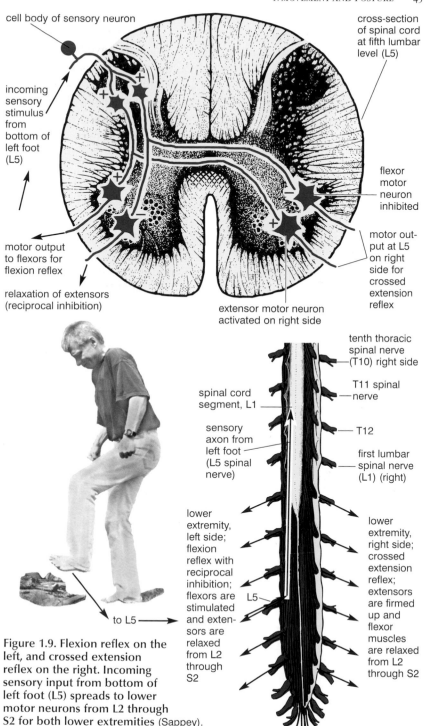

Figure 1.9. Flexion reflex on the left, and crossed extension reflex on the right. Incoming sensory input from bottom of left foot (L5) spreads to lower motor neurons from L2 through S2 for both lower extremities (Sappey).

The phenomenon is known as *reciprocal inhibition*, and it is an integral part of the flexion reflex (fig. 1.9).

Unlike stretch reflexes, flexion reflexes create effects well beyond the site of the stimulus. We can see this when a nurse pricks a child's index finger with a needle to draw blood. The child's entire upper extremity reacts, not just the flexors in the offended finger. A vehement jerk backward shows that the flexion reflex facilitates flexor motor neurons and inhibits extensor motor neurons for the entire upper extremity.

THE CROSSED-EXTENSION REFLEX

The *crossed-extension reflex* adds yet another ingredient to flexion reflexes— a supporting role for the opposite side of the body. Through the agency of this reflex, as the extremity on the injured side flexes, the extremity on the other side extends. This would happen if you stepped on a hot coal near a campfire. You don't have to think about either reflex; you lift your injured foot in a hurry, contracting flexors and relaxing extensors on that side— everything at the same time—toe, ankle, knee, hip, and even the torso. And as the injured foot lifts, the crossed-extension reflex contracts the extensors and relaxes the flexors on the opposite side of the body, strengthening your stance and keeping you from toppling over into the fire (fig. 1.9).

The crossed-extension reflex is accomplished by interneurons whose axons cross to the opposite side of the spinal cord and innervate motor neurons in a reverse pattern from that seen on the side with the injury—motor neurons for extensor muscles are facilitated, and motor neurons for flexor muscles are inhibited.

The flexion reflexes also serve many other protective functions. For example, if you sprain your ankle badly, the pain endings become more sensitive. The next time you start to turn your ankle, the higher centers in the brain associated with volition and consciousness allow the spinal flexion reflexes to act unencumbered and collapse the limb before your weight drops fully on the vulnerable joint. This prevents re-injury. A "trick" knee demonstrates the same mechanisms—an old injury, a sudden physical threat, unresistant higher centers, and unconscious flexion reflexes. Your bad knee buckles quickly, and you are saved from more serious injury.

RECIPROCAL INHIBITION AND A STIFF BACK

Since they restrain movement, flexion reflexes usually influence hatha postures negatively, but there are certain situations in which we can use them to our advantage. If you tend to be stiff and not inclined to forward bending, try this experiment early in the morning. First, for comparison, slowly lower into a standing forward bend with your fingers extended. Notice that you hesitate to come forward smoothly. This may happen even if you are in excellent condition but not warmed up. The deep back muscles are extensors of the back; they lengthen eccentrically as you bend, resisting forward movement and only allowing you come into the posture with creaks and wariness. Come up. Next, holding the elbows partially flexed, flex your fingers tightly, making fists. Holding that gesture, come forward again. You will immediately notice that you do this more easily and smoothly than when your fingers and elbows were extended. Come up, and repeat the exercise to your capacity.

Making your hands into fists reciprocally inhibits the motor neurons that innervate the deep back muscles. If you are in good condition this merely helps you come forward more smoothly and confidently, but if your back is tense from excess muscle tone in the deep back muscles you will be amazed at how much the simple act of flexing your fingers into fists eases you into a relaxed bend.

Why might your back be stiff from excess muscle tone? It is usually because of pain that causes the back muscles to become taut and act as a splint to prevent movement. This is fine for a while as a protective measure, but at some stage it becomes counterproductive and leads to other problems. When stiffness and mild back pain emerge, you need enough muscle tone to prevent re-injury, it is true, but you do not need enough to lock you up for months on end. The reciprocal inhibition caused by making your hands into fists helps relax the extensor muscles in the back and allows you to ease further into a forward bend. If your back is chronically stiff, but not in acute pain, you can benefit by doing this exercise several times a day.

THE VESTIBULAR SYSTEM, SIGHT, AND TOUCH

So far we have seen how motor neurons drive the musculoskeletal system, how association neurons channel our will to the motor neurons, and how sensory input from muscles, tendons, and pain receptors participate with motor neurons in simple reflexes. But that's only the beginning. Many other sources of sensory input also affect motor function. Some of the most important are the vestibular sense, sight, and touch.

THE VESTIBULAR SENSE

We have little conscious awareness of our *vestibular sense* even though it is critical for keeping us balanced in the field of gravity. Its receptors lie close to the organ of hearing—the *inner ear*—in little circular tubes called *semicircular canals* and in a little reservoir called the *utricle*, all of which are embedded in the bony region of the skull just underneath the external ear. The semicircular canals and the utricle are all involved with maintaining our equilibrium in space, but within that realm they are sensitive to different stimuli—the semicircular canals to rotary acceleration, and the utricle to linear acceleration and to our orientation in a gravitational field. They also participate in different reflexes: the semicircular canals coordinate eye movements, and the utricles coordinate whole-body postural adjustments.

Except for pilots, dancers, ice skaters, and others who require an acute awareness of equilibrium, most of us take the vestibular system for granted. We don't notice it because it does almost all of its work reflexly, feeding sensory information into numerous motor circuits that control eye and body movements.

Because the semicircular canals are sensitive to rotary acceleration, they respond when we start or stop any spinning motion of the body. One of their several roles is to help us maintain our equilibrium by coordinating eye movements with movements of the head. You can experience these if you sit cross-legged on a chair or stool that can rotate, tip your head forward about 30°, and have some assistants turn you around and around quickly for 30–40 seconds. Make sure you keep balanced and upright. Don't lean to the side or you will be pitched off onto the floor. Then have your assistants stop you suddenly. You eyes will exhibit little jerky movements known as *nystagmus*, and you will probably feel dizzy. Children play with this reflex when they spin themselves until they get dizzy and fall down. The sensation they describe as the world "turning" is due to nystagmus. The perception is disorienting at first but it slows down and stops after a while.

The receptors in the semicircular canals stop sending signals after about 30 seconds of spinning, which is why you have your assistants rotate you for that period of time. It is also why the reaction slows down and stops in 30 seconds after you are abruptly stopped. Third-party observers obviously cannot observe nystagmus during the initial period of acceleration while you are being spun around. To observe these eye movements in a practical setting, we must rely on what we call *post-rotatory nystagmus*, the eye movements that occur after you have been stopped suddenly.

The neurological circuitry for nystagmus is sensitive to excessive alcohol, and this is why highway patrol officers ask suspected drunks to get out of the car and walk a straight line. If the suspect is suffering from alcohol-induced nystagmus, the ensuing dizziness is likely to make walking straight impossible. Spontaneous (and continuing) forms of nystagmus that are not induced by drugs or alcohol may be symptomatic of neurological problems such as a brain tumor or stroke.

Occasionally students in hatha yoga classes are sensitive to dizziness when they do neck exercises. They may have had such problems from childhood or they may just not be accustomed to the fact that they are stimulating their semicircular canals when they rotate their head. And even otherwise healthy students who are just getting over a fever may be sensitive to dizziness. In any case, anyone who is sensitive should always do neck exercises slowly.

The second component of the vestibular organ, the utricle, detects two modalities: speeding up or slowing down while you are moving in a straight line, and the static orientation of the head in space. The rush of accelerating or decelerating a car is an example of the first case. As with the semicircular canals, stimulation ends after an equilibrium is established, whether sitting still or going 100 miles per hour at a constant rate on a straight road. The utricles also respond to the orientation of the head in the earth's gravitational field—an upright posture stimulates them the least and the headstand stimulates them the most. The receptors in the utricle adapt to the stimulus of an altered posture after a short time, however, which is why it is so important for pilots of small planes to depend on instruments for keeping properly oriented in the sky when visual feedback is absent or confusing. For example, a friend of mine was piloting a small plane and flew unexpectedly into a thick bank of clouds. Instantly lost and disoriented, and untrained in flying on instruments, he calculated that he would just make a slow 180° turn. Unfortunately, after having made the turn and exiting the clouds, he was shocked to see that he was headed straight toward the ground. Fortunately, he had enough airspace to pull out of the dive.

In ordinary circumstances on the ground, the receptors in the utricle do more than sense the orientation of the head in space: they trigger many whole-body postural reflexes that maintain our balance. This is the source of the impulse to lean into curves while you are running or cycling around a track. We also depend on the utricle for underlying adjustments of hatha yoga postures that we trigger when we tilt the head forward, backward, or to one side. Every shift of the head in space initiates reflexes that aid and abet many of the whole-body postural adjustments in the torso that we take for granted in hatha yoga.

The well-known righting reflexes in cats can give us a hint of how the vestibular system influences posture in humans. If you want to see these reflexes operate, drop an amicable cat, with its legs pointed up, from as little as a few inches above the floor. It will turn with incredible speed and land on all four feet, even if it has been blindfolded. Careful study reveals a definite sequence of events. The utricle first detects being upside down, and then it detects the falling sensation of linear acceleration toward the floor. In response to this the cat automatically rotates its head, which stimulates neck muscles that in turn leads to an agile twisting around of the rest of the body and a nimble landing on all four feet. The cat does all this in a fraction of a second. Comparable reflexes also take place in human beings, although they are not as refined as in cats.

SIGHT

When we are moving we are heavily dependent on vision, as anyone can attest who has stepped off a curb unawares or thought erroneously that one more step remained in a staircase. This is true to a lesser extent when we are standing still. If you stand upright with your feet together and your eyes open, you can remain still and be aware that only minuscule shifts in the muscles of the lower extremities are necessary to maintain your balance. But the moment you close your eyes you will experience more pronounced muscular shifts. For an even more convincing test, come into a posture such as the tree or eagle with your eyes open, establish your balance fully, and then close your eyes. Few people will be able to do this for more than a few seconds before they wobble or fall.

Visual cues are especially important while coming into a hatha yoga posture, but once you are stable you can close your eyes in most poses without losing your balance provided your vestibular system and joint senses are healthy. On the other hand, if you want to study your body's alignment objectively you can do it only by watching your reflection in a mirror. It is all too easy to deceive yourself if you depend purely on your muscle- and joint-sense to establish right-left balance.

THE SENSE OF TOUCH

The sense of touch brings us awareness of the pleasure and luxury of comfortable stretch, and because of this it is the surest authority we have for telling us how far to go into a hatha yoga posture. The vestibular reflexes and vision help with balance, and pain tells us how far not to go in a stretch. But the sense of touch is a beacon. It both rewards and guides.

The modality of touch includes *discriminating touch*, *deep pressure*, and *kinesthesis*. All three are brought into conscious awareness in the cerebral cortex, and along with stretch reflexes, vision, and the vestibular sense, they make it possible for us to maintain our balance and equilibrium. Discriminating touch is sensed by receptors in the skin, and deep pressure is sensed by receptors in fasciae and internal organs. Kinesthesis, which is the knowledge of where your limbs are located in space, as well as the awareness of whether your joints are folded, straightened, stressed, or comfortable, is sensed mostly by receptors in joints. If you lift up in a posture such as the prone boat and support your weight only on the abdomen, you can feel all three aspects of touch—contact of the skin with the floor, deep pressure in the abdomen, and awareness of extension in the spine and extremities.

Touch receptors adapt even more rapidly than receptors in the vestibular system, which means that they stop sending signals to the central nervous system after a few seconds of stillness. That's why holding hands with someone gets boring in the absence of occasional squeezing and stroking. Without movement, the awareness of touch disappears. Rapid adaptation to touch is extremely important in hatha yoga postures, relaxation, and meditation. If your posture is stable, the receptors for touch stop sending signals back to the brain and you are able to focus your attention inward, but as soon as you move the signals return and disturb your state of silence.

TOUCH AND THE GATE THEORY OF PAIN

If you bump your shin against something hard, rubbing the injured region alleviates the pain, and if your knee hurts from sitting for a long time in a cross-legged posture, the natural response is to massage the region that is hurting. There is a neurological basis for this—the *gate theory of pain*, according to which the application of deep touch and pressure closes a "gate" to block the synaptic transmission of pain in the spinal cord. Although it has not been possible to substantiate this theory as it was initially proposed, we all know experientially that somehow it works. So even though the mechanism is still uncertain, the general idea is widely accepted as self-evident—somewhere between the spinal cord and the cerebral cortex, touch and pressure pathways intersect with the ascending pathways for pain and either block or minimize its perception.

We use this principle constantly in hatha yoga. To illustrate, interlock your hands behind your back and press the palms together. Pull them to the rear so they do not come in contact with the back, and come into a forward bend. If you are not warmed up you may notice that you feel mild discomfort from the stretch. Now come up, press the forearms firmly against the back on either side of the spine, and come forward again. The contrast will be startling. The sensation of deep touch and pressure against the back muscles stops the discomfort immediately.

Is this good or bad? That is a vital question, and one of the challenges of hatha yoga is to learn how far this principle can safely be taken. If you underestimate the importance of the signals of pain, and diminish that pain with input from touch and pressure, you may injure joints and tissues. But if you baby yourself, you'll never progress. The answer, unfortunately, is that you may not know if you have gone too far until the next morning. If you are sore you know you misjudged.

CONNECTIVE TISSUE CONSTRAINTS

Our bodies are made up of four primary tissues: *epithelium*, *muscle*, *nervous tissue*, and *connective tissue*. Epithelia form coverings, linings, and most of the internal organs. Muscle is responsible for movement, and nervous tissue is responsible for communication. That leaves connective tissue—the one that binds all the others together. If you were able to remove all the connective tissue from the body, what was left would flatten down on the floor like a hairy, lumpy pancake. You would have no bones, cartilage, joints, fat, or blood, and nothing would be left of your skin except the epidermis, hair, and sweat glands. Muscles and nerves, without connective tissue, would have the consistency of mush. Internal organs would fall apart.

To understand epithelia, muscle, and nervous tissue we have to understand their cells, because it is the cells that are responsible for what the tissue does. Connective tissues are a different matter. With the exception of fat, the one connective tissue that is made up almost entirely of cells, it is the *extracellular* (outside of cells) *substance* in each connective tissue that gives it its essential character. The extracellular materials impart hardness to bone, resilience to cartilage, strength to tendons and fasciae, and liquidity to blood. And yet the extracellular components of connective tissues are entirely passive. Trying to relax a ligament or release fasciae with our power of will would be like trying to relax leather.

So are the connective tissues alive? Yes and no. Yes, in that living cells in the various connective tissues manufacture its extracellular components and organize the tissue. Also yes, in that the extracellular space in connective tissue is teeming with electrical activity. But no, in that the extracellular materials are nonliving. And one more no, in that the only way we can access them is through the agency of living cells. Only through neurons and their commands to muscle cells can we release tension in a tendon, execute weight-bearing activities that add bone salts to bone, and stimulate the laying down of additional connective tissue fibers in tendons and fasciae. And only with cells derived from epithelial tissues can we accomplish the absorption, manufacturing, and eliminatory functions that are needed for supporting the tissues of the body in general. In the end, our aim of molding and shaping the extracellular components of our connective tissues can only be accomplished indirectly.

The fact that the various connective tissues are so unlike one another is a reflection of the fact that their extracellular materials are diametrically different. Bone contains bone salts; tendons, ligaments, and fasciae contain dense accumulations of ropy fibers; loose connective tissue contains loose accumulations of the same fibers; elastic connective tissue contains elastic fibers; and blood contains plasma. So we can't work with connective tissues in general; we have to envision and work with each one individually.

Connective tissues not only give us shape, they also restrain activity. Bone butting against bone brings motion to a dead stop. Cartilage constrains motion, but more softly than bone. Ligaments constrain movements according to their architectural arrangements around joints. Sheets of fasciae, which are essentially layers of connective tissue, enclose and organize muscles and nerves, sometimes more restrictively than we would like. Finally, loose connective tissue helps bond the entire body together, constraining movement between fasciae and skin, adjacent muscle groups, and internal organs.

BONY CONSTRAINTS

Ligaments, muscles, and the joint capsule itself all aid in holding the elbow joint together, but underlying these supports, bony constraints ultimately limit both flexion and extension. Flexion is limited when the *head of the radius* and the *coronoid process* of the *ulna* are stopped in the *radial and coronoid fossae* in the lower end of the humerus, and extension is stopped when the hooked upper end of the ulna—the *olecranon process*—comes to a stop in a matching *olecranon fossa* in the humerus. Even though thin layers of cartilage soften the contact between the radius and ulna in relation to the humerus, the architectural plan limits flexion and extension as certainly as doorstops and provide us with clear examples of bony constraints to movement. It is not something we would want to alter (fig. 1.10).

In the spine we see another example of how one bone butting up against another limits movement. The lumbar spine can extend and flex freely, but matching surfaces of the movable intervertebral joints in this region are oriented vertically in a front-to-back plane that severely limits twisting (fig. 1.11). Because of this, almost all the twisting in a spinal twist takes place in the neck and chest, where the matching surfaces of comparable joints are oriented more propitiously (chapters 4 and 7). As with the elbow, we would not want to alter this design. If the lumbar region, isolated as it is between the pelvis and chest, could twist markedly in addition to bending forward and backward, it would be hopelessly unstable.

CARTILAGINOUS CONSTRAINTS

Cartilage has the consistency of rubber or soft plastic. It gives shape to the nose and external ears, and it forms a cushioning layer at the ends of long bones. Our main concern in this discussion, however, is not with these examples but with the joints called *symphyses*—the *intervertebral disks* between adjacent vertebral bodies (figs. 1.11, 4.10b, 4.11, and 4.13b), as well as the *pubic symphysis* between the two *pubic bones* (figs. 1.12 and 3.2). At all of these sites symphyses restrict movement, something like soft but thick rubber gaskets glued between blocks of wood that allow a little movement but no slippage. To that end the pubic symphysis is secure enough to bind the two halves of the pelvic bowl together in front and yet permit postural shifts and deviations; intervertebral disks bind adjacent vertebrae together tightly and yet permit the vertebral column as a whole to bend and twist.

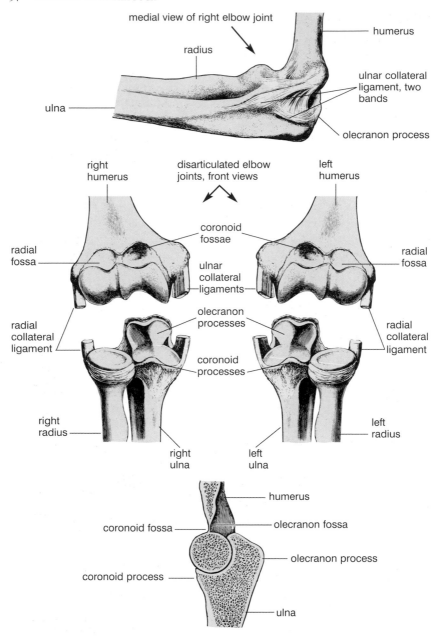

Figure 1.10. Bony stops for elbow flexion and extension, with the joint capsule pictured above, front views of the disarticulated right and left elbow joints shown in the middle, and a longitudinal cut through the joint and two of its three bones shown below. Extension is stopped where the olecranon process butts up against its fossa, and flexion is stopped where the head of the radius and coronoid process butt up against the radial and coronoid fossae (Sappey).

TENDONS AND LIGAMENTS

By definition, *tendons* connect muscles to bones, and *ligaments* connect bone to bone. They are both made up of tough, ropy, densely packed, inelastic connective tissue fibers, with only a few cells interspersed between large packets of fibers. Microscopically, tendons and ligaments are nearly identical, although the fibers are not packed as regularly in ligaments as in tendons. In a tendon the fibers extend from the belly of a muscle into the substance of a bone, lending continuity and strength to the whole complex. Ligaments hold adjoining bones together in joints throughout the body, often permitting small gliding motions, and usually becoming taut at the end of a joint's range of motion.

Ligaments and tendons can accommodate no more than about a 4% increase in length during stretching, after which tearing begins. This can be a serious problem. Because the extracellular connective tissue fibers in tendons and ligaments depend only on a few scattered living cells for repair and replacement, and because the tissue is so poorly supplied with blood vessels, injuries are slow to heal. The most common of these is *tendinitis*, which is caused by tears in the fibers at the interface between tendon and

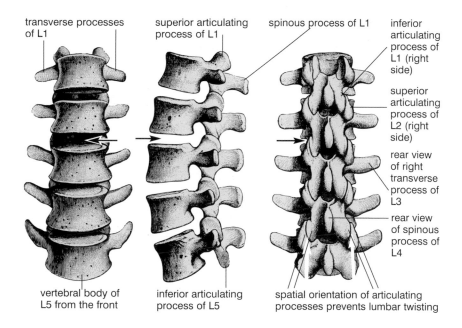

transverse processes of L1

superior articulating process of L1

spinous process of L1

inferior articulating process of L1 (right side)

superior articulating process of L2 (right side)

rear view of right transverse process of L3

rear view of spinous process of L4

vertebral body of L5 from the front

inferior articulating process of L5

spatial orientation of articulating processes prevents lumbar twisting

Figure 1.11. Lumbar vertebrae from the front, side, and behind. The vertical, front-to-back orientation of the articulating processes and their joint surfaces provides a bony stop that prevents lumbar twisting. Spaces that represent the location for the intervertebral disk between L2 and L3 are indicated by arrows (Sappey).

bone. If someone keeps abusing this interface with repetitive stress, whether typing at a computer keyboard, swinging a tennis racket, or trying compulsively to do a stressful hatha yoga posture, the injury can take a year to heal, or even longer.

The main purpose of ligaments is to restrain movable joints, and this becomes a major concern in hatha yoga when we want to stretch to our maximum. We might at first think of loosening them up and stretching them out so they do not place so many restraints on hatha postures. But ligaments don't spring back when stretched and lengthened (at least not beyond their 4% maximum), and if we persist in trying to stretch them beyond their limits we often do more harm than good. Once lengthened they become slack, and the joints they protect are prone to dislocation and injury. Ligaments have their purpose; let them be. To improve ranges of motion and flexibility, it is better to concentrate on lengthening muscles.

JOINT CAPSULES

Joint capsules are connective tissue encasements that surround the working surfaces of the class of joints known as *synovial joints*, including hinge joints, pivot joints, and ball-and-socket joints. Joint capsules for synovial

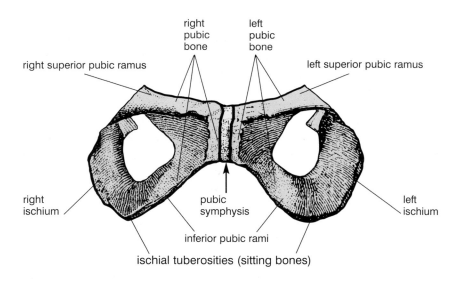

Figure 1.12. Pubic symphysis shown where it joins the two sides of the pelvis. This enlargement of the two pubic bones and ischia (front view) is taken from fig. 3.2, which shows the entire pelvis in perspective (Sappey).

joints have several roles: they provide a container for the slippery *synovial fluid* that lubricates the mating surfaces of the opposing bones; they house the *synovial membrane* that secretes the synovial fluid; they provide a tough covering of tissue into which ligaments and tendons can insert; and of special interest to us here, they and their associated ligaments provide about half the total resistance to movement.

The shoulder joint reveals an excellent example of a joint capsule. Like the hip joint, the shoulder joint is a ball and socket—the ball being the head of the humerus and the socket being the *glenoid cavity* of the scapula (fig. 1.13). The joint capsule surrounds the entire complex and accommodates tendons that pass through or blend into the joint capsule, as well as ligaments that reinforce it on the outside. To feel how it restricts movement, raise your arm overhead and pull it to the rear as far as possible: within the shoulder you can feel the joint capsule and its ligaments tightening up.

EXTENSILE LIGAMENTS

Extensile ligaments are not really ligaments; they are skeletal muscles held at relatively static lengths by motor neurons firing a continuous train of nerve impulses. They have greater elasticity than connective tissue ligaments because of their muscular nature, but other than that they function to maintain our posture like ordinary ligaments. What they don't do, by definition, is move joints through their full range of motion, which is what we usually expect from skeletal muscles. According to the conventional definition, extensile ligaments are mostly postural muscles in the torso, but it is arguable that for maintaining a stable meditation posture, every muscle in the body (excepting the muscles of respiration) becomes an extensile ligament.

Unlike connective tissue ligaments, the length of extensile ligaments can be adjusted according to the number of nerve impulses impinging on the muscle. And since every muscle associated with the torso and vertebral column is represented on both sides of the body, the matching muscles in each pair should receive the same number of nerve impulses per second on each side, at least in any static, bilaterally symmetrical posture. If that number is unequal, the paired muscles will develop chronically unequal lengths that result in repercussions throughout the central axis of the body. In hatha yoga, this condition is especially noticeable because it is the primary source of right-left musculoskeletal imbalances.

Axial imbalances can be spotted throughout the torso and vertebral column, but they are especially noticeable in the neck, where the tiny *suboccipital muscles* function as extensile ligaments to maintain head position (fig. 8.20). If your head is chronically twisted or tipped slightly to one side, it may mean that you have held the matching muscles on the two sides at unequal lengths over a long period of time. Motor neurons have become

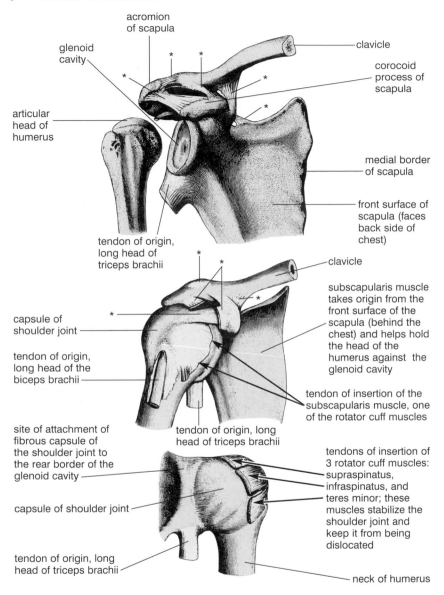

acromion
of scapula

glenoid
cavity

clavicle

corocoid
process of
scapula

articular
head of
humerus

medial border
of scapula

front surface of
scapula (faces
back side of
chest)

tendon of origin,
long head of
triceps brachii

clavicle

subscapularis muscle
takes origin from the
front surface of the
scapula (behind the
chest) and helps hold
the head of the
humerus against the
glenoid cavity

capsule of
shoulder joint

tendon of origin,
long head of the
biceps brachii

tendon of insertion of the
subscapularis muscle, one
of the rotator cuff muscles

site of attachment of
fibrous capsule of
the shoulder joint to
the rear border of the
glenoid cavity

tendon of origin, long
head of triceps brachii

tendons of insertion of
3 rotator cuff muscles:
supraspinatus,
infraspinatus, and
teres minor; these
muscles stabilize the
shoulder joint and
keep it from being
dislocated

capsule of shoulder joint

tendon of origin, long
head of triceps brachii

neck of humerus

Figure 1.13. Disarticulated right shoulder joint as viewed from the front (top image); right shoulder joint with its capsule, also from the front (middle image); and right shoulder joint with its capsule as viewed from behind (bottom image). Envision the chest as being located in front of the scapula and to the observer's right in the top two images (the surface of the scapula shown here faces the back of the chest). In the bottom image, envision the back of the scapula as being located to the observer's left; except for being a much deeper and more confined dissection, this view is similar to the one shown in figure 1.1. Asterisks indicate stabilizing ligaments, and arrows indicate rotator cuff tendons (Sappey).

habituated to long-established firing patterns, the bellies of the muscles themselves have become shorter on one side and longer on the other, and the connective tissue fibers within and surrounding the muscle have adjusted to the unequal lengths of the muscle fibers. Correcting such imbalances requires years of unrelenting effort; neither the bellies of the muscles or their connective tissue fibers can be lengthened or shortened quickly.

FASCIAE

Fasciae are sheets of connective tissue that give architectural support for tissues and organs throughout the body, holding everything together and providing for a stable infrastructure. They are crudely like leather gloves that form a boundary around your hands. Underneath the skin and subcutaneous connective tissue, fasciae organize and unify groups of muscles, individual muscles, and groups of muscle fibers within each muscle. They form a tough envelope around all the body cavities, and they surround the heart with a heavy connective tissue sack—the fibrous pericardium. We have superficial fascia just underneath the skin, and deep fasciae overlying muscle groups. The singular and plural terms are almost interchangeable—we can say deep fascia of the back, fascia of the body, or fasciae of the body.

Fascia is flexible if we keep moving, stretching, and breathing, but if we allow any part of the body to remain immobile, its fasciae become less flexible and eventually restrict our movements, like gloves that fit so tightly that you can't bend your fingers.

LOOSE CONNECTIVE TISSUE AND GROUND SUBSTANCE

Loose connective tissue is composed of *ground substance*, scattered fibers, and cells. It fills in the spaces between the three primary tissues that are mostly cellular—muscle, epithelia, and nervous tissue—and between all the other connective tissues, including bone and cartilage, blood and lymph, tendons and ligaments, joints and joint capsules, fasciae, fat, and lymphatic tissue. But loose connective tissue is more than a filler. Its ground substance is crudely comparable to glycerine—it lubricates and smoothes movement throughout the body. The ground substance permits slippage of adjacent structures as well as slippage of individual connective tissue fibers relative to one another in tendons and ligaments. Were it not for the connective tissue fibers and their submicroscopic attachments to muscle fibers, nerve fibers, and epithelia, ground substance would let everything slip and slide apart. This would be as unwelcome—by itself— as an oil spill on an icy road.

Ground substance is normally fluid, but it congeals and loses moisture if the surrounding tissues are inactive. And as it loses moisture it loses its lubricating properties. The entire body tightens down. Tendons, ligaments, and joint capsules become brittle, muscles lose much of their elasticity and ability to function smoothly, and the tissues become susceptible to injury. These reversals are the main reasons for morning stiffness, and they are a compelling argument for beginning every day with a session of hatha yoga. To rehydrate the ground substance a short, lazy practice is not as effective as a long and vigorous one, and you get what you pay for. The benefit is well-being; the payment is work and stretch.

STRETCHING

If you ask most people what it takes to lift a barbell overhead they will say muscles, bones, and joints. If you ask them what is needed for running a marathon they will say heart, lungs, and legs. And if you ask them what is required for dance or gymnastic performance they will say strength, grace, and agility. But if you ask them what is most important for increasing flexibility they will probably just give you a blank look. And yet those of us who practice hatha yoga know that improving flexibility is one of our greatest challenges. Even the simplest postures are difficult when one is stiff, and that is why instructors are always encouraging us to stretch. But what exactly do they mean?

Given what we have discussed so far, we know that we should be wary of increasing flexibility by trying to free up bony stops or loosening up cartilaginous restraints, joint capsules, tendons, and ligaments. What we can do is lengthen nerves and the bellies of muscles, the two kinds of extendable anatomical structures that run lengthwise through limbs and across joints.

THE BELLIES OF MUSCLES

Muscles have to be lengthened only a little to permit a respectable improvement in a joint's range of motion. But when we are stretching them and looking for long-term results, are we dealing with their individual muscle fibers or with their associated connective tissue fibers? The answer is both. The individual muscle fibers within a muscle can grow in length by the addition of little contractile units called *sarcomeres*. We know this from studies of muscles that have been held in casts in stretched positions. And by the same token, if a muscle is held by a cast in a foreshortened state, sarcomeres are lost and the muscle fibers become shorter.

It is not enough to increase the length of muscle fibers alone. A matching expansion of the connective tissue within and around the muscle is also needed, including the overlying fascia, the connective tissue that surrounds packets of muscle fibers, and the wrappings of individual fibers. And this is what happens during programs of prolonged stretching. The connective tissue gradually follows the lead of the muscle fibers, the muscle as a whole gets longer, and flexibility is improved. Hatha yoga stretches are a safe and effective way to bring this about. And in the occasional cases when we want to tighten everything down, all we have to do is stop stretching so much and concentrate on repetitive movements and short ranges of motion. The muscle fibers will quickly get shorter, and the connective tissues will soon follow suit.

NERVES

The issue of peripheral nerves is another matter. Nerves are sensitive to stretch but are not robust enough to limit it; they can accommodate to stretch only because they don't take a straight course through the tissues that surround them and because their individual nerve fibers meander back and forth within the connective tissue ensheathments of the nerve itself. During the course of stretching a limb, the gross path of a nerve through the surrounding tissues is first straightened, and as the stretch continues, the serpentine course of the individual fibers within the nerve is also straightened. And even after that, the enveloping connective tissue has enough elasticity to accommodate about 10–15% additional stretch without damaging the nerve fibers.

Without their connective tissue ensheathments nerves would be hopelessly vulnerable, not only to stretch but also to trauma and compression among tense muscles, bones, and ligaments. The protection is not fail-safe, however, because in extreme cases these ensheathments can accommodate to even more stretch than the 10–15% that is safe for their enclosed nerve fibers. The early warning signs are numbness, sensitivity, and tingling, and if these are ignored, sensory and motor deficits may develop. Your best protection is awareness and patience—awareness of why nerve stretch is a potential problem and the patience to work slowly when and if mild symptoms emerge. If nerve pain keeps turning up chronically, seek professional help.

IN THE LAST ANALYSIS

Research has shown beyond question that the length of muscle fibers can be increased as a result of prolonged stretching, or decreased as a result of chronic foreshortening. It is also clear that the connective tissue ensheathments of either muscles or nerves can be stretched too much. But there is another factor in the equation: the nervous system plays a pivotal role in causing muscles to either relax or tighten up, and this either permits stretch or limits it. So is it the active role of the nervous system or the passive role of the connective tissues that ultimately restrains movement? Since nerve impulses keep stimulating muscle cells during ordinary activities, there is only one way to find out for certain: to check someone's range of motion when they are under deep anesthesia, when the nervous system is not stimulating any skeletal muscle cells except those needed for respiration.

This has been done, inadvertently but many times. Any operating room attendant can tell you that when patients are anesthetized, their muscles become so loose that care has to be taken not to dislocate the joints, and this will happen even if the patient is extremely stiff in waking life. So why can't therapists take advantage of anesthesia-induced flexibility to increase the range of motion around joints? The answer is that without the protection of the nervous system the tissues tear—muscle fibers, connective tissue fibers, and nerves. And this proves that even though connective tissues provide the outermost limits to stretch, it is the nervous system that provides the practical limits in day-to-day life. When we have reached those limits the nervous system warns us through pain, trembling, or simply weakness that we are going too far, and most important, it warns us before the tissues are torn.

THREE POSTURES

Three hatha yoga postures illustrate the principles of movement we have been discussing. They are all simple to analyze and study because they exhibit bilateral symmetry, in which the two sides of the body are identical in structure and perform identical movements. Each one presents different challenges. We'll begin with the corpse posture.

THE CORPSE POSTURE

The corpse posture reveals several common problems that arise when people try to relax. Lie supine on a padded surface with the knees straight, the feet apart, the hands out from the thighs, and the palms up. Relax completely, allowing your body to rest on the floor under the influence of gravity (fig. 1.14). When you first lie down most of the motor neurons that innervate the skeletal muscles are still firing nerve impulses, but your breathing gradually becomes even and regular, and the number of nerve impulses per second to your muscles starts to drop. If you are an expert in relaxation, within a minute or two the number of nerve impulses to the muscles of your hands and toes goes to zero. Then, within five minutes the motor neuronal input to the muscles of your forearms, arms, legs, and thighs diminishes and also approaches zero. The rhythmical movement of the *respiratory diaphragm* lulls you into even deeper relaxation, finally minimizing the nerve impulses to the deep postural muscles of the torso. The connective tissues are not restraining you. Pain is not registered from any part of the body—the posture is entirely comfortable. This is an ideal relaxation.

In the early stages of practice any number of problems can interfere with the ideal. First, let's say you hurt your right shoulder playing basket-ball earlier in the day. Tension in that region is still high and stands out painfully in contrast to the relaxation in the rest of the limb and on the other side. In addition, you have an old back injury and the muscles around the vertebral column are holding it in a state of tension. You would like to lift your knees to relieve the stress, but you do not wish to seem unsporting. So you override the impulses of the flexion reflexes and continue to suffer with your knees straight.

This is absurd. All problems in the body tattle on themselves in one way or another, and you cannot relax your body because it is rebelling. You would not be in so much pain if you were walking around the block because the movement would keep you from noticing it, but when you try to relax you are aware of nothing else. The posture becomes increasingly irritating, and your mind, far from being still, is oscillating between awareness of the discomfort and longing for escape. If your instructor holds you in this pose for more than a minute or two you are in the wrong class. You are not yet ready for this work. You need to heal, move, and stretch—not lie still.

Those who are uncomfortable can sometimes improve the situation by simply moving into partially flexed positions—bending the knees, placing the hands on the chest, and supporting the head with a thick pillow. For restful sleep, it is not surprising that most people lie on their sides and curl up in an attitude of flexion.

THE PRONE BOAT

The prone boat posture demonstrates the simplest kind of movement against gravity. To experience this lie face down on the floor. Stretch your arms toward the feet, straight out to the sides, or overhead, as you prefer. Raise the arms, thighs, and head away from the floor all at once, keeping the knees and elbows extended (fig. 1.15). You are lifting into the posture with the muscles on the *posterior* (back) side of the body. The neck, back, hamstring, and calf muscles are all shortening concentrically and drawing the body up in an arc.

Figure 1.14. The corpse posture, for whole-body relaxation.

Although by most standards the prone boat is an easy posture, especially with the hands alongside the thighs, it can be challenging if you are in poor physical condition. A set of muscles is being used which is rarely exercised as a group in daily life, and if you keep your elbows and knees extended you may not be able to lift your hands and feet more than an inch or so off the floor. The combination of inflexibility and unfamiliarity keeps the antagonist muscles on the *anterior* (front) side of the body active, and this in turn restrains the lift. Whole-body extension is the essence of the prone boat, but the pull of gravity, lack of strength posteriorly, muscular resistance anteriorly, an abundance of flexion reflexes, and various connective tissue restrictions in the spine may all limit you. For beginners the activity of the nervous system is the main impediment to the posture.

Fascia is the main obstacle for intermediate students. The nervous system is commanding the posterior muscles to contract strongly and the anterior muscles to relax, but connective tissues and the design of the joints prevent marked extension. With time and practice the anterior muscles will relax and permit a full stretch. Finally, advanced students confidently lift to their maximum and play with the edges of neuronal control, tugging on their connective tissues with an educated awareness while at the same time keeping the breath even and regular without straining or faltering.

THE HAMSTRINGS-QUADRICEPS THIGH PULL

This standing forward bend demonstrates the interactions among agonist muscles, their antagonists, gravity, and the clasp knife reflex. Stand with the feet about 12 inches apart. Flexing the knees as necessary, bend forward and press the torso tightly against the thighs, which keeps the back relatively straight and prevents strain. Now, holding the chest and abdomen firmly in place, try to straighten the knee joint. The quadriceps femoris muscles on the front of the thighs try to accomplish this, and the hamstring

Figure 1.15. The prone boat. As you lift up into the posture, muscles on the back side of the body shorten concentrically; as you slowly lower yourself down, they lengthen eccentrically. Tension in muscles and connective tissues on the front side of the body increases as you lift up and decreases as you come down.

muscles on the back side of the thighs resist, but you can undercut the hamstring resistance by activating their clasp knife reflexes. Just massage the musculotendinous junctions of the hamstring muscles behind the knee joints while you are trying to press your hips up (fig. 1.16).

At this point the quadriceps femoris muscles are shortening concentrically, straightening the knees, and raising the body up against the force of gravity. At the same time the hamstring muscles, which are antagonists to the quadriceps, are actively even though unconsciously resisting. If you are in good condition your nervous system allows you to press upward to your personal maximum, but if you have recently hurt your knee or sprained your ankle, flexion reflexes responding to pain will limit you. As you press up you are making an isotonic movement. If you go to your maximum but then keep pressing, you are exercising isometrically.

If you were to bounce, which we do not want here, you would stretch the muscle spindles dynamically and stimulate the myotatic stretch reflexes. If you move slowly, you will be stimulating the Golgi tendon organs and eliciting the clasp knife reflex in both the quadriceps femoris and the hamstring muscles. Although this will tend to relax both sets of muscles, the focus of your will is to straighten the knee joint, with the result that the higher centers of the brain override the reflex in the quadriceps femori and allow it fuller rein in the hamstrings. Neurological circuits for reciprocal innervation also probably inhibit the motor neurons whose axons innervate the hamstring muscles.

The main resistance to lifting up comes from the hamstrings. If you are an advanced student and not feeling any trace of joint pain, you can try to relax the hamstrings and extend the knee joints more completely, contracting the quadriceps as much as your strength and health permit. This posture is different from the prone boat, in that your attention is more restricted. In the prone boat you are trying to relax the entire front side of the body; here you are trying to relax only the hamstrings.

Figure 1.16. Standing hamstrings-quadriceps thigh pull. The first priority is placing the torso solidly against the thighs in order to protect the lower back. Under those circumstances, trying to lift the hips forceably in combination with massaging the Golgi tendon organs in the hamstring tendons encourages deep relaxation and eventual lengthening of the hamstring muscles (simulation).

hamstring muscles

If you are relatively healthy, as you reach the limits of nervous system control, the fasciae begin to play an important role in limiting your efforts to straighten the knees and raise up. You reach a point at which the connective tissue fibers within and surrounding the hamstring muscles will not allow any more lifting. They are now like wires pulled taut, stretched to their limit. The only way to get more length in the system is to patiently lengthen the muscles and nerves with a long-term program of prolonged stretches.

PUTTING IT ALL TOGETHER

We have covered a great deal of territory in this chapter, but in so doing we have laid the foundation for everything that is to come. To sum it up: sensory input to the brain and the power of will both ultimately influence the motor neurons, which in turn preside over the actions of the musculo-skeletal system. The reflexes are in the background and out of our immediate awareness, but without them we would be in dire straits. Without the stretch reflexes our movements would be jerky and uncertain, like film portrayals of Frankenstein's monster. And without pain receptors and flexion reflexes we would soon be a battleground of burns and injuries. Without the reflexes from our vestibular system we would teeter about, uncertain of our balance and orientation. Without sensation from touch and pressure pathways we would lose most of the sensory input that gives us pleasure—and along with its loss, its guidance. In the end, the nervous system drives the musculoskeletal system, and these two in combination maintain and sculpt connective tissues, which in turn passively restrict movement and posture. All of this takes place within the field of gravity under the auspices of will and creates the practice of hatha yoga.

"Fortunately, science is self-correcting. Whatever the herd is doing, truth eventually becomes known and stops the stampede."
— Gershon, in *The Second Brain*, p.35.

CHAPTER TWO
BREATHING

"The lungs are placed in a recess so sacred and hidden that nature would seem to have specially withdrawn this part both from the eyes and from the intellect: for, beyond the wish, it has not yet been granted to any one to fit a window to the breast and redeem from darkness the profounder secrets of nature. For of all of the Parts of the body, the lungs alone, as if shrinking from observation, cease from their movement and collapse at once on the first entrance of light and self-revelation. Hence such an ignorance of Respiration and a sort of holy wonder. Still let me draw near to the inmost vitals, and concerning so obscure a matter, make at least a guess."

— John Mayow, in *Tractatus Quinque* (1674), quoted from Proctor's *A History of Breathing Physiology*, p. 153.

Yogis knew nothing of physiology, at least in terms that would have been helpful to 17th and 18th century European scientists and physicians like John Mayow, but for a long time they have made extraordinary claims about the value of studying the breath. They say flatly, for example, that the breath is the link between the mind and the body, and that if we can control our respiration we can control every aspect of our being. This is the endpoint, they tell us, that begins with simple hatha yoga breathing exercises. Every aspect of our being? That's a lot, by any standard. No matter: even though such comments may stimulate our curiosity, their pursuit is outside the scope of this book. Our objective here is to pursue studies in breathing as far as they can be tested objectively and experientially, and then to discuss some of the relationships between yoga and respiration that can be correlated with modern biomedical science: how different patterns of breathing affect us in different ways, why this is so, and what we can learn from practice and observation.

Breathing usually operates at the edge of our awareness, but will and volition are always at our disposal. Just as we can choose how many times to chew a bite of food or adjust our stride when we are walking up a hill, so can we choose the manner in which we breathe. Most of the time, however, we run on "automatic," allowing input from internal organs to manage the rate and depth of our breathing. Yogis emphasize choice. They have discovered the value of regulating respiration consciously, of breathing evenly and diaphragmatically, of hyperventilating for specific purposes, and of suspending the breath at will. But even though these aims might seem laudable, readers should be made aware that the classical literature of hatha yoga generally warns students against experimenting intemperately with breathing exercises. Verse 15 of Chapter 2 of the *Hatha Yoga Pradipika* is typical: "Just as lions, elephants, and tigers are gradually controlled, so the prana is controlled through practice. Otherwise the practitioner is destroyed." This sounds like the voice of experience, and we ought not dismiss it casually. We'll revisit the issue of temperance at the end of the chapter after having examined the anatomy and physiology of respiration. There are reasons for caution.

To understand the benefits of controlled breathing we must proceed step by step, beginning with a look at the overall design of the respiratory system, and then at the way skeletal muscles draw air into the lungs. Next we'll see how breathing affects posture and how posture affects breathing. After that we'll explore how the two major divisions of the nervous system—*somatic* and *autonomic*—interact to influence breathing. Then we'll turn to the physiology of respiration and examine how lung volumes and blood gases are altered in various breathing exercises. That will point us toward the mechanisms by which respiration is regulated automatically and at how we can learn to override those mechanisms when we want to. Finally we'll examine four different kinds of breathing—thoracic, paradoxical, abdominal, and diaphragmatic—and the relationships of each to yoga breathing practices. At the end of the chapter we'll return to the issue of moderation in planning a practice.

THE DESIGN OF THE RESPIRATORY SYSTEM

Every cell in the body needs to breathe—taking up oxygen, burning fuel, generating energy, and giving off carbon dioxide. This process, known as cellular respiration, depends on an exchange—moving oxygen all the way from the atmosphere to lungs, to blood, and to cells, and at the same time moving carbon dioxide from cells to blood, to lungs, to atmosphere. The body accomplishes this exchange in two steps. For step one we draw air into the lungs, where it comes in contact with a large wet surface area—the collective hundred million alveoli—into which oxygen can dissolve and from which carbon dioxide can be eliminated. For step two oxygen travels in the pulmonary circulation from the lungs to the heart and in the systemic circulation from the heart to the cells of the body. Carbon dioxide travels in the opposite direction, first from the cells of the body to the heart in the systemic circulation, and then from the heart to the lungs in the pulmonary circulation (fig. 2.1 and chapter 8).

Everything about the respiratory system is accessory to the movement of oxygen and carbon dioxide. Airways lead from the nose and mouth into the lungs (fig. 2.2). Air is pulled backward in the nose past the *hard and soft palates*, where it makes a 90° turn and enters a funnel-shaped region,

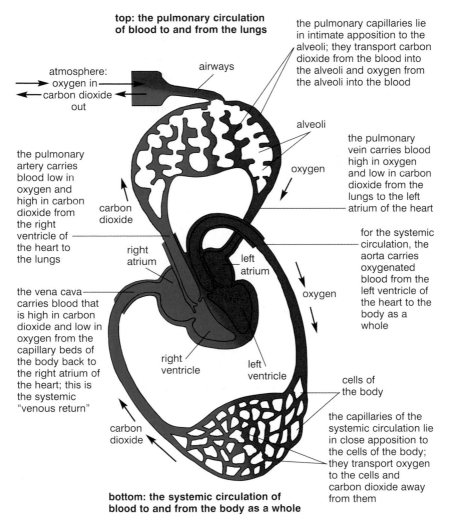

Figure 2.1. Cardio-respiratory system. As indicated by the arrows, oxygen is transported from the atmosphere to the cells of the body: from airways to lungs to the pulmonary circulation, heart, and finally to the systemic circulation. Carbon dioxide is transported in the other direction: from the cells to the systemic circulation, heart, pulmonary circulation, lungs, airways, and atmosphere (Dodd).

the pharynx. From there it continues downward into the *larynx*, which is the organ for phonation and whose vocal cords vibrate to create sound. Below the larynx air passes into the *trachea*, the right and left *primary bronchi*, and then into the two lungs, each of which contains 10 *bronchopulmonary segments* that are served individually by *secondary bronchi*. The secondary bronchi in turn divide into *tertiary bronchi* and smaller subdivisions (*bronchioles*) that collectively compose the *bronchial tree* (fig. 2.3). The terminal bronchioles of the bronchial tree in turn open into the tiny *alveoli*, giving a microscopic view

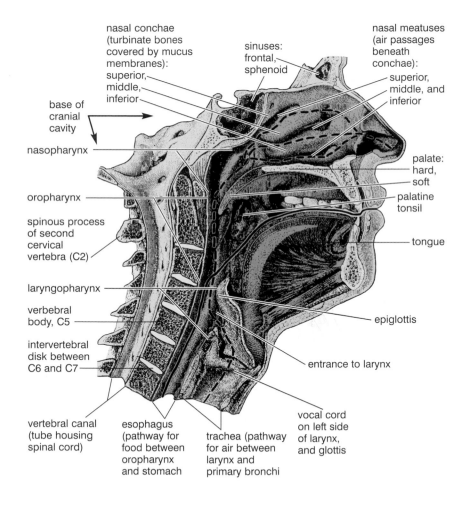

Figure 2.2. Nearly mid-sagittal cut (just to the left of the nasal septum) showing the left half of the head and neck, and revealing the crossing passageways for food (solid line from the oral cavity into the esophagus) and air (dashed lines from the nasal passages into the trachea). (from Sappey).

of the lungs the appearance of a delicate lacy network. The trachea and other large tubes in the airways are held open by incomplete rings of cartilage, and the alveoli remain open because a special surfactant on their walls limits their expansion during the course of a full inhalation and yet prevents surface tension from collapsing them during the course of a full exhalation.

The pharynx is a crossroads for the passage of air and food. Air passes down and forward from the *nasopharynx* into the *laryngopharynx* and then into the larynx and trachea. Food is chewed in the mouth, and from there it is swallowed backward into the *oropharynx* and across the pathway for air into the *esophagus*, which is located behind the trachea just in front of

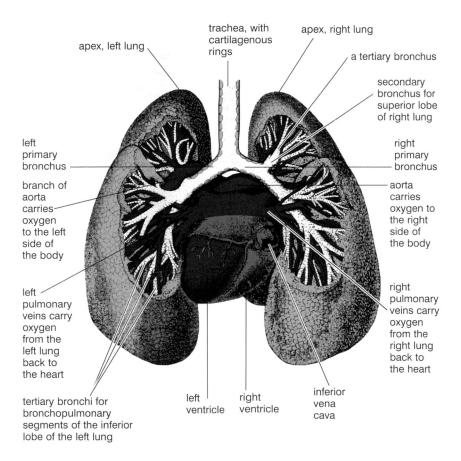

Figure 2.3. Isolated heart-lung preparation as viewed from behind. The aorta and superior vena cava are not visible from this perspective. The bronchial tree branches into right and left primary bronchi, 5 secondary bronchi (3 for the right lung and 2 for the left), and 20 tertiary bronchi to the bronchopulmonary segments (10 for each lung). Branches of the pulmonary arteries and veins are likewise associated with each of the bronchopulmonary segments (Sappey).

the vertebral column (fig. 2.2). The *glottis*, which is the narrowed aperture in the larynx at the level of the *vocal cords*, closes when we swallow. You can feel that happen if you initiate either an inhalation or an exhalation and then swallow some saliva. You will find, no matter what part of the breathing cycle you are in, that swallowing obstructs breathing. If it doesn't, food may "go down the wrong way," as children put it, and we choke.

The lungs are mostly composed of air: 50% air after full exhalation and 80% air after full inhalation. If you slap the side of your chest you'll hear a hollow sound; contrast this with the lower-pitched liquidy sound that comes from slapping your hand against your abdomen. A slippery membrane that is itself impervious to air covers the lungs, which can in turn be likened roughly to blown-up balloons that fill the rib cage, excepting that the "balloons" are not tied off at their necks. So why don't they deflate just like loosed balloons that fly away? The answer is fundamental to the design of the respiratory system. The lungs have an inherent elasticity, and they remain inflated inside the rib cage only because they faithfully track changes in the volume of the chest as it gets larger and smaller. How can this be? It is because nothing lies between the outer surfaces of the lungs and the chest wall except a potential space, the *pleural cavity*. This cavity contains no air, only a vacuum which holds the lungs tightly against the inner surface of the chest wall, along with a small amount of lubricating fluid that permits the lungs to expand and contract as the chest expands and contracts through the agency of the surrounding muscles of respiration (figs. 2.4, 2.6, and 2.9).

TWO EMERGENCY SITUATIONS

Two emergency situations will put all this in perspective. First, if your rib cage were penetrated on one side in a traumatic injury, air would rush into the pleural cavity and cause the lung on that side to collapse. This is called *pneumothorax*. How quickly it develops depends on the size of the injury. With a large enough opening, the lung collapses almost like letting the air out of the neck of a balloon, as might be surmised from Mayow's astute observations more than 300 years ago.

More perilously than pneumothorax for one lung, if both sides of the rib cage are grossly penetrated, both lungs collapse to their minimum size and shrink away from the chest wall. With the pleural cavities filled with air, the muscles of respiration cannot get purchase on the external surfaces of the lungs to create an inhalation, and unless someone holds your nose and blows directly into your mouth to give you artificial respiration, you will die in a few minutes.

[Technical note: There is one other alternative. If you were thinking fast enough and not too distracted by the injury, you could balloon out your cheeks and quickly (two times per second) "swallow" air into your lungs 10–15 times for inhalation, closing the glottis after each swallow. To exhale you simply open the glottis. For an interesting exercise in awareness, and to feel for yourself how the lungs effortlessly get smaller when you open your glottis to exhale, hold your nose and breathe this way for 2–3 minutes.]

The remedy for pneumothorax in a hospital setting also illustrates the architecture of the system. It's simple, at least in principle. Tubes are sealed into the openings of the chest wall, and the air is vacuumed out of the pleural cavities. This pulls the external surfaces of the lungs against the inner wall of the chest and upper surface of the diaphragm, and the muscles of respiration can then operate on the inflated lungs in the usual manner.

A second emergency situation involves an obstruction in the airway, perhaps a big chunk of food that has dropped into the larynx instead of the esophagus. If it is too big to get all the way through into the trachea, the obstruction may get stuck in the larynx, block the airway, and prevent you from breathing. In such cases the natural reaction for most people is to try

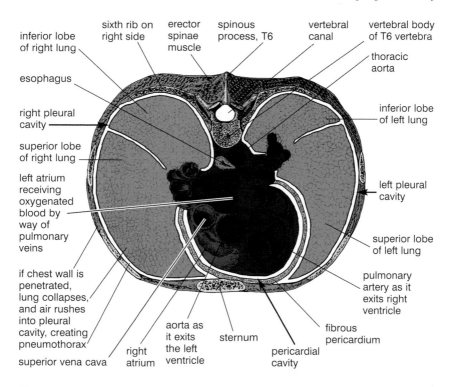

Figure 2.4. Cross-sectional view through the upper chest, looking from above at a section through the lungs and pleural cavities, and at a three-dimensional view of the upper portion of the heart with great vessels, pericardial cavity, and fibrous pericardium. Pericardial and pleural cavities are greatly exaggerated (Sappey).

to inhale more forcefully, but this will almost certainly reinforce rather than relieve the obstruction. Trying to exhale may be more productive. Or a second person, someone who knows first aid, could try the Heimlich maneuver, pulling sharply in and up on the abdominal wall from behind to create enough intra-abdominal and intra-thoracic pressure to force the object from the laryngopharynx back into the oropharynx, where it can either be coughed out externally or swallowed properly.

The emergency surgical remedy for a complete obstruction of the larynx is a tracheotomy, making a midline incision between the larynx and the pit of the throat, quickly separating the superficial muscles, and opening the exposed trachea with another midline incision just below the thyroid gland. This allows inhalation and exhalation to take place below the obstruction.

In the case of pneumothorax, when the chest wall has been penetrated and the lungs are collapsed, the muscles of respiration can expand and contract the chest, but the effort is all for nothing since the requisite contact between the inner surface of the chest wall and the outer surface of the lungs has been lost. It's like a car stuck in the snow—the wheels turn but they can't move you forward. The second case is like a car with its drive wheels immobilized in concrete—the blockage in the airway completely frustrates the action of the muscles of respiration. In both situations we are trying to pull air into the lungs by using our force of will but we are unable to support our inner needs with our external efforts.

THE MUSCLES OF RESPIRATION

Inhalations can take place only as a result of muscular activity. Exhalations are different: the lungs have the capacity to get smaller because their elasticity keeps pulling them, along with the rib cage, to a smaller size. And as already mentioned, the size of the lungs follows the size of the chest in lockstep: anything that expands and contracts the chest also expands and contracts the lungs, whether it is lifting or compressing the rib cage, lowering or raising the dome of the respiratory diaphragm, releasing or pressing inward with the abdominal muscles, or allowing the elasticity of the lungs to draw in the chest wall.

The way in which the muscles of respiration accomplish breathing is more complex than the relatively simple way a muscle creates movements around a joint. Three main sets of muscles are active when you breathe normally: the *intercostal muscles*, the *abdominal muscles*, and the *respiratory diaphragm*. We'll start our discussion with the intercostal muscles.

THE INTERCOSTAL MUSCLES

When we breathe, and in particular when we emphasize chest breathing, the short intercostal (between the ribs) muscles operate as a unit to expand

and contract the chest (figs. 2.5 and 2.9). Two sets of these muscles, one under the other, act on the rib cage. The *external intercostal muscles* run between the ribs in the same direction as the most external sheet of abdominal muscles (figs. 2.7, 2.9, 3.11–13, and 8.8); they lift and expand the rib cage for inhalation, like the movement of an old-fashioned pump handle as it is lifted up from its resting position. The *internal intercostal muscles* run at right angles to the external layer; they pull the ribs closer together as well as down and in for exhalation (usually a forced exhalation). If you place your hands on your chest with the fingers pointed down and medially (toward the midline of the body), this approximates the orientation of the external intercostal muscles, and if you place your hands on your chest with the fingers pointing up and medially, this approximates the orientation of the internal intercostal muscles (fig. 2.5). The external intercostal muscles do not always act concentrically to lift the rib cage; during quiet breathing they

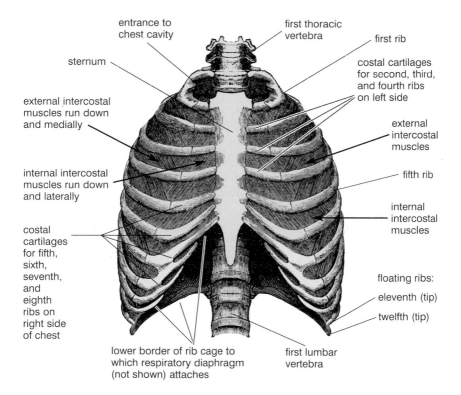

Figure 2.5. Surface view of chest. The internal intercostal muscles are visible in front near the sternum where they are not overlain by the external intercostals, and they are also visible laterally where the external intercostals have been dissected away (between the fifth and sixth ribs). As a group, the external intercostal muscles lift the rib cage up and out to support inhalation, and the internal intercostal muscles pull it down and in to complete a full exhalation (from Morris).

also act isometrically to keep the rib cage from collapsing inward when the respiratory diaphragm (see below) creates the vacuum that draws air into the lungs.

THE ABDOMINAL MUSCLES

In breathing, the abdominal muscles (figs. 3.11–13, 8.8, 8.11, and 8.13–14) function mainly in deep and forced exhalations, as when you try to blow up a balloon in one breath. For that task the muscles shorten concentrically, pressing the abdominal wall inward, which in turn pushes the abdominal organs up against the relaxed (or relaxing) diaphragm. In combination with the action of the internal intercostal muscles, this forcibly decreases the size of the chest cavity and pushes air out of the lungs. You can also feel the action of the abdominal muscles by pursing the lips and forcing the breath out through the tiny opening. In yoga the abdominal muscles are important for what yogis refer to as even breathing, and they are also key elements for many breathing exercises.

THE ANATOMY OF THE DIAPHRAGM

Because the respiratory diaphragm is completely hidden inside the torso, most people have only a rudimentary notion of what it looks like or how it operates. The simplest way to describe it is to say that it is a domed sheet of combined muscle and tendon that spans the entire torso and separates the chest cavity from the *abdominal cavity* (figs. 2.6–9). Its rim is attached to the base of the rib cage and to the lumbar spine in the rear.

The diaphragm is shaped like an umbrella, or an upside-down cup, except that it is deeply indented to accommodate the vertebral column. It consists of a *central tendon*, a *costal portion*, and a *crural portion*. The central tendon forms the top surface of the dome, which floats there freely, attached only to the muscle fibers of the costal and crural portions of the diaphragm. It is thus the only "tendon" in the body which does not attach directly to the skeleton. The largest part of the diaphragm is its costal component, whose muscle fibers fan down from the central tendon and attach all around to the lower rim of the rib cage (figs. 2.7–9). The crural portion of the diaphragm consists of the *right crus* and *left crus*, which attach to the forward arch of the lumbar spine (figs. 2.7–8). These are separated from one another by the aorta as it passes from the *thoracic cavity* into the abdominal cavity. The architecture of the diaphragm thus permits it to move the central tendon of the dome, the base of the rib cage, the lumbar spine, or any combination of the three.

You can note the site of the costal attachment of the diaphragm by hooking your fingers under the rib cage and tracing its lower margin. It is high in front where it attaches to the *sternum*, and lower where you trace it laterally (to the side), but you can't feel it behind because the deep back muscles are in the way. You can also occasionally feel the region where the *crura* (plural form) of the diaphragm attach to the lumbar vertebrae, especially in someone slender who is lying flat on the floor, because the lumbar region sometimes arches forward to within an inch or so of the surface of the abdominal wall. This vividly illustrates how far forward the diaphragm can be indented by the spinal column.

The diaphragm has to be one of the most interesting and complex muscles in the body. Because it is a thin sheet, its shape bears the impressions of its immediate surroundings—the rib cage, the heart and lungs, and the abdominal organs, and it is dependent on the existence and anatomical arrangements of these structures for its function. The diaphragm's extensive relationship with

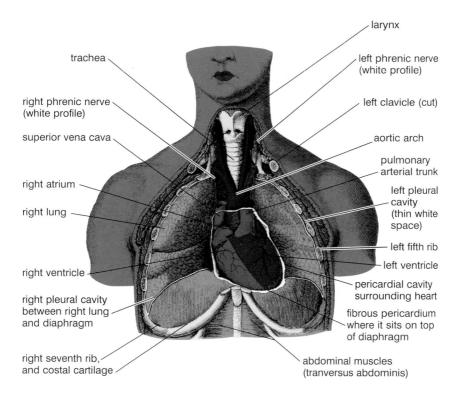

Figure 2.6. Front view of the chest, with the first six ribs, clavicles, and sternum cut away to reveal the internal organs, which include: the larynx, trachea, lungs and pleural cavities; the heart, great vessels (aorta, vena cava, pulmonary artery, and pulmonary vein, not all shown), pericardial cavity, and fibrous pericardium; the upper front portion of the respiratory diaphragm; and the right and left phrenic nerves. The pleural cavities are represented by the thin white spaces between the lungs and the body wall, and between the lungs and diaphragm (Sappey).

the chest wall is a case in point. Even though the costal portion of the diaphragm extends to the base of the rib cage, the lungs are never pulled that far inferiorly (toward the feet), and for much of its area the costal portion of the diaphragm is in direct contact with the inner surface of the rib cage, with only the potential space of the pleural cavity separating the two. This region into which the lungs never descend is called the *zone of apposition* (fig. 2.9); without its slippery interfaces, the outer surface of the diaphragm could not slide easily against the inner aspect of the rib cage, and the dome of the diaphragm could not move up and down smoothly when we breathe.

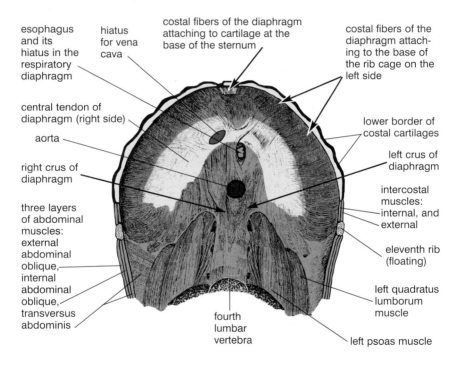

Figure 2.7. A view of the respiratory diaphragm looking at its underneath side from below. It's like a rejected upside-down bowl that a potter pushed in on one side. The pushed-in place is the indentation for the vertebral column, and the bottom of the bowl contains hiatuses (openings) for the esophagus, aorta, and inferior vena cava. The central tendon of the diaphragm is represented by the large, lightly contrasted central arch. The muscle fibers of the diaphragm are disposed radially from the central tendon: the costal fibers attach to the base of the rib cage most of the way around (approaching the viewer in the third dimension); and the right and left crura attach to the lumbar vertebrae below (between and in front of the origins of the psoas muscles). (from Morris)

THE FUNCTION OF THE DIAPHRAGM

To analyze the origins and insertions of a muscle that is shaped like an indented umbrella is a bit daunting, but that is what we must do if we want to understand how the diaphragm functions in breathing and posture. We'll begin with the simplest situation, which is found in supine postures. Here the base of the rib cage and the lumbar spine act as fixed origins for the diaphragm, and under those circumstances the central tendon has to

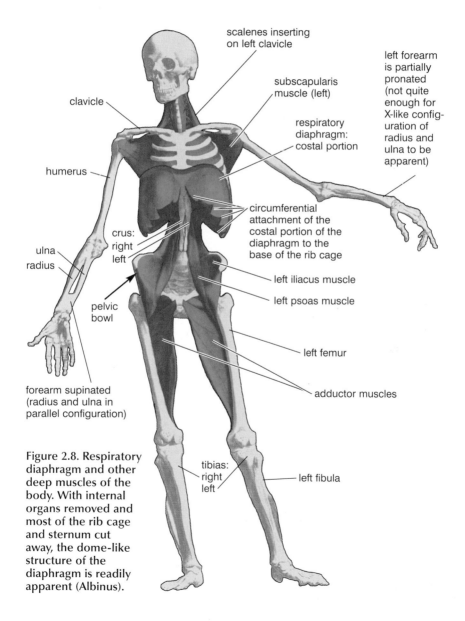

scalenes inserting on left clavicle

subscapularis muscle (left)

left forearm is partially pronated (not quite enough for X-like configuration of radius and ulna to be apparent)

clavicle

respiratory diaphragm: costal portion

humerus

circumferential attachment of the costal portion of the diaphragm to the base of the rib cage

crus: right left

ulna radius

left iliacus muscle

left psoas muscle

pelvic bowl

left femur

forearm supinated (radius and ulna in parallel configuration)

adductor muscles

Figure 2.8. Respiratory diaphragm and other deep muscles of the body. With internal organs removed and most of the rib cage and sternum cut away, the dome-like structure of the diaphragm is readily apparent (Albinus).

tibias: right left

left fibula

act as the movable insertion. The dome of the "cup," including the central tendon, descends and flattens during inhalation, putting pressure on the contents of the abdomen and creating a slight vacuum in the chest that draws air into the lungs. By contrast, the dome of the diaphragm is drawn upward during exhalation by the inherent elasticity of the lungs, and as that happens air escapes into the atmosphere.

Whenever the chest and spine are fixed, as typically occurs during relaxed breathing in a supine position, the top of the dome of the diaphragm is pulled straight downward during inhalation, like a piston, with the chest wall acting as the cylinder. During a supine inhalation the fibers of the diaphragm shorten concentrically and pull the central tendon inferiorly. During a supine exhalation its fibers lengthen eccentrically as the central tendon is both pushed from below and pulled from above— pushed by gravity acting on the abdominal organs and pulled by the elastic recoil of the lungs. The abdominal wall remains relaxed. It stretches out

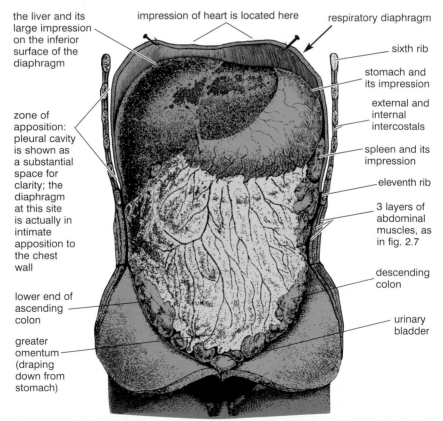

the liver and its large impression on the inferior surface of the diaphragm

impression of heart is located here

respiratory diaphragm

sixth rib

stomach and its impression

external and internal intercostals

zone of apposition: pleural cavity is shown as a substantial space for clarity; the diaphragm at this site is actually in intimate apposition to the chest wall

spleen and its impression

eleventh rib

3 layers of abdominal muscles, as in fig. 2.7

descending colon

lower end of ascending colon

greater omentum (draping down from stomach)

urinary bladder

Figure 2.9. Abdominal organs in place, with the diaphragm and lower half of the rib cage cut to illustrate the extensive zone of apposition, into which the lungs do not descend even during the course of a maximum inhalation (Sappey).

anteriorly (forward) as the dome of the diaphragm descends during inhalation, and it moves back posteriorly (toward the back of the body) as the diaphragm relaxes and rises during exhalation. Only in supine and inverted postures do we see the diaphragm act with such purity of movement.

This kind of breathing is carried out in its entirety by the diaphragm, but it is often referred to as *abdominal breathing*, or *belly breathing*, because this is where movement can be seen and felt. It is also known as *deep diaphragmatic breathing* in recognition of its effects in the lower abdomen. Finally, we can call it *abdomino-diaphragmatic breathing* to indicate that the downward movement of the dome of the diaphragm not only draws air into the lungs, it also pushes the lower abdominal wall anteriorly.

Another type of diaphragmatic breathing operates very differently. Amazingly, its principal mechanical features were accurately described by Galen (a first century Roman physician and the founder of experimental physiology) almost two thousand years ago, even though his concept of why we breathe was pure fantasy. During inhalation the primary action of this type of breathing is not to enlarge the lungs by pulling the dome of the diaphragm inferiorly, but to lift the base of the chest and expand it laterally, posteriorly, and anteriorly. It works like this. If there is even mild tension in the lower abdominal wall, that tension will impede the downward movement of the dome of the diaphragm. And since the abdominal organs cannot be compressed, they can act only as a fulcrum, causing the diaphragm to cantilever its costal site of attachment on the rib cage outwardly, spreading the base of the rib cage to the front, to the rear, and to the sides, while at the same time pulling air into the lower portions of the lungs. In contrast to the pump handle analogy for intercostal breathing, *diaphragmatic breathing* has been likened to lifting a bucket handle up and out from its resting position alongside the bucket (see Anderson and Sovik's *Yoga, Mastering the Basics* for illustration and further explanation). Without the resistance of the abdominal organs, the diaphragm cannot create this result. The intercostal muscles serve to support the action of the diaphragm, not so much to lift and enlarge the chest but to keep it from collapsing during inhalation.

[Technical note: Precise language does not exist, at least in English, for describing in a single word or phrase how the respiratory diaphragm operates to expand the rib cage in diaphragmatic breathing. A "cantilever truss," however, from civil engineering, describes a horizontal truss supported in the middle and sustaining a load at both ends, and this comes close. In the special case of the human torso, the abdominal organs and intra-abdominal pressure provide horizontal support for the dome of the diaphragm, and the lift and outward expansion of the base of the rib cage is a load sustained at the perimeter of the base of the rib cage.]

The origins and insertions of the diaphragm for abdominal inhalations are different than for diaphragmatic inhalations, and understanding the subtleties of these functional shifts will further clarify the differences between the two types of breathing. For abdominal breathing in the corpse and inverted postures, both the costal attachment to the rib cage and the crural attachment to the spine act as stationary origins; the only part of the diaphragm that can move (the insertion, by definition) is the central tendon in the dome, which moves inferiorly during inhalation and superiorly (toward the head) during exhalation. By contrast, for diaphragmatic breathing, the central tendon is held static by the relative tautness of the abdominal wall and serves mainly as a link between the spinal attachments of the crura, which now act as the stationary origin, and the costal attachment to the base of the rib cage, which now acts as the movable insertion.

To summarize, diaphragmatic breathing occasions an expansion of the rib cage from its lower border. To differentiate it from abdomino-diaphragmatic breathing, in which the rib cage remains static, we can call it *thoraco-diaphragmatic breathing*. It should be mentioned that the terms abdominal breathing, belly breathing, deep diaphragmatic breathing, and diaphragmatic breathing have all been in casual, although generally noncritical, use for a long time, but the terms "abdomino-diaphragmatic" and "thoraco-diaphragmatic" have not appeared in the literature before now.

HOW BREATHING AFFECTS POSTURE

The way breathing affects posture and the way posture affects breathing will be continuing themes throughout the rest of this book. The importance of these issues have long been recognized in yoga, but most commentaries are vague and imprecise. Here I am aiming for simplicity: photographic records of exhalations and inhalations, and superimpositions of computer-generated tracings of inhalations (since these are always larger) on the exhalations. As seen in both this chapter and in chapters 3 and 5, such images provide a source of raw data not only for how inhalations result in movements of the chest and abdomen but also for how they affect the body from head to toe. The single most important key to understanding all such effects is the operation of the respiratory diaphragm, and to introduce the subject, we'll explore two exercises that will help you become aware of its anatomy and understand two of its main roles in movement other than those for respiration itself.

A VARIATION OF THE COBRA

Lie face down on the floor and interlock your arms behind your back, grasping your forearms or elbows. Or you can simply place your hands in the standard cobra position alongside the chest. Strongly tighten all the muscles from the hips to the toes, and use the neck and deep back muscles to lift the head, neck, and chest as high as possible. You are not making any particular use of the diaphragm to come into this position. Now inhale and exhale deeply through the nose. Notice that each inhalation raises the upper part of the body higher and that each exhalation lowers it (fig. 2.10). Because you are keeping the back muscles engaged continuously during both inhalation and exhalation, the lifting and lowering action is due entirely to the muscles of respiration.

In this variation of the cobra pose we hold the hips, thighs, and pelvis firmly, which stabilizes the lower back and the spinal attachment of the crus of the diaphragm. Inhalation creates tension at all three of the diaphragm's attachments: one on the vertebral column, one on the base of the rib cage, and the third on the central tendon. But because the hip and thigh muscles have been tightened, the spinal attachment is stabilized, excepting only a slight lifting effect that is translated to the hips. What happens in the torso illustrates clearly how respiratory movements influence posture: with the abdomen pressed against the floor, the contents of the abdominal cavity cannot easily descend, and this restricts the downward movement of the central tendon, which now acts as a link between the two muscular portions of the diaphragm. With the crural attachments stabilized, the only insertion that can be mobilized without difficulty is the one at the base of the rib cage. This attachment therefore expands the chest from its base, draws air into the lungs, and lifts the upper body. If you are breathing smoothly and deeply you will feel a gentle, rhythmic rocking movement as the head, neck, and chest rise and fall with each inhalation and exhalation. This is a perfect illustration of thoraco-diaphragmatic breathing.

In this exercise the action of the diaphragm during inhalation reinforces the activity of the deep back and neck muscles and thus deepens the back-ward bend. During exhalation the muscle fibers of the diaphragm lengthen eccentrically as they resist gravity. When they finally relax at the end of exhalation, the backward bend in the spine is maintained only by the deep

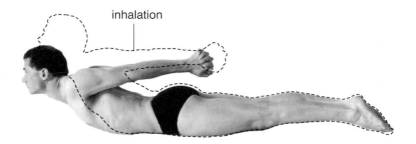

Figure 2.10. Cobra variation with tightly engaged lower extremities. Diaphragmatic inhalation (dotted line) lifts the upper half of the body over and above what can be accomplished by the back muscles acting alone (halftone). Contrast with the diaphragmatic rear lift in figure 2.11.

muscles of the back and neck. This is an excellent exercise for strengthening the diaphragm, because after you have lifted to your maximum with the deep back muscles, you are using the diaphragm, aided by the external intercostal muscles acting as synergists, to raise the upper half of the body even higher—and this is a substantial mass to be lifted by a single sheet of muscle acting as prime mover. Furthermore, if you keep trying as hard as possible to inhale deeply without closing the glottis, you will be creating the most extreme possible isometric exercise for this muscle and its synergists, the external intercostals. But be watchful. If this effort creates discomfort in the upper abdomen on the left side, please read the section in chapter 3 on *hiatal hernia* before continuing.

THE DIAPHRAGMATIC REAR LIFT

Next try a posture that we can aptly call the diaphragmatic rear lift. Again lie face down, placing your chin against the floor, with the arms along the sides of the body and the palms next to the chest. Keeping the chest pressed firmly against the floor, relax all the muscles from the waist down, including the hips. Take 10–15 nasal breaths at a rate of about one breath per second.

With the thighs and hips relaxed, and with the base of the rib cage fixed against the floor, the action of the diaphragm during inhalation can be translated to only one site: the spinal attachment of the crus. And because the deep back muscles are relaxed, each inhalation lifts the lower back and hips, and each exhalation allows them to fall toward the floor (fig. 2.11). Make sure you produce the movement entirely with the diaphragm, not by bumping your hips up and down with the gluteal (hip) and back muscles. Because the inhalations increase the lumbar curvature, this exercise will not be comfortable for anyone with low back pain.

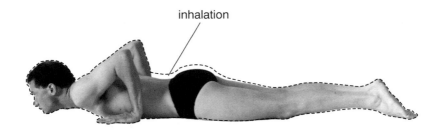

inhalation

Figure 2.11. Diaphragmatic rear lift. With the rib cage anchored against the floor, its lower rim acts as an origin for the diaphragm rather than an insertion (as happens in the cobra posture in fig. 2.10). If the gluteal region and lower extremities remain completely relaxed, the crural attachments of the diaphragm then lift the hips during inhalation and lower them back down during exhalation.

You can feel the diaphragmatic rear lift most easily if you breathe rapidly; the quick inhalations whip the hips up and away from the floor and the sudden exhalations drop them. But if you breathe slowly and smoothly you will notice that each inhalation gradually increases the pull and tension on the hips and lower back, even though it does not create much movement, and that each exhalation gradually eases the tension. When you are breathing slowly enough, you can also feel the muscle fibers of the diaphragm shorten concentrically during inhalation and lengthen eccentrically during exhalation as they control the gravity-induced lowering of the hips toward the floor.

The origins and insertions of the diaphragm are reversed in the diaphragmatic rear lift in comparison with the cobra variation, and this creates repercussions throughout the whole body. In the cobra variation we fix the hips and thighs, allowing the costal attachment of the diaphragm to lift the rib cage, and with the rib cage the entire upper half of the body. In the diaphragmatic rear lift we do just the opposite: we fix the rib cage, relax the hips and thighs, and allow the crural insertion of the diaphragm to lift the lumbar spine and hips.

These two postures also show us how important it is that the diaphragm is indented so deeply by the vertebral column that it almost encircles the spine. This enables it to act both from above and behind to accentuate the lumbar arch during inhalation, lifting the upper half of the body in the cobra variation, and lifting the sacrum and hips in the diaphragmatic rear lift.

THE SOMATIC AND AUTONOMIC SYSTEMS

The way we breathe affects far more than our posture, and we can best explore those ramifications by looking at the two great functional divisions of the nervous system—somatic and autonomic—and at the tissues and organs they each oversee. The *somatic nervous system* is concerned with everything from the control of skeletal muscle activity to conscious sensations such as touch, pressure, pain, vision, and audition. For the *autonomic nervous system*, think first of regulation of blood pressure, viscera, sweat glands, digestion, and elimination—in fact, any kind of internal function of the body that you have little or no interest in trying to manage consciously. This system is concerned with sensory input to the brain from internal organs—generally more for autonomic reflexes than for inner sensations—as well as for motor control of smooth muscle in the walls of internal organs and blood vessels, cardiac muscle in the wall of the heart, and glands (figs. 10.4a–b). Both systems are involved in breathing.

THE SOMATIC SYSTEM

Since breathing draws air into the lungs, and since the lungs are internal organs, we might suppose that the muscles of respiration are controlled by the autonomic nervous system. But they're not. The act of breathing is a somatic act of skeletal muscles. In chapter 1 we discussed the somatic nervous system, although without naming it, when we discussed the control of the skeletal muscles by the nervous system. Respiration makes use of this system, whether we want to breathe fast or slow, cough, sneeze, or simply lift an object while going "oomph." When we participate consciously in any of these activities we breathe willfully to support them, and we do so from the command post in the cerebral cortex that influences the lower motor neurons for respiration. If you are consciously and quietly using the diaphragm as you breathe, you are activating the lower motor neurons whose axons innervate the diaphragm by way of the *phrenic nerves* (figs. 2.6 and 2.12). If you are eight months pregnant the diaphragm can't function efficiently, and in order to breathe you will have to activate lower motor neurons whose axons innervate the intercostal muscles by way of the *intercostal nerves*. And if you are trying to ring the bragging bell at a state fair with a sledge hammer, you will make a mighty effort and a grunt with your abdominal muscles, again calling on motor neurons from the thoracic cord to transmit the cerebral commands to the muscles of the abdominal wall.

The cell bodies for the phrenic nerves are located in the spinal cord in the region of the neck (the *cervical region*), and the cell bodies for the intercostal nerves are located in the spinal cord in the region of the chest (the *thoracic region*). In the neck the spinal cord contains eight cervical segments (C1–8), and in the chest it contains twelve thoracic segments (T1–12; figs. 1.5 and 2.12). The diaphragm is innervated by the right and left phrenic nerves from spinal cord segments C3–5; the intercostal and abdominal muscles are innervated by the intercostal nerves from spinal cord segments T1–12 (figs. 1.5 and 2.12).

Both the phrenic and intercostal nerves are necessary for the full expression of breathing. If for any reason the intercostal nerves are not functional, leaving only the phrenic nerves and a functioning diaphragm intact, the diaphragm will support respiration by itself (fig. 2.12, site d). But in that event the external intercostal muscles will no longer maintain the shape of the chest isometrically, and every time the dome of the diaphragm descends and creates a vacuum in the lungs and pleural cavity, the chest wall will be tugged inward. On the other hand, if for some reason the phrenic nerves are not functional (see asterisks fig. 2.12), but the intercostal nerves and muscles are intact, the vacuum produced by activity of the external intercostal muscles will pull the dome of the flaccid diaphragm higher in the chest during the course of every inhalation.

Like all typical somatic motor neurons, those for respiration are controlled from higher centers in the brain, and life cannot be supported by spinal cord transections above C3 (fig. 2.12, site c). A transection at C6 is

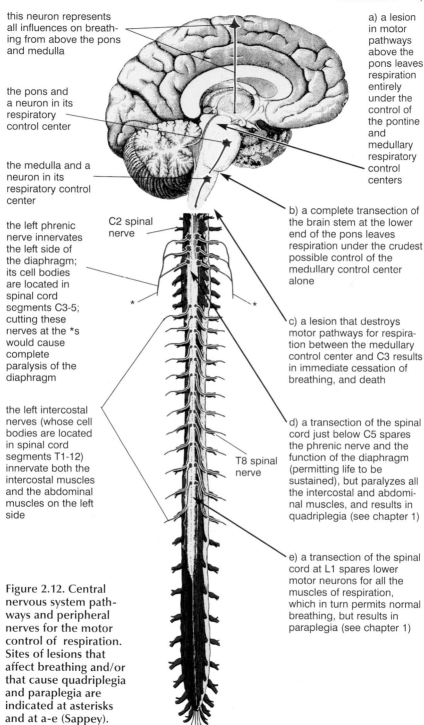

this neuron represents all influences on breathing from above the pons and medulla

the pons and a neuron in its respiratory control center

the medulla and a neuron in its respiratory control center

the left phrenic nerve innervates the left side of the diaphragm; its cell bodies are located in spinal cord segments C3-5; cutting these nerves at the *s would cause complete paralysis of the diaphragm

C2 spinal nerve

the left intercostal nerves (whose cell bodies are located in spinal cord segments T1-12) innervate both the intercostal muscles and the abdominal muscles on the left side

T8 spinal nerve

a) a lesion in motor pathways above the pons leaves respiration entirely under the control of the pontine and medullary respiratory control centers

b) a complete transection of the brain stem at the lower end of the pons leaves respiration under the crudest possible control of the medullary control center alone

c) a lesion that destroys motor pathways for respiration between the medullary control center and C3 results in immediate cessation of breathing, and death

d) a transection of the spinal cord just below C5 spares the phrenic nerve and the function of the diaphragm (permitting life to be sustained), but paralyzes all the intercostal and abdominal muscles, and results in quadriplegia (see chapter 1)

e) a transection of the spinal cord at L1 spares lower motor neurons for all the muscles of respiration, which in turn permits normal breathing, but results in paraplegia (see chapter 1)

Figure 2.12. Central nervous system pathways and peripheral nerves for the motor control of respiration. Sites of lesions that affect breathing and/or that cause quadriplegia and paraplegia are indicated at asterisks and at a-e (Sappey).

not quite as serious. It spares input from the brain to the somatic motor neurons whose axons travel in the phrenic nerve, and in that manner spares the function of the diaphragm, as mentioned above, but it eradicates input to the somatic motor neurons that innervate the intercostal and abdominal muscles, as well as to the rest of the skeletal muscles of the body from the neck down, resulting in quadriplegia (fig. 2.12, site d; also see chapter 1). If a complete transection occurs at L1—the first lumbar segment of the spinal cord—all input to all motor neurons for all muscles of respiration is spared and breathing is normal, although such a transection would result in paraplegia (fig. 2.12, site e; see also chapter 1).

THE RESPIRATORY CENTERS

Breathing goes on twenty-four hours a day. We can regulate it mindfully from the cerebral cortex if we want, in the same way that we can regulate our movement and posture, but most of the time our minds are occupied elsewhere and we rely on other motor centers to manage respiration. These respiratory control centers are located in the two lowest segments of the *brain stem* (the continuation of the spinal cord into the brain). A crude rhythm for respiration is generated in the lowest of these segments—the *medulla*—and this is fine-tuned by the next higher segment—the *pons* (fig. 2.12). Input from these centers to the motor neurons of respiration is unconscious. Willed respiration, of course, is directed from the cerebral cortex and can override the rhythms generated by the lower segments of the brain. But even if higher centers have been destroyed by a stroke or traumatic head injury (fig. 2.12, site a), the controlling centers for respiration in the pons or even just the medulla may still survive, allowing someone who is otherwise brain-dead to continue breathing indefinitely.

We depend on the respiratory centers to manage somatic aspects of breathing automatically, but sometimes the mechanisms do not work perfectly. In a rare form of sleep apnea—the *central hypoventilation syndrome*—the automatic control of ventilation is lost but the ability to breathe voluntarily is preserved. This is roughly similar to a circumstance immortalized in Jean Giraudoux's play *Ondine*. Ondine, a water nymph and an immortal, married Hans, a mortal, even though she knew that such a union was forbidden and that Hans was doomed to die if he was unfaithful to her. When the prophecy was fulfilled, Hans was deprived of his automatic functions. "A single moment of inattention," he tells Ondine, "and I forget to breathe. He died, they will say, because it was a nuisance to breathe." And so it came to be. This form of sleep apnea is now known as Ondine's curse.

Although the respiratory pathways in the brain stem support the most primitive form of rhythmic breathing, higher centers can either smooth this out or disrupt it. We all know that when we are in intense emotional states our breathing becomes jerky and irregular. Watch a baby struggle to breathe while it is preparing to cry, or think of how uncontrollable laughing affects a teenager's breathing. By contrast, when we are calm, the somatic motor circuits for respiration will be delicately balanced and our breathing will be smooth and even. Maintaining such even-tempered states is one of the aims of yoga.

THE AUTONOMIC NERVOUS SYSTEM

When you think of the autonomic nervous system, the first point is not to confuse the terms automatic and autonomic. We can breathe automatically courtesy of the somatic nervous system, but the word autonomic is derived from "autonomy," the quality of being independent. In the context of the two great divisions of the nervous system, the autonomic nervous system is largely independent of the somatic system; it consists of a vast auxiliary network of neurons that controls viscera, blood vessels, and glands throughout the body. It is not, however, completely autonomous, because it interacts with the somatic nervous system—it both feeds sensory information from within the body into the somatic systems of the brain and spinal cord (in this case our main concern is the respiratory centers), and is affected by the somatic motor systems in return.

We constantly depend on smooth interactions between the somatic and autonomic nervous systems. You race around the block using your skeletal muscles, which are controlled by the somatic nervous system, but you would not get far unless your autonomic nervous system sped up your heart, stimulated the release of glucose from your liver, and shunted blood from the skin to the skeletal muscles. And if, instead of running around the block, you sit down and read a book after dinner, you flip the pages using your skeletal muscles and depend on the unconscious operation of your autonomic nervous system to digest your meal. Respiration, as it happens, is the foremost function in the body in which signals from internal organs have a constant and continuing effect on somatic function, in this case the rate and depth of breathing, twenty-four hours a day.

If we look at an overview of how the autonomic nervous system operates, controlling autonomic influences from the central nervous system (the brain and spinal cord) are relayed to their visceral targets by two systems of autonomic motor neurons: sympathetic and parasympathetic. The *sympathetic nervous system* prepares the body for emergencies ("fight or flight") and the *parasympathetic nervous system* maintains the supportive functions of the internal organs. Between them, by definition, these two systems execute the autonomic motor commands from the brain and spinal cord. More of these interactions will be discussed in chapter 10, in which we'll be concerned with the importance of the autonomic nervous system in relaxation.

Here our concern is limited mainly to breathing, and the first thing to note is that the most important autonomic relationship involving the control of respiration is sensory. This does not mean sensory in regard to something you can feel; it refers to influences from sensory receptors that have an impact on breathing. Specifically, the sensory limb of the autonomic nervous system carries information on oxygen and carbon dioxide levels in the blood and cerebrospinal fluid to the respiratory control centers in the brain stem. You would see the important respiratory linkage between the autonomic and somatic systems in operation if you were suddenly rocketed from sea level to the top of Alaska's Mount Denali. You would immediately begin to breathe faster because your somatic respiratory control centers receive autonomic sensory signals that your blood is not getting enough oxygen, not because you make a conscious somatic decision that you had better do something to get more air.

There are also purely autonomic mechanisms that affect breathing in other ways. The most obvious example is familiar to those who suffer from asthma, or from chronic obstructive pulmonary disease (COPD) combined with bronchitis, and that is the difficulty of moving air through constricted airways. It is not very helpful to have healthy skeletal muscles of respiration if the airways are so constricted that they do not permit the passage of air. Although this is a complex and multifaceted problem, the autonomic nervous system involvement appears to be straightforward. In quiet times when there is less need for air, the parasympathetic nervous system mildly constricts the smooth muscle that surrounds the airways, especially the smaller bronchioles, and thereby impedes the flow of air to and from the alveoli. But in times of emergency or increased physical activity, the sympathetic nervous system opens the airways and allows air to flow more easily. Those who have chronic respiratory diseases have an acute awareness of how difficult it can be to medicate and regulate this system.

HOW BREATHING AFFECTS THE AUTONOMIC NERVOUS SYSTEM

All of our concerns so far have been with how the nervous system influences breathing. These are all widely recognized. What is not as well-known is that different methods of breathing can affect the autonomic nervous system and have an impact on the functions we ordinarily consider to be under unconscious control. Abnormal breathing patterns can stimulate autonomic reactions associated with panic attacks, and poor breathing habits in emphysema patients produce anxiety and chronic overstimulation of the sympathetic nervous system. By contrast, quiet breathing influences the autonomic circuits that slow the heartbeat and reduce blood pressure, producing calm and a sense of stability. Our ability to control respiration consciously gives us access to autonomic function that no other system of the body can boast.

2:1 BREATHING

One breathing technique that can produce a beneficent effect on the autonomic nervous system is 2:1 breathing—taking twice as long to exhale as to inhale. For those who are in good condition, 6-second exhalations and 3-second inhalations are about right, and if you can regulate this without stress, the practice will slow your heart down and you will have a subjective experience of relaxation. As with almost all breathing exercises in yoga, both inhalation and exhalation should be through the nose.

This connection between heart rate and breathing, known as *respiratory sinus arrhythmia*, involves reflex activity from the circulatory system to the brain stem that causes the heart to beat more slowly during exhalation than it does in inhalation. It is a natural arrhythmia, called "respiratory" because it is induced by respiration, and called "sinus" because the receptors that stimulate the shifts in heart rate are located in the aortic and carotid sinuses, which are bulbous enlargements in those great vessels. If you take longer to exhale than to inhale, especially when you are relaxing, the slowing-down effect of exhalation will predominate. This is an excellent example of how we can willfully intervene to produce effects that are usually regulated by the autonomic nervous system.

There are limits on both ends to the effects of 2:1 breathing. If you are walking briskly, exhaling for two seconds and inhaling one second, you will not get this reaction, and if you take it too far in the other direction, which for most people means trying to breathe fewer than five breaths per minute (8-second exhalations and 4-second inhalations), the exercise may become stressful and cause the heart rate to increase rather than slow down. The golden mean—that which is entirely comfortable—is best.

There is one well-known practical consequence of respiratory sinus arrhythmia. For decades doctors have known empirically that *pursed-lip breathing* against moderate resistance is helpful for those with obstructive lung disease. What is not generally realized is that the practice is helpful mainly because it lengthens exhalations, slows the heart rate, decreases the amount of air remaining in the lungs after exhalation, and reduces fear and anxiety. Knowledgeable yoga teachers realize that the same end can be accomplished through a different approach—lengthening exhalations by pressing in gently with the abdominal muscles while at the same time breathing through the nose.

THE PHYSIOLOGY OF RESPIRATION

Different hatha yoga breathing exercises affect respiration in different ways, but before we can understand how they do this we need a little more background. We'll start our discussion with a look at the amount of air found in the lungs and airways at different stages of the breathing cycle.

These values—the lung volumes, capacities, and anatomic dead space—vary according to stature, age, sex, and conditioning, so to keep things simple we'll always use round numbers that are characteristic for a healthy young man. The numbers are generally smaller for women, for older men and women, and for those in poor physical condition. That's not so relevant to us here. Our main interest is not in how the lung volumes and capacities vary in different individuals; it is in how they vary with different breathing practices and postures. The numerical representations in fig. 2.13, as well as in all the charts on respiration, are only simulations, but they will be a useful starting point for more rigorous inquiry.

LUNG VOLUMES, CAPACITIES, AND THE ANATOMIC DEAD SPACE

There are four *lung volumes* (fig. 2.13). We'll begin with the *tidal volume*, which is the amount of air that moves in and out in one breath. Textbooks state that in our healthy young man it amounts to one pint, or about 500 ml (milliliters) during relaxed breathing, but this volume is obviously circumstantial—when we are climbing stairs it will be greater than when we are sitting quietly. The *inspiratory reserve volume*, about 3,300 ml (3 1/2 quarts), is the additional air you can inhale after an ordinary tidal inhalation. The *expiratory reserve volume*, about 1,000 ml, is the additional air you can exhale after a normal tidal exhalation. The *residual volume*, about 1,200 ml, is the amount of air that remains in the lungs after you have exhaled as much as possible.

Lung capacities, of which there are also four, are combinations of two or more lung volumes (fig. 2.13). First, the *vital capacity* is the total amount of air you can breathe in and out; it totals 4,800 ml and is the combination of the tidal volume plus the inspiratory and expiratory reserve volumes. This is the most inclusive possible definition of the yogic "complete breath," and is an important clinical value.

Second, the *total lung capacity* is self-explanatory. In a healthy young man is amounts to about 6,000 ml and is the sum of all four lung volumes, or alternatively, the sum of the vital capacity and the residual volume.

Third, the *inspiratory capacity* is the total amount of air you can inhale at the beginning of a normal tidal inhalation. This is a restrictive definition of the yogic "complete breath," which is the combination of the tidal volume and inspiratory reserve volume (about 3,800 ml).

Fourth, the *functional residual capacity*, 2,200 ml, is the combination of the residual volume and the expiratory reserve volume. As its name implies, this is an especially practical quantity—the amount of air in the lungs at the end of a normal exhalation that will be mixed with a fresh inhalation. This usually amounts to a lot of air—more than four times as much as an ordinary tidal volume of 500 ml. One point of pursed-lip breathing, discussed earlier, is to drastically decrease this value so that the fresh air that you inhale is mixed with a smaller volume of oxygen-poor air.

Lung volumes and capacities differ markedly in different hatha yoga postures and practices. For example, agni sara (chapter 3) almost obliterates the expiratory reserve volume and increases the tidal volume from 500 ml to possibly 1,600 ml (figs. 3.31–33); inverted postures (chapters 8 and 9) decrease the expiratory reserve volume and shift the tidal volume closer to the residual volume; and the bellows breath, which will be discussed at length later in this chapter, minimizes the tidal volume.

The *anatomic dead space* is another extremely important clinical value—the air-filled space taken up by the airways, which include the nasal passages, pharynx, larynx, trachea, right and left primary bronchi, and the branches of the bronchial tree that lead to the alveoli. It is called a dead space because it does not, unlike the alveoli, transport oxygen into the blood and carbon dioxide out. This space ordinarily totals about 150 ml, so for a tidal volume of 500 ml, only 350 ml of fresh air actually gets to the alveoli. You can get an immediate idea of its significance when you are snorkeling. If you breathe through a snorkel tube with a volume of 100 ml, the practical size of the anatomic dead space increases from 150 ml to 250

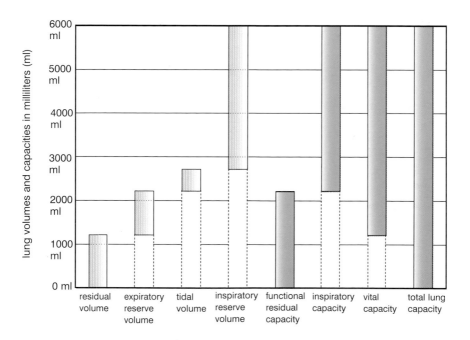

Figure 2.13. The four lung volumes(on the left) and the four lung capacities (on the right). The latter are combinations of two or more lung volumes. All the values are simulations for a healthy young man.

ml, you have to inhale 600 ml through the end of the tube just to get 350 ml to the alveoli, and you might have a few moments of panic before you adjust to the need for deeper breaths. Clinical concerns with the anatomic dead space are often grim: in terminal emphysema patients its volume sometimes approaches and exceeds the vital capacity.

ALVEOLAR AND MINUTE VENTILATION

When we consider how much air we inhale and exhale over a period of time, the first thing we think of is the *minute ventilation*, the amount of air we breathe in and out over a period of 60 seconds. This is what we feel—the touch of the breath in the nostrils, in and out, over a period of one minute. All you have to do to calculate your minute ventilation is measure your tidal volume and multiply that value times the number of breaths you take per minute. According to textbooks, this would be 500 ml per breath times 12 breaths per minute, and this equals 6,000 ml per minute.

The minute ventilation does not tell us everything we need to know, however, because what is most important is not the amount of air that moves in and out of the nose or mouth, but the amount of air that gets past the anatomic dead space into the alveoli. This is also measured over a period of one minute and is called, logically enough, the *alveolar ventilation*. It is our primary concern when we want to know how breathing affects the content of oxygen and carbon dioxide in the blood, and that is our main interest in yoga breathing exercises. To calculate the alveolar ventilation, subtract the size of the anatomic dead space from the tidal volume before multiplying by the respiratory frequency. For example, 500 ml of tidal volume minus 150 ml of anatomic dead space equals 350 ml per breath, and 350 ml per breath times 12 breaths per minute yields an alveolar ventilation of 4,200 ml per minute.

The values given for lung volumes and capacities, as well as for minute and alveolar ventilation, are only textbook examples—it is not uncommon to breathe more rapidly and take in a smaller tidal volume for each breath. If you watch a dozen people closely in casual situations, such as when they are sitting on a bus with their arms folded across their chests, you can easily count the breaths they take per minute, and it is usually faster than the textbook standard of 12 breaths per minute: 24–30 breaths per minute is a lot more common. This is of no great consequence because everyone simply adjusts their tidal volume so that their alveolar ventilation stays within a normal range (fig. 2.14). In meditation the rate of breathing generally seems to slow down, but it can still vary widely and may either be faster or slower than the standards cited in the medical literature on respiration. Here too, you adjust the rate of breathing and the tidal volume so that the alveolar ventilation comes in line with the metabolic requirements of the practice.

ATMOSPHERIC, ALVEOLAR, AND BLOOD GASES

The whole point of breathing is to get oxygen from the atmosphere to the cells of the body and carbon dioxide from the body into the atmosphere, and to understand how this happens we need to know how diffusion and pressure differentials drive those processes. Here's how it works: A gas moves from a region of high concentration to one of low concentration, just as a drop of dye placed in a glass of water gradually diffuses throughout, sooner or later equalizing the mixture until it has colored all the water in the glass uniformly. Very crudely, something similar happens in the body. There is much concentration of (or pressure from) oxygen in the atmosphere, less in the alveoli, less than that in the arterial blood, and less yet in the cells of the body that are using the oxygen. By the same token, there is much concentration of (or pressure from) carbon dioxide in the vicinity of the cells that are eliminating it, somewhat less in the veins and alveoli, and almost none in the atmosphere.

The standard measure of pressure we use for gases is millimeters of mercury (mm Hg), which is the height of a column of mercury that has the same weight as a column of gas that extends all the way out to the stratosphere. In other words, if we think of ourselves as bottom-dwellers in a sea of air, which we assuredly are, the weight of a column of air above us at sea level is the exact equivalent of the weight of a column of mercury of the same diameter that is 760 mm in height. We use this unit for measuring

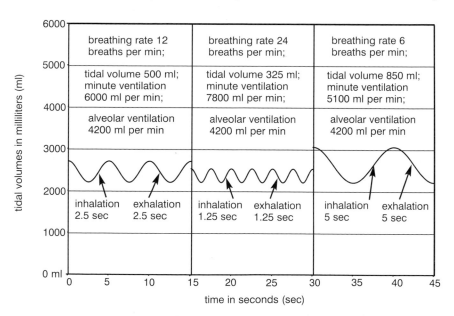

Figure 2.14. Three modes of breathing with identical alveolar ventilations. The numerical values are simulations for a healthy young man.

many values: total atmospheric pressure; the atmosphere's itemized content of nitrogen, oxygen, and other gases; the decreased oxygen and increased carbon dioxide in the alveoli; and the content of oxygen and carbon dioxide in the blood.

Atmospheric pressure decreases with increasing altitude. At sea level it is 760 mm Hg, and of this total, the oxygen share is about 150 mm Hg, the nitrogen share is about 580 mm Hg, and water vapor is about 30 mm Hg, depending on the humidity. At the summit of Pike's Peak in Colorado atmospheric pressure is 450 mm Hg (oxygen 83 mm Hg), and at the summit of Mount Everest in the Himalayas it is 225 mm Hg (oxygen 42 mm Hg). Going in the other direction to a depth of 165 feet under water (which is considered by diving experts a prudent maximum depth for breathing atmospheric air that has been pressurized by the depth of the water), atmospheric pressure is 4,500 mm Hg and oxygen is 900 mm Hg.

Returning to more ordinary circumstances, let's limit ourselves for the moment to what we would see inside and outside the body at sea level. If we are quietly breathing atmospheric air at our favorite seaside resort, where the oxygen content is about 150 mm Hg, we'll end up with oxygen levels of about 104 mm Hg in the alveoli, which is reduced from 150 mm Hg because of the transfer of oxygen from the alveoli into the blood. Passing on down the pressure gradient, arterial blood contains slightly less oxygen, about 100 mm Hg. Venous blood, or blood that has just released its oxygen in the tissues, contains dramatically less, about 40 mm Hg. Carbon dioxide decreases in the other direction from the blood to the atmosphere, from a high of 46 mm Hg in venous blood to 40 mm Hg in arterial blood and the alveoli, and finally to a negligible 0.3 mm Hg in the atmosphere.

The numbers for atmospheric, alveolar, and blood gases can all be compared conveniently in table 2.1. The ones we are especially concerned with when we look at pulmonary ventilation and breathing exercises in yoga are the pressures for oxygen and carbon dioxide in atmospheric air, alveoli, arterial blood, and venous blood.

Too little alveolar ventilation is *hypoventilation*, and too much is *hyperventilation*. Both conditions will have repercussions in the alveoli, arterial blood, and venous blood, as well as on tissues throughout the body. Hypoventilation will result in reduced levels of oxygen and increased levels of carbon dioxide at all those sites, and hyperventilation drives the figures in the opposite direction (see table 2.2).

HYPOVENTILATION

Everyone has an intuitive understanding that we have to have oxygen to live, and most people have experienced an undersupply of oxygen at one time or another, if only from holding the breath. What's not always recognized, at least from personal experience, is that the momentary discomfort of smothering is a warning of something more serious: that the cells of the brain and spinal cord are acutely sensitive to oxygen deficits, that a severe deprivation of oxygen will cause temporary damage to the tissue in less than a minute, and that neurons totally deprived of oxygen for about five minutes (as in the case of stroke) will die.

gases	dry air, any temperature	moist air, 75°, 80% relative humidity	warm (98.6°) wet air	alveolar gaseous pressures	arterial blood gases	venous blood gases
oxygen	159.1	150	**149.2**	**104**	**100**	**40**
carbon dioxide	0.3	0.3	**0.3**	**40**	**40**	**46**
water vapor	0.0	30	47	47	47	47
nitrogen	600.6	579.7	563.5	569	573	627
totals, in mm Hg	760	760	760	760	760	760

Table 2.1. The above chart shows pressures in mm Hg (millimeters of mercury) expected during the course of relaxed breathing at sea level ; the most important eight values are shown in boldface. Nitrogen is inert: its values are determined solely by altitude and the summed specific pressures for oxygen, carbon dioxide, and water vapor.

Hypoventilation, or underbreathing, is a related matter, and another condition that is familiar to people with respiratory problems. They call it shortness of breath. Hypoventilation is not usually a serious matter for anyone who is in good health, for whom a few deep breaths will usually step up the alveolar ventilation enough to bring the oxygen and carbon dioxide levels into balance. This is also the aim of several hatha yoga breathing exercises that increase ventilatory capacity, especially the bellows breath.

gases	alveolar gases			arterial blood gases			venous blood gases		
relative ventilation	hypo-vent.	normal vent.	hyper-vent.	hypo-vent.	normal vent.	hyper-vent.	hypo-vent.	normal vent.	hyper-vent.
oxygen	90	**104**	140	85	**100**	120	32	**40**	60
carbon dioxide	50	**40**	15	51	**40**	15	56	**46**	30

Table 2.2. The above chart shows simulations of alveolar and blood gases in mm Hg for hypoventilation, normal breathing, and hyperventilation. The six figures in boldface are the norms, repeated from table 2.1.

But vigorous practice of bellows breathing brings up the question of hyperventilation, or overbreathing, and this, paradoxically, can create a deficit in the supply of oxygen for the cells of the central nervous system where we need it the most.

HYPERVENTILATION

Let's say you are hyperventilating during the course of an extreme bellows exercise. If this involves breathing in and out a tidal volume of 500 ml three times per second, you will end up with an alveolar ventilation of 180 breaths per minute times 350 ml per breath, which equals 63,000 ml per minute, or fifteen times the norm of 4,200 ml per minute. If you were in world-class athletic condition and running full speed up forty flights of stairs, this would be fine. During heavy exercise your body will use all the oxygen it can get, and it will also need to eliminate a heavy overload of carbon dioxide. It's not, however, a good idea for an ordinary person to breathe in this way. Extreme hyperventilation when you are not exercising strenuously skews the blood gases too much.

Our first thought is that hyperventilation must drive too much oxygen into your tissues, but this is inaccurate. Except for a few special circumstances, such as breathing 100% oxygen for prolonged periods, or breathing oxygen at high pressure in deep-sea diving, you can't get too much, and the increased oxygen in the blood that results from hyperventilation is certainly not harmful.

The problem with hyperventilation is not that it increases arterial oxygen but that it decreases arterial carbon dioxide, and that can have an unexpected side effect. What happens is that a substantial reduction in arterial carbon dioxide constricts the small arteries and arterioles of the brain and spinal cord. The way this happens, or at least the end result, is very simple: an arteriole acts crudely like an adjustable nozzle on the end of a garden hose that can open to emit a lot of water or clamp down to emit only a fine spray. As carbon dioxide in the blood is reduced, the arterioles clamp down and the blood supply to the tissue is restricted until there is so little blood flowing to the brain that it doesn't matter how well it is oxygenated. Not enough blood (and therefore not enough oxygen) can get through the arterioles to the capillary beds and adequately support the neurons.

Hyperventilating vigorously enough to dramatically lower blood carbon dioxide doesn't necessarily result in death or even obvious clinical symptoms, but it can cause more general complaints such as fatigue, irritability, lightheadedness, panic attacks, or the inability to concentrate. It's not illogical that the folk remedy for panic attacks, which is still routinely administered by triage nurses in emergency rooms, is to have someone who is in such a state breathe into a paper bag. Rebreathing our exhaled carbon dioxide increases carbon dioxide levels in the blood and opens the cerebral circulation. There are better solutions, however, and triage nurses who have also had some training in relaxed yogic breathing practices would be more imaginative, perhaps suggesting something as simple as having the patient lie supine and breathe abdominally with their hands or a moderate weight on the abdomen.

Extremely low blood levels of carbon dioxide can cause you to pass out. Children at play sometimes hyperventilate, hold their breath after a deep inhalation, and then strain against a closed glottis. If they do this for only 3–4 seconds they will drop to the floor like stones. Increasing intrathoracic pressure from straining will have diminished the venous return to the heart (and thus the cardiac output) immediately after the cerebral circulation has been partially occluded by hyperventilation, and these two ingredients combined cut off enough of the blood supply to the brain to cause an immediate but temporary loss of consciousness. The danger of passing out from constricted brain arterioles is also why lifeguards do not allow swimmers to hyperventilate vigorously before swimming underwater. Hyperventilating followed by holding the breath after a deep inhalation is not harmful to children on a grassy lawn who will begin to breathe normally as soon as they lose consciousness, but it is deadly under water.

One of the most demanding tests of aerobic capacity is mountain climbing without bottled oxygen at altitudes higher than 25,000 feet. Superbly conditioned athletes are able to meet this standard and reach the summit of Mount Everest by hyperventilating the oxygen-poor atmosphere (42 mm Hg at 29,000 feet) all the way to the top. They can jam enough oxygen into their arterial blood to survive (about 40 mm Hg), and that's good; but the hyperventilation also drives their alveolar carbon dioxide down to less than 10 mm Hg, and that's not so good. They have to train rigorously at high altitudes to adapt the cerebral circulation to such extremely low levels of carbon dioxide. If most of us were transported unprepared to such an altitude (as would happen if we suffered a sudden loss of cabin pressure in an airliner cruising at 29,000 feet), we would experience so much reflex hyperventilation and subsequent constriction of the cerebral circulation that without supplemental oxygen we would pass out in about two minutes and die soon thereafter.

Beginning hatha yoga students who practice the bellows breath excessively may experience some adverse symptoms of hyperventilation, especially irritability. But if they continue the practice over a period of time, the cerebral circulation gradually adapts to decreased levels of carbon dioxide in the blood, and they can intensify their practice and safely gain the benefits of alertness and well-being associated with higher levels of blood oxygen.

CHEMORECEPTORS

The levels of oxygen and carbon dioxide in the blood and cerebrospinal fluid are monitored by *chemoreceptors*, specialized internal sensors of the autonomic nervous system. Sensory nerve endings associated with these receptors then transmit nerve impulses coded for distorted levels of oxygen and carbon dioxide directly to the circuits of the somatic nervous system that regulate breathing (fig. 2.15). Accordingly, the chemoreceptors are important keys to linking the autonomic and somatic systems.

There are two classes of chemoreceptors: peripheral and central. The *peripheral chemoreceptors*, which are located in the large arteries leading away from the heart, react quickly to substantial reductions of arterial oxygen and strongly stimulate respiration. If you restrict your breathing, or if

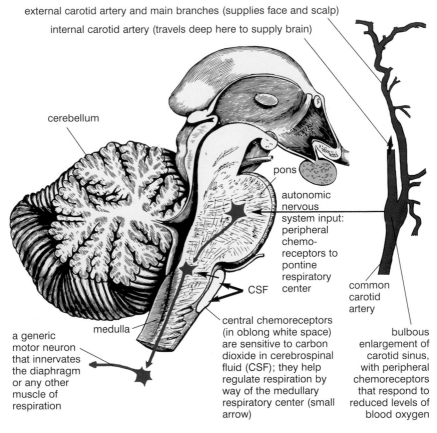

external carotid artery and main branches (supplies face and scalp)

internal carotid artery (travels deep here to supply brain)

cerebellum

pons

autonomic nervous system input: peripheral chemoreceptors to pontine respiratory center

CSF

common carotid artery

a generic motor neuron that innervates the diaphragm or any other muscle of respiration

medulla

central chemoreceptors (in oblong white space) are sensitive to carbon dioxide in cerebrospinal fluid (CSF); they help regulate respiration by way of the medullary respiratory center (small arrow)

bulbous enlargement of carotid sinus, with peripheral chemoreceptors that respond to reduced levels of blood oxygen

Figure 2.15. Brain stem and cerebellum on the left (with central chemoreceptors near the front surface of the medulla), and on the right, the carotid sinus (with peripheral chemoreceptors) just below the bifurcation of the common carotid artery into the internal and external carotid arteries (Quain).

you are at an altitude that cuts your arterial oxygen in half (that is, from 100 mm Hg to 50 mm Hg), the input of the peripheral chemoreceptors to the brain stem respiratory centers will quadruple your alveolar ventilation from a norm of 4,200 ml per minute to about 16,000 ml (16 liters) per minute. Even if you are well enough conditioned to walk up a 30° grade at sea level with only moderate increases in alveolar ventilation, you will find yourself panting when you hike up that same grade at a high altitude.

Although the peripheral chemoreceptors respond to large decreases in blood oxygen, they do not respond significantly to small decreases. If you are only somewhat short of oxygen you may simply lose the edge of your alertness and just feel like yawning and taking a nap, which is the point at which yoga breathing exercises are indicated.

Central chemoreceptors, which are located on the surface of the brain stem immediately adjacent to the somatic respiratory control centers, stimulate the rate and depth of respiration in response to increased levels of carbon dioxide, and dampen respiration if levels of carbon dioxide fall. They are more sensitive to small changes than the peripheral chemoreceptors, but they are slower to react because the cerebrospinal fluid in which they are bathed is isolated from the blood supply and does not respond instantly to changes in blood carbon dioxide.

The differing sensitivities of the peripheral and central chemoreceptors sometimes results in their working at cross-purposes. For example, at high altitudes decreased oxygen stimulates the peripheral chemoreceptors to increase ventilation, but this also lowers carbon dioxide, and when that happens the central chemoreceptors start to retard ventilation. You may require the extra air for the sake of the oxygen, but the response to decreased carbon dioxide confounds that need. Training the system to adapt to such conflicting signals is part of the process of high-altitude acclimation.

THE ROLE OF WILL

Dozens of physical, mental, and environmental factors cooperate to influence respiration, and some of these work at odds with one another. Our will can override most of them. You can counter the state of being bored and sleepy by practicing bellows breathing. If you are bicycling behind a smelly bus you can hold your breath, at least momentarily, to escape the fumes. If you have the habit of breathing irregularly you can learn even meditative breathing. If you are upset you can breathe slowly and evenly to calm down. Most important, you can learn to observe healthier breathing patterns while you are doing hatha yoga postures; then you can carry the refined habits over into your daily life. To see specifically how this works in the practical environment of yoga postures, we'll look at four different kinds of breathing: thoracic, paradoxical, abdominal, and diaphragmatic.

THORACIC BREATHING

Specialists in holistic therapies often condemn *thoracic,* or chest, *breathing,* but there are two possible scenarios for this mode of breathing that should be considered separately: one is empowering and has an honored role in hatha yoga, and the other is constricting and can create physical and mental health problems if it is done habitually. First, we'll look at the beneficial version.

EMPOWERED THORACIC BREATHING

To get a feel for the best of *empowered thoracic breathing* (fig. 2.29a), stand up, interlock your hands behind your head, pull your elbows to the rear as much as possible, bend backwards moderately, and inhale, expanding the chest maximally. Lift your elbows and expand the chest until you feel the intercostal muscles reach their outermost limits of isometric tension.

[Technical note: Although the diaphragm is not as obviously involved in this method of breathing as the intercostal muscles, it supports inhalation synergistically. How? Its muscle fibers resist lengthening by keeping the dome of the diaphragm from being pulled freely toward the head as inhalation proceeds (unlike what we'll soon see for paradoxical breathing), and at the peak of inhalation, it holds momentarily in a state of isometric tension.]

Next, let your hands hang down and pull your elbows slightly to the rear, again while bending back moderately and inhaling as much as you can. If you observe carefully you'll see that you can slightly increase your inspiratory capacity with the arms in this more neutral position. How can you prove this? Go back to the first posture, inhale as much as possible, then hold your breath at the glottis at the end of your fullest possible inhalation. Still holding your breath, assume the second position with your hands hanging and elbows back, and you will immediately confirm that you can inhale a little more. Then as a control experiment, just to be certain, try it the opposite way, first a maximum inhalation with the hands down and elbows back, and second with the hands behind the head and the elbows strongly lifted and pulled to the rear. You'll find that coming into the latter position secondarily (after locking the glottis in the first position) mandates a release of air once you open the glottis. These are not yoga practices, of course, but experiments to test the effects of particular arm positions on your inspiratory reserve volume during the course of empowered thoracic breathing. You can also experiment with any number of other standing postures. If, for example, you grasp your elbows tightly behind your back with your opposite hands, or come into a forward bend supporting your hands on the thighs just above the knees, you will find that these arm positions markedly limit your inhalation.

In general, there are three major reasons for variations in inspiratory capacity that are due to posture. One is obvious: sometimes the position of the upper extremities compresses the chest and limits inhalation mechanically.

The other two are more subtle: many of the muscles of the upper extremities serve either as synergists or as antagonists to the external intercostal muscles for enlarging the chest. The relationships are straightforward: any position that favors the synergistic effects will increase inspiratory capacity, and any position that favors the antagonistic effects will decrease it.

One of the most effective training exercises for increasing your inspiratory capacity takes its cue from a standard barbell exercise. In this case you can simply swing a broomstick or a light barbell without added weights from your thighs to 180° overhead, doing 10–15 repetitions while keeping your elbows extended. Exhale maximally as you bring the broomstick or barbell to your thighs, and inhale maximally as you bring it overhead. As a barbell exercise, this is designed to develop and stretch chest muscles such as the *pectoralis major* (fig. 8.8–9), but many of the muscles needed for moving the barbell through its arc also act synergistically with the external intercostals to facilitate inhalation. This is also a great exercise for children with asthma, who often tend to be parsimonious when it comes to using their chests for breathing. If their asthma is typically induced by exercise, they should of course use a broomstick instead of a barbell, and be sensitive to their capacity.

In hatha yoga generally, inhaling as much as you can is an excellent chest exercise any time you are doing simple whole-body standing back-ward bends (fig. 4.19), diaphragm-assisted backbends (fig. 5.7), cobra postures (especially those shown in figs. 2.10 and 5.9–12), the upward-facing dog (figs. 5.13–14), prone boats and bow postures (fig. 5.20–23), variations of the cat pose in which the lumbar region is arched forward (figs. 3.30 and 3.34b), or possibly best of all, any one of several variations of the fish posture (figs. 3.19a, 5.28, and 9.19). In fact, whenever an instructor suggests taking the deepest possible inhalations, this can only mean placing an emphasis on empowered thoracic breathing, and it works well in any relatively easy posture in which it is natural to thrust the chest out.

CONSTRICTED THORACIC BREATHING

Constricted thoracic breathing (fig. 2.29b) is typically shallow, rapid, and irregular. It is commonly associated with stress and tension, and our main interest in analyzing it is to understand why it is inadvisable to breathe that way habitually. Whenever someone criticizes chest breathing, this is what they are talking about.

To help students understand why constricted thoracic breathing is undesirable, ask them to lie in the corpse posture (figs. 1.14 and 10.2), placing the left hand on the abdomen and the right hand on the chest. First of all they should concentrate on moving only the front surface of the abdomen when they breathe; the right hand should be stationary and the left hand should rise toward the ceiling during inhalation and come back down during exhalation. Ask them to notice that this is natural and comfortable. Then, to do thoracic breathing, ask them to breathe so that the left hand is stationary and the right hand is lifted toward the ceiling. This feels so unnatural, at least in the supine position, that many students in a beginning class won't be able to do it. You will probably have to demonstrate and explain that you are not teaching a relaxed or empowered yogic breathing practice; you simply want students to experience this form of thoracic breathing so they can contrast it with other options.

In thoracic breathing the hand on the abdomen is stationary because rigid abdominal muscles prevent the dome of the diaphragm from moving, and the only way you can inhale is to lift and expand the upper part of the chest. This is not a relaxing breathing pattern, and some people will know in advance that the exercise will be stressful—don't insist that everyone do it.

When you breathe thoracically while standing (fig. 2.16), you can feel the external intercostal muscles expand the rib cage, especially during a deep inhalation, and you can feel them resist its tendency to get smaller during

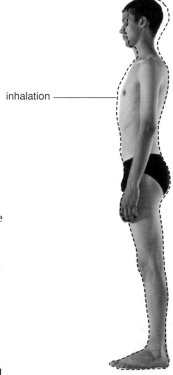

inhalation ————

Figure 2.16. Thoracic breathing. The dotted line reveals the profile for a moderately empowered thoracic inhalation, with the sternum lifted up and out in accordance with the "pump-handle" analogy. The abdomen and respiratory diaphragm remain relatively fixed in position, and the head is pulled to the rear. The halftone profiles a normal exhalation.

exhalation. This is fine for empowered thoracic breathing, but it feels out of place in the supine posture.

The role of the internal intercostals, whether standing or supine, is not so obvious. In the first place they do not become fully active except in forced exhalation, even in someone who has healthy breathing patterns. Second, habitual chest breathers are generally compulsive about inhalation, as though they are afraid to exhale, and because of this they may not make much use of their internal intercostal muscles under any circumstances.

Other muscles in the neck, chest, and shoulders also support thoracic breathing as a side effect to some other action. The scalenes (figs. 2.8 and 8.13), which take origin from the cervical spine and insert on the clavicle (the collarbone) and first rib, have their primary effect on the neck, but they also lift the chest during the course of a complete inhalation. We call this *clavicular breathing* to differentiate it from lifting the chest with the intercostal muscles. In addition, as mentioned earlier, most of the muscles that stabilize the scapula and move the arms also have indirect effects on breathing for the simple reason that they attach to the chest.

PROBLEMS WITH CONSTRICTED THORACIC BREATHING

During constricted chest breathing both inhalation and exhalation are hesitant and tentative. This breathing pattern is not common among experienced yoga students, who have a large repertoire of more useful forms of breathing, but you see it occasionally in beginning classes. And once in a while during the course of a classroom demonstration you'll even hear someone say "That's how I always breathe!" The abnormal upper body tension associated with this form of breathing is palpable—both literally and figuratively—in faces, necks, and shoulders.

Habitual chest breathing not only reflects physical and mental problems, it creates them. It mildly but chronically overstimulates the sympathetic nervous system, keeping the heart rate and blood pressure too high, precipitating difficulties with digestion and elimination, and causing cold and clammy hands and feet. In common usage chest breathing is known as "shallow" breathing, and if you watch people breathe in this fashion for any length of time you will notice that every once in a while they will sigh, yawn, or take a much deeper breath to bring in more air.

If you really want to understand shallow breathing you have to experiment with yourself. In either a supine or upright posture, try taking 20–30 constricted thoracic breaths, lifting only the upper part of the chest. Be careful not to move the abdomen, and try to keep the lower part of the chest from moving. To do this you have to keep the abdominal wall rigid and hold the lower part of the sternum and the lower ribs still. If you are healthy this will give you an unusual and unsettling feeling, and pretty soon you'll have an irresistible urge to take a deep breath—if not two or three. You'll wonder how anyone could possibly develop this breathing pattern as a lifetime habit.

Chest breathers often feel short of breath because constricted thoracic breathing pulls most of the air into the upper portions of the lungs. But when we are upright it is the lower portions of the lungs that get most of the blood supply. Why? The pulmonary circulation to the lungs is a low-pressure, low-resistance circuit in which the average pressure in the pulmonary arteries is only 14 mm Hg. By contrast, the pressure in arteries of the systemic circulation averages about 100 mm Hg (chapter 8). The 14 mm Hg pulmonary arterial pressure is more than enough to perfuse blood into the lower parts of the lungs, but it is inadequate to push the blood into the upper parts of the lungs. This means that when you are taking constricted thoracic inhalations, you are bringing the bulk of the air into the parts of the lungs that are most poorly supplied with blood. You can't make efficient use of the extra ventilation to the upper parts of the lungs because of the poor circulation, and yet you get scanty ventilation to the lower parts of the lungs that are getting the bulk of the blood supply. It's no wonder those who breathe thoracically need to take occasional breaths that will fill their lungs from top to bottom.

The disadvantages of constricted chest breathing are ordinarily empha-sized, but this mode of breathing is occasionally necessary. If you should happen to overindulge in a holiday meal and then follow it up with a rich dessert, try taking a walk. You will notice that the restricted form of thoracic breathing is the only comfortable way you can breathe. A five-mile walk can be useful, but the last thing you'll want to do *en route* is to press against your stomach with your diaphragm (figs. 2.9 and 2.29b).

THE NEED FOR EMPOWERED THORACIC BREATHING

In addition to certain postures in hatha yoga, thoracic breathing works beautifully in aerobic exercise, in which a freer and more vigorous style of thoracic breathing is combined with increased cardiac output. The aroused heart creates pulmonary arterial pressures high enough to perfuse the entire lungs with blood at the same time they are being ventilated from top to bottom. In hatha yoga this also happens in a series of briskly executed sun salutations or in any other postures that stress the cardiorespiratory system, such as triangles (chapter 4) or lunging postures (chapter 7), espe-cially when performed by beginners. In hatha yoga we also frequently use an empowered and healthy form of thoracic breathing for the complete breath (which we'll discuss later in this chapter) and in most other circumstances in which you are taking fewer than two breaths per minute.

PARADOXICAL BREATHING

Empowered chest breathing carried to extremes is *paradoxical breathing* (fig. 2.29c). Try inhaling so deeply that the abdominal wall moves in during inhalation rather than out. Or imagine a situation which shocks you. Let's say you dart into a shower thinking that the water will be warm, and instead find it ice cold. You will probably open your mouth and suck in air with a gasp. Try breathing this way three or four breaths under ordinary circumstances and notice how you feel. This is paradoxical breathing, so-named because the abdominal wall moves in rather than out during inhalation, and out rather than in during exhalation (fig. 2.17). Unless someone is in a state of considerable anxiety, we rarely see this in the corpse posture—it is more common while sitting or standing.

During a paradoxical inhalation, the external intercostal muscles enlarge and lift the rib cage, lift the abdominal organs and the relaxed diaphragm, and suck in the abdominal wall. During a paradoxical exhalation, the abdomen moves back out because the rib cage relaxes and releases the vacuum on the diaphragm and *abdominopelvic region.*

Paradoxical breathing stimulates the sympathetic nervous system even more than thoracic breathing. In an average class only a few students will have

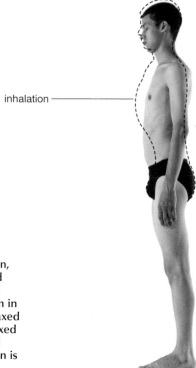

inhalation

Figure 2.17. Paradoxical breathing. During inhalation, the external intercostal and other accessory muscles of respiration create a vacuum in the chest that pulls the relaxed diaphragm up and the relaxed abdominal wall in. The end stage of a natural exhalation is profiled by the halftone.

the confidence to try it enthusiastically, and those who do it for 10–15 deep breaths may get jumpy and nervous. This is its purpose: preparation for fight or flight. Paradoxical breathing gives you an immediate jolt of adrenaline. The problem is that some people breathe like that much of the time, making life a constant emergency. Our bodies are not built for remaining this keyed up, and keeping the sympathetic nervous system in a constant state of arousal is hard on the supportive systems of the body. Digestion, circulation, endocrine function, sexual function, and immune function are all either put on hold or are stressed by continual sympathetic discharge.

SUPINE ABDOMINAL BREATHING

The antidote for chronic thoracic and paradoxical breathing is *abdominal breathing*, or *abdomino-diaphragmatic breathing* (fig. 2.29d). It is simple, natural, and relaxing—especially in the supine position. To try it, lie in the corpse posture, and again place the right hand on the upper part of the chest and the left hand on the upper part of the abdomen. Breathe so that the left hand moves anteriorly (toward the ceiling) during inhalation and posteriorly (toward the floor) during exhalation. The right hand should not move. Take the same amount of time for exhalation as inhalation. Notice that inhalation requires moderate effort and that exhalation seems relaxed. This is abdominal breathing. As discussed earlier, it is accomplished by the respiratory diaphragm.

Because the contents of the abdominal cavity have a liquid character, gravity pushes them to a higher than usual position in the torso when you are lying down. The diaphragm acts as a movable dam against this wall of abdominal organs, pressing them inferiorly (toward the feet) during inhalation and restraining their movement superiorly (toward the head) during exhalation. As the diaphragm pushes the abdominal organs inferiorly during inhalation, the abdominal wall is pushed out, thus pressing the left hand anteriorly.

We perceive the gravity-induced exhalation as a state of relaxation, but careful observation will reveal that the diaphragm is actually lengthening eccentrically throughout a supine exhalation. In other words, it is resisting the tendency of gravity to push the diaphragm superiorly. You can feel this for yourself if you breathe normally for a few breaths, making the breath smooth and even, without jerks, pauses, or noise. Then, at the end of a normal inhalation, relax completely. Air will whoosh out faster, proving that some tension is normally held in the diaphragm during supine exhalations. You can relax the diaphragm suddenly if you like, but exhalations that are restrained actively are more natural, at least for anyone who has had some training in yoga.

LUNG VOLUMES AND ALVEOLAR VENTILATION

Since the abdominal organs and the dome of the diaphragm ride to a higher than usual position in the chest in a supine posture, less air than usual is left in the lungs at the end of a normal exhalation. This is reflected in a decreased expiratory reserve volume. You can prove this to yourself if you breathe abdominally, first sitting upright and then lying down supine, and subjectively compare the two expiratory reserve volumes. What you do is come to the end of a normal exhalation in each case and then breathe out as much as possible—all the way down to your residual volume. It will be obvious that the supine position decreases the amount of air you can breathe out to about one-half of your upright expiratory reserve volume, let's say from 1,000 ml to about 500 ml (fig. 2.18).

Supine abdominal breathing is both natural and efficient. Using the above figures, if you were to maintain a tidal volume of 500 ml when you are supine, you will be mixing that tidal volume with only 1,700 ml of air instead of the 2,200 ml in your functional residual capacity when you are upright. And because your tidal volume for each breath is getting mixed with a smaller functional residual capacity, you will not need to breathe as

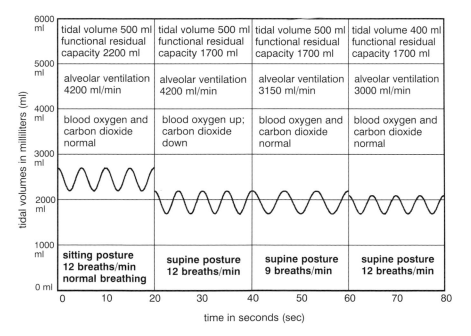

Figure 2.18. Tidal volume simulations for abdominal breathing in a sitting posture (far left), and for three conditions of abdominal breathing in a supine position, the first with alveolar ventilation identical to the sitting posture, second with the breathing rate slowed down, and third with the tidal volume decreased.

deeply or as fast. In fact, if you were to keep your alveolar ventilation constant at 4,200 ml/minute as a textbook norm, the improved efficiency of the alveolar exchange would soon be reflected in increased blood oxygen and decreased blood carbon dioxide. What happens, of course, is that you either slow your rate of respiration or decrease your tidal volume (or both), and that keeps blood oxygen and carbon dioxide within a normal range.

SMOOTH, EVEN BREATHING

The corpse posture is a good place to learn one of the most important skills in yoga: smooth, even breathing. When you are relaxed and breathing nasally and abdominally, it is easy to inhale evenly, smoothly merge the inhalation into the exhalation, and smoothly exhale. You may pause at the end of exhalation, but if you do so for any length of time the diaphragm will have relaxed completely during the pause and you may find that you are starting your next inhalation with a jerk. The best prevention for that disturbance is to begin your inhalation consciously just as exhalation ends.

USING A SANDBAG

The movements of the diaphragm are delicate and subtle, and not always easy to experience, but when you are supine you can place a sandbag that weighs 3–15 pounds on the upper abdomen just below the rib cage, and you will immediately notice the additional tension needed for inhalation and controlled exhalation. Make sure the chest does not move and that the weight is light enough to push easily toward the ceiling (fig. 2.22a). The exercise is valuable both for training and strengthening. It helps students learn to sense the activity of the diaphragm by increasing the amount of work and tension needed for inhalations (concentric shortening of the muscular parts of the diaphragm) and for controlled exhalations (eccentric lengthening of the muscular parts of the diaphragm). The cobra variant and the diaphragmatic rear lift (figs. 2.10–11) give the diaphragm more exercise by requiring it to lift large segments of the body, but a light sandbag brings the student more in touch with the delicacy of its function.

Since breathing evenly with a sandbag increases neuromuscular activity in the diaphragm, this makes you aware of the challenges involved in moving it up and down without starts, stops, and jerks. And developing the control necessary to accomplish this is an important aid to learning even breathing. First try it with a sandbag weighing 10–15 pounds to feel a pronounced increase in muscular activity, and then try it with a book or much lighter sandbag weighing 1–3 pounds. After you have practiced with a lighter weight for a while, you will have become so sensitive to the subtle activity of the diaphragm that you will be able to sense its delicate eccentric resistance during exhalation without any weight at all. After about twenty deep breaths with a sandbag you'll also notice that it is natural to stop breathing for a few seconds at the end of an exhalation, and that this yields a moment of total relaxation. Here again, once you have experienced this with a sandbag you will notice that the same thing can happen with free relaxed abdominal breathing.

Caution: Don't pause the breath habitually. It's unnatural while inhaling and exhaling, or at the end of inhalation, so those times are not usually a problem, but at the end of exhalation, it's tempting. Don't do it except as an experiment in understanding the operation of the diaphragm. The medical lore in yoga (the oral tradition) is that the habit of pausing the breath at the end of exhalation causes heart problems.

THE INTERCOSTAL MUSCLES

If you are not using a sandbag, the extent to which the intercostal muscles are active during supine abdominal breathing is an open question. They may be serving to maintain the shape of the rib cage isometrically during inhalation (as in upright postures), but this may not be the case toward the end of a long and successful relaxation in the corpse posture. At that time the tidal volume and the minute ventilation are reduced so markedly that little tension is placed on the rib cage by breathing, and the intercostal muscles may gradually become silent. It would require electromyography using needle electrodes placed directly in the intercostal muscles to settle the point.

There will be no doubt about the activity of the intercostal muscles if you use a sandbag for this exercise. Now the diaphragm has to push the sandbag toward the ceiling, and as its dome descends its costal attachments pull more insistently on the base of the rib cage than would otherwise be the case. This pull can be countered only by isometric tension in the intercostal muscles; you can feel it develop instantly if you make a before-and-after comparison, first without a sandbag and then with one.

ABDOMINAL BREATHING IN SITTING POSTURES

We discussed abdominal breathing in the supine corpse posture first because in that pose we find the simplest possible method of breathing: the diaphragm is active in both inhalation and exhalation, the intercostal muscles act only to keep the chest stable, and the abdominal muscles remain completely relaxed. Abdominal breathing in sitting postures is quite different. First of all, when we are upright, gravity pulls the abdominal organs inferiorly instead of pushing them higher in the torso, and this is what causes the shift in expiratory reserve volumes from approximately 500 ml in the supine posture to about 1,000 ml in the upright posture. It also means that the diaphragm cannot act as purely like a piston as it can in supine and inverted postures.

The other major difference between supine and upright abdominal breathing is that when we are upright we can choose between exhaling actively or passively. We can simply relax as we do when we sigh, allowing the elasticity of the lungs to implement exhalation, or we can assist exhalation with the abdominal muscles, which we do in many yoga breathing exercises and for all purposeful actions such as lifting a heavy weight or yelling out a command. A quiet breathing pattern with relaxed exhalations is simpler, so we'll look at that first.

ABDOMINAL BREATHING WITH A RELAXED ABDOMEN

Breathing abdominally with a relaxed abdomen is a prelude to meditative breathing because it gives one an opportunity to understand the subtle problems involved with breathing quietly. To begin, sit straight in a chair. Don't slump but don't pitch yourself forward with an arched *lumbar lordosis*, either. Make sure the lower abdomen is not restrained by tight clothing. Because the abdominal muscles wrap around to the rear it is better not to lean against the back of the chair. Now breathe so that the lower abdomen moves outward during each inhalation and comes passively inward during each exhalation. Breathe evenly and nasally, making sure the chest does not move. The abdominal muscles have to be completely free. If they are even mildly tensed you will not be doing abdominal breathing. Notice, even so, that the abdominal movement is minimal and that the rest of the body is stable except for a slight backward movement of the head during inhalation (fig. 2.19).

When you are sitting, the two most critical moments for relaxed, even breathing are at the transitions—one between inhalation and exhalation, and the other between exhalation and inhalation. These are the times when the breath is more likely to jerk or become uneven. But if you imagine that your breath is making a circular pattern it is easier to accomplish these transitions smoothly. Pretend you are on a Ferris wheel. Going up is inhaling; coming down is exhaling. The upward excursion smoothly decelerates to zero as you circle up to the top; the downward excursion smoothly accelerates from zero as you start coming down. At the bottom, just the opposite happens: a downward deceleration (exhalation) merges smoothly into an upward acceleration (inhalation).

If you are riding a real Ferris wheel with your eyes closed you know you have reached the top and bottom of its circular movement by feel—the only time there is a jerk is when it stops to let someone off. And therein lies the nub of the matter. What we want from relaxed, even breathing is no jerks—just the sensation that you are making a transition from inhalation to exhalation and from exhalation to inhalation. The actual pattern of breathing is elliptical rather than circular, but the image of a Ferris wheel is still useful, especially for beginners. The main point is that even though no air is moving in or out at the ends of inhalation and exhalation, you can merge inhalation with exhalation (and exhalation with inhalation) without effort if you focus on smooth movement along the ellipse. There will be different challenges at each junction, so we'll look at them separately.

THE JUNCTION OF INHALATION WITH EXHALATION

The end of inhalation is the least troublesome. Nerve impulses keep impinging on the muscle fibers of the respiratory diaphragm even after exhalation begins, and this operates to smooth the transition between the end of inhalation and the beginning of exhalation. Picture your inhalation as you feel it. If you make the transition from inhalation into exhalation in slow motion, initiating your exhalation ever so slowly, you will feel a slight hesitation as you start to exhale, which reflects the continuing flow of nerve impulses into the diaphragm as its dome begins to ascend. If you have healthy breathing habits little effort is needed to tune this mechanism delicately and make an even transition from inhalation into exhalation, but if you find yourself holding your breath at the end of inhalation it is better to first concentrate on breathing evenly in bending, twisting, and inverted postures—the poses themselves correct bad habits.

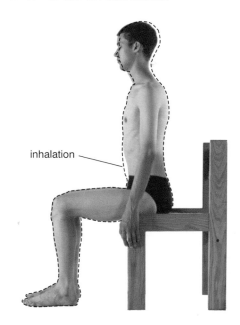

Figure 2.19. Abdominal, or abdomino-diaphragmatic breathing. During inhalation, the lower abdomen comes forward and the dome of the diaphragm descends. There is little movement or enlargement of the rib cage, although the external inter-costal muscles are active enough to keep the chest from collapsing inward as the dome of the diaphragm is pulled downward. The head and neck are pulled slightly to the rear during inhalation.

inhalation

[Technical note: In an upright posture the diaphragm continues to receive nerve impulses as its dome starts to rise during exhalation, but referring to its muscular components as lengthening eccentrically during that time would be pushing the use of the term eccentric too far. The phrase eccentric lengthening is customarily applied *only to a muscle's resistance to the force of gravity.* There is no doubt that eccentric lengthening of the diaphragm occurs during exhalation in a supine posture and even more obviously in inverted postures (in other words, gravity aids exhalation and the diaphragm resists as its muscular components lengthen), but I am not using the term eccentric here because the primary cause of the upward excursion of the diaphragm during relaxed exhalations in an upright posture is the elasticity of the lungs, not gravity. On the contrary, under these circumstances gravity actually has the opposite effect: rather than aiding exhalation and resisting inhalation, in upright postures it resists exhalation and aids inhalation. Why? The liver is firmly adherent to the underside of the diaphragm (this association is shown artificially dissected and pulled apart in fig. 2.9), and the heart is situated just above the diaphragm. Under these conditions the force of gravity tends to pull all three (the liver, the dome of the diaphragm, and the heart) down at the same time, mildly aiding inhalation and restricting exhalation rather than the other way around.]

THE JUNCTION OF EXHALATION WITH INHALATION

As exhalation in an upright posture continues, the diaphragm finally relaxes, and toward the end of exhalation its motor neurons have largely ceased to fire. This makes it difficult to negotiate a smooth transition between the end of exhalation and the beginning of inhalation because the motor neurons create a jerk in the system when they start firing again, something like starting a cold car that cranks in fits and coughs before it runs smoothly. In an average beginning class, two or three times as many students will find it more difficult to avoid a discontinuity at the junction of exhalation and inhalation than at the junction of inhalation and exhalation.

ABDOMINAL BREATHING WITH ACTIVE EXHALATIONS

It is very easy to remedy the jerk at the beginning of inhalation. All you have to do is maintain tension in the abdomen throughout exhalation, especially toward the end, and merge that tension into the cycle of inhalation. If you are uncertain of how to do this, first learn to emphasize exhalation in a contrived situation. Purse the lips so that only a small amount of air can escape, and blow gently as if you are blowing up a balloon. Notice that the abdominal muscles are now responsible for the exhalation. Keep blowing as long as you can. After you reach your limit notice that inhalation is passive, especially at its start. Why? If you have exhaled almost to your residual volume, the chest will spring open passively and the abdominal wall will spring forward of its own accord, at least until you have inhaled your normal expiratory reserve volume. Then, as you begin to inhale your normal tidal volume, the diaphragm begins its active descent.

After you have gotten a feel for exaggerated exhalations, try even abdominal breathing while you are sitting in a chair. Again, your clothing must be loose so there are no restrictions on the movement of the lower abdomen. Begin by taking 2-second exhalations and 2-second inhalations. Imagine the ellipse, the exhalation going down and the inhalation going up, one count each second: down, down, up, up, down, down, up, up. Then create an image in your mind in which you are actively pressing in the abdominal muscles during exhalation and releasing them forward during inhalation. Still riding the ellipse, think "travel down" the ellipse and "push in" the abdomen for exhalation, and think "travel up" the ellipse and "ease out" the abdomen for inhalation: down and in, up and out, down and in, up and out. Assisting exhalations with the abdominal muscles does two things: it masks any jerks and discontinuities that come from starting up the contraction of the diaphragm, and even more important, it keeps alive your intent to breathe evenly.

THE IMPORTANCE OF POSTURE

No breathing technique will work unless you are sitting correctly, as two simple experiments will show. First sit perfectly straight and breathe evenly, remaining aware of the elliptical nature of the breathing cycle and making sure that you are not creating pauses or jerks at either end of the ellipse. Now slump forward slightly and allow the lumbar lordosis to collapse. Notice three things: inhalation is more labored, exhalation starts with a gasp, and it is impossible to use the abdominal muscles smoothly to aid exhalation. Breathing evenly is impossible and meditation is impossible. The lesson is obvious: Don't slump.

Now sit on the edge of a chair. Keep the lumbar lordosis maximally arched but lean forward, making an acute angle between the torso and the thighs. Watch your breathing. The abdominal muscles now have to push strongly against a taut abdomen to aid exhalation. Then, at the beginning of inhalation, if you relax your respiration, air rushes into the airways. Try restraining inhalation and notice that active abdominal muscles are required to prevent the sudden influx of air. The lesson here? Don't lean forward, even with a straight back.

THE BELLOWS BREATH AND KAPALABHATI

The bellows breath (*bhastrika*) and *kapalabhati* are both highly energizing abdominal breathing exercises. In their mild form they are excellent for beginners, because they require only that students be acquainted with even abdominal breathing. The bellows breath imitates the movement of the blacksmith's bellows, and kapalabhati requires sharp exhalations and passive inhalations. The chest should not move very much in either exercise even though the intercostal muscles remain isometrically active.

To do the bellows exercise, sit with your head, neck, and trunk straight and unsupported by a wall or the back of a chair. To begin, exhale and inhale small puffs of air rapidly and evenly through your nose, breathing abdominally and pacing yourself to breathe in and out comfortably about 30 times in 15 seconds. Keep your shoulders relaxed and your chest still.

The blacksmith's bellows operates by pulling air into a collapsible chamber and then blowing it forcibly into a pile of glowing coals. In the bellows breath the diaphragm pulls air into the lungs and the abdominal muscles force it out. And just as the blast of extra oxygen from the air in the blacksmith's bellows kindles additional combustion in the coals, so does the additional oxygen pulled into the lungs by the bellows exercise increase the potential for combustion throughout the body.

The bellows is an easy and rewarding exercise if it is not overdone. The main problem beginning students encounter is coordinating the actions of the diaphragm and abdominal muscles without breathing thoracically or paradoxically. The secret is to start with active, even, abdominal breathing. Watching yourself in a mirror, breathe evenly, using 2-second exhalations and 2-second inhalations (15 breaths per minute). Then gradually increase your speed, taking 1-second exhalations and 1-second inhalations (30 breaths per minute), making sure not to move your shoulders or heave your chest up and down. Notice that the entire body is stable during inhalation except for the abdomen (fig. 2.20a). Then take one breath per second, then two breaths per second, then three, then possibly four. Make sure you give equal emphasis to both phases of the cycle. When and if you lose control, slow down.

Kapalabhati is similar to bellows breathing except that it consists of a sharp inward tap with the abdominal muscles, a quick pressing in that results in a sharp exhalation. To inhale, just relax. Inhalation is passive and requires only releasing tension in the abdominal wall (fig. 2.20b). Try the exercise for one breath per second at first, and gradually increase your speed as you get more confidence and experience.

Kapalabhati is one of the six classic cleansing exercises in hatha yoga, and it is especially effective in lowering alveolar carbon dioxide in the lower segments of the lungs. Like the bellows, kapalabhati is not only energizing, it develops strength and stamina, and it teaches you to coordinate the abdominal muscles for skillful use in other exercises such as *agni sara*, *uddiyana bandha*, and *nauli*, all of which we'll discuss in chapter 3.

The physiological correlates of bellows breathing and kapalabhati differ (fig. 2.21). If we assume a tidal volume of 200 ml for the bellows and 300 ml for kapalabhati, we'll get 50 ml of alveolar ventilation per breath for the bellows and 150 ml of alveolar ventilation per breath for kapalabhati. At a rate of three breaths per second, your alveolar ventilation for the bellows would be 180 breaths per minute times 50 ml per breath, or 9,000 ml per minute. By contrast, if you take two breaths per second for kapalabhati, your alveolar ventilation would be 120 breaths per minute times 150 ml per breath, or 18,000 ml per minute.

Looking carefully at the graph (fig. 2.21), notice not only that kapalabhati creates more alveolar ventilation than the bellows, but that its functional residual capacity is smaller. The reason for this is that each sharp exhalation in kapalabhati begins before you have inhaled up to the point at which an ordinary tidal inhalation would begin in the bellows. On the other hand, kapalabhati is nearly always practiced for a shorter period of time than the bellows. So even though the projected alveolar ventilation is 50% greater in kapalabhati, doing the bellows for longer times can easily make up the difference.

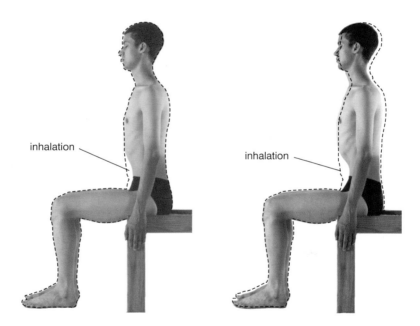

Figure 2.20a. The bellows breath, which is accomplished by breathing abdominally at the rate of 1–4 breaths per second, with inhalations and exhalations equally emphasized and equally active. The exercise mimics a blacksmith's bellows, with a tidal volume of about 200 ml.

Figure 2.20b. Kapalabhati, an abdominal breathing exercise in which exhalations are emphasized by sharply pressing in with the abdominal muscles. Inhalations are mostly passive and the tidal volume is about 300 ml. The exhalations thrust the head, neck, and chest slightly forward even as they drive the dome of the relaxed diaphragm toward the head and the abdominal wall to the rear.

Which exercise should be learned first? The simplicity and ease of a moderately paced bellows (two breaths per second) argues for concentrating on it first, but if you compare a few seconds of kapalabhati with one minute of a fast-paced bellows breath, kapalabhati will be the milder exercise and less likely to result in hyperventilation. Either choice is fine. After a little experience students naturally adjust the rate, extent, and depth of their respiration so that both exercises are comfortable.

For beginners the most common challenge of these two exercises is to stay relaxed and not breathe diaphragmatically, thoracically, or paradoxically. The chest and shoulders should remain still except for the moderate impact on the chest of movements that originate from the lower abdomen. It is easier to accomplish this with kapalabhati because all of the emphasis is on the lower abdomen. In the bellows, if it is difficult to keep the chest still, the only solution is to return to even abdominal breathing and start over. Go slowly enough to maintain control, even if you have to slow down to 30–60 breaths per minute.

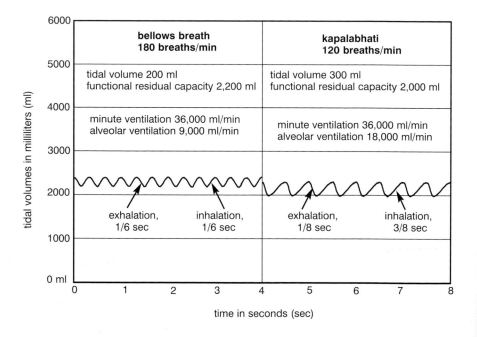

Figure 2.21. Bellows and kapalabhati, simulated comparisons. The bellows exercise is usually faster, but kapalabhati ordinarily makes use of a larger tidal volume and a decreased functional residual capacity.

In both exercises it is important not to maintain any tension in the lower abdomen during inhalation, for if you do, it will impede the downward displacement of the dome of the diaphragm and force a lateral expansion of the chest (thoraco-diaphragmatic breathing), or even frankly thoracic breathing. This is impractical because the chest is a cage—rigid in comparison to the abdominal wall—and except for speedy thoraco-diaphragmatic and thoracic breathing in aerobic exercise, it is unreasonable physiologically and unsettling neurologically to breathe by enlarging and contracting such an enclosure quickly. The most efficient way to breathe rapidly in hatha yoga exercises such as the bellows breath and kapalabhati is to create most of the motion in the softest tissues—and that means the lower abdomen. This is easier to regulate in kapalabhati than in the bellows because of the extra emphasis on exhalation.

Once you have mastered the technique of quickening the bellows to one breath per second with even abdominal breathing, it isn't too difficult to increase the speed to over 120 breaths per minute. Serious students can speed up gradually to 180–240 breaths per minute, and advanced practitioners approach 300 breaths per minute. It's fun, and the faster the better. But beware of hyperventilating: build your capacity slowly but surely.

THE LONG VIEW OF LEARNING AND TEACHING

Bad breathing habits are likely to be insidious, but they are not intractable. Even though they go on 24 hours a day year in and year out, change is still possible because the respiratory motions are entirely controlled by somatic motor neurons—you have the potential of thinking the actions through and controlling them willfully.

How to proceed? Whether bad breathing habits involve constricted chest breathing, reversing the movement of the abdomen in paradoxical breathing, jerking the breath, or pausing between inhalation and exhalation, anyone who has such problems should first master abdominal breathing. Thoraco-diaphragmatic breathing is not advisable at first; it will be especially confusing to chronic chest breathers. The best solution for such students is a regular practice of a variety of postures in hatha yoga, and the corpse posture is the place to start. In the supine position almost everyone can learn to breathe in a way that allows a hand or sandbag on the abdomen to move smoothly toward the ceiling. As soon as problem breathers have that mastered, they should work with abdominal breathing while sitting up straight in a chair, first just inhaling and exhaling naturally. This means: first, making sure that the lower abdomen relaxes completely and protrudes during inhalation, and that the chest does not lift up and out; and second, allowing the abdominal muscles to remain passive during exhalation, thus permitting the abdominal wall to sink back in. It may help to make tiny sighs to insure that each exhalation is entirely passive.

After mastering abdominal breathing with passive exhalations in an upright posture, students should learn to use the abdominal muscles to aid exhalation and cultivate even breathing. This will lead naturally to the bellows breath and kapalabhati. Both of those exercises should be approached with a sense of experimentation, observation, and play. Rushing yourself or someone else into developing new breathing habits will only create anxiety and disrupt rather than benefit the nervous system.

DIAPHRAGMATIC BREATHING

Yogis are not the only ones who know about *diaphragmatic breathing* (fig. 2.29e). Martial artists, public speakers, and musicians are all united in its praise. But even those who practice it have a hard time describing precisely what they do and how they do it. This is not surprising—it's a difficult concept. We'll approach it here by looking at how diaphragmatic breathing differs from abdominal and chest breathing and take note of how it feels and where you feel it in the body.

Abdominal breathing, or abdomino-diaphragmatic breathing, brings your attention to the lower abdomen. If you sit with it for a while in meditation you will be relaxed, but your attention will be drawn to the pelvis and the base of the torso. It is a good technique for beginners, but in the long run it results in a depressed, overly relaxed sensation. Thoracic and paradoxical breathing go to the other extreme. They bring your attention to the upper chest and spin you off into realms that are not wanted for meditation: heady sensations for thoracic breathing, and tangents of anxiety and emergency for paradoxical breathing. Diaphragmatic breathing, or *thoraco-diaphragmatic breathing*, is the perfect compromise. It brings your attention squarely to the middle of the body, to the borderline between the chest and the abdomen, and from there it can balance and integrate the opposing polarities.

Diaphragmatic breathing is also the most natural way to breathe in everyday life. Whenever you gear up mentally and physically for any activity, the additional concentration is reflected in diaphragmatic breathing. And in the yoga postures that call for it, the effort to maintain the required tension in the abdominal muscles will bring you more control and awareness of the torso than any other type of breathing.

DIAPHRAGMATIC BREATHING IN THE CORPSE POSTURE

We'll start with diaphragmatic breathing in the corpse posture. To begin, lie supine and breathe abdominally for five or six rounds, allowing the lower abdomen to relax and protrude during inhalation and to drop back toward the floor during exhalation. Then, to create thoraco-diaphragmatic breathing, hold enough muscle tone in the abdominal muscles as you inhale to prevent the lower abdomen from moving anteriorly during that phase of the cycle. You can feel what happens next. Since the tension in the abdominal muscles does not allow the abdominal wall to protrude as the central tendon starts to descend, the diaphragm can act only at its costal insertion to lift and expand the rib cage. This draws air into the lungs and at the same time enlarges the upper abdomen, as opposed to the lower. As in abdominal breathing, the external intercostal muscles remain active; you can feel them lengthen actively against the resistance of the lungs' elasticity as the chest wings out during inhalation, especially toward the end of inhalation. Diaphragmatic breathing in the corpse posture requires more attention than abdominal breathing, and because of this it is useful as a concentration exercise and for the deep inhalations and long exhalations in 2:1 breathing.

SANDBAG BREATHING

In the corpse posture, sandbags of various weights will strengthen and further educate the diaphragm, intercostal muscles, and abdominal muscles. As mentioned earlier, a sandbag weighing 3–15 pounds is best for training in abdominal breathing because it can be comfortably pressed toward the ceiling with each inhalation, and its fall can be comfortably restrained during exhalation. The chest is stable, and both the upper and lower abdomen are thrust anteriorly (along with the sandbag) by inhalation (fig. 2.22a).

To intensify the exercise and create diaphragmatic breathing, increase the weight of the sandbag to the point at which it is a bit awkward to press it toward the ceiling. This much weight, about 20–30 pounds for a healthy young man with good strength, makes it more convenient to breathe diaphragmatically than abdominally. If you adjust the amount and placement of the weight perfectly, the tension on your upper abdomen will cause the diaphragm to flare the rib cage out from its base. You have to play with the resulting sensations and analyze the movements carefully. In this case (fig. 2.22b), if a 25-pound bag of lead shot is placed just beneath the rib cage, inhalation lifts the chest and upper abdomen up and forward, but the movement of the lower abdomen is checked, at least in comparison with abdominal inhalation shown in fig. 2.22a.

If you increase the weight even more, to 30–50 pounds or so, you will create so much tension in the abdomen that the dome of the diaphragm is unable to descend at all. In that event the only way you will be able to breathe comfortably will be by lifting the upper part of the rib cage and breathing thoracically (fig. 2.22c). Placing two 25-pound bags of lead shot on the abdomen creates two results: it requires that there will be a substantial increase in the anterior displacement of the upper chest during inhalation, and it holds the middle and lower portions of the abdomen fixed in position.

BREATHING IN THE CROCODILE POSTURE

If you still have trouble sensing the ways in which abdominal and diaphragmatic breathing operate and differ from one another, the distinctions will become more clear if you try breathing in two variations of the

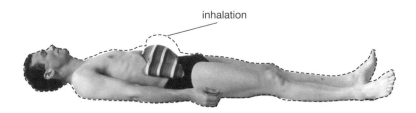

Figure 2.22a. Abdominal, or abdomino-diaphragmatic breathing, with a 14-pound sandbag. The diaphragm pushes against the abdominal organs, ultimately pressing the abdominal wall and sandbag toward the ceiling.

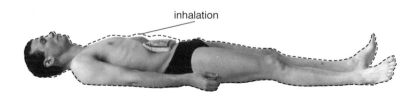

Figure 2.22b. Diaphragmatic, or thoraco-diaphragmatic breathing with a 25-pound bag of lead shot. The extra weight is somewhat more difficult to lift than the 14-pound sandbag, and this creates more of a tendency for the rib cage to be enlarged from its base than for the weight to be pushed toward the ceiling.

Figure 2.22c. Thoracic breathing, as required by 50 pounds of weight placed on the abdomen and lower border of the chest. After a modest downward excursion of the dome of the diaphragm (inhalation), its muscle fibers remain in a state of isometric contraction and the brunt of inhalation must be borne by the external intercostals.

crocodile posture. First, to experience abdominal breathing, lie prone, with the feet apart, the elbows flexed, and the arms stretched out in front. Your hands should be pulled in enough for the forehead to rest on the bony part of the wrist. This is the most relaxed variation of the crocodile (fig. 2.23). The position of the arms restricts thoracic breathing, the position of the chest against the floor restricts diaphragmatic breathing, and the position of the lower abdomen against the floor restricts what we conventionally think of as abdominal breathing. Still, in a modified form, abdominal breathing is what this is, with the hips and lower back rather than the front of the abdomen responding to the rise and fall of the dome of the diaphragm.

Abdominal breathing in this sleepy, stretched-out crocodile requires a more active diaphragm than abdominal breathing in the supine position. Why? The weight of the entire torso against the floor in the prone position restrains inhalation more than the weight of the abdominal organs by themselves in the supine position—it feels something like breathing in the corpse posture with a lead apron spanning your entire chest and abdomen. If you make a nominal effort to breathe evenly, the diaphragm also has to work more strongly to restrain exhalation. At the end of exhalation, of course, it can relax completely, just as it does in the corpse posture.

Next, to experience an unusual form of diaphragmatic breathing, lie in the more traditional easy crocodile with the elbows flexed and the arms at a 45–90° angle from the torso. The hairline should rest against the forearms. Adjust the arms so that the lower border of the chest is barely touching the floor. This arches the back and creates a mild backbending posture (fig. 2.24). Now we are entering complex and unexplored territory. The lower abdomen still cannot protrude because it is against the floor; thoracic breathing is restricted by the extreme arm position even more than in the previous posture; and the attempt of the diaphragm to descend is checked because the base of the rib cage and upper abdomen is still held in position. The only parts of the body that appear to yield for inhalation

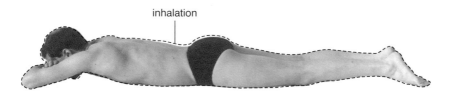

inhalation

Figure 2.23. A specialized type of abdominal breathing created by a stretched-out crocodile posture. The floor cannot yield to the descending dome of the diaphragm during inhalation, so the lower back and hips are lifted by default.

are the lower back and hips (fig. 2.24), just as in the stretched crocodile. Isolated comparisons of the dashed superimpositions for inhalations in these two postures is not helpful to our analysis, however, because the experience of breathing in them is completely different. Although it is not reflected in the photographs, inhalation in the beginner's crocodile creates a characteristic tension at the base of the rib cage which is absent in the stretched-out posture. For that reason we can—indeed we must—classify breathing in the beginner's crocodile as thoraco-diaphragmatic breathing.

DIAPHRAGMATIC BREATHING IN SITTING POSTURES

To experience the center-of-the-trunk sensation that characterizes diaphragmatic breathing in sitting postures, sit upright in a chair and first review abdominal breathing as a basis for comparison. Then to breathe diaphragmatically, inhale gently while holding just enough tension in the abdominal muscles to make sure that the lower abdomen is not displaced anteriorly during inhalation. There is a sense of enlargement in the lower part of the chest and a feeling of expansion in the upper part of the abdomen just below the sternum. The lateral excursion of the rib cage (fig. 2.25a) is more pronounced than the anterior movement (fig. 2.25b), but you may have to take a few slow, deep inhalations to confirm this.

All of these observations will be lost on chest breathers because the difference between the mild lower abdominal tension that creates diaphragmatic breathing and the frank rigidity of the entire abdominal wall that is associated with constricted thoracic breathing is far too subtle for them to feel and comprehend. They will get mixed up every time. As discussed earlier, anyone who has the habit of chronic chest breathing should not try to do thoraco-diaphragmatic breathing until they have become thoroughly habituated to abdominal breathing. Their first goal must be to break the habit of constricted chest breathing forever.

inhalation

Figure 2.24. Objectively, this beginner's crocodile posture again appears to lift the lower back and hips as in abdomino-diaphragmatic breathing, but appearances can be deceiving. The subjective feel of the posture is that the mild back-bending position severely restricts lifting of the lower back; more emphasis is felt at the base of the rib cage. For that reason, and because the extreme arm position also restricts thoracic breathing, this posture is admirably suited for training in thoraco-diaphragmatic breathing.

HOW DIAPHRAGMATIC BREATHING AFFECTS POSTURE

If you examine your body carefully when you are breathing diaphragmatically in the easy crocodile (fig. 2.24), the cobra (fig. 2.10), or the diaphragmatic rear lift (fig. 2.11), you will notice that inhalation raises your posture up and back, and that exhalation lowers it down and forward. This principle also holds true when you are standing, sitting straight, or even lounging in a soft chair. During inhalation in all such postures the head moves back, and during exhalation it comes forward. During inhalation the *cervical lordosis* (the forward arch in the neck) decreases, thus raising the head; during exhalation it increases, lowering the head. The shoulders move back during inhalation and forward during exhalation. The *thoracic kyphosis* (the posterior convexity in the chest) decreases during inhalation and increases during exhalation. Finally, if you are sitting straight the *lumbar lordosis* increases during inhalation and decreases during exhalation.

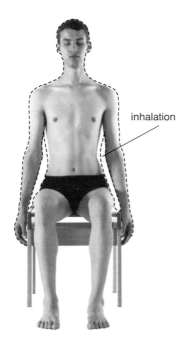

inhalation

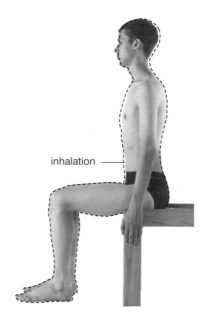

inhalation

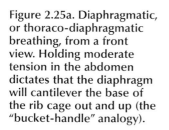

Figure 2.25a. Diaphragmatic, or thoraco-diaphragmatic breathing, from a front view. Holding moderate tension in the abdomen dictates that the diaphragm will cantilever the base of the rib cage out and up (the "bucket-handle" analogy).

Figure 2.25b. Diaphragmatic breathing from the side, illustrating the forward movement of the chest when moderate tension is held in the abdominal muscles during inhalation.

The movements are subtle, but if you purposely try to make them in the opposite direction, you will see instantly that they are contrary to the normal pattern.

An understanding of these principles is of practical value to meditators because they can take advantage of the slight postural changes caused by breathing to adjust and improve their sitting postures. Try it. Sit comfortably on the edge of a straight chair and breathe evenly and diaphragmatically in a cycle of 4–6 seconds for each round of inhalation and exhalation. Resolve not to make noticeable movements for the next five minutes. Now, with each inhalation lift your posture, allowing the inhalation to pull the head back, flatten the thoracic kyphosis, and increase the lumbar lordosis. These adjustments should be so slight that they are barely perceptible, even to the practitioner. Press the abdomen in actively during each exhalation so as not to lose ground. Pretend that the breath is acting like a ratcheting mechanism on a pulley that is lifting a weight. With each inhalation you gain a single cog, and during exhalation the ratchet prevents the weight from falling. You can also imagine that the breath is a thread which lifts the posture during inhalation and then holds it from falling during exhalation. The resolve not to move in this exercise is critical, so good concentration is required. If you make adjustments that are externally visible, the body accepts the habit of moving, and the posture deteriorates when concentration lapses.

Next check the effects of diaphragmatic breathing when you are slouched. You will notice the same problems you encountered with abdominal breathing in a slouched posture: labored inhalations, an inability to start exhalations without gasping, and the difficulty of using the abdominal muscles to aid exhalation. The entire torso is lifted up and back with each breath, but each exhalation drops it forward. You can see an extreme example of this if you dip your head forward while you are slouched. Each inhalation rolls the body up, and each exhalation rolls it down.

Now try sitting perfectly straight (but without arching your torso forward from the hips). Notice that the posture itself defines diaphragmatic breathing. Unless you are too flabby, the abdomen is held taut enough by the posture to make abdominal breathing inconvenient. You can play with the edges of this. Hold the posture less rigorously, and you will see that you begin to breathe abdominally. Sit straight, and the taut abdomen will force you to breathe diaphragmatically. Carrying this to an extreme, if you bend forward from the hips markedly while maintaining a prominent lumbar lordosis, the abdomen gets so taut that inhalation becomes very laborious. You will then either have to resort to chest breathing or make excessive effort to breathe diaphragmatically.

EVEN DIAPHRAGMATIC BREATHING

Many of the principles underlying even abdominal breathing apply to even diaphragmatic breathing as well. Make sure there are no jerks in your breath. This is more difficult in diaphragmatic breathing than it is in abdominal breathing because the process is more complex and you are constantly monitoring the tension in your abdomen. Until you get accustomed to doing this, it may create slight disruptions during inhalations.

Be careful that you are not creating a pause at the end of inhalation. This is less of a problem in diaphragmatic breathing than it is in abdominal breathing because the additional tension in the abdomen (as well as the focus of mental attention at the junction of the chest and abdomen) keeps the diaphragm in a state of tension well into exhalation. Be even more watchful that you are not creating a pause at the end of exhalation. As with abdominal breathing, it is important to assist exhalation with the abdominal muscles, causing that part of the cycle to flow smoothly and naturally into the inhalation. As inhalation proceeds, however, there is an important difference between abdominal and diaphragmatic breathing: during abdominal breathing, the abdominal muscles facilitate even breathing only at the beginning of inhalation, but during diaphragmatic breathing, they remain active throughout inhalation so that their isometric tension can force the diaphragm to spread its costal attachment laterally and enlarge the rib cage.

Breathe through your nose, and try not to create noise. If your breathing is noisy, you may have to work with cleansing, diet, allergies, and breathing exercises to solve the problem, but this is essential. Noisy breathing will distract your mind as long as it lasts.

Observe in your mind's eye the elliptical nature of the breathing cycle. Smoothly decelerate your rate of inhalation and merge it into exhalation exactly as you would round off an ellipse at the top of a chalkboard. Smoothly accelerate your exhalation under the control of your abdominal muscles as you draw the chalk down the ellipse; smoothly decelerate your exhalation and merge it into the inhalation as you carry your mark around the bottom of the ellipse.

Until you have mastered even breathing don't try to lengthen your inhalations and exhalations. A 2-second inhalation and a 2-second inhalation is fine, or a little faster or slower. The longer you try to make the cycle, the more difficult it is to make it even. So be completely natural at first without thinking of trying to accomplish anything.

After several months of practice you can slowly work up to making your breaths longer, so long as you are still not jerking, pausing, or making noise. If you are taking fewer than six breaths per minute, you will be adding a thoracic component to diaphragmatic breathing, which means that you are activating the external intercostal muscles concentrically, especially toward the end of inhalation. You will also be pressing more insistently with the abdominal muscles to lengthen the exhalation. And if you carry this to an extreme, going slowly, you will finally approach breathing your vital capacity with each cycle of exhalation and inhalation. This is the complete breath, our next topic.

THE COMPLETE BREATH

The complete breath is one of the simplest and yet most rewarding of all the yoga breathing exercises. To begin, breathe in and out a few times normally and then exhale as much as possible, all the way down to your residual volume. Then for the complete breath inhale as much as possible, which will be your vital capacity (fig. 2.26). Continue by exhaling and inhaling your vital capacity as many times as you want.

This is a lot of ventilation even if you breathe slowly. If you inhale and exhale your vital capacity three times in one minute, your minute ventilation will be 14,400 ml per minute (4,800 ml per breath times three breaths per minute), and your alveolar ventilation will be 13,950 ml per minute (4,650 ml times 3 breaths per minute). After just six such breaths your blood gases will have shifted perceptibly—arterial oxygen will have moved from perhaps 100 mm Hg to 120 mm Hg and arterial carbon dioxide from perhaps 46 mm Hg to 35 mm Hg (fig. 2.27). For this reason the complete breath is both cleansing and energizing, but if you do it slowly and evenly it will also produce a sense of calm and stability.

You can practice the complete breath when you are sitting, standing, or lying down, but it is most commonly done in a supine position with the

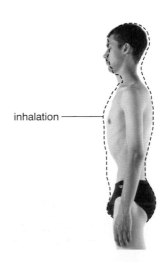

inhalation

Figure 2.26. The complete breath, or inhalation and exhalation of the vital capacity. The halftone shows a profile of the fullest possible exhalation, and the dotted outline shows the fullest possible subsequent inhalation.

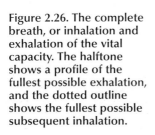

hands stretched overhead, usually at the end of a class or at the end of a series of sun salutations. Most instructors will suggest filling the lungs from below—expanding the lower, then the middle, and finally the upper parts of the lungs.

A common and less extreme variation of the complete breath is to simply inhale and exhale your inspiratory capacity instead of your vital capacity, and unless the instructor specifically asks you to exhale as much as you possibly can before starting the complete breath, inhaling and exhaling the inspiratory capacity is what most people will do naturally.

[Technical note: In addition to the proven anti-aging effects of a calorie-restricted and high-nutrition diet in experimental animals, the ability to quickly inhale a commodious vital capacity appears to be one of the most reliable predictors of longevity in humans. Whether this argues for the principle of trying to increase your inspiratory and vital capacity is not so certain, but it certainly can't hurt anyone who is in good enough health to do the postures. In chapter 3, we'll concentrate on exercises that focus on exhalation rather than inhalation—increasing your vital capacity by developing the ability to exhale your full expiratory reserve volume and minimize your residual volume.]

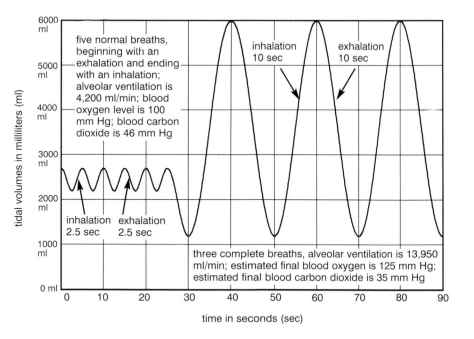

Figure 2.27. Simulations of three complete breaths (in this case inhaling and exhaling the vital capacity) following an initial exhalation of the expiratory reserve volume. Even though the subject is only taking three breaths per minute, breathing in and out the full vital capacity a few times is expected to markedly increase blood oxygen and decrease blood carbon dioxide.

ALTERNATE NOSTRIL BREATHING

One of the best breathing exercises for calming the nervous system is alternate nostril breathing, or *nadi shodanham*. This is a concentration as well as a breathing exercise, and it is possibly the single most important preparation for meditation in hatha yoga. There are dozens of variations to suit differing needs, abilities, and temperaments. At one extreme, mental patients, flighty or hyperactive children, or anyone who has difficulty concentrating can simply sit up straight, rest their elbows on a desk, press the right nostril shut with the right index finger, and exhale and inhale three times. Then they can press the left nostril shut with the left index finger and again exhale and inhale three times. This simple exercise can be repeated for 5 minutes at a pace of 1- to 2-second exhalations and 1- to 2-second inhalations (15–30 breaths per minute) using abdominal breathing. It trains concentration because it requires sitting straight, counting the breaths, switching nostrils at the proper moment, and, most important of all, breathing evenly with no noise, jerks, or pauses.

A slightly more complex version of alternate nostril breathing begins with making the classical *mudra* (gesture) with the right hand, curling the index and middle fingers in toward the palm. Closing the right nostril with the thumb, exhale and inhale once through the left nostril (fig. 2.28a). Then, closing the left nostril with the ring (fourth) finger (fig. 2.28b) exhale and inhale once through the right nostril. Go back and forth like that for 5 minutes. Breathe abdominally or diaphragmatically as your abilities permit.

Figure 2.28a. Alternate nostril breathing, closing the right nostril with the right thumb.

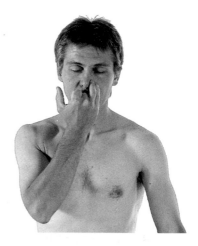

Figure 2.28b. Alternate nostril breathing, closing the left nostril with the right fourth finger.

The following version of this exercise is the one ordinarily taught in hatha yoga classes even though it is more elaborate and demanding of concentration than the previous exercises. Using the same hand mudra as for the second exercise, exhale through the left nostril and inhale through the right three breaths, then exhale through the right and inhale through the left three more breaths. Then breathe three breaths evenly with both nostrils open. Next, exhale through the right and inhale through the left three times, then exhale through the left and inhale through the right three times, and again take three even breaths with both nostrils open. That's 18 breaths. Repeat this three times, for 54 breaths total. As before, your concentration will be on posture, abdominal or diaphragmatic breathing, and, above all, on even breathing. If you can avoid sacrificing even breathing, you can slow down to 3-second exhalations and 3-second inhalations, or ten breaths per minute.

If you do this practice three times a day, it centers the attention and calms the mind, and it is therefore ideal for anyone who wishes to remain balanced and focused. For a more advanced practice, students will gradually slow down the pace of breathing until they are finally taking 20-second exhalations and 20-second inhalations. Ultimately they will practice pranayama, or breath retention (which, as will be discussed shortly, should never be undertaken except under the supervision of a competent instructor).

A TRADITIONAL WARNING

Cautions to be judicious and respectful of breathing exercises abound in the literature on hatha yoga. And it does indeed seem from anecdotal reports of explorers in this field that the rhythm and record of our respiration resonates throughout the body. It seems to accentuate whatever is in the mind, whether it be benevolence or malevolence, harmony or disharmony, virtue or vice. On the negative side, experienced teachers report that quirkiness of any sort gets accentuated in students who go too far. It might be an abusive streak, laughing inappropriately, speaking rudely, flightiness, twitchiness, or nervous tics. Right to left physical imbalances also become exaggerated. Unfortunately, novices often close their ears to warnings: having become addicted to their practice, they will not be denied. Competent teachers of hatha yoga will be watchful of these simple matters and wary of tutoring refractory students. Even the beginning exercises discussed in this chapter should be treated with respect.

Apart from psychological concerns, the special physiological hazards of breathing exercises is that they can cause problems without giving us traditional signals warning us against doing something harmful. In athletics, the practice of asana, experiments with diet, or just tinkering with any object in the physical world, we depend on our senses to tell us that we are exceeding our capacity or doing something inadvisable. But breathing exercises are different. In that realm we are dealing with phenomena that our senses, or at least our untutored senses, are often unable to pick up, even though they can still affect the body. And because of this, advanced exercises should be undertaken only by those who are adequately prepared. Given such preparation, and given that one is enjoying a balanced life of cheerful thoughts, positive feelings, and productive actions, the yoga breathing exercises have the potential for producing more powerful and positive benefits than any other practice in hatha yoga. Again, that's a big claim, but experienced yoga instructors will agree.

"With respect then to the use of respiration, it may be affirmed that an aerial something essential to life, whatever it may be, passes into the mass of the blood. And thus air driven out of the lungs, these vital particles having been drained from it, is no longer fit for breathing again."

— John Mayow, in *Tractatus Quinque* (1674), quoted from Proctor's *A History of Breathing Physiology*, p. 162.

an empowered thoracic inhalation

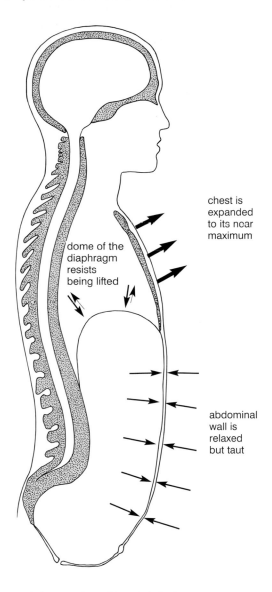

chest is
expanded
to its near
maximum

dome of the
diaphragm
resists
being lifted

abdominal
wall is
relaxed
but taut

Figure 2.29a. Empowered thoracic breathing: inhalation.
1) The dome of the diaphragm resists being pulled toward the head, and thereby supports inhalation indirectly.
2) The abdominal wall is relaxed but taut.
3) The external intercostal muscles actively lift the chest up and out.
4) The rib cage expands to its near maximum.
5) The mental state is celebratory.

a constricted thoracic inhalation

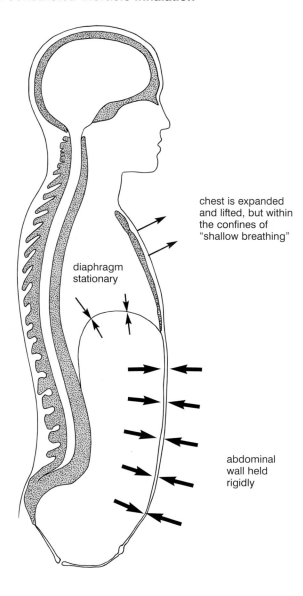

chest is expanded
and lifted, but within
the confines of
"shallow breathing"

diaphragm
stationary

abdominal
wall held
rigidly

Figure 2.29b. Constricted thoracic breathing: inhalation.
1) The diaphragm is relaxed and almost immobile.
2) The abdominal wall is held rigidly.
3) The external intercostal muscles actively lift the chest up and out.
4) The rib cage expands to within self-imposed and constricted limits.
5) The mental state can become anxious.

a paradoxical inhalation

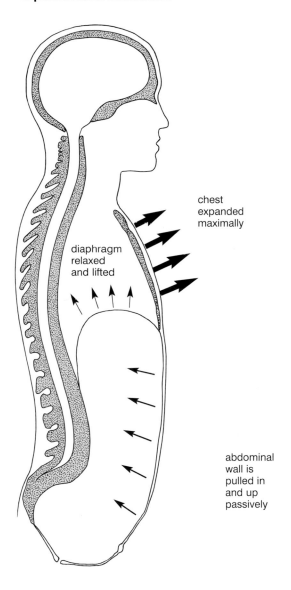

chest
expanded
maximally

diaphragm
relaxed
and lifted

abdominal
wall is
pulled in
and up
passively

Figure 2.29c. Paradoxical breathing: inhalation.
1) The diaphragm is completely relaxed and lifted by the chest.
2) The abdominal wall is pulled in and up passively.
3) The external intercostal muscles actively lift the chest up and out.
4) The rib cage expands maximally.
5) Overdone, the mental state can become anxious and panicky.

an abdominal (abdomino-diaphragmatic) inhalation

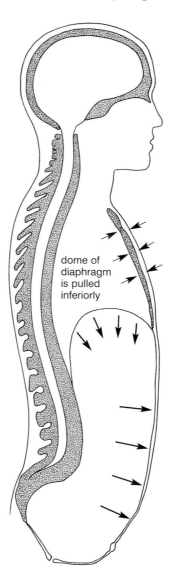

intercostal muscles
are held in a gentle
state of isometric
contraction to keep
the chest from
collapsing inward
during inhalation

dome of
diaphragm
is pulled
inferiorly

abdominal wall
is relaxed and
pushed forward
by the action of
the diaphragm

Figure 2.29d. Abdominal (abdomino-diaphragmatic) breathing: inhalation.
1) The dome of the diaphragm moves down in a fairly simple piston-like action.
2) The lower abdominal wall is relaxed and pushed forward by the diaphragm.
3) The intercostal muscles actively hold the chest wall in a stable position.
4) The rib cage remains about the same size in all parts of the breathing cycle.
5) The mental state is relaxed and may get sleepy.

a diaphragmatic (thoraco-diaphragmatic) inhalation

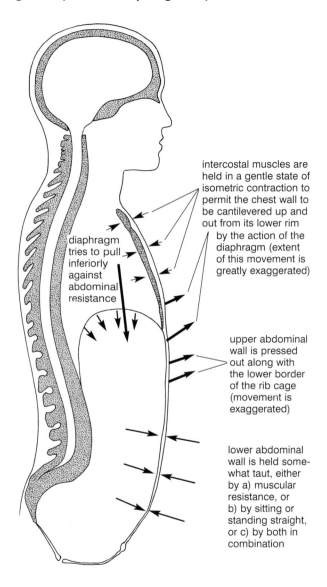

intercostal muscles are held in a gentle state of isometric contraction to permit the chest wall to be cantilevered up and out from its lower rim by the action of the diaphragm (extent of this movement is greatly exaggerated)

diaphragm tries to pull inferiorly against abdominal resistance

upper abdominal wall is pressed out along with the lower border of the rib cage (movement is exaggerated)

lower abdominal wall is held somewhat taut, either by a) muscular resistance, or b) by sitting or standing straight, or c) by both in combination

Figure 2.29e. Diaphragmatic (thoraco-diaphragmatic) breathing: inhalation.
1) The diaphragm presses down against the abdominal organs.
2) The abdominal wall is relatively taut, in part from muscular activity.
3) The intercostal muscles maintain the overall integrity of the chest wall.
4) The rib cage is flared at its base by the costal attachment of the diaphragm.
5) The mental state is clear and attentive, and is focused on the mid-torso.

CHAPTER THREE
ABDOMINOPELVIC
EXERCISES

"Agni sara means 'energizing the solar system,' the area of the body associated with digestion. Agni sara also benefits the bowels, bladder, nervous system, circulatory system, and reproductive system. Of all the exercises this one is the most beneficial, and if time is very short, it may be performed alone."

— Sri Swami Rama, in *Exercise Without Movement*, p. 53.

A sure way to develop what yogis call inner strength is to tone the abdominal region. If energy in the arms and shoulders is weak, a strong abdomen can give you an extra edge, but if the abdomen is weak, look out, because even the strongest arms and shoulders are likely to fail you. One of the most memorable boxing matches of the century (Muhammad Ali vs. George Foreman, Oct. 30, 1974) is a perfect example. Ali knew he had lost his edge for dancing around the ring "like a butterfly," and that he probably couldn't win unless he adopted unconventional tactics. Suspecting that Foreman would not have the stamina for a long bout, Ali had prepared a steely but resilient belly with thousands of repetitions of sit-ups and other abdominal exercises. He called on this secret strength early in the match, repeatedly going to the ropes and letting his opponent punch himself to exhaustion. Reality hit in the eighth round—with a few precise strokes Ali brought Foreman down for the count.

The structural foundation for abdominopelvic strength and energy (this is obviously a literary rather than a scientific use of the term "energy," something like saying someone has "a lot of pep," or "a lot of guts") is the pelvis and abdomen, a complex region whose architecture can be understood most easily by studying two simple and familiar exercises: crunches and sit-ups. Then we'll be able to make more sense of the general design of the abdominopelvic region in relation to the chest and lower extremities. This in turn will enable us to discuss leglifts, sit-ups, the boat postures, and the peacock. These seemingly diverse exercises not only strengthen the torso, they stimulate abdominal energy by using the abdomen as a fulcrum for manipulating large segments of the body in relation to one another in the field of gravity.

In the second half of the chapter we'll shift our attention to the *anatomical perineum* and discuss practices that work with that region and with the abdomen and pelvis in relative isolation, in contrast to the abdominal exercises, which demand use of the body as a whole. The practices in the second half of the chapter include *ashwini mudra*, *mula bandha*, *agni sara*, *uddiyana bandha*, and *nauli kriya*. Last, we'll take a critical look at contraindications and benefits.

CRUNCHES AND SIT-UPS

If you asked the instructor at your local health club to show you the best abdominal exercise, you would probably be told to do crunches. You would lie down supine, draw the feet in, bend the knees, interlock the fingers behind the head, and then pull the upper half of your body into a fourth of a sit-up, just enough to lift your shoulders well off the floor. Then you would lower yourself back down and repeat the movements as many times as you want. This is not a bad exercise. It strengthens the abdominal muscles and stretches the back in one of the safest possible positions. Sit-ups are a different matter. In high school gym classes from years gone by, students used to count the number of rapid-fire sit-ups (jerk-ups, actually) they could do in a minute with the knees extended and the hands interlocked behind the neck. If you are strong and under eighteen this probably won't hurt you, but if you are older and have a history of back problems it is likely to make them worse.

The muscles responsible for crunches and sit-ups include both abdominal muscles and hip flexors. The abdominal muscles encircle the abdomen and extend from the chest to the pelvis. The hip flexors, which are located deep in the pelvis (and thus hidden from view), flex the *femur* at the hip joint. They include the *iliacus* and *psoas* muscles (or the *iliopsoas*, considering the two of them together as a team). They run from the pelvis to the upper part of the femur in the case of the iliacus, and from the lumbar spine to the femur in the case of the psoas (figs. 2.8, 3.7, and 8.13).

Crunches are relatively safe because the knees are bent and the lumbar region is rounded posteriorly (to the rear). Under these circumstances, the abdominal muscles pull you up and forward, and the iliopsoas muscles aid that movement as synergists by bracing the ilia and the lumbar region (fig. 3.1). By contrast, if you do sit-ups with the knees straight, the psoas muscles first pull the lumbar spine into a more fully arched position anteriorly (to the front), and then they pull the torso up and forward. If you have back problems, it is this initial pull on the lumbar arch that can create problems. Later in this chapter (fig. 3.21a–b) we'll see several ways to approach sit-up exercises more safely.

THE FOUNDATION OF THE BODY

To understand how crunches, sit-ups, and leglifts operate mechanically, as well as to lay the groundwork for discussing standing, backward bending, forward bending, twisting, and sitting postures in later chapters, we must look at the pelvis and its relationships with the spine and thighs in detail.

THE HIP BONES AND SACRUM: THE PELVIC BOWL

We'll first examine the *pelvic bowl*, which is formed from the combination of the two *pelvic bones* (the *hip bones*) and the *sacrum*—the lowest of the four main segments of the spine. The pelvic bones have two roles: one is to link the vertebral column with the thighs, legs, and feet; the other is to define (in combination with the sacrum) the base of the torso and provide a skeletal framework for the pelvic cavity and the organs of elimination and reproduction.

In the fetus each hip bone is made up of three segments: the *ilium*, the *ischium*, and the *pubis*. We often speak of them individually, but in adults they are fused together into one piece, with one hip bone on each side. To the rear, the iliac segments of the pelvic bones form right and left *sacroiliac joints* with the sacrum (fig. 3.2–4).

To understand the three-dimensional architecture of the pelvic bowl, there is no substitute for palpating its most prominent landmarks. You can start by feeling the *crests* of the ilium on each side at your waistline. Then locate the *ischial tuberosities* (the "sitting bones") behind and below; these are the protuberances upon which your weight rests when you sit on a bicycle seat or on the edge of a hard chair.

To continue your exploration, locate the two *pubic bones* in front, just above the genitals. They join one another at the *pubic symphysis*, a fibro-cartilaginous joint which keeps the two sides of the pelvis locked together in front (figs. 1.12 and 3.2–4); their *rami* (ramus means "branch") connect with the *ilia* and ischia on each side (figs. 3.2–4). First trace the upper margin of each pubic bone laterally. What you are feeling are the *superior pubic rami*, bony projections that extend into the groin toward the ilium on each

iliopsoas muscles (hip flexors) act as synergists to brace the pelvis

Figure 3.1.
Crunch exercise,
safely lifting up
and forward
with bent knees.

abdominal muscles act as agonists (prime movers)

side. An inch or so lateral to the pubic symphysis, these projections are overlain by the iliacus and psoas muscles passing out of the pelvis to their combined insertion on the front of the femur. And beyond the softness of these muscles, the superior pubic rami connect with the ilia, which are again easily palpable.

Next locate the *inferior pubic rami*, which connect to the ischia (figs. 3.2–4). To find them, stand with your feet wide apart and locate the bones that extend from the base of the pubic region inferiorly, laterally, and posteriorly. They form a deep upside-down V. About halfway back each inferior pubic ramus merges into the next component of the hip bone, the ischium. It's hard to locate the lateral border of the inferior pubic ramus because the tendons of the adductor muscles (figs. 2.8, 3.8–9, and 8.13–14) are in the way. And in the male it is also difficult to palpate the inside, or medial, border of the inferior pubic rami because the penis is rooted in the converging arms of the V. In the female the medial borders of these bones are more accessible. In either case, following them posteriorly will finally lead you to the ischial tuberosities.

Returning to the ilium, which continues laterally from each superior pubic ramus, you will find a prominent bony point, the *anterior superior iliac spine*, and just below this protuberance, the less obvious *anterior inferior iliac spine* (figs. 3.2–4). If your abdomen is not in the way, you will become aware of the right and left anterior superior iliac spines when you lie prone on a hard surface. From these landmarks, trace the crests of the ilia laterally along the waistline. If you are slender and not heavily muscled, you can poke your thumb inside the iliac crest and feel the top half inch or so of the inside of the pelvic bowl from which the iliacus muscle originates. Then, as you follow the crest of the ilium around to the back, you will come to a solid mass of muscle, the *erector spinae*, below which the ilium articulates with the sacrum.

THE SACROILIAC JOINTS AND THE SPINE

The two pelvic bones connect with the rest of the torso through the sacrum at the two sacroiliac joints (figs. 3.2–4), which are formed on each side of the sacrum at the junction of two rough but matching surfaces (figs. 3.3 and 6.2)—the lateral surface of the sacrum and the medial surface of the pelvic bone. Even though these are movable synovial joints whose mating surfaces are bathed in synovial fluid, and even though their matching L-shaped groove-and-rail architecture permits some movement in children and healthy young adults, heavy bands of deep fasciae and well defined *sacroiliac* and *iliolumbar ligaments* (fig. 3.4) bind the joints together on the outside and restrain their movement in most people over the age of 25. Athletic young women are notable exceptions; their sacroiliac joints are generally more mobile than those for men in comparable condition. We'll explain the nature of the complex movements that are possible at the sacroiliac joint in chapter 6.

Anatomical differences account for some of the variations in sacroiliac mobility between men and women, along with the female hormones estrogen, progesterone, and relaxin. The latter all become especially important in the last month of pregnancy for loosening up the sacroiliac joints, along with the pubic connections in front. All must yield to permit the passage of the baby through the birth canal.

Internally, the sacroiliac joints sometimes become *ankylosed*, which means they have formed a partial or complete bony union. Older men are particularly apt to develop this condition, and once it begins, their sacroiliac components can slip relative to one another only with considerable difficulty and unpleasantness. Such slippage usually happens as a result of a fall, but any impact that disturbs the partially locked relationship between the two sides of the joint will traumatize the opposing surfaces and probably cause extreme pain. *Sacroiliac sprains* (tears) of the binding ligaments are yet another problem: in this case they are a common cause of lower back pain.

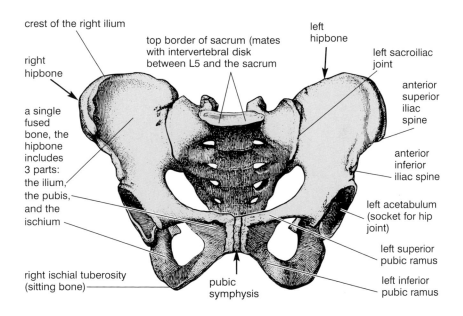

Figure 3.2. The female pelvis, with sacrum and two hipbones. The sacrum articulates in the rear with the ilia at the sacroiliac joints, and the two hipbones articulate with one another in front (by way of the right and left pubic bones) at the fibrocartilagenous pubic symphysis. Also see fig. 1.12 (Sappey).

Because the sacroiliac joints in adults bind the pelvic bones so firmly to the sacrum, every tilt, rotation, and postural shift of the pelvis as a whole affects the vertebral column, and with the vertebral column, the entire body. If you rotate the top of the pelvis posteriorly (which is by definition a *posterior pelvic tilt*, or colloquially, a pelvic "tuck"), the top of the sacrum is carried to the rear, and this causes the lumbar curvature to flatten and lose its lordosis (forward arch), or in the extreme to become rounded

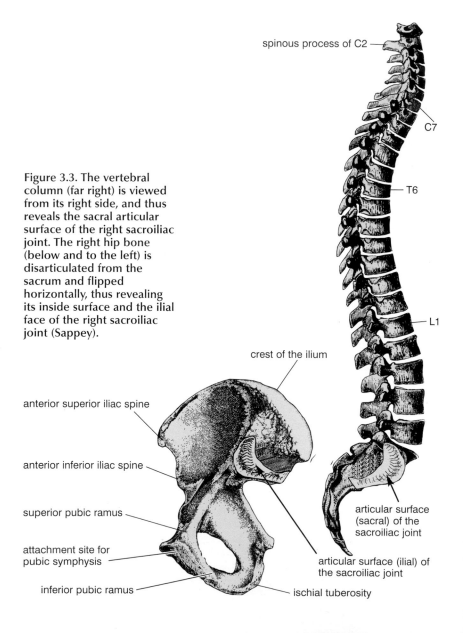

Figure 3.3. The vertebral column (far right) is viewed from its right side, and thus reveals the sacral articular surface of the right sacroiliac joint. The right hip bone (below and to the left) is disarticulated from the sacrum and flipped horizontally, thus revealing its inside surface and the ilial face of the right sacroiliac joint (Sappey).

spinous process of C2

C7

T6

L1

crest of the ilium

anterior superior iliac spine

anterior inferior iliac spine

superior pubic ramus

attachment site for pubic symphysis

inferior pubic ramus

articular surface (sacral) of the sacroiliac joint

articular surface (ilial) of the sacroiliac joint

ischial tuberosity

posteriorly. On the other hand, pulling the top of the pelvis forward, which is defined as an *anterior pelvic tilt*, increases the depth of the lumbar lordosis. And if you stand on one foot the tipped pelvis will create side-to-side deviations of the spine.

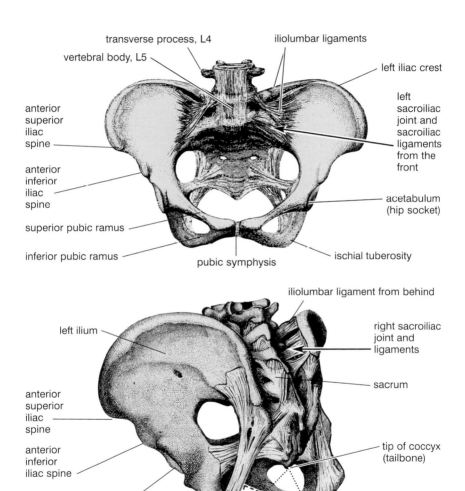

Figure 3.4. Pelvic restraining ligaments from the front (above) and from the side and behind (below). The borders of the diamond-shaped anatomical perineum are shown below, and include the anal triangle behind (dotted line), and the urogenital triangle in front (dashed line), with a shared border (solid line) connecting the two ischial tuberosities (Sappey).

THE HIP JOINTS AND THEIR PRIME MOVERS

Most people understand the hip joints intuitively so long as they are dealing with a simple imperative such as "bend forward from the hips," or understanding that a simple "hip replacement" involves replacing the head of the femur with a steel ball that will fit into the hip socket. Questioned beyond that, most people will fall silent; they have no notion of what makes up the socket or how movements take place. But now we have begun to develop a distinct image of the pelvic bowl. We have seen how the two pelvic bones are united in front at the pubic symphysis and how the pelvic bones articulate with the sacrum behind, and we have palpated several bony landmarks on each side. We only need a few more details to complete the picture.

The *acetabulum* (socket) for each hip joint is located at the lateral and inferior aspects of the pelvic bowl (figs. 3.2 and 3.4–5). You can't feel the acetabulum, but you can feel the bony protuberance just below the joint that sometimes bumps into things—the *greater trochanter* of the femur (figs. 3.5–6). If you stand up and locate this landmark near where your hands fall alongside your thighs, you will notice that it moves around as you swing your thigh back and forth.

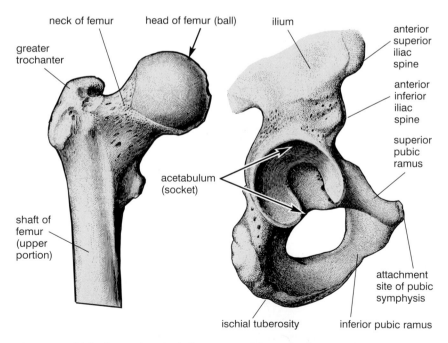

Figure 3.5. Right femur (on the left) as viewed from the front, and right hip bone (on the right) as viewed from the side. The head of the femur fits snugly into the acetabulum, forming a ball-and-socket joint (Sappey).

The pelvic bowl is the foundation for all movements of the thighs at the hip joints, including flexion, extension, abduction, adduction, and rotation. To flex the thigh in a leglift (figs. 3.15–17) you contract the psoas and iliacus muscles (figs. 2.8, 3.7, and 8.13), which, as we have seen, run from the pelvis to the upper part of the femur in the case of the iliacus, and from the lumbar spine to the femur in the case of the psoas. For activities such as lifting each knee (as in running in place), or for stepping forward (as in walking), the origins of these muscles are on the torso and their insertions are on the thighs, but for sit-ups and crunches (fig. 3.1), the origins and insertions are reversed—the thighs are fixed and the entire body is pulled up and forward.

To extend the thigh actively in a posture such as the locust (figs. 5.15–19) you tighten the *gluteus maximus* muscle (figs. 3.8, and 8.9–10), which takes origin from the posterior surface of the ilium and which has two insertions,

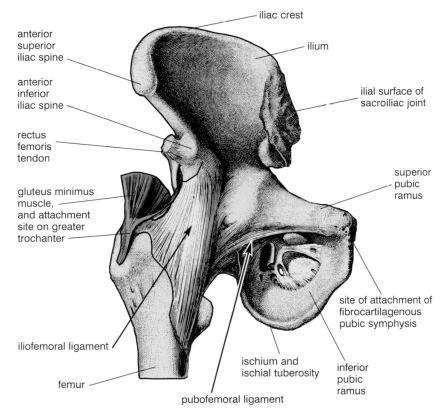

iliac crest

anterior superior iliac spine

ilium

anterior inferior iliac spine

ilial surface of sacroiliac joint

rectus femoris tendon

gluteus minimus muscle, and attachment site on greater trochanter

superior pubic ramus

iliofemoral ligament

site of attachment of fibrocartilagenous pubic symphysis

femur

ischium and ischial tuberosity

inferior pubic ramus

pubofemoral ligament

Figure 3.6. Right hip bone. femur, and joint capsule, with the iliofemoral and pubofemoral ligaments visible in front, and the ischiofemoral ligament hidden behind. These three ligaments in combination become taut during hip extension, and loose during hip flexion (for example, when the knee is lifted); (from Sappey).

one on the femur (fig. 3.10b), and the other in a tough band of connective tissue—the *iliotibial tract*—that runs all the way down past the knee to the leg (figs. 3.8–9 and 8.12). You can feel the activity of the gluteus maximus become pronounced if you stand up and pull the thigh to the rear while pressing against the gluteal region with your hand. By contrast, many other postures such as the camel (figs. 5.34–35) hyperextend the hip joint passively, and this is resisted both by the psoas and iliacus muscles (figs. 2.8, 3.7, and 8.13), and by the *rectus femoris* component of the quadriceps femoris muscle (figs. 3.9, 3.11, and 8.8–9).

To abduct the thigh, which you do when you lift the foot straight out to the side, you tighten the *gluteus medius* and *gluteus minimus* muscles (figs. 3.8, 3.10a–b, 8.9–10, and 8.12), which take origin from beneath the gluteus maximus and insert on the greater trochanter. To adduct the thighs, which you do by pulling them together, you tighten the adductor muscles, which take origin from the inferior pubic rami and insert below on the femurs and *tibias* (figs. 2.8, 3.9, and 8.13–14).

If the muscles of the hips and thighs are strong and flexible, and if you are comfortable extending the thighs fully in any standing, kneeling, or prone posture, you'll finally encounter resistance to extension in a deep spiral of ligaments that surround the ball and socket hip joint—the

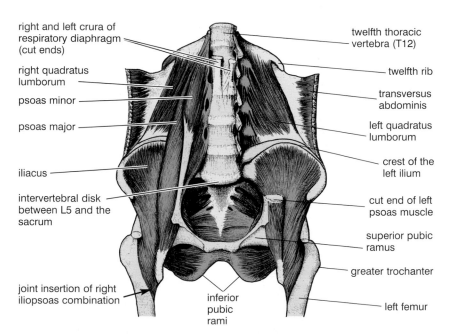

right and left crura of respiratory diaphragm (cut ends)

right quadratus lumborum

psoas minor

psoas major

iliacus

intervertebral disk between L5 and the sacrum

joint insertion of right iliopsoas combination

inferior pubic rami

twelfth thoracic vertebra (T12)

twelfth rib

transversus abdominis

left quadratus lumborum

crest of the left ilium

cut end of left psoas muscle

superior pubic ramus

greater trochanter

left femur

Figure 3.7. Deep dissection of the pelvis and lower abdomen revealing the psoas and iliacus muscles and their conjoined insertions on the femurs. Their contraction lifts the thighs, thus bringing about hip flexion (Sappey).

iliofemoral, ischiofemoral, and *pubofemoral ligaments* (fig. 3.6). You won't feel this spiral unless you know it is there, but it will become increasingly taut as the thighs are extended. When that happens, the head of the femur is driven into the acetabulum of the pelvic bone in a near-perfect fit, and the thigh will extend no more. The spiral will unwind as the thighs are flexed. If this spiral is removed and the hip joint opened up, the head of the femur and the acetabulum become visible (fig. 3.5).

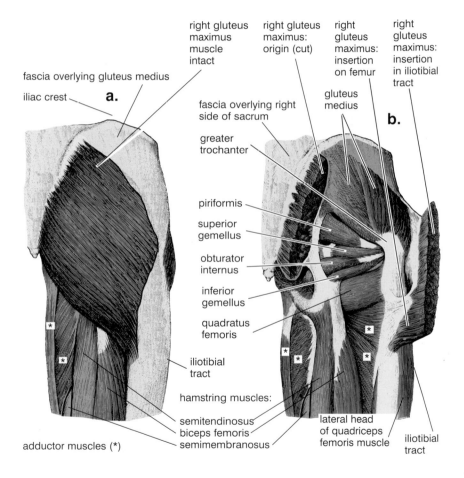

Figure 3.8. Right gluteal region and upper thigh from behind, with superficial dissection on the left (a) and deeper dissection on the right (b). The partial cutaway of the gluteus maximus on the right (b) exposes deeper muscles of the hip, as well as a clear picture of the dual insertion of the gluteus maximus to he iliotibial tract and the femur (Sappey).

THE QUADRICEPS FEMORIS MUSCLE

The quadriceps femoris is the largest muscle on the front of the thigh (figs. 1.2, 3.9, and 8.8–9) and the foremost anti-gravity muscle in the body. Three of its four components, or "heads," take origin from the femur and act on the tibia by way of the patellar tendon. Its fourth head, the rectus femoris (figs. 3.9, 3.11, and 8.8–9), takes origin from the front of the pelvis (the anterior inferior iliac spine, figs. 3.2–6) and joins the other three components below. The quadriceps femoris is the muscle, more than any other, that stands you up from a squatting position. You can test its strength by standing in a 90° bent-knee position for 30 seconds with your back flat against a wall and then slowly rising. For those who are older and in a weakened condition,

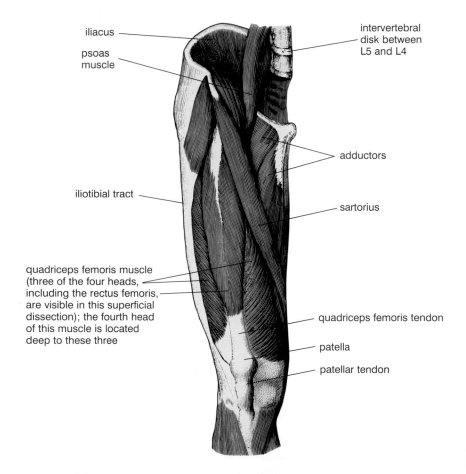

Figure 3.9. Right side of pelvis (deep dissection), right thigh, and right knee, as viewed from the front (Sappey).

this is the muscle that gives them pause when they want to climb up or down stairs without holding onto a handrail. It is also Waterloo for inexperienced skiers who are trying to negotiate a bowl of deep powder for the first time: they are firmly (even though wrongly) convinced that they have to keep their weight back and their ski tips visible to avoid toppling over into the snow. Although that can indeed happen—it's called a "face plant" or "header"—most novices overcompensate for the possibility and quickly pay for their error: quadriceps femoris muscles that are soon burning with pain.

THE HAMSTRING MUSCLES

On the back sides of the thighs are the *hamstring muscles*, most of which have their origin on the ischial tuberosities. Like the quadriceps femoris muscles, the hamstrings insert below the knee joint, in this case both medially and laterally (figs. 3.10, 8.10, and 8.12). Tight hamstrings are the bane of runners—thousands of repetitive strides make these muscles shorter and shorter until they are barely long enough to permit full extension of the knees.

As two-joint muscles that pass lengthwise across two joints instead of one (from the ischial tuberosities of the pelvis all the way to the proximal ends of the tibias and fibulas), the hamstrings contribute both to extension of the thighs at the hip joints and to flexion of the legs at the knee joints. This architectural arrangement facilitates walking and running beautifully, but it creates a problem in hatha yoga. Since the hamstrings reside on the back sides of two joints—the knee and the hip—each of which is crucial in its own way for forward bending, these muscles are major obstacles to such movements. It's obvious that you could relieve tension on the hamstrings in forward bends by easing up either on hip flexion or knee extension, but releasing flexion of the hips would be contrary to the whole idea. What everyone does naturally is to flex their knees slightly, insuring that the hamstring muscles don't tug so insistently on the base of the pelvis as one attempts to bend forward. This was the principle involved in chapter 1 when we bent the knees before pulling the torso down against the thighs in the standing hamstrings-quadriceps thigh pull, and this is why we keep the knees bent in crunches. It is also why the knees should be bent if you insist on doing high-speed sit-ups. Otherwise the hamstring muscles tug on the ischial tuberosities from below and create too much tension in the lower back as you jerk yourself up and forward.

THE SPINE AND ABDOMINAL WALL

The pelvic bowl is not merely the link between the thighs and the upper half of the body; it is also the foundation for the torso. Knowing this, if you look at a skeleton, even with one glance, you will sense an immediate cause

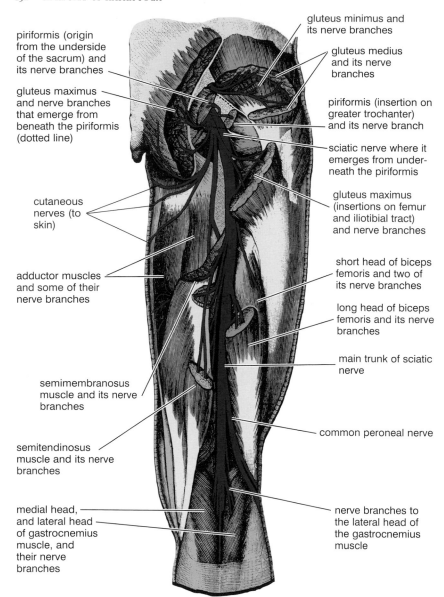

piriformis (origin from the underside of the sacrum) and its nerve branches

gluteus maximus and nerve branches that emerge from beneath the piriformis (dotted line)

cutaneous nerves (to skin)

adductor muscles and some of their nerve branches

semimembranosus muscle and its nerve branches

semitendinosus muscle and its nerve branches

medial head, and lateral head of gastrocnemius muscle, and their nerve branches

gluteus minimus and its nerve branches

gluteus medius and its nerve branches

piriformis (insertion on greater trochanter) and its nerve branch

sciatic nerve where it emerges from underneath the piriformis

gluteus maximus (insertions on femur and iliotibial tract) and nerve branches

short head of biceps femoris and two of its nerve branches

long head of biceps femoris and its nerve branches

main trunk of sciatic nerve

common peroneal nerve

nerve branches to the lateral head of the gastrocnemius muscle

Figure 3.10a. Nerves to muscles of the back of the hip and thigh originate from spinal segments L4, L5, S1, and S2, and run down the back of the thigh on the extensor side of the hip joint. The large sciatic nerve and associated branches to the gluteus maximus emerge from just underneath the piriformis muscle (shown intact in fig. 3.8b, in two parts connected by the dotted lines here in fig. 3.10a, and removed except for its tendon of insertion in fig. 3.10b). Nerves to the gluteus medius, gluteus minimus, and piriformis are shown above, and nerves to the hamstrings, gastrocnemius, and adductors are shown below. A superficial branch of the common peroneal nerve swings around to an anterior, subcutaneous, and vulnerable position just below the knee (chapter 10); (from Sappey).

for alarm: there are many bones and much skeletal density in the pelvis and lots of ribs and vertebrae in the upper torso, but there are only five lumbar vertebrae connecting the two regions (figs. 4.3–4). This arrangement could not provide adequate support to the torso if it were acting alone. It needs the help of the soft tissues, especially sheets of muscle and fasciae. To that end the skeleton is supported by a "tube" containing the abdominal organs, a tube that is bounded in front and on the sides by the abdominal muscles, braced posteriorly by the spine and deep back muscles, capped by the respiratory diaphragm, and sealed off below by the *pelvic diaphragm*. The tube runs all the way from the sternum to the pubis in front but is quite short laterally.

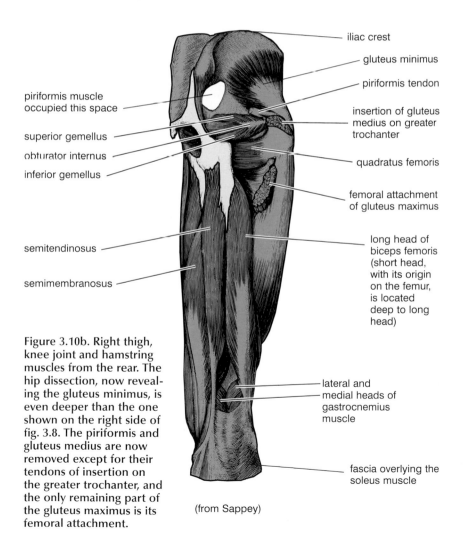

iliac crest

gluteus minimus

piriformis tendon

piriformis muscle occupied this space

insertion of gluteus medius on greater trochanter

superior gemellus

obturator internus

inferior gemellus

quadratus femoris

femoral attachment of gluteus maximus

semitendinosus

long head of biceps femoris (short head, with its origin on the femur, is located deep to long head)

semimembranosus

Figure 3.10b. Right thigh, knee joint and hamstring muscles from the rear. The hip dissection, now revealing the gluteus minimus, is even deeper than the one shown on the right side of fig. 3.8. The piriformis and gluteus medius are now removed except for their tendons of insertion on the greater trochanter, and the only remaining part of the gluteus maximus is its femoral attachment.

lateral and medial heads of gastrocnemius muscle

fascia overlying the soleus muscle

(from Sappey)

We have four pairs of *abdominal muscles* (figs. 2.7, 2.9, 3.11–13, 8.8, 8.11, and 8.13). Three of these form layers that encircle the abdomen, and the fourth is a pair of longitudinal bands. The *external abdominal oblique* layer runs diagonally from above downward in the same direction as the external intercostal muscles. If you place your hands in the pockets of a short jacket with your fingers extended, the fingers will point in the direction of the external abdominal oblique muscle fibers. The *internal abdominal oblique* layer is in the middle. Its fibers also run diagonally but in the opposite direction, from laterally and below to up and medially in the same direction as the internal intercostal muscles. The innermost third layer, the *transversus abdominis*, runs horizontally around the abdominal wall from

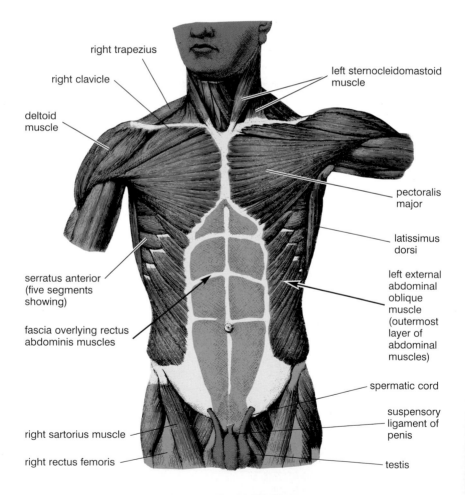

Figure 3.11. Torso, with superficial muscles of the chest and neck, fascia covering the rectus abdominis muscles, and the external abdominal oblique muscles (Sappey).

back to front. These three layers together act as a unit, helping to support the upper body and contributing to bending, twisting, and turning in a logical fashion. They are also necessary for coughing, sneezing, laughing, and various yoga breathing exercises.

The fourth pair of abdominal muscles, the *rectus abdominis* muscles (rectus means "straight"), run vertically on either side of the midline between the pubic bone and the sternum. As discussed earlier in this chapter, the rectus abdominis muscles are the prime movers (agonists) for flexion of the spine in crunches, while the hip flexors serve as synergists for bracing the pelvis and lumbar region. The roles are then reversed for old style sit-ups, in which the hip flexors become the prime movers for jerking the torso up and forward at the hip joints, and the rectus abdominis muscles serve as synergists for bracing the spine.

THE CAVITIES AND INTERNAL ORGANS

Within the "tube" of the torso are the thoracic, abdominal, and pelvic cavities, as well as most of the internal organs. The heart, lungs, and esophagus lie within the *thoracic cavity*, which is bounded externally by the rib cage and inferiorly by the respiratory diaphragm (figs. 2.6–9). The

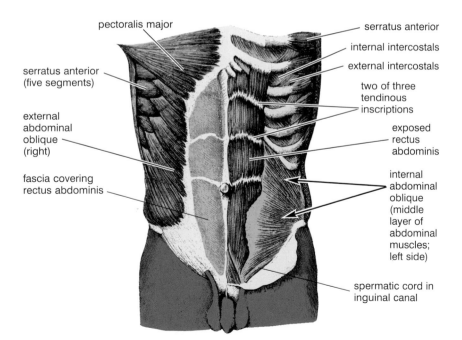

Figure 3.12. External abdominal oblique and rectus abdominis fascia on the torso's right side, and internal abdominal oblique and exposed rectus abdominis muscle on the torso's left side (Sappey).

stomach, intestines, liver, pancreas, spleen, and kidneys are contained within the *abdominal cavity* (figs. 2.9 and 3.14), which is separated from the chest by the diaphragm, protected posteriorly by the spine and deep back muscles, and surrounded anteriorly and laterally by the abdominal muscles. The urinary bladder, the terminal end of the colon, and portions of the reproductive systems lie in the *pelvic cavity* (figs. 2.8 and 3.7) and open to the external world by way of passages through the pelvic diaphragm at the base of the pelvic bowl (figs. 2.29a–e, 3.14, and 3.24–26). The pelvic cavity is defined above by the upper limits of the bony pelvis and below by the pelvic diaphragm, but otherwise it is confluent with the abdominal cavity. Thus, we refer to them together as the *abdominopelvic cavity* (fig. 3.14, illustration on the right).

Most of the internal organs are not fixed in position but can slide around by virtue of slippery external surfaces: pleural and pericardial membranes in the chest, and peritoneal membranes in the abdomen and pelvis. Within the thoracic cavity, the *pericardial membranes* surround the heart and enclose the *pericardial cavity* while the *pleural membranes* surround the lungs and enclose the *pleural cavities* (figs. 2.4 and 2.6). Within both the

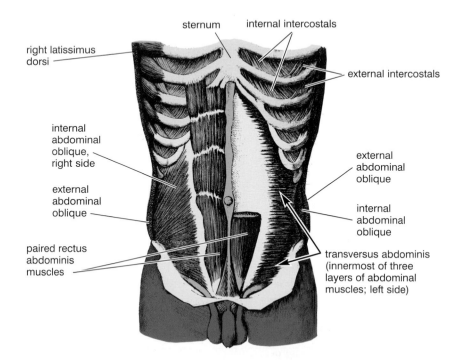

Figure 3.13. Internal abdominal oblique and exposed rectus abdominis on this torso's right side, and transversus abdominis and sectioned rectus abdominis muscle on the torso's left side (Sappey).

abdominal and pelvic cavities, the *peritoneal membranes* surround the abdominal and pelvic organs and enclose the *peritoneal cavity*. Like the pleural and pericardial cavities, the peritoneal cavities are potential spaces only, as illustrated by a schematic midsagittal section through this region (fig. 3.14). These spaces contain only a small amount of fluid which allows the organs to move relative to one another. The most famous trick question in a medical gross anatomy course is: Name all the organs in the pleural, pericardial, and peritoneal cavities. The correct answer is: None.

INTRA-ABDOMINAL AND INTRATHORACIC PRESSURE

Lubricating fluids in the peritoneal cavity impart a liquid character to the internal organs in the abdominopelvic cavity and allow that region to act as a *hydraulic* (having to do with liquid) *system*. This means that if something presses against the abdominal wall, hydraulic pressure is transmitted throughout the entire region just like squeezing a capped tube of toothpaste at one site will cause the tube to bulge out everywhere else. The

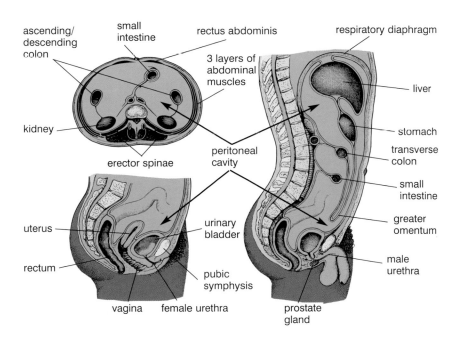

Figure 3.14. Schematic drawings of the peritoneal cavity and a few abdominal and pelvic organs: male on the right, female on the lower left, and gender-neutral cross section on the upper left. White spaces indicate the peritoneal cavity (greatly exaggerated) in all three drawings. The entirety of the abdominopelvic cavity (within which reside the abdominopelvic organs and the peritoneal cavity) is illustrated in the mid-sagittal section on the right (Sappey).

abdominal wall is the soft part of the tube, and the respiratory and pelvic diaphragms seal it at either end. A separate unit, the chest, is bounded by the rib cage and the respiratory diaphragm. The glottis can seal the air within the chest, with the result that the chest can act as a *pneumatic* (having to do with air) *system*. Such a system remains at atmospheric pressure any-time the glottis is open, but if you inhale and close the glottis, the system can be compressed (and is indeed often compressed) by the action of the abdominal muscles and external intercostals.

Even though the thoracic and abdominopelvic regions are anatomically independent, the former functioning as a pneumatic system and the latter as a hydraulic system, the trunk as a whole operates as a cooperative unit. For example, if you bend over from an awkward position to pick up a heavy object, and if you have to do that with your knees straight, your spine is vulnerable to injury from too much stress on the lumbar region. If you were to try that maneuver from a bent-forward position when you are breathing freely—or even worse, if you were to hold your breath after an exhalation—the weight of the object would create a frighteningly efficient shearing effect on all the intervertebral disks between the chest and the sacrum. Were it not for our ability to supplement skeletal support with the hydraulic and pneumatic pressures within the abdominopelvic and thoracic cavities, the intervertebral disks in the lumbar region would quickly degenerate and rupture. To protect yourself you will have to increase intra-abdominal pressure, and you can do this with or without the aid of compressed air in the chest.

You can protect your spine with respect to how you use your hydraulic and pneumatic systems in one of three ways. First, before you lift you can inhale, close the glottis, and hold your breath. Then you can tighten your abdomen, pelvic diaphragm, and internal intercostal muscles all at the same time so that the pneumatic pressure in the chest comes into equilibrium with the hydraulic pressure in the abdominopelvic cavity. This allows the respiratory diaphragm to remain relaxed and increases pressure in the torso as a whole. That increased pressure then supports the action of the back muscles in two ways: it creates a taut, reliable unit from which to lift the object, and it produces a lengthening effect on the spine which spreads the vertebrae apart and eases strain on the intervertebral disks. (It should be noted that for older people, especially those who might be vulnerable to cardiovascular problems, this is an emergency measure only, because it will result in an immediate increase in blood pressure.)

A second way to protect your back, if you have a strong respiratory diaphragm and know how to use it, is to keep the glottis and airway open as you lift, and at the same time press down with the diaphragm, in with the abdominal muscles, and up with the pelvic diaphragm. This is a very different situation from the first one. Here it is the respiratory diaphragm rather than the glottis that seals the top of the tube and counters the action of the abdominal muscles and pelvic diaphragm. Just the same, it eases strain on the intervertebral disks in the critical lumbar region. The main difference between the two techniques is that now the thoracic region is not involved because the airway is open and intrathoracic pressure is not increased.

The third way to protect the spine, and one that comes naturally to most of us, is to mix and match the options. Prepare yourself with an inhalation, partially close the glottis, press down with the diaphragm, and coordinate your lifting effort with a heavy grunting sound, which is a signal that the glottis has been partially closed. What happens exactly? You start with an intent to use method number two—increasing intra-abdominal hydraulic pressure alone—but augment that effort by increasing pneumatic pressure in the chest at the precise moment that maximum protection for the back is needed. This is the choice of championship weightlifters, who continue to breathe during the easier portions of the lift, and then emit a mighty grunt to complete it.

In all hatha yoga postures that involve bending forward and then lifting back up in a gravitational field (for practical purposes this means anywhere but in a swimming pool), it is increased intra-abdominal pressure far more than the action of individual muscles that protects and braces the back. If you want to strengthen the abdominopelvic region to the maximum, and if you want this region to link the upper and lower halves of the body in the most effective and efficient manner, you will have to exercise the second option—keeping the glottis open—when you do the exercises and postures that follow. This means always placing the burden for creating intra-abdominal pressure on the respiratory diaphragm, the abdominal muscles, and the pelvic diaphragm. The first option, holding the breath at the glottis, should be used only as an emergency measure for extricating yourself safely from a posture that is beyond your capacity.

ABDOMINOPELVIC ENERGY

Yoga is concerned first and foremost with the inner life, and the abdomino-pelvic exercises are no exception. On the most obvious level yoga postures strengthen the abdominal region and protect the back. But when you do them you also come alive with energy that can be felt from head to toe. Leglifts, sit-ups, the sitting boat postures, and the peacock all create these effects through manipulating the limbs and torso in a gravitational field while you are using the abdominal region as a fulcrum for your efforts. And the harder you work the more energizing the exercise.

SUPINE LEGLIFTS

"Safety first" is a wise slogan, and the frailties (or challenges) of the human frame offer us many opportunities to practice it. If you have not had much experience with abdominopelvic exercises, please read the section on contraindications at the end of this chapter before doing them. Even the simplest leglifts and sit-up exercises should be approached with caution. The first rule: Until you know your body and its limitations well, your first line of protection is to keep the lower back flattened against the floor as you do these exercises. With a little training you can confidently make skillful use of the hydraulic nature of the abdominopelvic region, and after you are certain of yourself you can explore other options.

SUPINE SINGLE LEGLIFTS

Supine single leglifts are the safest beginning leglifting exercises because they are not likely to strain an inexperienced or sensitive back. Start with the thighs adducted, the knees extended, the feet extended (toes pointed away from you), and the hands alongside the thighs, palms down. Slowly raise one foot as high as possible (fig. 3.15) and then slowly lower it back to the floor. Repeat on the other side. Keep breathing. If you are comfortable you can try several variations of this exercise. One is to keep the knee extended, pull the flexed thigh as close as possible to the head (using the hip flexors, not the upper extremities), and hold it there for 30–60 seconds. And after you have come most of the way down you can hold the foot isometrically an inch or two away from the floor.

Figure 3.15. Single leglift. This posture is a safe hamstring stretch for the flexed thigh because the pelvis and lower back are stabilized against the floor.

What happens in single leglifts is that the psoas and iliacus muscles flex the hip while the quadriceps femoris muscle keeps the knee extended. The rectus femoris (the straight head of the quadriceps femoris) assists in both roles: it aids flexion of the hip because of its origin on the anterior inferior iliac spine, and it assists the rest of the quadriceps in keeping the knee extended (fig. 3.9). The posture itself creates the stable conditions that make single leglifts an easy exercise. First, the hamstring muscles of the side being lifted pull inferiorly on the ischial tuberosity on that same side, which keeps the pelvis anchored in a slightly tucked position and the lumbar spine flattened against the floor, and second, keeping the opposite thigh and leg flat on the floor improves the stability of the pelvis and lumbar spine even more. The combination permits flexion of one thigh with little or no stress on the lower back.

THE BICYCLE AND OTHER VARIATIONS

To further prepare for more difficult yoga postures, and to get both lower extremities into the picture but still without placing a great deal more stress on the lower back than is occasioned by single leglifts, flex both knees, draw them toward the chest, and bicycle your feet around and around. Next, and a little harder, lower your feet closer to the floor and pump them back and forth horizontally. Intensifying even more, straighten your knees and press your feet toward the ceiling. This is easy if you have enough flexibility to keep your thighs and legs perpendicular to the floor. If you can't do that, bend your knees slightly, and from that position, keep one leg lifted and slowly lower the opposite foot to within an inch of the floor, straightening the knee on the way down; then raise it back up and repeat on the other side. Keeping one foot up while lowering and raising the other is almost as easy as keeping one leg flat on the floor. You can also create a scissoring motion, with the feet meeting midway or near the highest position. And any time you need to create less pull on the underside of the pelvis from the hamstrings, bend the knees.

When you do variations that are more demanding than the simple bicycling motion, you'll find that your abdominal muscles tighten, increasing intra-abdominal pressure and pressing the lower back against the floor in cooperation with the respiratory diaphragm. This assumes, of course, that your airway is open; if you lock it at the glottis, the diaphragm will remain relaxed and you will miss one of the main points of the exercise.

THE FIRE EXERCISE

As soon as you are comfortable doing single leglifts and their variations for 5–10 minutes, you can try the fire exercise, named from its energizing effects on the body as a whole. To get in position for this one, sit on the floor, lean back, support yourself on the forearms, and place the hands under the hips or slightly behind them, palms down, or up, if that feels easier. Keeping the feet together, extend the toes, feet, and knees, and draw the head forward while keeping the back rounded. Exhale, and at the same time slowly lift the feet as high as possible, drawing the extended knees toward the head (fig. 3.16). Slowly come back down, not quite to the floor if your strength permits. Come up and down as many times as you can without strain, inhaling as required and always breathing evenly. If coming all the way up and down is too difficult, simply tighten the muscles, lift the feet an inch or so, and hold in that position isometrically. After a few days you may have enough strength to do the full exercise.

The fire exercise is intended for breath training as well as for building abdominal strength, and if you watch your breathing carefully, you'll notice that the posture feels more powerful when you exhale. As is true for many day-to-day activities, inhalation is mostly a preparation for the intensity associated with exhalation. In the case of the fire exercise, it's a matter of muscle mechanics: to support the posture efficiently with intra-abdominal pressure, the diaphragm must be continuously active, and to do this it must operate within a fairly narrow range with its muscle fibers moderately stretched and its dome high in the torso. For this reason you will find your-self exhaling almost as much as you can and taking small inhalations. If you take a deep inhalation as an experiment, you will immediately sense a loss of abdominal and diaphragmatic strength. As always for exercises such as these, unless you are faced with an unexpected emergency keep the airway open, supporting the posture only with hydraulic pressure in the abdominopelvic unit.

Figure 3.16. Fully lifted position for the fire exercise. Its key feature is that the back remains rounded posteriorly. An alternative and slightly easier hand position is to place them under the pelvis, palms up.

Like health club crunches, two features of this practice make the fire exercise safe: the back is rounded posteriorly, and the psoas muscles help lift the thighs from a stabilized origin on the inner curvature of the lumbar spine (fig. 3.16). If you start with the back straight or less firmly rounded to the rear, the psoas muscles will destabilize the lumbar region by pulling it forward before they begin to flex the hips; this is fine if your back is strong and healthy, but too stressful if it is not.

You will immediately sense the difference between the straight and the curved-to-the-rear positions of the spine if you do the following experiment. First round your back and try the fire exercise in its standard form. Sense your stability. Then (provided your back is sound) try lifting your feet after lowering your head and shoulders to the rear and letting the lumbar region relax and come forward. The instability of the second starting position will shock you. The lesson: if you are unable to maintain a stabilized posterior curvature, don't do the fire exercise. Instead, work with crunches and the single leglift variations until you are strong enough to keep the back rounded to the rear.

THE BASIC SUPINE DOUBLE LEGLIFT

No matter what kind of leglift you try, if you do not do it while keeping your lower back flattened against the floor in the supine position or rounded to the rear when the head and upper back are lifted, it has to be considered an advanced practice. The supine double leglift is a case in point. Lying flat with the legs extended, a small amount of space will usually be found between the lumbar region and the floor, and if this is allowed to remain when the psoas muscles flex the thighs, those muscles will not be pulling from a stabilized lower back that is pressed to the rear, but from a wavering and inconstant lumbar lordosis. It is therefore essential, before starting the supine double leglift, to press this region to the floor with a posterior pelvic tilt and hold it there for the duration of the exercise.

To begin the supine double leglift, lie down with your thighs adducted, knees extended, feet and toes extended, and hands alongside the thighs, palms down. Next, in order to establish enough intra-abdominal pressure to dominate the lumbar region decisively, strongly engage the abdominal muscles along with the respiratory and pelvic diaphragms, and holding that position tenaciously, slowly lift the feet (by flexing the hips) as high as possible (fig. 3.17) and then lower them to within an inch of the floor. Come up and down for as many repetitions as you want, breathing evenly throughout the exercise.

Since the muscular leverage for pressing the lumbar region to the floor comes from the abdominal muscles, especially from the rectus abdominis, learning to activate those muscles is the most important part of the exercise.

If you cannot get the feel of tightening them when you are lying flat on the floor, which is the case for most beginning students who have never been very athletic, lift your head and shoulders while holding your hands against your abdomen for feedback—it is impossible to lift your upper body without engaging the abdominal muscles. Then try to generate that same feeling as a preparation for the leglift, but without lifting the upper body.

Double leglifts are difficult not only because they depend on strong abdominal muscles that are acting in a manner to which they are not accustomed, but also because the knees must be kept fully extended. To flex the hips with the knees straight, a tremendous force has to be exerted on the insertions of the iliopsoas muscles at the proximal (near) end of the femur, and this is like trying to lift a board by gripping it with your fingers at one end. The rectus femoris muscles aid leglifts as synergists because they are pulling from the front of the pelvis to their insertions on the *patella* (kneecap) instead of from the pelvis to the proximal portion of the femur, but even with help from these muscles, the exercise is still a test of strength for many students. The endeavor is further complicated by the fact that keeping the knees straight during the leglift stretches the hamstring muscles, which are antagonists to the iliacus, psoas, and rectus femoris muscles. That stops a lot of people in a hurry.

TRAINING ALTERNATIVES

There are fewer sights more unnerving to a yoga instructor than watching a group of beginners struggle with double leglifts, permitting their lumbar regions to lift off the floor as they start to raise their feet, and at the same time holding their breath at the glottis. If you are teaching a class in which

Figure 3.17. End position for the supine double leglift. The key requirement of this posture is to actively keep the lower back flattened against the floor using the abdominal muscles.

several people are struggling, you should stop everything and demonstrate the proper technique. Try this: Lie down and ask two volunteers, one on either side of you, to press their right hands against your abdominal wall and place their left hands under your lower back. Then tighten your abdominal muscles. This will push their right hands toward the ceiling, and at the same time it will flatten your lumbar region to the floor against their left hands. Then do a double lift showing first how the lower back should be kept down, and second, letting it lift inappropriately away from the floor. Keep up a stream of conversation to prove that you are supporting the effort with your diaphragm and not holding your breath.

If students have the knack but not the strength to keep their backs against the floor during the double leglift (which is very common), another trick is to try this exercise: before beginning the lift tell them to bend the knees enough to raise the thighs to a 30–45° angle, then lift the feet off the floor and straighten the knees. This will make it easier to keep the back against the floor and make it possible to complete the leglift properly from the higher angle. They should come partially down in the same way, being sensitive to when they can no longer keep their back braced against the floor, at which time they should bend their knees and either lift back up or come all the way down.

BREATHING

The respective natures of the chest and the abdominopelvic regions of the torso are very different from one another: the abdominopelvic cavity is like an oblong rubbery egg filled with water, and the egg is topped by a cage of bone filled with air. Everyone doing leglifts should increase pressure only in the egg. You will have to squeeze down from above with the respiratory diaphragm, up from below with the pelvic diaphragm, and in with the abdominal muscles. To maintain this pressure the respiratory diaphragm has to be strong enough during both inhalation and exhalation to counteract the effects of the abdominal muscles, and this effort must be sustained throughout the exercise. So when do we breathe? All the time. And how? It depends. If you have excellent hip flexibility, you will be able to lift the thighs 90°, exhaling as you lift, and when you have reached that position you can relax and breathe any way you want. But if your hip flexibility is limited, and if you have to keep working against tight hamstring muscles even in the up position, your breathing will be intense and focused on exhalation all the time. You will be taking tiny inhalations whenever you can.

You can take breathing one step further and intensify the energizing effect of leglifting exercises, as well as their difficulty, by keeping as much air out of the lungs as possible, exhaling all the way to your residual volume and then taking small inhalations. If your residual volume is 1,200 ml, you might breathe in and out a tidal volume between 1,200 ml and 1,400 ml instead of between 1,400 ml and 1,600 ml, which approximates what would be most natural in the active stages of double leglifts (fig. 3.18). Breathing this way is more difficult because the already hard-working abdominal muscles (especially the rectus abdominis) now have to work even harder to keep air out of the lungs.

THE SUPERFISH LEGLIFT

We have seen that the lumbar region can be stabilized for leglifts either by rounding it posteriorly as in crunches or the fire exercise, or by keeping it flattened against the floor. But anyone with a sound back can also try leglifting with the lumbar region stabilized in an arched forward position. This is a variation of the fish posture (figs. 3.19a, 5.28, and 9.19)—a posture I'm calling a superfish leglift (fig 3.19b).

To do this posture place your palms up under the hips, stretch your feet out in front, and support all or most of your weight on your forearms. Let your head barely touch the floor. Now arch up maximally by lifting the chest and abdomen into the most extreme possible position. This will stabilize the lumbar arch (fig. 3.19a). Then keeping the feet together, and the toes, feet, and knees extended, slowly raise the heels away from the

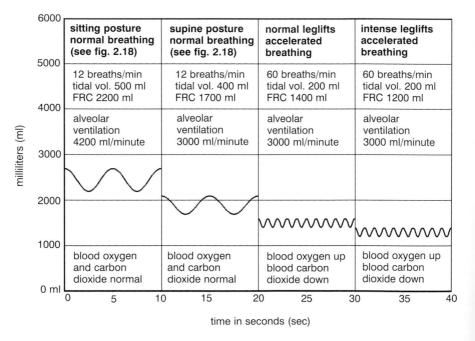

Figure 3.18. Simulated shifts in respiration during leglifting. The two conditions on the left are repeated from fig. 2.18. FRC= functional residual capacity.

floor. Raise up only as far as you can without degrading the arch in the back. If you have excellent hamstring flexibility you may be able to flex your thighs to a 30° angle (fig. 3.19b), or even more, but most people will only be able to raise their feet a few inches before the hamstrings start tugging so much on the base of the pelvis that they pull the lower back toward the floor. So you have a choice: either lift your feet up and down and allow the back to follow in reverse—back down feet up, followed by back up feet down—or lift your feet only until the lower back starts to lose its emphasized forward curve. The main benefit of the exercise comes not from how far you raise the feet but in experiencing the intense pull of the hip flexors on the accentuated lumbar arch. Keep breathing, but for this posture focus on inhalation rather than on exhalation, because emphasizing exhalation will press the lumbar region posteriorly and defeat your purpose.

THE SLOW LEGLIFT WITH A RELAXED ABDOMEN

This next leglifting exercise turns everything we have said so far on its head because it is carried out with relaxed abdominal muscles; for this reason it is only for advanced students with healthy, flexible, and adventuresome backs. The sequence of movements is not only an excellent strength-building exercise for the iliopsoas muscles, which will be doing

Figure 3.19a. Fish posture. The chest and abdomen are lifted as high as possible and the posture is supported mainly by the upper extremities.

Figure 3.19b. Superfish leglift. The thighs are flexed as much as possible without degrading the lumbar arch, which is stabilized in the forward position.

most of the work, it is also a golden opportunity to observe complex muscular action.

Begin in the supine position. Keep the abdominal muscles relaxed, and in slow motion develop enough tension in the iliopsoas muscles to prepare to lift your feet. Notice that as tension develops, the lumbar arch increases (fig. 3.20a). This is a formidable concentration exercise. What you are doing is diametrically opposed to the standard double leglift, and keeping the abdominal muscles relaxed as you increase tension in the hip flexors goes against every natural inclination.

It is important to sense that the iliopsoas muscles are raising the lumbar arch maximally before they lift the heels off the floor. Follow the movement of the lumbar region to its limit. As soon as that point is reached the arch will be stabilized and the hip flexors will finally begin to lift the thighs (fig. 3.20b). At that precise moment focus your concentration on not, repeat not, tightening the rectus abdominis muscles. Although this is counter to your natural predilections, any tension in those muscles pulls the lower back toward the floor. As with the superfish leglift, unless you have long hamstrings and exceptional hip flexibility you will not be able to lift up very far and at the same time maintain the deep lumbar lordosis. Nearly everyone will find that their hamstrings start pulling the lower back toward the floor before they can even get their thighs flexed 45°, much less 90°.

Figure 3.20a. For the first stage of a double leglift with a relaxed abdomen, tension in the iliacus and psoas muscles lifts the lumbar arch forward as the abdominal muscles remain completely relaxed.

Figure 3.20b. Completion of slow leglift with relaxed abdomen. Its key feature is keeping the lumbar arch stabilized in the forward position before and while the feet are lifted slightly off the floor.

AN ADVANCED BREATHING EXERCISE

In a second variation of the leglift with relaxed abdominal muscles, instead of focusing primarily on exhalation as we do for the fire exercise and ordinary double leglifts, inhale slowly as you develop tension for raising the lumbar region and for starting to lift the feet off the floor. This facilitates arching the lumbar region forward. Then to continue the leglift, exhale as you flex the thighs to 90° while your lower back is being pulled down against the floor by the hamstrings and abdominal muscles. Breathe to suit yourself while resting at 90° of hip flexion. Then brace yourself and exhale while lowering the feet back to a few inches away from the floor. Next, inhale as you cautiously relax the abdominal muscles, which allows the lumbar arch to become re-established. Then lower the feet the rest of the way, exhale and rest with your feet on the floor, and inhale again to begin a new lifting cycle.

This method of breathing helps you coordinate the challenging musculoskeletal requirements of the exercise. When you start the sequence, the thighs are the fixed origins for the iliopsoas muscles, and the lumbar region and pelvis serve as the insertions. Then, as soon as the lumbar region is lifted to its maximum, the origins and insertions reverse: the lumbar region and pelvis serve as origins and the thighs become the insertions. Coordinating the breath with all of this while you are watching the activity of the rectus abdominis muscles and the hamstrings, and at the same time keeping in mind everything else that is going on, will make you aware of the architecture of the abdominopelvic region more than any other exercise.

YOGA SIT-UPS

Yoga sit-ups are a far cry from the fast, jerky exercises in a high school gym class. For one thing, they should always be done in slow motion. For another, they should always be done with full awareness of the spine as you roll up into a sitting position "one vertebra at a time," as hatha yoga teachers like to say. Yoga sit-ups also differ fundamentally from leglifts in that for sit-ups you are rolling up the part of the body (the torso) that controls the movement itself, while in leglifts you are raising up a part of the body that is merely connected to the lifting unit.

The initial position for sit-ups is lying supine, keeping the thighs together, flexing the feet and toes, extending the knees, and pressing the lower back to the floor. Then, with the hands pointed toward the feet and the lower back held against the floor, flex the head toward the chest. Breathing evenly, continue to roll up one vertebra at a time (fig. 3.21a) until you are in a sitting position. Concentrate on the action of the abdominal muscles, and stretch the hands forward as much as possible. Come down from the posture in reverse order, slowly rolling down, first the sacrum, then the lumbar region, chest, and finally the head and neck, breathing evenly all the way. If you are unable to lift up significantly, just squeeze up as much as is comfortable, hold the position isometrically for a few seconds, and slowly roll back down. You will still benefit from the posture. Work on it every day, and you will soon be lifting up with ease. When you have developed enough strength to do sit-ups with the hands pointing toward the feet, you can work with progressively more difficult hand and arm positions—placing the fists in the opposite armpits (figs. 6.13–14), catching the opposite earlobes, interlocking the hands behind the head, and stretching the arms overhead.

Holding your back flat against the floor while initiating a sit-up powerfully activates the abdominal muscles, and this enables them to act as prime movers for rolling you up and forward, but if you start with the lower back arched forward, beware. The abdominal muscles will be relaxed and less effective, and the psoas muscles will create excess tension at the lumbar lordosis, exactly as in old-style sit-ups. Do not let that happen. If you

Figure 3.21a. Intermediate position for a slow easy yogic sit-up.

iliacus and psoas muscles first brace, then pull forward actively to complete the posture

rectus abdominis muscles act as prime movers for initiating sit-up

rectus femoris muscles pull forward on pelvic bowl

quadriceps femoris muscles keep knees extended and thighs braced

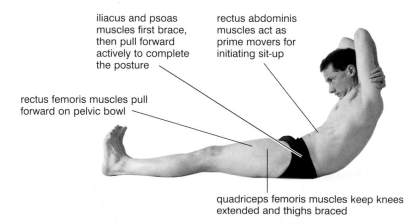

Figure 3.21b. Locations of muscles involved in slow sit-ups.

don't have enough control to keep the back against the floor, bend the knees before you do the sit-up just as you would in crunches.

Sit-ups in yoga, whether done with extended or flexed knees and hips, complement leglifts because they involve some of the same muscles. But there the similarities end. Leglifts simply flex the hips, but for sit-ups, muscles from head to toe on the front of the body act first to brace and then to bend the torso up and forward like an accordion. The iliopsoas and rectus femoris muscles first act as synergists, bracing the pelvis and lumbar region and merely supporting the action of the rectus abdominis. Then as the upper body is pulled further up and forward, the hip flexors take a more active role. Picturing the locations of all three hip flexors plus the rectus abdominis muscle from the side makes it obvious that the rectus abdominis is the only one of the four that has a good mechanical advantage for initiating the sit-up, especially when the knees are straight and the thighs are flat against the floor (fig. 3.21b).

Even if you are careful to keep the lower back against the floor as you start the sit-up, the exercise still compresses the spine and should be done for only a few repetitions. If you are looking for an athletic abdominal exercise that can be repeated hundreds of times, all modern trainers recommend that you do sit-ups by first bending your knees and pulling your heels toward your hips. When the hips are partially flexed as in fig. 3.1, the iliopsoas and the rectus femoris are able to act more powerfully as synergists from the beginning to support rolling up and forward, keeping the pelvis stabilized at the crucial moment the sit-up is being initiated by the rectus abdominis muscles.

THE SITTING BOAT POSTURES

Leglifts and sit-ups are dynamic exercises that feature isotonic movements, whereas sitting boats are classic yoga postures that are held isometrically. The latter resemble boats when viewed from the side, and are even as tippy as boats because of the way you must balance your weight on the pelvic bowl. And since the sitting boat postures are ordinarily held for 30–60 seconds, they require more coordination and balance than leglifts or sit-ups. We'll explore two variations: one makes use of a rounded back—a flat-bottom boat; and the other makes use of a straight back—a boat with a keel.

THE FLAT-BOTTOM BOAT

The boat posture with the back rounded should be mastered first because it is safer and more elementary, and because it doesn't require nearly as much hip flexibility as the posture with the straight back. Start from a sitting position with the fingers interlocked behind the head or neck, the feet and toes extended (pointed away from the head), and the knees extended.

Round the back posteriorly and slowly lean to the rear. You will be lengthening the iliopsoas and abdominal muscles eccentrically as gravity pulls your head and shoulders closer to the floor. Then lean back even further while flexing the thighs, and right after that pull the torso forward with the abdominal muscles. Finally, flex the thighs as much as possible with a combination of the psoas, iliacus, and rectus femoris muscles. Keep the knees extended and hold the pose isometrically for 10–60 seconds (fig. 3.22a). If you start shaking, you've gone beyond your capacity and should come back down. Your back should be rounded enough for your weight to be supported on the relatively flat surface of the sacrum—the flat bottom of the boat—so balance ought not be a serious problem, but if it is, sit on a softer surface or a pillow.

The other way to come into this posture is to raise up from a supine position. With the toes, feet, and knees extended, press the lower back to the floor using the abdominal muscles. Holding that position, raise the head and pull the shoulders up and forward, and when that movement is partially underway, tighten the psoas and iliacus muscles concentrically to flex the thighs. If you have developed the requisite strength from leglifts and sitting-up exercises, the entire sequence should be easy. Try it slowly to analyze its components. You will probably find yourself minimizing your tidal volume and focusing on exhalation, exactly as you did with the fire exercise and most of the other leglifts.

THE BOAT WITH A KEEL

The back is kept straight in the second sitting boat posture, and this may not be easy. From a sitting position with the feet together and with the toes and knees extended, sit ramrod straight, which includes arching the lower back forward. Next, stretching your hands out in front of you, lean backward, keeping the hips flexed at a 90° angle, which of course lifts the feet. You will be supporting the posture with the psoas, iliacus, and abdominal muscles, and especially (in contrast to the round-bottom boat), with the rectus femoris muscles (figs. 3.9, 8.9, and 8.11), which are prime movers for keeping the front of the pelvis pulled forward and for maintaining the 90° angle between the pelvis and the thighs. Hold the pose isometrically for 10–60 seconds (fig. 3.22b). It is harder to balance in this posture than in the previous one because you are poised on your sitting bones (the keel of the boat) instead of the flat of the sacrum. Again, use a pillow if balancing is too difficult.

If you have limited hip flexibility because of tight hamstrings, you will find yourself struggling to keep your back straight. The problems are comparable to those faced by students trying to do 90° leglifts while keeping their lumbar regions arched forward in the superfish leglift, as well as in trying to lift up in a double leglift while keeping the abdomen relaxed. All such postures, including this straight-back boat pose, are impossible if hip flexibility is poor. And even intermediate-level students find it difficult to resist the hamstring stretch and at the same time summon the strength to calmly hold the posture in its ideal form.

THE PEACOCK

Peacocks are said to have extraordinary powers of digestion and assimilation, and that is one reason the posture has been given this name. The hatha yoga literature tells us that the peacock pose so enlivens the abdominopelvic region that if you have mastered it and hold it regularly for three minutes a day, you can ingest poison without harm. It might be wise to take that with a grain of salt, but the peacock is certainly the supreme posture for developing abdominopelvic energy. What is more, the completed posture looks like a male peacock as it struts its stuff with a long plume of colorful feathers trailing behind.

There are several ways to approach and complete the posture. Here's one: To come into the preparatory position, you first kneel with the thighs abducted and the toes flexed. Then you lower the top of the head to the floor, and place the palms on the floor between the knees with the fingers

Figure 3.22a. Flat-bottomed boat. This is a beginner's rounded-back posture, and is especially valuable for those with poor hip flexibility.

Figure 3.22b. Straight-back keel boat, for advanced students who have good strength and hamstrings long enough to permit 90° of hip flexion.

pointing behind you. Bring the wrists and elbows together tightly, and pull the hands toward the head until the forearms are perpendicular to the floor and the elbows are in contact with the abdomen (fig. 3.23a). The wrists will be extended about 90°. If this is a problem because of previous wrist injuries or wrist inflexibility, you may not be able to do the peacock until the situation has been corrected with other stretches. Most women will have to squeeze their breasts between the arms above the meeting point for the elbows. If you try to create more room for the breasts by allowing the elbows to come apart, one or both elbows will slip to the side and off the abdomen when you attempt to complete the posture.

Holding this position, take the knees back as far as possible and then straighten them, sliding the toes as far back as you can, supporting your weight on the top of the head, the hands, and the feet (fig. 3.23b). This may be all you can do. If so, remain in this position for 20–60 seconds to build your capacity.

Still keeping the elbows in position, lift the head. Then slowly take your weight forward by extending the elbows, supporting most of your weight on the hands and some of your weight on the extended feet (fig. 3.23c). Again, you may find it useful to remain in this position for 20–60 seconds rather than go further and fall forward.

Now, while bracing the back and thighs to keep the body as straight as possible, pitch your weight forward by extending the elbows until you are balancing all of your weight on the hands, paying special attention to the

Figure 3.23a. Preparatory position, peacock.

Figure 3.23b. Second position, peacock, with elbows flexed about 90°.

fingertips. You have to keep the body rigid enough for the toes to lift off the floor (fig. 3.23d), and the back muscles have to be very powerful to accomplish this, especially if you want to keep the back relatively straight in the final pose. Although we'll delay detailed comments on the design of the upper extremities until chapter 8, the muscles that stabilize the two scapulae (the shoulderblades) are also crucial to this effort, especially one—the *serratus anterior*—that keeps the scapula flat against the back and pulled to the side (figs. 3.11–12 and 8.9).

Assuming that your abdominal muscles, back muscles, and scapular supporting muscles are strong enough to support the posture, and assuming that you have been able to keep the elbows in position, the main problem for most people is developing enough strength in the flexors of the forearms to permit a slow and controlled eccentric extension of the elbows. To complete the posture, the forearm flexors have to support the entire weight of the body. They lengthen eccentrically as you bring your weight forward, and as you try to come into the final isometric position you may exceed their limits. One of three reactions is typical: you may fall forward on your nose as the flexors suddenly relax and give way under the influence of inhibitory input to motor neurons from Golgi tendon organs; you may fall

Figure 3.23c. Third position for the peacock, with weight supported only between the feet and the hands.

forearm flexors

Figure 3.23d. Peacock posture completed. As the weight comes further forward, the forearm flexors lengthen eccentrically to support the posture.

to one side, usually as the weaker arm slips off the abdomen; or your motor pathways may just deliver up a resounding objection because they sense at some level that you will not have the strength to support the final posture. In this last case students often do something silly, like tossing their feet into the air as if they were trying to levitate. Their feet, of course, fall back to earth just like anything else that is tossed in the air. You can complete this posture, at least using the approach described here, only by bringing your weight forward.

A successful peacock pose depends to a great extent on your body type and weight distribution. If you have a big chest and small hips and thighs the bulk of your weight will be forward and you will not have to extend the elbows a lot to complete the posture. But if you have a small chest, big hips, and heavy thighs and legs, a greater proportion of your weight will be to the rear and you will have to extend your elbows more fully. As you do that, however, the forearm flexors start losing their mechanical advantage and the pose becomes more difficult to complete and hold. This is easy to prove. If it happens that you can complete the posture easily you'll not have any trouble supporting a 25-pound weight on your midback without additional extension of the elbows. But if someone were to place a 5-pound weight on your feet, the increased elbow extension needed to bring your weight forward to a point of balance will probably drop you to the floor like lead.

It is commonly said that the peacock is more difficult for women than for men because they have less upper body strength, but the main reason that women have more difficulty is that a greater proportion of their weight is distributed in the lower half of the body. It follows, then, that to make the posture easier, all they need to do is to fold in some of their lower body weight. The classic solution: do the posture with the legs folded up in the lotus pose so the elbows will not have to extend as much to support it. This is possible, of course, only if you are comfortable in the lotus.

The peacock develops more intra-abdominal pressure than any other posture because the abdominopelvic unit (which is bounded by the respiratory diaphragm, the pelvic diaphragm, and the abdominal muscles) is support-ing the weight of the body through the elbows and arms. And because the diaphragm is working so hard, you can breathe only under duress. Never-theless, you should always keep breathing. It is tempting to hold the glottis shut and equalize intra-abdominal pressure with intrathoracic pressure, but that is impractical because you can hold your breath for only so long. In addition, the substantial increase in intrathoracic pressure could be dangerous to the heart and circulation. It is much better to keep the air-way open and limit the increase in pressure to the abdominopelvic cavity. Obviously the peacock is only for those who are in splendid athletic condition.

THE PELVIS AND THE ANATOMICAL PERINEUM

A famous conductor, rehearsing the chorus for the Verdi *Requiem*, once stopped the music and shouted to the performers, "No! No! Squeeze it in—push it up!" He may not have known it, but he was telling them to seal off and control the anatomical perineum—the base of the pelvis—and thereby cultivate what we have been calling abdominopelvic energy. All trained singers have learned that the purest and richest sound originates from this region. In the language of singers, the base of the body "supports" the voice.

The perineum and pelvis not only establish a foundation for creating an intensely lyric sound, they form the lowermost portion of the abdominopelvic unit, support the weight of the abdominal and pelvic organs, and bear their full share of intra-abdominal pressure. As an experiment, next time you sense an impending sneeze or a fit of violent coughing, notice that you prepare for the sharp increases in intra-abdominal pressure by pulling the base of the body in and up with an intensity that will match the expected explosiveness of the expected sneeze or cough.

For singers and public speakers who are engaging an audience, a tripartite muscular effort within the torso is apparent. The anatomical perineum pushes up against the pelvic organs, the abdominal muscles squeeze in from the front, the sides, and behind, and the muscle fibers of the respiratory diaphragm lengthen against resistance, slowly submitting to the ascent of the dome of the diaphragm during exhalation. All three act together to oversee a whole-body regulation of the passage of air past the vocal cords in the larynx.

THE PELVIS AND THE PERINEUM

The pelvis and the perineum contain the pathways for elimination, serve as focal points for sensual pleasure, and accommodate all aspects of procreation; but even though these functions are all enormously significant to us personally and are treated at length by all esoteric traditions, our main purpose here is to understand how the pelvis and perineum are important in postures and breathing exercises. To that end we'll simply outline their anatomy and concentrate on several important practices that enable the student to sense their architecture experientially and lay the groundwork for more advanced study.

The pelvis and perineum are difficult terms to comprehend because the words "pelvis," "pelvic," and "perineum" each have more than one meaning. First consider the pelvis. The way we used this word in the first half of the current chapter was in reference to the pelvic bowl, which, in addition to the two pelvic bones, includes the pubic symphysis plus the sacrum, the sacroiliac joints, and all the pelvic restraining ligaments (fig. 3.4). But lay people occasionally refer to the pelvis more generally as including the region of the body between the upper portion of the thighs and the lower abdomen, and indeed, in the section of this chapter on intra-abdominal pressure, we referred to the pelvic cavity as a part of the combined abdominopelvic region. And finally, we commonly make reference to the contents of the pelvic bowl, that is, to the pelvic organs.

The word "perineum" also has more than one meaning. Gross anatomy textbooks usually include a chapter titled "The Anatomical Perineum" that describes the contents of a diamond-shaped region that forms the base of the pelvis and that contains the anus, the genitals, and their supporting muscles. The more common definition of the perineum, however, refers to a much smaller region, not inclusive of the anus and the genitals, but between them. This is the site hatha yoga teachers are referring to when they tell you to place one heel in the perineum, and this is the region the obstetrician slices through to do an episiotomy. These variations in usage are rarely a problem, however, for anyone who has awareness of both possibilities, because the meaning of the term is nearly always clear from its context.

CONTENTS OF THE PELVIC BOWL

The contents of the pelvic bowl are best seen in a dissection in which a mid-sagittal cut has been made from the waist down (figs. 3.24–25). The reason for this is simple: most of the pelvic structures in which we are interested either lie in the midline or are visible from the perspective of a front-to-back cut that runs straight down the middle of the body. In both sexes, the skeletal framework of the lumbar, sacral, and coccygeal spine are visible behind, along with the pubic symphysis in front. Also in both male and female, the rectum, anus, bladder, and urethra are plainly seen. The penis, prostate gland, and scrotum are found exclusively in the male (fig. 3.24), and the uterus, vagina, labia, and clitoris are found exclusively in the female (fig 3.25).

THE ANATOMICAL PERINEUM

The anatomical perineum is shaped roughly like a diamond. It is defined by four points: the inferior border of the pubic symphysis, the tip of the *coccyx* (the tailbone), and the two ischial tuberosities. The diamond is made up of two triangles: the *urogenital triangle* anteriorly, and the *anal triangle* posteriorly (figs. 3.4 and 3.27). These two triangles share a common base, which is an imaginary line between the two ischial tuberosities, but except for this one line, the two triangles lie in different planes. The anal triangle extends up and back to the coccyx, and the urogenital triangle extends up and forward along the inferior pubic rami to the pubic symphysis (fig. 3.4). The line connecting the two ischial tuberosities is lower than any other part of the perineum, which is obvious to any slender person sitting upright on a hard seat.

THE PELVIC DIAPHRAGM

We can best understand the structures included within the anatomical perineum if we build our understanding from the inside out. The deepest layer, the *pelvic diaphragm*, is a broad, thin sheet of muscle and fasciae that spans the entire diamond-shaped region, encircling the anus posteriorly and lying deep to the genitals anteriorly. Seen in three dimensions it is shaped like a deep hammock. Stand up and envision such a hammock at the base of the body. It is suspended between the pubic bones in front and the sacrum behind, and it supports the internal structures of the pelvic

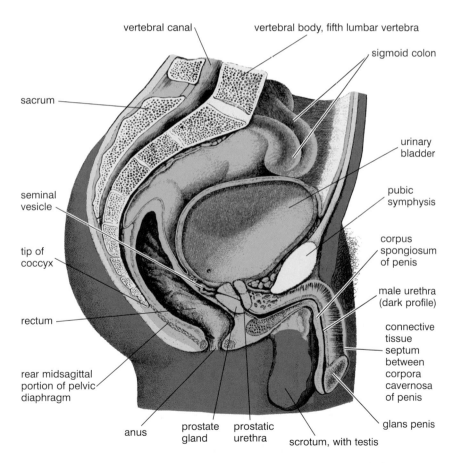

Figure 3.24. Midsagittal (longitudinal, front-to-back) section through the male pelvis revealing midline parts of the reproductive system, the midline terminal portions of the digestive and urinary systems, and the rear portion of the pelvic diaphragm (front portion is interrupted by the genitals). (Sappey)

cavity just as a hammock in your back yard supports the human frame. The midsagittal segment of the hammock that runs from the anus to the coccyx is visible to the rear in figs. 3.24–25. The pelvic diaphragm is interrupted by the anus and its sphincters to the rear, and by the midline structures of the genitals in front. A frontal section (male) through the prostate gland and urethra illustrates how the borders of the hammock extend up and to either side (fig. 3.26). In a superficial dissection of either male or female only the rear half of the funnel comes into view because the genitals cover it in front (figs. 3.28–29); in a deeper dissection (female) with the genitals removed (fig. 3.27), it becomes obvious that the pelvic diaphragm forms a sling around cross sections of the vagina and urethra.

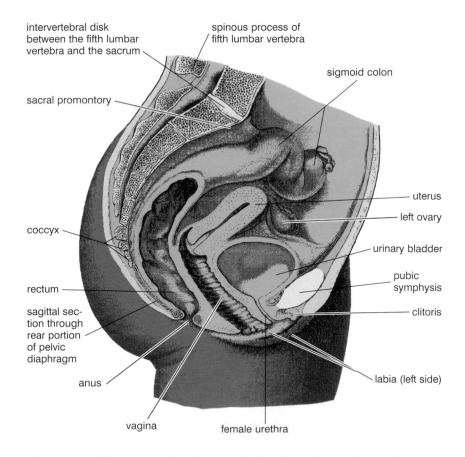

Figure 3.25. Midsagittal section through the female pelvis revealing midline parts of the reproductive system, the midline terminal portions of the digestive and urinary systems, and the rear segment of the pelvic diaphragm (Sappey).

In hatha yoga the pelvic diaphragm is activated consciously by two practices: *ashwini mudra* and *mula bandha*. There are subtle and not-so-subtle differences between the two. We'll begin with ashwini mudra.

ASHWINI MUDRA

Ashwini is the Sanskrit word for "mare" ("horse" would be an appropriate translation except that the word *mudra*, which means "gesture," is a feminine noun in Sanskrit and requires a feminine modifier), and ashwini mudra in hatha yoga is named for the movement of the pelvic diaphragm in a horse after it has expelled the contents of its bowel. During the expulsion phase, the cone-shaped pelvic diaphragm moves to the rear, and after the contents of the bowel are dropped, the muscles of the pelvic diaphragm pull strongly inward. In so doing they cleanse the anal canal. In human beings the same thing happens—you first bear down, opening the anus and expelling the contents of the bowel, and then the pelvic diaphragm pulls inward and upward while contracting the anal sphincter. The pulling inward motion, which we also do reflexly from moment to moment during the day, is ashwini mudra. This is not as obvious as it is in a horse, because

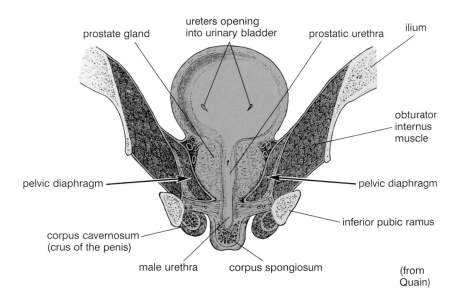

Figure 3.26. Frontal (longitudinal, side-to-side) section through the male pelvic diaphragm, urinary bladder, prostate gland, prostatic urethra, and corpus spongiosum, as well as the paired corpora cavernosa, inferior pubic rami, and ilia. The pelvic diaphragm forms a deep hammock that extends lengthwise from the pubis to the coccyx and that supports the internal pelvic organs. Here we see a section through the sides of the hammock, and in fig. 3.27 we see it as a whole.

in humans the whole region is enveloped in loose connective tissue and covered with the superficial structures of the perineum, but it is the same gesture.

As a natural movement, ashwini mudra is often forceful, especially when it is associated with keeping the base of the abdominopelvic cavity sealed during sharp or extreme increases in intra-abdominal pressure, or when it is used as a last-ditch means for retention (think of restraining diarrhea). As a yoga practice, ashwini mudra is not so intense, but it still acts as a perineal seal, fortified in this case by tightening the gluteal muscles along with the pelvic diaphragm and anal sphincter. The mudra is applied for a few seconds, released, and repeated. Ideally, only the gluteals, the pelvic diaphragm, and the anal sphincter are activated, but the proximity of muscles overlying the genitals anteriorly sometimes makes this difficult, and you will often feel them tighten along with the rest when you try to create the gesture.

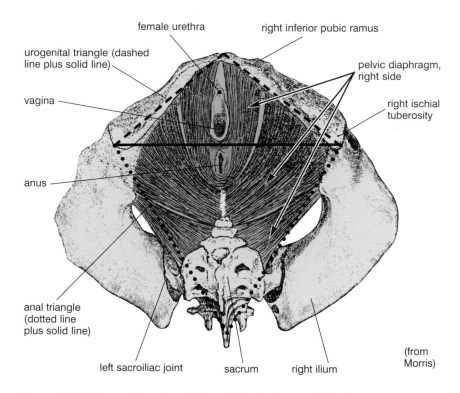

Figure 3.27. A deep dissection of the pelvic diaphragm of the female (view from below). The superficial muscles and external genitals (see figs. 3.28-29) have all been removed, revealing the underlying hammock-like pelvic diaphragm (see figs. 3.24-26), as well as the urogenital and anal triangles (see also fig. 3.4). A comparable male dissection is almost identical, except that a cross-section of the penis and male urethra is substituted for the vagina and female urethra.

Some postures make a pure ashwini mudra easy, and others make it difficult. If you stand with the feet well apart and bend forward 20–30°, you will find it awkward to contract the anus and pelvic diaphragm, and almost impossible to contract them without activating the muscles around the genitals as well. Now stand upright with the heels and toes together and try it again. This is easier. If you don't tighten too vigorously you may be able to isolate the pelvic diaphragm and the gluteals from the muscles of the genitals. Next, bend backward gently, keeping the heels together and the thighs rotated out so the feet are pointed 90° away from one another. Keep the knees extended. Then tighten gently behind and try to release in front. This is one of the easiest upright postures in which to accomplish a pure ashwini mudra. Last, bring the toes together and rotate the heels out. This again makes it difficult.

These simple experiments illustrate the general rule: any posture that pulls the hips together will make ashwini mudra easier, and any posture that pulls the hips apart will make it more difficult. That, as it happens, is one problem with all cross-legged sitting postures. Try it. When the thighs are flexed with respect to the spine and abducted out to the sides, it is almost impossible to contract the gluteals and only a little less difficult to isolate the anus and pelvic diaphragm from the genitals. But if you try the mudra in the shoulderstand or headstand with the heels together and the toes out, you will find that it is easy because gravity is already pulling the pelvic diaphragm toward the floor. Little or no effort is needed to achieve a fully pulled-in feeling, and that effort need not involve the genitals. Now lie supine on the floor and notice that you can easily tighten up in the rear without recruiting muscles around the genitals. Prone, it is more difficult, at least in men, in whom the muscles associated with the genitals are stimulated by contact with the floor.

One of the best postures for ashwini mudra is the upward-facing dog. As long as the pelvis is lifted slightly off the floor (figs. 5.13–14), it is impossible to do this pose without activating the pelvic diaphragm, yet it does not stimulate the muscles in the urogenital triangle in the least. The down-facing dog (figs. 6.17 and 8.26), not surprisingly, creates the opposite effect: this posture is one of the easiest poses for recruiting the muscles of the urogenital region in isolation, but a pose in which it is almost impossible to isolate the muscles associated with ashwini mudra.

MULA BANDHA

In ashwini mudra we strongly activate the pelvic diaphragm, the anus, and the gluteals. Mula bandha is more delicate. Here we mildly activate the pelvic diaphragm plus—more strongly—the overlying muscles of the urogenital triangle, which includes the muscles associated with the genitals and the urethra. Therefore, to understand mula bandha we have to examine the anatomical disposition of these muscles.

THE MUSCLES OF THE UROGENITAL TRIANGLE

Looking at a superficial dissection, we see that three pairs of muscles overlie the genitals. In both male (fig. 3.28) and female (fig. 3.29), the *superficial transverse perineal muscles* course laterally in the shared border of the urogenital and anal triangles, extending laterally from a heavy band of centrally located connective tissue—the *central tendon of the perineum*—to the ischial tuberosities. The *bulbospongiosus* muscles in the male encircle the base of the penis; in the female those same muscles encircle the vagina and urethra. The *ischiocavernosus* muscles in both the male and female lie superficial to the erectile tissues of the *corpora cavernosa*, which themselves course from the inferior pubic rami to the body of the penis in the male and to the clitoris in the female. In a slightly deeper plane of the urogenital diaphragm (in both male and female), the *deep transverse perineal muscles* spread out laterally in sheets that attach to the inferior pubic rami, and the urethral sphincters encircle the urethrae.

MULA BANDHA, THE ROOT LOCK

Unlike ashwini mudra, which is often a response to sharp and sudden increases in abdominopelvic pressure, mula bandha (the root lock) is a gentle contraction of the pelvic diaphragm and the muscles of the urogenital triangle. It does not counter intra-abdominal pressure so much as it seals urogenital energy within the body, controlling and restraining it during breathing exercises and meditation (again, this is a literary rather than a scientific use of the term "energy"). What actually happens is more easily sensed than described, so we'll begin with a series of exercises.

First try sitting in a hard chair covered with a thin cushion. In a neutral position, neither perfectly upright or slumped, try to blow out but without letting any air escape. Try hard. Notice that the pelvic region contracts and lifts up involuntarily enough to counter the downward push from the chest and abdominal wall. Now try the mock blowing maneuver again, but this time keep the pelvic region relaxed, and notice that it feels like straining for a bowel movement. Try it one last time, but this time lift the entire anatomical perineum consciously, and you will quickly sense that these efforts bring both the pelvic diaphragm and the muscles of the urogenital region into play.

Next sit really straight, arching the lower back forward. Exhale, pressing in with the abdominal muscles, and notice that it is natural to find a focus for your attention at a point between the anus and genitals. You may sense a slight tension in the muscles of the genitals, but little or none in the anus, and certainly none in the gluteal muscles. This describes the root lock. You don't have to make extreme efforts. The cushion on which you are sitting places enough pressure on the muscles of the urogenital triangle to focus your awareness on the lock.

Now try the same exercise in a slumped posture with the back rounded to the rear. This changes everything. It shifts your attention from the front of the anatomical perineum to the rear, and it elicits a mild ashwini mudra instead of mula bandha because you are tipping backward toward the plane of the anal triangle and away from the plane of the urogenital triangle.

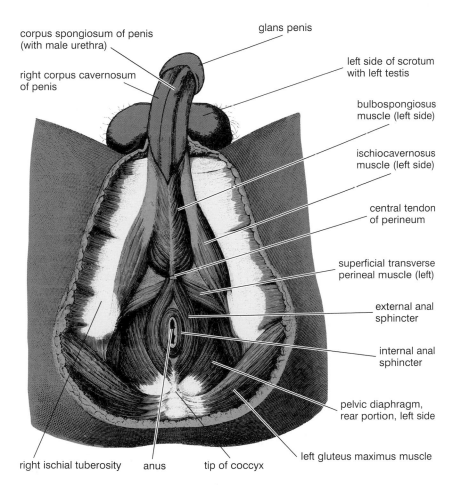

corpus spongiosum of penis
(with male urethra)

glans penis

right corpus cavernosum
of penis

left side of scrotum
with left testis

bulbospongiosus
muscle (left side)

ischiocavernosus
muscle (left side)

central tendon
of perineum

superficial transverse
perineal muscle (left)

external anal
sphincter

internal anal
sphincter

pelvic diaphragm,
rear portion, left side

left gluteus maximus muscle

right ischial tuberosity   anus   tip of coccyx

Figure 3.28. Male anatomical perineum. The anal portion of the pelvic diaphragm is shown below (in the anal triangle). The external genitals and their associated muscles are shown above (in the urogenital triangle), where they are superimposed over (and thus hide) the front portion of the pelvic diaphragm (from Sappey).

(The angle between these two planes is shown clearly in fig. 3.4). Sitting straight rocks you up and forward so that contact with the cushion favors the root lock. The lesson: sit straight if you wish to apply mula bandha.

If this is still confusing, it will be helpful to first experience a gross version of the root lock. The best concentration exercise for this is to sit upright and try breathing in concert with slowly increasing and decreasing tension in the perineum. With the beginning of exhalation gradually tighten the muscles of the pelvic diaphragm and genitals, aiming for maximum contraction at the end of exhalation. As inhalation begins, slowly relax. Repeat the cycle for ten breaths several times a day. At first it may be difficult to tighten the muscles without also tightening the gluteal muscles, but if you are careful to sit straight it will become easy.

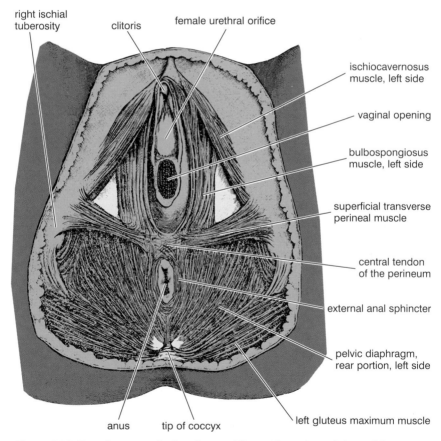

Figure 3.29. Female anatomical perineum. The anal portion of the pelvic diaphragm is shown below (in the anal triangle). The external genitals and their associated muscles are shown above (in the urogenital triangle), where they are superimposed over (and thus hide) the front portion of the pelvic diaphragm (from Morris).

With practice you will be able to sense the contraction of successive layers of muscles from the outside in. Starting superficially and with a minimal effort, you can feel activity in the ischiocavernosus, bulbospongiosus, and superficial transverse perineal muscles. And with a little more attention you can activate the deep transverse perineal muscles and the urethral sphincter. And with yet more effort you can activate the pelvic diaphragm.

The central tendon of the perineum, which as discussed previously is located at the dividing line between the anal and urogenital triangles, appears to be the key structure around which the more delicate versions of mula bandha are organized. This is an extremely tough fascial region into which the superficial and deep transverse perineal muscles insert. If you can learn to focus your attention on this tiny region while creating minimal physical contraction of the nearby muscles, you will be feeling the root lock. (Yoga teachers who speak of placing awareness on the perineum are referring to this region.) Concentrate on the sensation, and in time mula bandha will feel natural and comfortable. With experience you can hold the lock constantly, which is what yogis recommend for meditation.

A MODIFIED CAT STRETCH

In addition to the down-facing dog (figs. 6.17 and 8.26), which was just mentioned in the section on ashwini mudra, one of the best yoga postures for helping you come in contact with the delicacy and precision of the root lock is a modified cat pose. From a kneeling position, bring your chin to the floor, swing your elbows out, and bring the upper part of the chest as low as possible, arching your back deeply and mimicking a cat peering under a couch (fig. 3.30). Then tighten the perineal region generally. You will immediately notice that the exposed anus in this position brings the sensations toward the front of the diamond-shaped perineum rather than behind, and that even if you squeeze vigorously the gluteal muscles remain relaxed. After you have practiced this pose several times and gotten accustomed to its associated sensations, you can try to find the same feelings when you apply the root lock in sitting postures.

Figure 3.30. Modified cat stretch, for sensing mula bandha in preference to ashwini mudra.

AGNI SARA

Agni sara, or "fanning the fire," is a breathing exercise, an abdominal exercise, and a powerful stimulus to abdominopelvic health. When it is done with full attention and for an adequate span of daily practice, it stokes the fire of the body like no other exercise. But before trying it we'll first do a training exercise for active exhalations, and then we'll work with a moderate practice—A and P breathing—that is accessible to everyone.

ACTIVE EXHALATION

During the course of relaxed, casual breathing, you make moderate efforts to inhale and you usually relax to exhale, but all of the exercises that follow make use of active exhalations, in some cases breathing out all the way down to your residual volume. To get an idea of what is involved, try the following exercise: Inhale moderately through the nose, purse the lips, and exhale as if you were trying to blow up a balloon in one breath. Try this several times. If you slowly breathe out as much air as possible through the resistance of the pursed lips, you'll notice that exhalation is accompanied by a tightening of the muscles throughout the torso, including the abdominal muscles, the intercostal muscles in the chest, and the muscles in the floor of the pelvis. At first you will notice the abdominal muscles pressing the relaxed diaphragm up (and pushing the air out) with the chest in a relatively fixed position; then you'll notice the chest being compressed inward; and finally, toward the end of exhalation, you will notice the contraction of the pelvic diaphragm. This sequence of events will also take place if you breathe out normally, but creating resistance through the pursed lips makes the muscular efforts much more obvious.

A AND P BREATHING

This preliminary exercise to agni sara, called *akunchana prasarana*, or A and P breathing for short, involves active exhalations and relaxed inhalations. The literal meaning of the phrase is apt: "squeezing and releasing." Stand with the trunk pitched forward, the hands on the thighs just above the knees, the elbows extended, the feet about a foot and a half apart, and the knees slightly bent. Much of the weight of the torso is placed on the front of the thighs. Breathe in and out a few times normally, and observe that the posture and the angle of the body pulls the abdominal organs forward and creates a mild tension against the abdominal wall. Notice that countering the tension produced by the force of gravity requires that a mild effort be made even at the beginning of exhalation, and the greater the forward angle the greater the effect.

To do A and P breathing, assume the same posture as in the trial run, and press in slowly (squeezing) from all sides with the abdominal muscles as you exhale, all the while bolstering the effort with the chest. Your first impulse is to emphasize the upper abdomen. Try it several times, observing exactly where the various effects and sensations are felt. Notice that the effort in the upper abdomen is accompanied by a slight feeling of weakening in the lower abdomen. The lower region may not actually bulge out physically, but it feels as if it might. Now try to exhale so that the upper abdomen, the lower abdomen, and the sides are given equal emphasis, as though you are compressing a ball. Exhalation might take 6–7 seconds and inhalation 3–4. Inhalation is mostly passive (releasing) and manages itself naturally. Take 10–15 breaths in this manner.

Much of inhalation is passive in A and P breathing because the chest springs open and the abdominal wall springs forward of their own accord. The strong emphasis on exhalation means that you are breathing in and out a tidal volume which is the combination of your normal tidal volume for an upright posture plus part of your expiratory reserve. Your revised tidal volume for A and P breathing might be about 900 ml for each breath rather than the textbook tidal volume of 500 ml, and along with this, your new expiratory reserve volume would become about 600 ml rather than 1,000 ml (fig. 3.32, left-hand panel). In any case, A and P breathing boosts your energy by increasing blood oxygen and decreasing blood carbon dioxide. It is a simple exercise, but one that is both relaxing and invigorating.

THE CLASSIC AGNI SARA

A and P breathing can be done by anyone, but agni sara and its more complex variations are intense practices that require training and conditioning. Their effects on the body are powerful enough for them to be contraindicated by several medical conditions (see the end of this chapter). They should also be done only on an empty stomach and after evacuating the bowels.

Like A and P breathing, the classic agni sara is usually done standing with the torso at a 60–70° angle from upright, the feet apart, the knees slightly bent, the hands on the thighs just above the knees, the elbows extended, and the arms supporting the torso. And like A and P breathing, the practice focuses on exhalation. But instead of utilizing a mass contraction of abdominal muscles, agni sara requires a step-by-step muscular effort. To do the practice, focus your attention on the area just above the pubis, and press the abdomen in at that site before pressing in with the middle region of the abdomen. Then continue to exhale, gradually recruiting muscles higher in the abdominal wall and ending with the internal intercostal muscles of the rib cage—all the time holding tension below. Exhale as much as possible. Notice that exhalation not only presses the abdomen and chest in, it also presses the back to the rear (fig. 3.31). For inhalation reverse the process, relaxing the chest and upper abdomen first and lower abdomen last.

If you watch yourself from the pubis to the sternum when you have nothing on, it will be easier to learn, but even then it may take several weeks of daily practice and concentration before you can do the exercise with confidence.

[Technical note: There's no unambiguous language for indicating in a simple phrase the source of the movement involved in exhalation. Sometimes teachers say "push the abdomen in," but muscles, of course, never push. Others say "pull the abdomen in," but this sounds as though something other than the abdominal muscles themselves might be responsible, as we'll soon see is the case for uddiyana bandha. Although "press the abdomen in" isn't perfect, it's at least general enough not to be misleading.]

The rectus abdominis muscles are not single muscles extending from the pubis to the sternum, but a series of short muscles that are isolated from one another by horizontal lines of connective tissue called *tendinous inscriptions*, which are responsible for the muscular segmentation and washboard look in the abdomen of a bodybuilder. The wave of abdominal contraction in agni sara is possible only because each segment is separately innervated and can be controlled individually.

In agni sara the modifications of the lung volumes and capacities are more extreme than in A and P breathing. Here exhalation combines your normal tidal exhalation (500 ml) with your entire expiratory reserve volume (1,000 ml), creating a tidal volume of 1,500 ml, an expiratory reserve volume of zero, and a functional residual capacity which is now equal to your residual capacity of 1,200 ml (fig. 3.32). As with A and P breathing, inhalation takes care of itself and is passive except for the last 500 ml. Inhalation ordinarily takes about half to three-fourths the time as exhalation and is accompanied by the gradual relaxation of the intercostal and abdominal muscles.

Figure 3.31. Agni sara. The halftone reveals the profile for full exhalation, and the dotted line reveals the profile for inhalation. Exhalation is taken all the way through the expiratory reserve volume (in other words, to the residual volume), but inhalation doesn't extend into what would ordinarily be the inspiratory reserve volume.

For those who are interested in numbers, minute and alveolar ventilations are easily calculated (fig. 3.32). A reasonable practice is to breathe out and then in 4 times per minute, which would yield a minute ventilation of 6,000 ml (1,500 ml per breath times 4 breaths per minute), and an alveolar ventilation of 5,400 ml per minute (1,350 ml per breath times 4 breaths per minute). With practice and self-control you can slow down even more, to as little as two breaths per minute, yielding a minute ventilation of 3,000 ml and an alveolar ventilation of 2,700 ml per minute. You might be concerned that 2,700 ml per minute will not supply enough fresh air to the alveoli in comparison to the standard 4,200 ml per minute mentioned in chapter 2, but in that case we were mixing 350 ml of fresh air with a functional residual capacity of 2,200 ml, and here we are mixing 1,350 ml of fresh air in each breath with only 1,200 ml of residual volume. It's plenty.

During agni sara the diaphragm remains generally passive. It is relaxed throughout most of exhalation, although it probably resists lengthening toward the very end of exhalation as it opposes the upward movement of the abdominal organs. And likewise, during inhalation the dome of the diaphragm moves downward passively as you breathe in what is ordinarily your expiratory reserve volume. If you have controlled the release into this

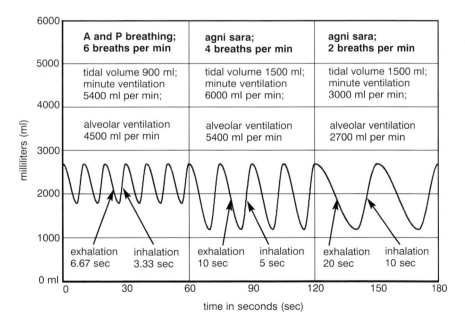

Figure 3.32. Simulated comparisons of agni sara with A and P breathing in a young man. Compare these three modes of breathing with normal breathing in fig. 2.14. Even though taking only two agni sara breaths per minute (above right) drops the alveolar ventilation precipitously (to 2700 ml per minute), exhaling all the way to the residual volume is adequate for maintaining the blood gases.

phase of inhalation slowly, you'll notice that it is not the diaphragm but the abdominal muscles that are in command, lengthening eccentrically to restrain the gravity-induced drop of the abdominal organs until you reach what would have been your normal tidal range. At that point they release more completely and allow the diaphragm to preside over an ordinary tidal inhalation of the last 500 ml.

Although the chest does not at first glance seem to play a prominent role in agni sara, the internal intercostal muscles do become activated for compressing it inward during exhalation, especially in the last stages when you are approaching your residual volume. Then, as you begin to inhale, the chest springs open passively, restrained only by the abdominal muscles and internal intercostals. Since the emphasis of agni sara is on exhalation, the chest shows only a modest enlargement during the period of what would have been an ordinary tidal inhalation in normal breathing.

The airway is open in agni sara, and intra-abdominal pressure remains in approximate equilibrium with atmospheric pressure, so it is not necessary to apply mula bandha. And it is not only unnecessary to apply ashwini mudra—in the bent-forward position it is impractical. Beginners, however, may find it helpful to establish the root lock while they are learning to activate the wavelike recruitment of abdominal muscles from below. Focusing their attention on the front of the perineum may help them develop and refine the practice. After that they should stop paying special attention to the root lock. It is not a part of this practice.

We are not looking for speed in agni sara, but for control. A common mistake is to whoosh air out too quickly at the beginning of each exhalation. Try to make the rate approximately equal throughout so that you still have a third of your agni sara tidal volume left when you have a third of your time to go. Approach the end of exhalation by recruiting the highest of the abdominal muscles, and keep pressing. Your time for exhalation and inhalation will lengthen with practice, working up quickly to 6–10 seconds for exhalation and 3–5 seconds for inhalation; with a little more practice it is easy to manage three or even two breaths per minute. In the beginning it helps to set a clock on the floor and watch the second hand, but this will soon become a distraction. Discard the clock after a few days and concentrate on the sensations.

For an even more intense practice of agni sara, and for an occasional change of pace, you can exhale as usual, and then instead of releasing fully into inhalation, take a minimal chest inhalation and then immediately re-establish an even more powerful exhalation. Repeat this several times before inhaling as in the traditional agni sara discussed above. This exercise, which can only be repeated a few times before you have to inhale fully, dramatically increases the power of agni sara.

AGNI SARA IN OTHER POSITIONS

If you have a problem doing agni sara standing, you can try it in other postures. First you can kneel, with the body lifted up off the heels, the hands on the thighs, and the torso pitched forward (fig. 3.33). Or you can rest the buttocks on the heels and drop the hands or elbows either to the thighs or to the floor just in front of the knees. Or you can place the elbows on the knees and touch the abdomen with one hand to give yourself feedback and encouragement when you try to exhale and inhale in a wavelike motion. It is also comfortable to do agni sara in cat stretch variations, arching the back up as much as possible during exhalation (figs. 3.34a–b, halftone images). And for a potent combination of breathing practices, the cat stretch poses are also excellent for going back and forth between agni sara exhalations and empowered thoracic inhalations (chapter 2) that take you to the outermost limits of your inspiratory capacity for these postures (figs. 3.34a–b, superimposed dotted lines).

If these standard poses for agni sara are inconvenient for you, you can still do the practice in a chair or in a meditative sitting posture. It works best to sit up straight, arch the lumbar lordosis, and pitch yourself forward at a slight angle. This will enable you to feel the countering tension of the abdominal muscles in each region of the abdominal wall.

If you are chronically short of breath you will have a fine personal practice if you aim for 10–15 breaths per minute for only one minute. Every bit of extra exhalation induced by agni sara is a blessing. Even if your alveolar ventilation remains exactly the same as it is in ordinary relaxed breathing, it will be more efficient for bringing in oxygen and removing carbon dioxide from the blood, just as we saw for kapalabhati (chapter 2). If you have chronic obstructive pulmonary disease and do agni sara several times each day you will please and surprise yourself, your family, and your doctors.

Figure 3.33. Agni sara kneeling. Again, the halftone reveals the profile for full exhalation, and the dotted line reveals the profile for inhalation. The tidal inhalation and exhalation combines an ordinary tidal volume with the entire expiratory reserve volume.

If you are in ordinary good health a meaningful practice of agni sara will require at least 10 minutes at the rate of 3–4 breaths per minute, and as you gradually increase your time you will achieve an enriched sense of well-being. If you get up early in the morning and do agni sara before doing hatha, it will give you a burst of energy and enthusiasm. And if you are unable for one reason or another to practice hatha yoga postures, agni sara

Figure 3.34a. Hands-and-knees pose. The maximum inhalation (dashed line) with head up and maximum lumbar lordosis is superimposed on the maximum exhalation with head down and the back arched toward the ceiling (halftone).

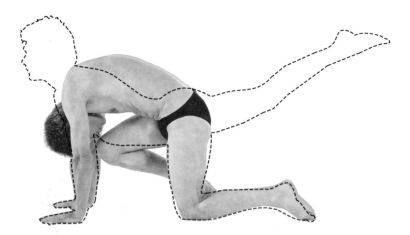

Figure 3.34b. Cat stretch. The maximum inhalation (dashed line) with the head and right foot up in combination with the maximum lumbar lordosis, is super-imposed on the maximum exhalation with the head down and knee to nose (halftone).

is probably more important. Finally, if you really want to learn agni sara, you have to both stay within your capacity and at the same time explore your limits. You'll not sense the power of this practice unless you do it 30–40 minutes a day for 10 days in a row.

AGNI SARA WITH OTHER HATHA YOGA PRACTICES

If you are an advanced student, you can use agni sara (or A and P breathing) to intensify the hatha yoga postures in which you are comfortable. You will have to breathe faster than usual, of course, because the postures will increase your needs for oxygen and carbon dioxide exchange. You will also have to modify the patterns of exhalation according to the demands of the posture. For example, in a deep standing forward bend you can both see and feel what is happening, but a standing backbend or spinal twist permits little obvious movement in the abdomen. That's fine. Either way, it's the attempt to press in from below that generates the surge of energy. And for all standing postures in which you are emphasizing an empowered thoracic inhalation (chapter 2), you can not only increase your inspiratory reserve volume by trying to inhale more deeply, you can use agni sara to exhale most or all of your expiratory reserve volume, thus inhaling and exhaling your vital capacity (the volume of which is specific to the particular posture) with every breath.

After you have worked successfully with agni sara, bellows breathing, and kapalabhati for some time, you can experiment with an agni sara type of movement during the exhalation phases of the bellows breath and kapalabhati. Exhalation will produce an upward-moving wave of contraction—a pushing in and up sensation, rather than a mass contraction of the abdominal muscles. You can feel this if you span your hand across your abdomen with the middle finger on the navel, the thumb and index fingers above, and the ring finger and little finger below. You will feel the little finger on the lower abdomen moving inward during exhalation, and little or no movement where the thumb is located on the upper abdomen. Using this technique for the bellows and kapalabhati creates a mild scooping-up sensation. It requires more control than the standard techniques for bellows and kapalabhati, so you will need to slow down—perhaps to as few as 60 breaths per minute. You will never be able to do it as fast as the standard technique, but it is still a powerful abdominopelvic exercise and is excellent for training the abdominal muscles for more advanced practices.

UDDIYANA BANDHA, THE ABDOMINAL LIFT

Mula bandha seals the anatomical perineum, and agni sara teaches us special skills for using the abdominal muscles. When you have become proficient in both, you are ready to learn the second great lock in hatha yoga: *uddiyana bandha*, or the abdominal lift. To do it you must exhale, hold your breath out (as it's said in yoga), and create a vacuum in your chest that sucks your diaphragm and abdominal organs to a higher than usual position in the torso. This can happen only if the body is sealed above and below—above at the glottis and below at the perineum. Without these seals air would be drawn into the larynx and lungs above, and into the eliminatory and reproductive organs below. You hold the root lock reflexly and without having to think about it, but the glottis has to be held shut voluntarily.

UDDIYANA BANDHA

The best time to practice the abdominal lift is early in the morning, certainly before breakfast, and ideally after having evacuated the bowels. The same contraindications apply as in agni sara (see the end of this chapter). To begin, stand with your knees slightly bent and your hands braced against the thighs. As with agni sara, this stance lowers the abdominal organs downward and forward. Exhale to your maximum. Notice that you do this by pressing in first with the abdomen and then with the chest. Then do a mock inhalation with the chest, closing the glottis to restrain air from entering the lungs, and at the same time relax the abdomen. You should feel the chest lift. Holding the glottis closed for a few seconds, try harder to inhale, keeping the abdomen relaxed. The upper abdomen will form a deep concavity that extends up and underneath the rib cage. This is uddiyana bandha (fig. 3.35). If you get confused about how to prevent air from entering the lungs, forget about the abdominal lift for a week or so and simply practice trying to inhale after full exhalations while you are blocking your mouth and nose with your hands.

Figure 3.35. Uddiyana bandha: A maximum exhalation is followed by a mock inhalation with locked glottis and chin lock in combination with relaxed abdominal muscles.

To make holding the glottis shut feel more natural and comfortable, a third lock, jalandhara bandha (the chin lock), can be established by flexing the head forward so the chin is tucked into the *suprasternal notch*, the little concavity above the *sternum* at the pit of the throat. It is possible to do the abdominal lift without the chin lock, but its addition will make the closure of the glottis feel more secure, and many teachers consider it absolutely necessary. Fixing the eyes in a downward position also complements both uddiyana bandha and jalandhara bandha. Try looking up as you try them and you'll quickly sense the efficacy of looking down.

Come out of uddiyana bandha in two stages. First, while still holding the glottis shut, ease the vacuum in the chest by relaxing the external intercostal muscles, which will lower the dome of the diaphragm and the abdominal organs to a lower position in the trunk. Then, as soon as the abdominal wall is eased forward, press inward strongly with the chest and abdomen until the pressure above and below the glottis is equalized. You have to compress inward just as forcefully as when you first exhaled for uddiyana bandha; if you don't, air will rush in with a gasp when you open the glottis. As soon as the pressure is equalized, open the glottis and breathe in gently.

Where does the vacuum come from? In uddiyana bandha we are trying to inhale without inhaling, and this makes the thoracic cage larger, expanding it from side to side and from front to back. And since no air is allowed in, the air pressure inside the chest has to decrease, which in turn creates enough of a vacuum to pull the diaphragm up (provided it is relaxed) in proportion to the expansion of the rib cage. Coming down from uddiyana bandha, the side-to-side and front-to-back expansions of the chest are first relaxed and then compressed back into their starting positions of full exhalation, and the dome of the diaphragm and abdominal organs move inferiorly.

Uddiyana bandha is the only practice in hatha yoga that frankly stretches the respiratory diaphragm. It's true that you get a mild stretch of the diaphragm when you exhale as much as possible in agni sara and for the exhalation stage of uddiyana bandha, both of which push the dome of the diaphragm (from below) to the highest possible position the abdominal muscles can accomplish. But uddiyana bandha goes beyond this because the vacuum in the chest that is superimposed on full exhalation pulls the diaphragm (from above) to an even higher position. We can surmise that regular practice of uddiyana bandha will stretch, and in time lengthen the diaphragm's muscle and connective tissue fibers, as well as keep the zone of apposition (fig. 2.9) between the diaphragm and the chest wall healthy and slippery. You will be able to exhale more completely as you gradually lengthen the muscle fibers, and you will be able to breathe more comfortably and efficiently as you increase the diaphragm's mobility.

PROBLEMS

Many people, including yoga teachers, surprisingly, seem to have a great deal of difficulty learning uddiyana bandha. It is partly a matter of poor body awareness in the torso, but the most constant factor is simply your history. Many youngsters grow up doing the abdominal lift in play, often combining it with other manipulations such as rolling the rectus abdominis muscles from side to side or up and down. In a typical group of children, almost half of them will be able to do uddiyana bandha after only a few seconds of instruction and demonstration, and in a beginning hatha class for adults, those who did uddiyana bandha in play as children will usually be able to learn the yoga version immediately.

If you're having trouble, you are doing one of three things wrong. First, you may not be exhaling enough at the start. The less you exhale, the less convincing will be the lift. You have to exhale the entire expiratory reserve volume—only the residual volume of air should remain in the lungs. The second possibility is a corollary to the first. You may be letting in a little air on your mock inhalation. You have to try to inhale without doing so. That is the whole point of locking the airway at the glottis. The third, and usually the most intractable, problem is that you are not relaxing the abdomen during the mock inhalation. You must learn to distinguish between pressing in with the abdominal muscles, which we want only for the preliminary exhalation, and allowing the abdominal wall to be pulled in passively by the vacuum in the chest. Many students hold their abdominal muscles rigidly or even try to keep pushing in with them during the lifting phase of the practice, and this prevents the abdominal organs and abdominal wall from being sucked in and up. It is also common for students to relax their abdominal muscles momentarily but then get mixed up and try to assist the inward movement with an active contraction. It won't work. You have to relax the abdominal muscles totally and keep them relaxed to do this exercise.

Figure 3.36. A modified cat stretch encourages keeping the abdominal muscles relaxed during the mock inhalation phase of uddiyana bandha.

ANOTHER MODIFIED CAT STRETCH

If you consistently have trouble relaxing the abdominal muscles for uddiyana bandha in a standing position, try it in a cat stretch, similar to the one we used for exploring mula bandha (fig. 3.30), except more relaxed. Rest on the knees and forearms and lower the forehead down against the crossed hands. Press the shoulders toward the floor and increase the lumbar arch as much as possible. This position pitches the abdominal and pelvic organs forward and toward the chest. Now all you have to do is exhale as much as possible (which rounds your back posteriorly) and hold your breath at the glottis. Now relax, allowing the lower back to arch forward again, and notice that in this position it is unnatural to hold the abdominal muscles firmly. Uddiyana bandha comes effortlessly as your chest cooperates with gravity in pulling the abdominal organs to a higher position in your torso (fig. 3.36). Finally, continuing to hold your breath and keeping the abdomen relaxed, slowly lift your head and shoulders. Come up on your hands, walk them toward your knees and onto your thighs ever so delicately, and carefully come into an upright kneeling position without tightening the abdomen. If you are successful you will be doing uddiyana bandha.

FIRE DHAUTI

Here is a simple exercise that some texts call agni sara, and others refer to as fire dhauti. Come into uddiyana bandha (standing), and continuing to hold your breath, alternately lower and again lift the abdominal organs by decreasing and increasing the size of the chest cage with the intercostal muscles. Each time the abdominal organs are lowered, the abdominal wall is pushed out, and each time the abdominal organs are again lifted into the typical uddiyana bandha position, the abdominal wall is pulled in. It's a pumping action, and it is sometimes done fast, up to two times per second, but more frequently it is done about once per second.

Doing fire dhauti, keep in mind that you see and feel most of the action in the belly, but that the control of the maneuver depends on the chest as well as holding your breath after a full exhalation. The abdominal muscles themselves remain passive: they are pulled up passively by uddiyana bandha, and they are pressed back out by gravity and by the action of the chest. You keep holding the breath at the glottis, but the vacuum in the chest is diminished and even converted momentarily into a positive pressure as the diaphragm and abdominal organs are pressed inferiorly. You can do the pumping action, of course, only for the length of time that you can hold your breath. This practice is an excellent training exercise for those who are having trouble releasing the abdominal muscles in uddiyana bandha, because its vigorous up and down motion has the effect of freeing you from the habit of holding the abdominal muscles rigidly.

THE ABDOMINAL LIFT SUPERIMPOSED ON AGNI SARA

Agni sara pushes the abdominal organs from below, and uddiyana bandha lifts them from above. After mastering these two practices individually, advanced students will benefit from combining them. Start in the standard stance for agni sara and the abdominal lift, supporting the body with the hands resting on the thighs. Exhale completely by pressing in the abdominal muscles in a wave from the pubis to the rib cage. As soon as you have exhaled to the maximum, lower the chin to the top of the sternum, close the glottis, and do an abdominal lift. Relax the abdomen and diaphragm completely, and form as deep a concavity below your rib cage as you can manage. The moment you reach your limit of lifting, start lowering. Then as soon as you start the dropping action, initiate a gradual mass contraction of the abdominal muscles.

During the lowering phase you must develop just as much abdominal contraction as you felt before you initiated the lift. This is difficult to accomplish smoothly. The abdominal organs are like a baton that is being passed. The abdominal muscles press them toward the chest, which then grabs and lifts them as high as possible with the external intercostal muscles. In preparation for the releasing phase, the external intercostals lower the abdominal organs eccentrically back to the point at which the abdominal muscles retrieve them gracefully. As soon as the abdominal muscles and internal intercostal muscles team up and re-compress the chest, which equalizes the pressure on both sides of the glottis, you can open it knowing that air will not rush into the lungs. This is important. If you don't quite equalize the pressure you will hear a slight click in your throat as the glottis opens. Finally, after opening the airway, release the chin lock and the abdominal muscles as though you were coming out of agni sara, first just below the rib cage, then in the middle of the abdomen, and last in the lower abdomen and perineum. The whole exercise should be done noiselessly and with infinite smoothness, so much so that an observer who is not familiar with the practice will not realize that you are doing an abdominal lift or restraining inhalation.

A practice of 2–3 cycles per minute is reasonable for those who are in good condition. For example, if you are doing a complete cycle in 30 seconds, or 2 breaths per minute, you can take about 15 seconds for exhalation, 5 seconds for the abdominal lift, and 10 seconds for inhalation. You should develop a rhythm that you can maintain without obvious effort; otherwise you are trying to do too much. You can do about 10 cycles per day for the first week, and increase by 10 each week as long as you are comfortable.

UDDIYANA BANDHA WITH OTHER POSTURES

If you are well-practiced in both uddiyana bandha and hatha yoga postures, you can combine the two to create a more aggressive practice. As with agni sara, when you are bending forward you will feel the effects as well as notice the movements externally, and when you are bending backward or twisting, you will feel the effects without seeing them. As in the case of combining agni sara with postures, this is not a practice for beginners, because the vacuum developed in the chest and abdomen completely nullifies the intra-abdominal pressure we depend on to protect the back in stressful positions. You should be confident in both the abdominal lift and your hatha yoga postures before trying it.

NAULI

Nauli, which means "churning," is one of the most rewarding, if not the crown jewel, of the various abdominopelvic practices. To do it you must first do uddiyana bandha and then contract the rectus abdominis muscle, first on one side and then the other, creating a wavelike, side-to-side motion in the abdomen. The other abdominal muscles remain relaxed, leaving concavities lateral to the rectus abdominis on each side. To learn the exercise, most people first learn to isolate both rectus abdominis muscles at the same time after having established uddiyana bandha. Then, still holding uddiyana bandha, they learn to contract each rectus abdominis muscle individually, and finally they learn to coordinate the side-to-side motion for the final practice.

NAULI MADHYAMA

Uddiyana bandha creates a deep concavity in the abdomen because all of the abdominal muscles are relaxed. Looking at someone in profile, the abdominal wall has the appearance of an upside-down J, with the short limb of the J attached to the sternum, the arc of the J sucked up by uddiyana bandha, and the long limb of the J attached to the pubis. The rectus abdominis muscles, which run vertically on either side of the midline between the rib cage and the pubis, are included in this inverted J. Contracting them selectively—that is, superimposing their contraction onto uddiyana bandha—pulls the inverted J into a straight line, leaving the rest of the muscles relaxed (fig. 3.37a, middle image). This is *nauli madhyama*, the first step to learning nauli. It is another one of those practices children sometimes learn in play, and anyone who has done that can easily learn nauli after a month of so of trial and error.

Several exercises may be necessary for learning to isolate the rectus abdominis muscles. The most straightforward method is to apply uddiyana bandha in the standard position with the hands braced on the thighs. Then bring your attention to the lower abdomen and create in that region what teachers sometimes describe as a forward and downward push. Although muscles don't push, and in this case are merely pulled taut, a "push" is what the untutored observer sees, and the image seems useful for many students.

The most common problem with trying to isolate the rectus muscles is that the external abdominal oblique, internal abdominal oblique, and transversus abdominis muscles also tend to become recruited. But the whole effort to pull the rectus muscles forward is meaningless unless you keep the rest of the abdominal muscles relaxed. If the nauli madhyama technique doesn't work, don't struggle, but try it once a day on an empty stomach or whenever you feel adventuresome or energetic.

Another trick is to apply uddiyana bandha when you are in a supine position: press your fingers into the abdominal wall lateral to the rectus abdominis muscles on each side, and then lift your head up an inch or so off the floor. As you start to make an effort to lift your head, you will feel the rectus abdominis muscles contract just before the rest of the superficial muscles. Adjust your effort so that you engage the rectus muscles while keeping the others relaxed. If you make too much effort and lift the shoulders along with the head, you will engage the abdominal muscles *en masse*, which defeats the purpose of the exercise. Lift the head up and down to explore this. Mechanical feedback also helps, for which there are three possibilities. You can give yourself some little punches laterally to remind your other muscles to stay relaxed, you can squeeze your hands against the rectus muscles from the side as you feel them become engaged, or you can roll the overlying skin up and out to encourage the rectus muscles to contract in isolation.

If all else fails, you may be able to learn how to isolate the rectus muscles by coming into the cat stretch with the deepest possible uddiyana bandha, and then arching your back up. The rectus abdominis muscles are the prime movers for creating the arched posterior curvature, while at the same time partially straightening out the inverted J, so they will be activated selectively. Then all you have to do is concentrate on keeping the rest of the abdominal muscles relaxed. Try supporting yourself with one hand so the other hand is free to feel the preferential contraction of the rectus muscles.

NAULI

If you can learn to superimpose the contraction of the rectus abdominis muscles onto uddiyana bandha in various positions, you will sooner or later be able to do so in a standing position with your hands on your thighs. And if you cultivate this until it is second nature you will be able to learn nauli in short order. Here's how. Holding nauli madhyama, place both hands on the thighs just above the knees. Press diagonally in strong lunges first to one side and then the other to activate the individual rectus abdominis muscles. Do this one-two lunge at the rate of about 1–2 times per second. If you need to waggle the hips at the same time, that's fine. It will look silly for a while, but after a month or so of practice you will be able to produce the rolling movement that is characteristic of nauli, and be able to do so without much hip motion or side-to-side lunging. Refining the practice, you can go ten times from right (fig. 3.37b) to left, and ten times from left (fig. 3.37c) to right on one round of holding your breath. You can start practicing for one minute a day and add a minute each week. As with uddiyana bandha, many teachers consider it important to hold the chin lock for this practice. In any case, if you build up to 20 minutes a day you will find out why nauli is valued so highly in hatha yoga.

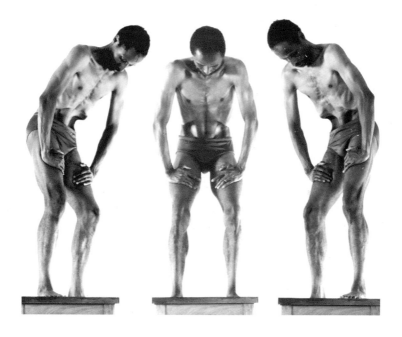

Figure 3.37b. Nauli right, with contraction of the right rectus abdominis muscle superimposed on uddiyana bandha.

Figure 3.37a. Nauli madhyama, with contraction of both rectus abdominis muscles superimposed on uddiyana bandha.

Figure 3.37c. Nauli left, with contraction of left rectus abdominis muscle superimposed on uddiyana bandha.

CONTRAINDICATIONS

For practices in the first half of the chapter, contraindications for leglifts, sit-up exercises, and the boat postures are obvious: lower back problems. And it ought not surprise anyone that the peacock, as well as other exercises that greatly increase intra-abdominal pressure, should be explored with caution, at least by anyone who is not already quite athletic. It is a myth, although a common one, that women in general should not do the peacock. For practices in the second half of the chapter, there are no contraindications for ashwini mudra and mula bandha, but agni sara, uddiyana bandha, and their derivatives are powerful exercises whose indiscriminate use is not recommended.

HIGH BLOOD PRESSURE

If you have high blood pressure, even the mildest of abdominopelvic exercises should be approached gingerly. Even if you are on medication that successfully lowers your blood pressure, all intense abdominopelvic exercises should be avoided. Holding your breath at the glottis after inhalation is always contraindicated. Holding your breath after exhalation, as in uddiyana bandha, is less dangerous but also inadvisable because we would expect it to quickly increase venus return, that is, the flow of blood back to the heart.

ULCERS

Intense abdominopelvic practices are all contraindicated for everyone with stomach and duodenal ulcers except in the case of practices recommended by a holistic physician who is willing to advise you.

HIATAL HERNIA

The esophagus passes through the respiratory diaphragm through the *esophageal hiatus* (fig. 2.7), and under certain conditions the upper part of the stomach may herniate through this region of the diaphragm into the thoracic cavity. This is called *hiatal hernia*. If you have occasional discomfort in that region after eating, or if you have acute discomfort just under the left side of the rib cage while trying the peacock, uddiyana bandha, or vigorous versions of the cobra, it may be that the differential between intra-abdominal pressure (which is higher) and intra-thoracic pressure (which is lower) is causing the problem. It is important to seek medical counsel from someone who is conversant with hatha yoga before continuing with any posture or exercise that causes such symptoms.

INGUINAL HERNIA

The *inguinal canal*, through which the testis passes around the time of birth on its way to the scrotum, is another region of weakness in which abdominal organs, or more commonly a little fatty tissue, usually from the greater omentum (fig. 2.9) can herniate out of the abdominal cavity. This condition—an *inguinal hernia*—can also occur in women, although it is less common than in men. If a little outpouching of soft tissue appears on one or both sides of the groin when you are upright, and if that outpouching disappears back into the abdominal cavity when you are lying down, it is almost certainly an inguinal hernia.

Inguinal hernias will become more pronounced in any standing posture and in all exercises such as the peacock that increase intra-abdominal pressure. Bicycling, walking, running, and sun salutations also commonly make inguinal hernias more prominent. But they are unpredictable: they can get worse quickly or remain about the same for months or years. If the condition is not repaired surgically, a support (truss) that presses against the hernia from the outside may be effective in keeping the contents of the abdomen out of the inguinal canal, but in the absence of such a device, strenuous upright postures and the peacock should be avoided.

MENSTRUATION AND PREGNANCY

No exercise involving breath retention should be practiced during menstruation or pregnancy, but the regular and enthusiastic practice of abdominopelvic exercises appears to be helpful in preventing premenstrual symptoms and cramping. During pregnancy, most practices in hatha yoga are contraindicated, especially those that increase intra-abdominal pressure but also those few that decrease it, such as uddiyana bandha. Ashwini mudra and mula bandha are fine and are even recommended during pregnancy, but agni sara is contraindicated because of its intensity. One caution for expert hatha yogis who have just given birth: the fascia that connects the two rectus abdominis muscles in the midline may have become weakened by pregnancy and childbirth, and women who were able to do the peacock easily before having children are sometimes unable to do so afterwards because the rectus abdominis muscles are now pulled uncomfortably apart in the effort to come into the posture.

BENEFITS

Everyone knows that developing strength, improving aerobic capacity, and increasing flexibility is important for physical conditioning. The question of how to accomplish these goals is less certain, but yogis insist that these are the benefits of leglifts, the peacock, agni sara, uddiyana bandha, and nauli. Why that happens is still something of a mystery, but we can call on our experience to make some reasonable guesses. If you are hungry and tired, but feel great after doing twenty leglifts and ten minutes of agni sara instead of eating and taking a nap, something obviously worked—and anyone who has a little knowledge of anatomy and physiology can make intelligent guesses about what, where, and how: you increased your blood oxygen and decreased your blood carbon dioxide; you stimulated the adrenal glands to release *epinephrine* (*adrenaline*) and steroids; you stimulated the release of *glucagon* from the *islets of Langerhans* in the pancreas; and the liver released extra glucose into the general circulation, cutting your appetite and preparing you for getting on with your day.

We can also look at the physiology of any specific practice and comment on events that are certain to result. For example, we can note that any activity that increases intra-abdominal pressure while the airway is being kept open will force blood more efficiently than usual from the venous system in the abdominal region up into the chest. Quantities can be debated, measurements taken, and opinions stated regarding how and why that might be beneficial, but there can be no argument about the reality of the effects.

> *"It may take many months to acquire the control and stamina necessary to perform this exercise (agni sara) correctly. Do not become discouraged. Your efforts will be rewarded with excellent health."*

— Swami Rama, in *Exercise Without Movement*, p. 55.

CHAPTER FOUR
STANDING POSTURES

"The first bipedalists were not semihuman creatures. They were animals opting to walk on their hind legs. It was a costly option for them to take up, and we are still paying the instalments. The mammalian spine evolved over a hundred million years and reached a high degree of efficiency, on the assumption that mammals are creatures with one leg at each corner and that they walk with their spine in a horizontal plane. Under those conditions the blueprint is one that would command the admiration of any professional engineer . . . . Such a mammal resembles a walking bridge."

— Elaine Morgan, in *The Scars of Evolution*, p. 25.

Creatures with an upright two-legged posture appeared along the coastal regions of Africa 4–6 million years ago. How this came to be is still controversial, but the posture is one of the defining characteristics of the modern human form. Another is that we are able to stand erect with minimal muscular activity in our thighs, hips, and backs. By contrast, the stance and gait of a dog or cat, or of the occasional monkey who chooses to walk upright at times on two legs, is dictated by joints in the supporting extremities that are always bent. This enables them to pounce or run at a moment's notice, but it also requires them to use muscular activity just to stand upright. The secret of our stance is simple—we can relax when we stand because we can lock our knees and balance on our hip joints without much muscular activity.

Most of us are only vaguely aware that we can balance our weight on top of the relaxed thighs, but everyone learns about knees in junior high school cafeteria lines when someone sneaks up behind you and buckles your knee as you are leaning on one leg. Your ensuing collapse shows you clearly that you were depending on the locked knee joint to hold you up and that your tormentor caught your relaxed muscles off guard.

"Locking the knees" is a phrase that has two implications: one is that the hamstrings will be relaxed, and the other is that additional extension will be stopped by ligaments. Instructors in dance, athletics, and the martial arts generally caution against this, arguing instead that the backs of the knees should never be thrust to the rear in a completely locked and hyper-extended stance. Although this thinking is widely accepted in the movement disciplines, and although it is certainly sound advice for all fields of study in which whole-body standing movements must flow freely, weary mountain climbers gratefully learn about a slow, choppy, "rest step," in which they stand for 2–4 seconds or even longer on a locked knee joint—just bones and ligaments—to save muscular effort before lifting their opposite foot onward and upward. And assuming that you are not preparing to pounce on someone at a social gathering, locking one knee is hard to fault for standing and engaging in quiet conversation first on one leg and then the other. This is a uniquely human gesture—a natural consequence and indeed the culmination of the evolution of our upright posture. An all-encompassing condemnation of the practice is ill-advised if not downright foolish.

Hatha yoga directs our attention to the knees in many postures—the sitting boats (figs. 3.22a–b), the superfish leglift (fig. 3.19b), sitting forward bends generally (chapter 6), the celibate's pose (fig. 8.25), and the fullest expressions of many inverted poses (chapters 8 and 9), just to mention a few—in which generating tension in the hamstrings or releasing tension in the quadriceps femoris muscles to permit frank bending of the knees would alter the funda-mental nature of the posture. In such cases there is nothing inherently wrong with simply saying "lock the knees." On the other hand, movement therapists are correct in noting that such a directive all too frequently gives students permission to absent themselves mentally from the posture. Rather than experimenting with the nuances of partially relaxing the ham-string muscles, and of alternating this with tightening both the quadriceps femoris muscles and hamstrings at the same time, students often take the lazy way out by simply locking their knees. They might remain unthink-ingly in a sitting forward bend for several minutes using a combination of tight quadriceps femoris muscles and relaxed hamstrings, or they might hyperextend their knees in a standing forward bend and support the posture with no more than bony stops and ligaments. The result: they end up in a few minutes with a sense of vague discomfort in their knees. Therefore, throughout the rest of this book, I'll acknowledge the current preferences in movement studies by referring not to locking but to extension of the knees, and I'll suggest accompanying this at selected times with relaxed hamstrings—essentially locking the knees without using that trouble-some phrase.

That we can stand with knees locked is obvious; sensing how our weight is balanced over the hip joints is more subtle. Feel the softness in your hips with your fingertips as you stand erect. Then bend forward 3–5° from the hips and notice that tension immediately gathers in the gluteal muscles to control your movement forward. Next, slowly come back up and feel the gluteals suddenly relax again just before your weight is balanced upright.

Our relatively relaxed upright posture is possible because a plumb line of gravity drops straight down the body from head to foot, passing through the cervical and lumbar spine, behind the axial center of the hip joint, in front of the locked knee joint, and far enough in front of the ankle joint to keep you from rolling over backward onto your heels (fig. 4.1). (Because the ankle joints do not lock, keeping your balance will require you to hold some tension in the calf muscles.) This architectural arrangement allows you to balance your weight gracefully from head to toe and accounts for why you can stand on your feet without much muscular effort.

The fact that we can remain in standing poses when we are relaxed, tensed, or anywhere in between often prompts spirited discussion among hatha yoga teachers. One instructor says to relax in standing postures; another says don't for a second relax in standing postures. Both can be correct, and we'll explore how and why later in this chapter. Putting first things first, however, we'll begin with the skeleton. We'll follow that with the general

Figure 4.1. A plumb line of gravity drops perpendicular to the gravitational field of the earth from the crown of the head to the feet in a frontal plane of the body. This plane passes through the cervical spine, the lumbar spine, behind the axial center of the hip joints, in front of the axial center of the knee joints, and in front of the ankle joints. The disposition of this plumb line of gravity allows us to balance upright in a relaxed posture except for enough tension in the leg muscles (front and back) to keep the line perpendicular to earth's gravitational field.

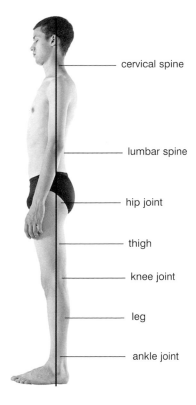

cervical spine

lumbar spine

hip joint

thigh

knee joint

leg

ankle joint

principles that underlie standing postures, looking first at a few simple exercises: the mountain pose, a side-to-side stretch, an overhead stretch, a twist, three backbends, six forward bends, and four side bends. Then we'll study the more complex dynamics of triangle postures in detail. Last, we'll look at two balancing postures: the eagle and the tree.

THE SKELETAL SYSTEM AND MOVEMENT

Every artist who wants to do figure drawing must first learn about the skeleton—the shape and placement of the skull, rib cage, pelvis, and scapula, as well as the rest of the bones of the extremities. Every curve, bump, indentation, and bulge in the body is superimposed on the underlying bones, and life-drawing instructors could hardly conceive of a better way to start their courses than to ask students to draw a skeleton covered with translucent plastic shrink-wrap—just skin and bones. Looking within such a model, students would see two distinct skeletal units: the *appendicular skeleton* and the *axial skeleton*. The former refers to the bones of the appendages (the upper and lower extremities), and the latter refers to the bones that lie in the central axis of the body—the *skull*, the *vertebral column*, and the *rib cage*, including the *sternum*. The appendicular skeleton, as the name suggests, is appended to the axial skeleton—the upper extremities are attached to the sternum at the *sternoclavicular joints*, and the lower extremities are attached to the *sacrum* at the *sacroiliac joints*. Taken together, the two units form the frame upon which the entire body is organized.

[Technical note: It is noteworthy and perhaps surprising to many that the hip joints, which are the sites for flexing, extending, and rotating the thighs, do not form axial-appendicular junctions. The reason is simple: both the femur and the pelvic bone are part of the appendicular skeleton, and it is the pelvic bone alone that articulates with the axial skeleton.]

We'll discuss the skeletal system and movement in more detail, but first we need to review some common anatomical terms that we'll be using routinely from this point on. "*Anterior*" refers to the front of the body; "*posterior*" refers to the back of the body; "*lateral*" refers to the side; "*medial*" means toward the midline; "*superior*" means above; "*inferior*" means below; "*proximal*" means closer to the torso; and "*distal*" means away from the torso. *Sagittal planes* lie from front to back, top to bottom; the one and only *midsagittal plane* is the sagittal plane that bisects the body in two right-left halves, and *parasagittal planes* include all sagittal planes that are lateral and parallel to the midsagittal plane. The *coronal* or *frontal planes* lie from side to side, top to bottom, such as a plane that runs through the ears, shoulders, torso, and lower extremities. The *transverse* or *cross-sectional planes* lie from side to side, front to back (fig. 4.2).

THE APPENDICULAR SKELETON

The appendicular skeleton for the lower extremities forms the foundation for standing postures. From top to bottom, it includes the *pelvic bones*, the *femur, patella, tibia, fibula,* and the bones of the ankles and feet (fig. 4.3). The pelvic bones and the sacrum comprise the *pelvic bowl* (figs. 2.8, 3.2, and 3.4), which is thus an axial-appendicular combination of three bones. The femur is the single bone in the long axis of the thigh, and the patella is the "kneecap." The tibia and fibula are in the leg. The anterior border of the tibia—the shin—is familiar to everyone as the front surface of the leg that is so vulnerable to painful bumps and bruises. The fibula is located laterally, deep to the calf muscles. The tibia and fibula remain in a fixed position parallel to one another, the tibia medial to the fibula from top to bottom. The bones of the ankle and foot include the *tarsals, metatarsals,* and *phalanges.*

The appendicular skeleton for the upper extremities is used for manipulating objects in our environment and is often an important accessory for bracing difficult standing poses. The bones of the upper extremity include

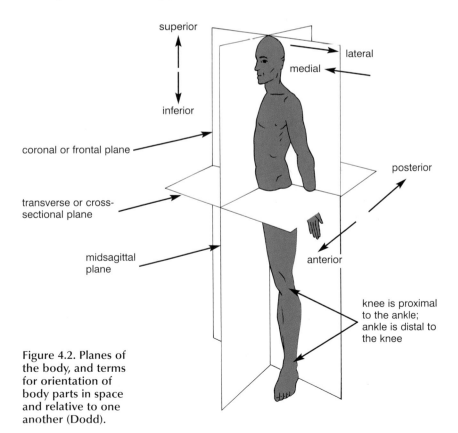

Figure 4.2. Planes of the body, and terms for orientation of body parts in space and relative to one another (Dodd).

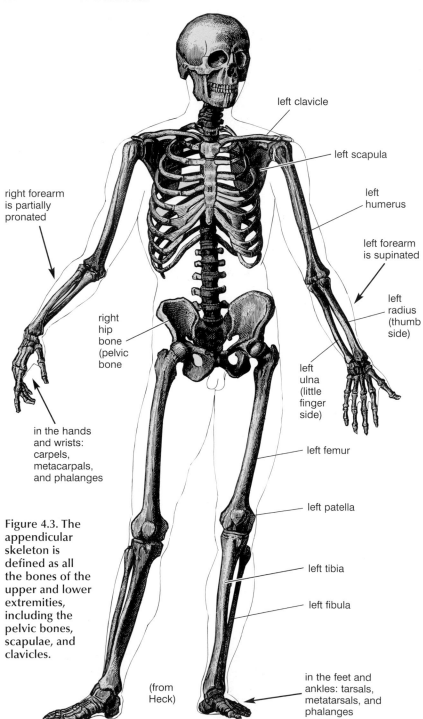

left clavicle

left scapula

left
humerus

left forearm
is supinated

left
radius
(thumb
side)

left
ulna
(little
finger
side)

left femur

left patella

left tibia

left fibula

right forearm
is partially
pronated

right
hip
bone
(pelvic
bone

in the hands
and wrists:
carpels,
metacarpals,
and phalanges

Figure 4.3. The
appendicular
skeleton is
defined as all
the bones of the
upper and lower
extremities,
including the
pelvic bones,
scapulae, and
clavicles.

(from
Heck)

in the feet and
ankles: tarsals,
metatarsals, and
phalanges

the *clavicle* (the collarbone), the *scapula* (the shoulderblade), the *humerus*, *radius*, *ulna*, and bones of the wrist and hand (fig. 4.3). The clavicle is the only bone of the upper extremity that forms a joint (the sternoclavicular joint) with the axial skeleton. It also happens to be the most commonly broken bone in the body. If you fall on the point of your shoulder, the dead weight of your upper body can snap the clavicle like a dry branch.

The humerus is the single bone of the arm, and the radius and ulna are the two bones of the forearm. If you stand upright, palms facing forward in the "*anatomical position*," the radius and ulna are parallel, with the ulna on the medial side near the hips and the radius on the lateral side. This position for the forearm is called *supination*; its opposite is *pronation*. You supinate your right forearm when you drive a wood screw into a plank clockwise. If you rotate the same screw out of the plank, turning your right hand counter-clockwise, you are pronating the forearm. During pronation the radius and ulna shift to form a long, skinny X so that the distal part of the radius is rotated to an inside position, and the distal part of the ulna is rotated to an outside position. Here (fig. 4.3) the left forearm is shown supinated and the right forearm is shown partially pronated. The bones of the wrist and hand include the *carpels*, *metacarpels*, and *phalanges*.

THE AXIAL SKELETON

The axial skeleton forms the bony axis of the body (fig. 4.4). In addition to the vertebral column (the spine), it includes the skull, the rib cage, and the sternum, or breastbone. Looking at the vertebral column from its right side reveals that the spine forms a reversed double S, with one reversed S on top of the other (viewed from the left side, envision a plain S on top of the other). The top curve faces right, the next one left, the third one again right, and the bottom one again left (figs. 4.10a–b). This reversed double S represents the four curves of the vertebral column. From the top down, the first and third convexities face anteriorly, and the second and fourth convexities face posteriorly. These are also the sites of the four main regions of the vertebral column: cervical, thoracic, lumbar, and sacral (figs. 2.29a–e, 4.4, and 4.10). A curve facing anteriorly is a *lordosis*, and a curve facing posteriorly is a *kyphosis*. The curves alternate: *cervical lordosis*, *thoracic kyphosis*, *lumbar lordosis*, and *sacral kyphosis*. Each region contains a specific number of *vertebrae*: seven in the cervical region (C1-7), twelve in the thoracic region (T1-12), and five in the lumbar region (L1-5). The sacrum is a single fixed bone.

In the fetus the entire spine is curved posteriorly, as mimicked by the child's pose (or fetal pose) in hatha yoga (fig. 6.18). When an infant begins to crawl, and later walk, secondary curvatures that are convex anteriorly—the cervical lordosis and lumbar lordosis—develop in the neck and lumbar

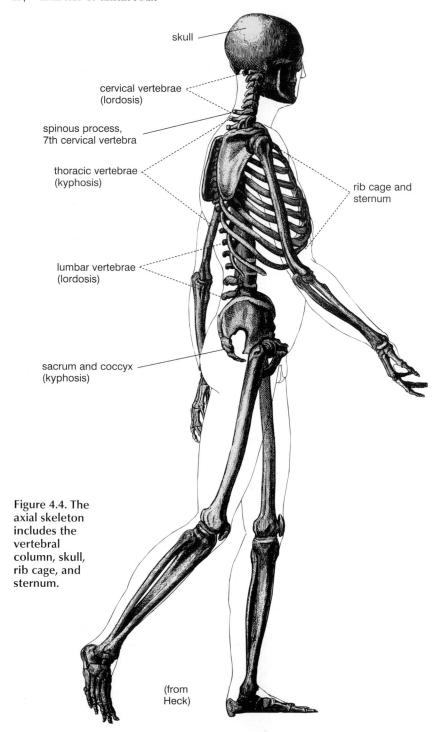

skull

cervical vertebrae
(lordosis)

spinous process,
7th cervical vertebra

thoracic vertebrae
(kyphosis)

rib cage and
sternum

lumbar vertebrae
(lordosis)

sacrum and coccyx
(kyphosis)

Figure 4.4. The
axial skeleton
includes the
vertebral
column, skull,
rib cage, and
sternum.

(from
Heck)

regions, while the posterior convexities—the thoracic kyphosis and the sacral kyphosis—are retained in the adult. The four curvatures act together as a spring for cushioning the upper body from the impact of running and walking. In standing postures the curvatures are easy to keep in their optimal configuration, neither too straight nor too pronounced, but in cross-legged meditative sitting postures the lack of hip flexibility makes this more challenging.

ANATOMY OF THE SPINE

Just as the spine of a book forms an axial hub around which pages turn, so does the human spine form the axial support around which the body moves. It forms the skeletal core of the torso, and it is the axial support for all hatha yoga postures. We cannot understand hatha yoga without understanding its structure and function. To begin, we'll examine the individual bones of the spine—the vertebrae—and then look at how they join together to form the vertebral column as a whole.

TYPICAL VERTEBRAE

We'll begin by looking a typical lumbar vertebra (L4) from above and from its left side. Viewing its superior surface from front to back (fig. 4.5a), anteriorly we see the upper surface of a cylinder, the *vertebral body*; posteriorly we see a *vertebral arch* that surrounds a space, the *vertebral foramen*, in which the spinal cord resides. Pointing backward from the rear of the vertebral arch is a bony projection, the *spinous process*. If you lie on your back and draw your knees up against your chest, you feel the spinous processes of the lumbar vertebrae against the floor. And because this is what everyone notices, laypeople often mistakenly refer to the tips of the spinous processes as "the spine."

Still looking at the upper surface of L4, we see that the foundation for the vertebral arch is composed of two columnar segments of bone, the right and left *pedicles*, which project backward from the posterior border of the vertebral body. From there, flatter segments of bone, the right and left vertebral *laminae*, meet in the midline to complete the vertebral arch posteriorly and provide the origin for the midsagittal spinous process. When surgeons have to gain access to the spinal cord or to *intervertebral disks* in front of the spinal cord, they do a *laminectomy* to remove the posterior part of the vertebral arch, including the spinal process. Near the junction of the pedicles and the laminae, the superior *articulating processes* are seen facing the viewer, and robust *transverse processes* point laterally in a transverse plane, as their names imply. It is possible to feel the latter from behind in the lumbar region, but it requires a determined, knowing fingertip.

A lateral view of L4 from its left side (fig. 4.5b) shows two articulating protuberances: the left *superior* and *inferior articulating processes*. The bilateral superior articulating processes of L4 form synovial joints with the inferior articulating processes of L3, and the bilateral inferior articulating processes of L4 form synovial joints with the two superior articulating processes of L5. In this manner the articulating processes connect the vertebral arches to one another from the neck to the sacrum. The left transverse process is visible extending toward the viewer, and a side view of the spinous process juts to the rear. Looking from the left side of L4 also confirms the cylindrical shape of the vertebral body. In this view the vertebral foramen is hidden by the left side of the vertebral arch.

We saw in chapter 2 that the lumbar spine deeply indents the respiratory diaphragm, and that its front surface (L4 and L5) can be palpated through the abdominal wall in someone slender who has a normal or overly prominent lumbar lordosis—the abdominal organs just slip out of the way as you probe. These two views of L4 reveal this anterior-most surface of the lumbar spine, and if you ever have occasion to palpate it directly, the term "axial" will gain new meaning. The spine really does form the axis of the body.

Moving up the spinal column, the top view of the seventh thoracic vertebra (fig. 4.6) reveals a smaller vertebral body than we see in the

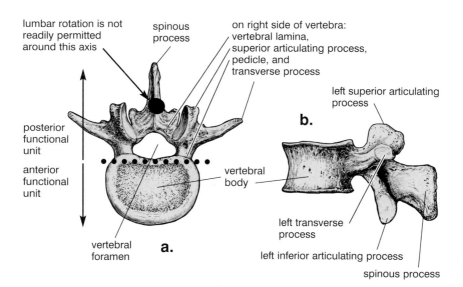

Figure 4.5. Superior view (from above) of fourth lumbar vertebra (a) and a view from its left side (b). The heavy dotted line in "a" indicates at the level of a single vertebra the separation of the anterior functional unit from the posterior functional unit. Notice also in "a" how the parasagittal orientation of the superior and inferior articulating processes, along with an axis of rotation that is located near the base of the spinous process (large dot), will be expected to inhibit lumbar twisting (from Morris).

lumbar region. And near the ends of the transverse processes we also see tiny *facets* (smooth articular regions) that represent sites for articulation with the ribs on each side (figs. 4.6a and 5.33). A side view from the left of the thoracic vertebrae (fig. 4.7b) reveals sharp spinous processes pointing downward rather than straight back, as well as vertebral bodies that are composed of shorter cylinders than those found in the lumbar region. This view also shows several other features: spaces for intervertebral disks; sites (from the left side) where the superior and inferior articulating processes meet; and spaces between the vertebral arches, the *intervertebral foramina* (again from the left side) through which spinal nerves T1–12 pass on their way to the chest wall.

Further up the spinal column, the cervical vertebrae have yet smaller vertebral bodies than those found in the thoracic vertebrae, but their vertebral arches are wider than you might imagine (fig. 4.8). If you feel your neck from the side and press deeply, you can feel the underlying hardness of the transverse processes of the cervical vertebrae.

The top two cervical vertebrae are specialized for articulation with the skull. C1 sits just underneath the skull, and C2 provides the axis around which C1 plus the skull rotate when the head is turned. We'll take a closer look at those two vertebrae and their articulations in chapter 7.

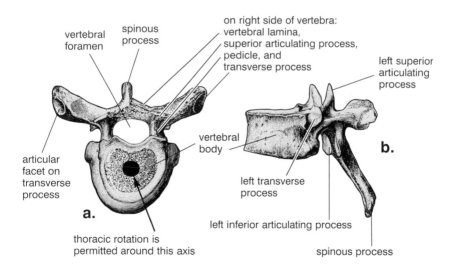

Figure 4.6. Superior view of the seventh thoracic vertebra (a) and a view from its left side (b). Notice how thoracic twisting will be permitted by the frontal orientation of the facets on the superior and inferior articular processes. Rotation of one vertebral body with respect to the next is also encouraged by an axis of rotation (large dot in "a") in the center of the vertebral body (Sappey).

At the base of the vertebral column is the sacrum, composed of what were originally five vertebrae that became fused together into one bone during fetal development (fig. 4.9). It is pointed at its lower end and has an articular surface on each side that mates with the ilia for forming the sacroiliac joints. You can feel the relatively flat posterior surface of the sacrum against the floor in the corpse posture. And if you are able to

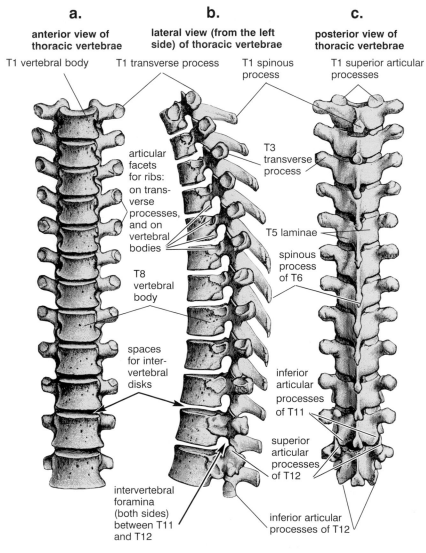

a.
anterior view of thoracic vertebrae

T1 vertebral body

b.
lateral view (from the left side) of thoracic vertebrae

T1 transverse process

c.
posterior view of thoracic vertebrae

T1 spinous process

T1 superior articular processes

articular facets for ribs: on transverse processes, and on vertebral bodies

T8 vertebral body

spaces for intervertebral disks

intervertebral foramina (both sides) between T11 and T12

T3 transverse process

T5 laminae

spinous process of T6

inferior articular processes of T11

superior articular processes of T12

inferior articular processes of T12

Figure 4.7. Thoracic vertebral column, from an anterior view (a), lateral view (b), and posterior view (c). Notice in the middle thoracic region how the spinous processes point sharply downward (Sappey).

palpate someone's L5 vertebral body through their abdominal wall, you can also feel the *promontory* (the top front border) *of the sacrum.* Below this promontory, the sacral kyphosis rounds so much to the rear that you can't feel its anterior surface. The *coccyx* (the "tail bone") is attached to the tip of the sacrum (figs. 4.4 and 4.10).

THE ANTERIOR FUNCTIONAL UNIT

The vertebral column is composed of two distinct functional units: an *anterior functional unit* composed of a stack of vertebral bodies and intervertebral disks that together form a flexible rod, and behind the anterior functional unit, a tubular *posterior functional unit* composed of a stack of vertebral arches and associated ligaments. We'll look first at the rod.

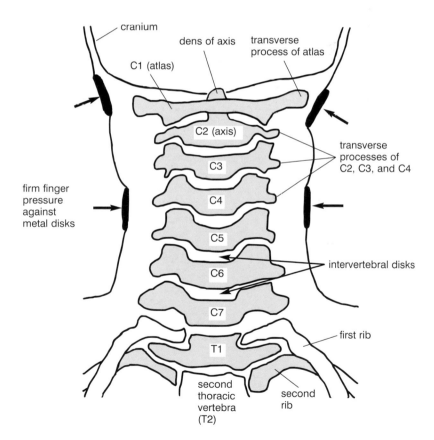

Figure 4.8. Skeletal structures of the neck in relation to finger pressure against metal disks from the side, drawn from two roentgenograms (X-rays). The images were taken from the front. Compare with the atlas and axis in fig. 7.2. Credits to Dr. Donald O. Broughton for the roentgenograms.

If we were to locate an isolated spine and saw off all the vertebral arches at the base of the pedicles, we would be left with a stack of vertebral bodies and intervertebral disks (fig. 4.10a). Each intervertebral disk forms a *symphysis* between adjacent vertebral bodies that allows a small amount of compression, expansion, bending in all directions, and twisting. The flexibility of the anterior functional unit, at least in isolation, is therefore restrained only by the integrity of the intervertebral disks and by *anterior and posterior longitudinal ligaments* that support the complex in front and back (figs. 4.12b and 4.13a–b).

Each intervertebral disk (fig. 4.11) has a semi-fluid core, the *nucleus pulposus*, which is surrounded by a tough but elastic connective tissue exterior, the *annulus fibrosis*. The nucleus pulposus comprises only about 15% of the total mass, but that's enough liquid to allow the disk to act hydraulically— every time you shift the angle of one vertebral body with respect to its neighbor, the nucleus pulposus shifts accordingly, bulging out the elastic annulus fibrosis on one side, and every time you twist, the nucleus pulposus presses the annulus fibrosis outward all around. The tough fibroelastic connective tissue comprising the annulus fibrosis fuses each intervertebral disk to the vertebral body above and below. They even run continuously from the disk into the bone. That is why an intervertebral disk never "slips." It can only rupture, exuding some of the gel-like nucleus pulposus through a weakened annulus fibrosis, or degenerate.

As with other connective tissues, intervertebral disks contain living cells which require nutrients for their survival and which produce metabolic

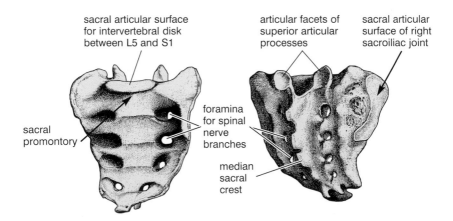

Figure 4.9. Anterolateral view (on the left) and posterolateral view of the sacrum (on the right), the latter also showing the sacral surface of the right sacroiliac joint. The sacrum is concave anteriorly and convex posteriorly, an architectural feature which is even more apparent in fig. 4.10a (Sappey).

waste products that have to be disposed of. But one thing is missing, at least after we have reached our mid-twenties, and that is blood vessels. The capillary beds that serviced the intervertebral disks during our youth are lost during the natural course of aging. So in older people, how do the living cells receive nourishment, and how are waste substances eliminated?

a.   **b.**   **c.**

side view of the four curvatures
and of the anterior and posterior
functional units of the spine

anterolateral
view of spine

posterior
view of spine

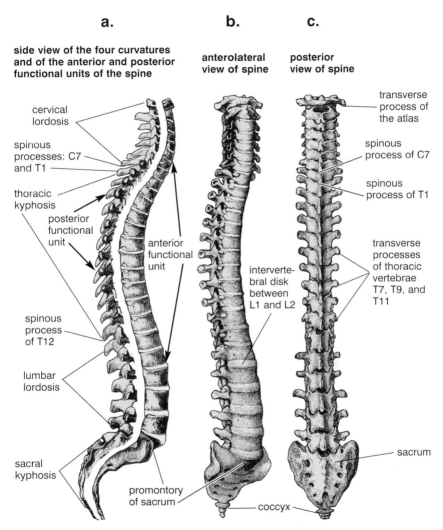

cervical
lordosis

spinous
processes: C7
and T1

thoracic
kyphosis

posterior
functional
unit

anterior
functional
unit

spinous
process
of T12

lumbar
lordosis

sacral
kyphosis

promontory
of sacrum

intervertebral disk
between
L1 and L2

coccyx

transverse
process of
the atlas

spinous
process of C7

spinous
process of T1

transverse
processes
of thoracic
vertebrae
T7, T9, and
T11

sacrum

Figure 4.10. Three views of the spinal column: a lateral view from the right side (a), an anterolateral view from right front (b), and a posterior view from behind (c). The middle image (b) is enough from the side that the four spinal curvatures start becoming apparent. The anterior and posterior functional units are shown separated from one another in the figure on the left (a). The tubular nature of the posterior functional unit, however, is apparent only in views of individual vertebrae from above (fig. 4.5) or from below (Sappey).

The biomedical literature suggests that nutrients are "imbibed," or absorbed, into the intervertebral disks from the vertebral bodies, which themselves are well supplied with blood, but little is known about this process except that healthy intervertebral disks contain 70-80% liquid and that the spine gets shorter during the day and longer during the night. Taken together, these two pieces of information suggest that when the intervertebral disks are compressed by gravity and muscular tension during the day, liquid is squeezed out, and that when tension is taken off the vertebral column at night, the vertebral bodies can spread apart, allowing the intervertebral disks to absorb nutrients. This would help explain why it has long been thought that one of the most effective treatments for acute back pain is bed rest, and why a thoughtful combination of exercise and relaxation is therapeutic for most chronic back ailments.

THE POSTERIOR FUNCTIONAL UNIT

If we were to locate another isolated spine and saw off most of the anterior functional unit from top to bottom, we would be left with the posterior functional unit—a tube made up of the stack of vertebral arches, all the restraining ligaments between the arches, and just enough of the posterior borders of the vertebral bodies and intervertebral disks to complete the tube anteriorly (figs. 4.5a and 4.10a). The interior of the posterior functional unit is the vertebral canal, which houses the spinal cord and spinal nerves.

Each vertebral arch forms small synovial joints with its neighbors through the agency of little facets, or flattened joint surfaces, that are located on adjacent superior and inferior articular processes. Like other synovial joints, these movable *facet joints* are characterized by joint surfaces covered with cartilage on their articular surfaces, synovial membranes and fluid, and joint capsules that envelop the entire complex.

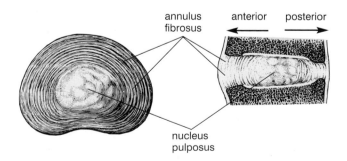

Figure 4.11. The intervertebral disk, in a cross-section from above (left), and in a mid-sagittal section between two vertebral bodies (right). In the figure on the right, the nucleus pulposus is being squeezed anteriorly by a backbend (Sappey).

Two inferior articular processes, one on each pedicle, form facet joints with matching surfaces on the superior articular processes of the vertebra below. Although the orientations of the articular processes restrict twisting of the spine in the lumbar region (fig. 1.11), their orientations permit it in the cervical and thoracic regions. You hear and feel the facet joints moving when a skilled bodyworker cracks your back. And many young athletes can twist their spine when they get up in the morning, and hear a sequence of pops, indicating that facet joints are being released one after another.

Just in front of the superior and inferior articular processes, and just to the rear and to each side of the intervertebral disks, are the intervertebral foramina, the openings in the vertebral column out of which the spinal nerves emerge (fig. 4.7b). In the lumbar region the locations of these foramina and their spinal nerves make them vulnerable to ruptured intervertebral disks, which may impinge against a spinal nerve and cause shooting pains down the thigh.

SPINAL STABILITY AND BENDING

The anterior and posterior functional units are described separately to clarify the concepts, but in fact they are bound together inextricably. The architectural arrangement that links them is propitious: it permits spinal movements and at the same time it insures spinal stability. For example, backward bending will compress the vertebral arches together posteriorly while spreading the front surfaces of the vertebral bodies apart anteriorly, and forward bending will pull the vertebral arches slightly apart while compressing the intervertebral disks in front. These movements, however, can be taken only so far because the entire complex is stabilized by ligaments (figs. 4.12–13). As just mentioned, the anterior and posterior longitudinal ligaments reinforce the flexible anterior functional unit, and the joint capsules for the facet joints help hold the vertebral arches together. In addition, there are yet more ligaments superimposed on the posterior functional unit—*interspinous ligaments* that run between adjacent spinous processes, a *supraspinous ligament* that connects the tips of the spinous processes, flat elastic ligaments known as *ligamenta flava* that connect adjacent laminae, and the cervical *nuchal ligament*, which is an elastic extension of the supraspinous ligament that reaches the head.

[Technical note: Quadrupeds such as dogs and cats have well-developed nuchal ligaments whose elasticity keeps their heads on axis without constant expenditure of muscular energy. Because of our upright posture, the significance of the nuchal ligament in humans is greatly reduced in comparison with that of quadrupeds, but it probably has a least a minor role in helping to keep the head pulled to the rear. It's no substitute for muscular effort, however, as evidenced by drooping heads in a room full of sleepy meditators.]

THE SPINE AS A WHOLE

We've seen representative vertebrae at each segment of the spine, and we've seen how their linked front portions form the anterior functional unit and how their linked vertebral arches form the posterior functional unit. Then we looked at how the combination of anterior and posterior functional units along with their restraining ligaments permits and yet limits bending. Now we need to examine the spine as a whole within the body (fig. 4.4). The depth of the four curvatures can be evaluated from the side (figs. 4.10a–b). If they are too flat the spine will not have much spring-like action when you walk and run, but if the curvatures are too pronounced, especially in the lumbar region, the spine will be unstable. Excess curvatures are more common—an orthopedist would tell you that a "*lordosis*" is an excess anterior curvature in the lumbar region and that a "*kyphosis*" is an excess posterior curvature of the chest. And sometimes people have what is called a reverse curvature in the neck, one that is convex posteriorly instead of anteriorly.

From either the front or the back, the spine should look straight. If it doesn't, it is usually because of an imbalance such as one leg being shorter

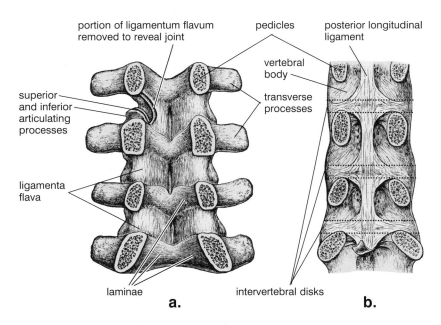

a. **b.**

Figure 4.12. Views of the ligaments in the lumbar portion of the vertebral canal from inside looking out. On the left (a) is an image looking posteriorly toward the vertebral laminae, ligamenta flava, pedicles (cut frontally), and transverse processes. On the right (b) is an image looking anteriorly toward the posterior longitudinal ligament, the posterior surfaces of the vertebral bodies, intervertebral disks (indicated by dotted lines), and pedicles (again in frontal section; Morris).

than the other, which causes the pelvis and spine to be angled off to the side. This creates a side-to-side curvature known as *scoliosis*, a condition which always includes a right-left undulation of compensatory curvatures higher up in the spine that ultimately brings the head back in line with the body. These compensatory curvatures develop because our posture adjusts itself to maintain the plumb line of gravity from head to toe with the least possible muscular effort, even if it results in distortions and chronic aches and pains.

THE DEEP BACK MUSCLES

We can't make a posture with just bones, joints, and ligaments: something has to move them. An engineer designing a plan for putting the body into motion might suggest using three layers of rope-and-pulley systems. The innermost layer would consist of miniature systems of ropes and pulleys connecting the smallest adjacent skeletal segments, the intermediate layer would consist of larger systems connecting bigger segments, and the outermost layer would consist of the largest and longest systems connecting the segments of the skeleton which are the furthest apart. And indeed, we

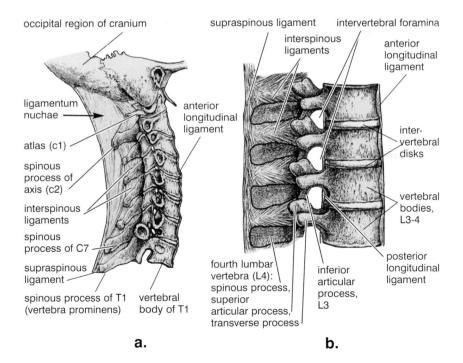

Figure 4.13. Side views (from the right) of the ligaments associated with the cervical portion of the vertebral column (a) and with the lumbar region (b). (from Morris).

can see elements of such a plan within the human body. The smallest and shortest muscles of the spine act between adjacent vertebrae. The middle layer of muscles—the deep back muscles—operates between the torso and the lower extremities to hold us upright. To see them (fig. 4.14 and 5.5) you would have to remove the upper extremities, including the third and outermost layer of muscles that act from the torso to manipulate the scapulae and arms (figs. 8.8–14).

The main component of the middle layer is the *erector spinae*, which runs between the neck and the pelvis on either side of the spinous processes. This muscle erects or extends the spine, as its name implies; it also restricts forward bending, aids side bending, and influences twisting of the torso in conjunction with the abdominal muscles. And in static postures such as sitting upright in meditative sitting postures, it acts as an extensile ligament, holding the spine straight with a mild isometric effort (chapter 10). Our main concern here is how the deep back muscles operate either to facilitate or restrict standing postures.

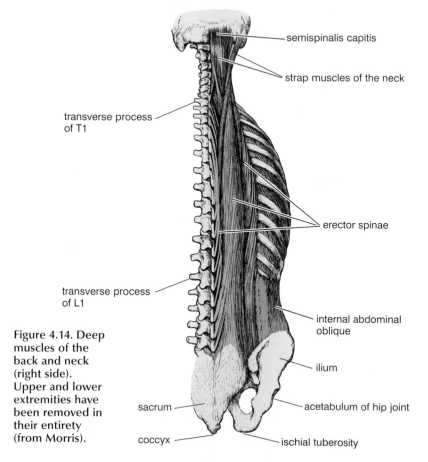

semispinalis capitis

strap muscles of the neck

transverse process of T1

erector spinae

transverse process of L1

internal abdominal oblique

ilium

Figure 4.14. Deep muscles of the back and neck (right side). Upper and lower extremities have been removed in their entirety (from Morris).

sacrum

coccyx

acetabulum of hip joint

ischial tuberosity

SYMMETRY AND ASYMMETRY

The axial skeleton, the appendicular skeleton, and muscles throughout the body all contribute to determining our bilateral symmetry. For perfect symmetry, every right-left member of every pair of bones, skeletal muscles, joints, and ligaments must be identical on both sides of the body—right and left knee joints, hip joints, femurs, and clavicles; and right and left erector spinae muscles, quadriceps femori, hamstrings, adductors, and gluteals.

To check out the symmetry in your own posture, look at yourself frontally in a full-length mirror, preferably in the buff. Place your feet about twelve inches apart and let your hands hang relaxed. Look carefully. Do the right and left extremities appear to be of equal length? Is one shoulder higher than the other? Do you lean slightly to one side? Do both forearms hang loosely, or is one elbow more bent? Does the waistline make a sharper indentation on one side than the other, creating extra space between the body and elbow on one side? Is the crest of the ilium higher on one side than the other? Is one nipple higher than the other? If you draw an imaginary line from the umbilicus to the center of the sternum, is it perpendicular to the floor, or slightly off?

Look down at your feet. Are they comfortable in a perfectly symmetrical position, or would it feel more natural if one or the other were rotated laterally? Do the toes all spread out and down squarely, or do some of them seem to clench in? You are not trying to change anything; you are just making observations. Don't despair if your body is not perfectly symmetrical; few are.

Most of us were born symmetrical, but our habitual activities have undermined our balance. Carrying a handbag on one shoulder, always lowering the chin to the same side against a telephone receiver, swimming freestyle and always turning the head in the same direction for breathing, and countless other right-left preferences create habitual tension on one side of the body that eventually results in muscular and skeletal misalignments and distortions.

So far we've been discussing only static anatomical symmetries and asymmetries. But these terms are also used in the context of movement. In that realm a symmetrical movement is one in which both sides of the body move at the same time and in the same way, while an asymmetrical movement is one in which each side of the body moves sequentially. As it happens, most of our everyday activities are accomplished asymmetrically. You don't hop forward two feet at a time—you walk, swinging your right hand forward in concert with your left foot, and swinging your left hand forward in concert with your right foot. Likewise, a boxer hits a punching bag with one hand and then the other, not with both hands at the same time. And every karate master knows that the power of a punch with one hand depends on simultaneously pulling the opposite elbow to the rear. In swimming we see both possibilities—the butterfly, the breast stroke, and the beginner's back stroke involve symmetrical movements; the freestyle and the back stroke crawl involve asymmetrical movements.

You can correct some right-left asymmetries with patience, persistence, and a well-thought-out practice plan. In fact, right-left balancing is an important quest in hatha yoga ("ha" is the Sanskrit word for "right," and "tha" means "left"). And apart from its importance in hatha yoga postures, bodily symmetry is beneficial to our overall health and comfort.

The best approach for correcting right-left imbalances is to concentrate on asymmetrical postures and activities, working first with one side and then the other, and watching for differences between the two. If you spot an imbalance you can do the same posture three times, starting and ending on the more difficult side, and over time this will tend to correct the situation. Symmetrical postures, by contrast, are often not very effective for correcting right-left imbalances. Both sides may get stronger and more flexible, but they will remain different. In certain cases the differences can even become exaggerated because making an identical effort on both sides may favor the side that is more flexible, and this leaves the constricted side even more out of balance.

If right-left imbalances are best remedied with asymmetric postures, front-to-back imbalances are best remedied with symmetric postures. Let's say you can't bend forward and backward very far, or that you perceive that your backward bends are more convincing than your forward bends, and you can't detect any difference in tightness when you compare the two sides of the body. The solution to this problem is to develop a personal program of symmetrical forward bends and backbends to redress the imbalance. But keep watching. These improvements are sooner or later likely to uncover asymmetrical limitations which until that time had been hidden: limitations to forward bending in the hip or hamstrings on one side, limitations in the hip flexors on one side for eccentric backward bending, or limitations to side bending on one side. Don't complain. Start over. Enjoy.

STANDING POSTURES

In addition to correcting muscular and skeletal imbalances, standing postures as a whole form a complete and balanced practice that includes twisting, forward bending, side bending, backward bending, and balancing. A standing forward bend even serves as a mild inverted posture for those who are flexible enough to bend all the way down. Before studying specific postures, however, we must examine some fundamental principles.

DEVELOPING A STRONG FOUNDATION

Should you stand relaxed, or should you purposely hold some tension in the hips and thighs when you are doing standing postures? This was our opening question, and the answer is not the same for everyone. Through long experience, advanced students know exactly when and where it is safe to relax, so they can do whatever they want. Beginners, however, who are embarking on a course of standing postures should be told straight out to plant their feet firmly and to hold the muscles of the hips and thighs in a state of moderate tension. The many muscles that insert in joint capsules keep them taut and establish a strong base for the posture. This not only reinforces the joints, it brings awareness to them and to the surrounding muscles—and where there is awareness there is safety. Tightening the muscles of the hips and thighs limits the range of motion, it is true, but it prevents torn muscles and injuries to the knee joints, sacroiliac joints, hip joints, and the lower back. In addition to these immediate benefits, developing a strong base over a period of years builds up the connective tissues in both the joints and their capsules. And as the joints become stronger, it becomes safer to relax the body more generally and at the same time intensify the stretches. Experts take this all for granted; they protect themselves without realizing it and are often not aware that beginners unknowingly place themselves in danger.

SETTING PRIORITIES

For novices, standing postures are the best training ground for experiencing the principle of learning to establish priorities from the distal to the proximal parts of the limbs. This means you should construct standing poses from your feet to your hips to your torso, and from your hands to your shoulders to your torso, rather than the other way around. That's desirable because your awareness of the body gets poorer and your ability to control the muscles diminishes as you move from distal to proximal, and if you first bend or twist the trunk and then manipulate the extremities, the latter movements take your attention away from the proximal structures of the body over which you have less awareness. By contrast, if you settle the distal portions of your limbs first, you can keep them stable with minimal effort while you place your attention on the central core of the body.

FOOT POSITION

The feet are the foundation for standing postures. This can be taken literally: small adjustments in how the feet are placed will affect your posture from head to toe. To see how this happens, stand with your feet together and parallel, draw lines straight down the front of your bare thighs with a marker, and imagine parasagittal planes through each of them. What we are going to see is that movements of the lines reflect rotation of the planes, and that rotation of the planes reflects rotation of the thighs. So keeping the knees straight, first rotate your feet so that the big toes remain together and the heels swing out 45° each (a 90° angle opening to the rear), and then rotate them in the opposite direction so that the heels are together and the toes are out (again a 90° angle, but now opening to the front). By definition, the thighs will have rotated medially in the first case and laterally in the second. We should note that in both situations the thighs account for only about two-thirds of the total rotation—30° at the hip and 15° at the ankle.

This experiment makes it clear that most of the rotation of the foot is translated to the thigh. If a foot slips out of position in a standing posture, it indicates weakness on that side, and to allow the weakness to remain indefinitely can only accentuate problems throughout the foundation of the body. The situation should be corrected, but don't force matters. Instead of hurting yourself by stressing the weaker side, ease up on the healthier side and resolve to take as long as necessary to make long-term adjustments. In any case, watch your foot position constantly.

FOUR SIMPLE STRETCHES

The best way to approach standing postures is to start simply, and the simplest standing stretches are those that do not require us to counteract gravity by tightening our lower extremities beyond what is needed to balance upright. This means that the torso is not bending backward, forward, or to the side. We'll start with the mountain pose.

THE MOUNTAIN POSE

The mountain pose is the basic beginning standing posture (fig. 4.15), from which all others are derived. To begin, stand with the feet together and parallel, and the hands alongside the thighs with the forearms midway between supination and pronation (the thumbs toward the front). Create a firm base by pulling the hips tightly together in ashwini mudra and by keeping the thighs tight all around. The quadriceps femoris muscles keep the kneecaps lifted in front, the hamstring muscles keep tension on the ischial tuberosities and the base of the pelvis, and the adductor muscles keep the thighs squeezed together. Keep the knees extended, but not hyper-extended beyond 180°. Find a relaxed and neutral position for the shoulders, neither thrown back artificially nor slumped forward. Just stand smartly erect. This is the mountain pose. It will keep the abdomen taut without any special effort and produce diaphragmatic breathing.

[Technical note: Most students do not have to be worried about hyperextension of the knees provided they keep some tension in the hamstrings. The few individuals who can hyperextend their knees beyond 180° should be watchful not to lock them, but to maintain a balancing tension all around their thighs—especially between their hamstrings and their quadriceps femoris muscles—which keeps their lower extremities on axis. It should also be mentioned that some instructors, perhaps a minority, suggest keeping the knees "soft" for the mountain posture, by which they mean keeping them ever so slightly bent. What's most important is awareness. Do whatever you want but be attentive to the results.]

THE SIDE-TO-SIDE STRETCH

Next we'll look at a simple side-to-side standing stretch (fig. 4.16). Stand with your feet a comfortable distance apart and tighten the muscles of the hips and thighs to make a solid pelvic base. Raise the arms to shoulder height. Now stretch the hands out to the side, palms down, with the five fingers together and pointing away from the body. Observe the sensations in the upper extremities. At first you may tend to clench the muscles, trying to force the hands out, but that's too extreme. Just search out regions, especially around the shoulders, which, when relaxed, will allow the fingers more leeway for reaching. You are still using muscular effort for the side stretch, but the muscles you are relaxing are now allowing others fuller sway. Gradually, delicate adjustments and readjustments will permit your fingertips to move further and further to the sides.

Figure 4.15. Mountain pose: the basic standing posture, from which all others are derived.

If you suspect that there is something mysterious about this, that some force other than your own muscular effort is drawing your fingertips out, a simple experiment will bring you back to reality. Stand in the same stretched posture and ask two people to pull your wrists gently from each side while you relax. As the stretch increases, the feeling is altogether different from the one in which you were making the effort yourself. Stretching once again in isolation will convince you that nothing but muscular effort is doing this work.

Like the mountain pose, the side-to-side stretch is excellent training for diaphragmatic breathing because the posture itself encourages it. The arm position holds the lower abdominal wall taut and the upper chest restricted, and this makes both abdominal and thoracic breathing inconvenient. You would have to make a contrived effort to allow the lower abdomen to relax and release for an abdominal inhalation, and you would have to make an unnatural effort to force the chest up and out for a thoracic inhalation. Students who tend to get confused when they try to breathe diaphragmatically in other positions learn to do it in this stretch in spite of their confusion. All the instructor has to do is point out what is happening.

Figure 4.16. Side-to-side stretch.

Right-left imbalances also become obvious in the side-to-side stretch. If students watch themselves in strategically placed mirrors, they will be acutely aware if one shoulder is higher than the other, if extension is limited more on one side than the other, or if there are restrictions around the scapula, often on one side. And with awareness begins the process of correction.

THE OVERHEAD STRETCH

Next try a simple overhead stretch (fig. 4.17). Stand this time with your heels and toes together, and with your base again firmly supported by contracted hip and thigh muscles. Bring the hands comfortably overhead with the fingers interlocked, the palms pressed together, and the elbows extended. Stretch up and slowly pull the arms to the rear, lifting the knuckles toward the ceiling. You can feel some muscles pulling the arms backward,

Figure 4.17. Overhead stretch, a simple and superb posture for learning to use the distal portions of the extremities to access proximal parts of the extremities and the core of the body.

and others resisting. Now you have to watch the elbows. It is easy to keep them extended in the first position, but as the arms are pulled back, one of them may begin to reveal weakness or restriction in its range of movement, or one forearm may show weakness that permits the interlocked hands to angle slightly off toward the weaker side. Take care to keep the posture as symmetrical as possible.

As you lift with more focused attention, you will feel the effects of this stretch first in the shoulders, then in the chest, abdomen, back, and finally the pelvis. As in the side-to-side stretch, the posture requires selective relaxation. Many students find this difficult and will keep all their muscles clenched, but any excess tension in the neck, shoulders, or back will make it difficult to feel the effects all the way down to the pelvis. If you only feel the posture affecting the upper extremities, you need to make more conscious efforts to relax selectively.

In the overhead stretch the extensors of the upper extremities will all be in a state of moderate tension and the flexor muscles will be in a state of relaxed readiness, simply countering the extensors. The posture's effectiveness will depend on how naturally this takes place. If the extensor muscles in the arm and shoulder are noticeably limited by their antagonists, you may not be able to straighten your forearms at the elbow joint. Or even if the forearms can be fully extended you may not be able to pull the arms backward. And if you feel pain in the arms and shoulders, you will not be able to direct much energy and attention to the trunk. But if you practice this stretch regularly, you will gradually notice that your efforts are affecting the central part of the body as well as the extremities.

The overhead stretch is also one of the best postures for learning how to work from distal to proximal because the proximal parts of the body are affected so clearly by each successively more distal segment. As with the side-to-side stretch and the mountain pose, the overhead stretch encourages diaphragmatic breathing. It doesn't restrict abdominal breathing as much as the side-to-side stretch (at least not unless you bend backward in addition to stretching up), but it restricts thoracic breathing even more.

THE STANDING TWIST

The last and most complicated of these simple stretches, and the only one that is asymmetrical, is the simple standing twist. First, to understand the dynamics of the pose, try it while holding as little tension in the hips and thighs as possible. We'll call this a relaxed standing twist. With the medial borders of the feet about twelve inches apart, twist to the right, leading with the hips, with the arms hanging. As you twist, the opposite hip projects backward and you dip forward almost imperceptibly, as though you were planning a twisted forward bend from the hips. This may not feel natural:

in fact, it shouldn't, because the healthy norm is to hold some tension in the hips and thighs when you twist. But this relaxed standing twist is a concentration exercise in doing just the opposite. In this manner you can get a feel for what not to do. Paradoxically, students who are not very body-oriented can do this exercise with little or no prompting, in contrast to the athlete or hatha yogi who finds it odd to relax and allow the opposite hip to move posteriorly.

Now twist again, but this time first plant the feet, hold the knees extended, and tense the gluteal muscles on the side opposite the direction of the twist. When you twist to the right and contract the left gluteus maximus, at least three things are noticeable: the left thigh becomes more extended, the left side of the hip is pulled down, and the torso straightens up. Now tense the quadriceps femoris muscles on the front of the thighs, paying special attention to the side toward which you are twisting. This complements the action of the opposite gluteus maximus. Last, tense the adductors on the medial sides of the thighs. The entire pelvis is now strongly supported by muscular activity (fig. 4.18). This is the correct feeling for a standing twist; it comes naturally to those who are in good musculoskeletal health but it feels artificial to those who are not.

Figure 4.18. Standing twist. In this and all other asymmetrical stretches, the text descriptions refer to what is seen in (and felt by) the model since that is ordinarily a student's frame of reference in a class. All such postures should be done in both directions.

BACKWARD BENDING

It is logical to examine standing backward bends next because they are easy, simple, symmetrical, and natural. For beginners they are simply an extension of standing up straight.

WHOLE-BODY BACKWARD BENDING

Whole-body backward bending, more than any other posture, demonstrates the principle of setting priorities from distal to proximal. Try this experiment: stand, lift your hands overhead, and clasp your hands and fingers together firmly. Press the palms together solidly, making sure that the hands do not angle to the right or left. Now extend the forearms at the elbows. Lock them firmly. With the heels and toes together, grip the floor gently with the feet. Tighten all the muscles of the thighs, lifting the kneecaps with the quadriceps femori, tensing the adductors medially and the hamstrings posteriorly, and squeezing the hips together. This stabilizes the knee joints and pelvis, creating a strong base. Pull your arms backward as far as possible and lift your hands toward the ceiling. If all of the priorities have been set in order, the lifting feeling in the posture can be felt progressively in the thorax, abdomen, and pelvis.

Now, with all the distal-to-proximal priorities established, you can place your attention on the totality of the posture. Bend backward in a whole-body arc, but without throwing your head back excessively. You will feel a whole-body bend as you access the core of the body. Hold your attention on the abdomen. Tissues are pulling on your torso from fingertips to toetips— fingers to hands, hands to forearms, forearms to arms, arms to shoulders, and shoulders to chest, abdomen, and pelvis; toes to feet, feet to legs, legs to thighs, and thighs to pelvis and vertebral column. Notice the whole-body tension, especially on the front side of the chest, abdomen, and thighs (fig. 4.19). Breathe evenly; do not come back so far that you have a desire to hold your breath at the glottis or that you hesitate to exhale. Within those limits, as soon as you are settled, pull back isometrically to exercise and strengthen the back muscles. With every breath, inhale deeply with thoracic inhalations that explore (within the limits of the posture) the fullness of your inspiratory capacity.

Numerous mechanisms protect you in a whole-body backward bend. Eccentric lengthening and finally isometric contraction of the iliopsoas muscles turn them into extensile ligaments, and these muscles, acting as short but powerful cables, resist excess extension at the hip joint. In the case of the iliacus they act between the femur and the pelvis, and in the case of the psoas they act between the femur and the lumbar region. They are not alone, of course. The spiraled ischiofemoral, pubofemoral, and iliofemoral ligaments back them up on each side, keeping the head of the femur pulled into congruence with the hip socket as you reach your limits of hip extension.

Tensing the quadriceps femoris muscles lifts the patellas and extends the knees, and this allows you to drop backward only so far. To come back further you would have to bend the knees and bring your lower body forward, but we do not want that here. In this standing whole-body backward bend we are thinking of beginning and intermediate students who should learn to protect their joints, muscles, ligaments, and tendons before trying more challenging postures.

The femoris quadriceps has another action that affects the backbend more subtly. As we saw in chapter 3, three of the four heads of this muscle take origin from the femur and do not have a direct effect on the pelvis, but the fourth, the rectus femoris (figs. 3.6, 3.9, and 8.8–9), attaches to the ilium at the anterior inferior iliac spine. The rectus femoris is a football-kicking muscle, for which it has two roles: acting as a hip flexor for helping the iliopsoas muscle initiate the kick, and then acting with the quadriceps

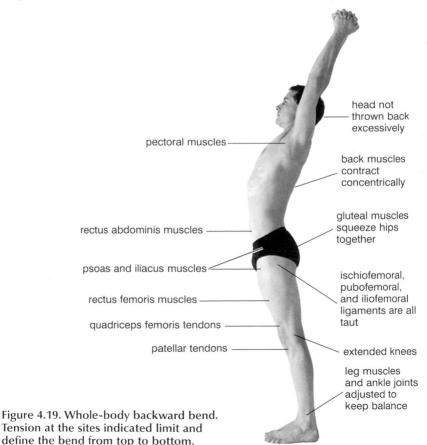

head not thrown back excessively

pectoral muscles

back muscles contract concentrically

gluteal muscles squeeze hips together

rectus abdominis muscles

psoas and iliacus muscles

rectus femoris muscles

ischiofemoral, pubofemoral, and iliofemoral ligaments are all taut

quadriceps femoris tendons

patellar tendons

extended knees

leg muscles and ankle joints adjusted to keep balance

Figure 4.19. Whole-body backward bend. Tension at the sites indicated limit and define the bend from top to bottom.

femoris muscle as a whole for extending the leg at the knee joint and assisting completion of the kick. The rectus femoris acts isometrically, however, in a standing backward bend, helping the iliopsoas muscles stabilize the pelvis and spine, and helping the rest of the quadriceps femoris stabilize the knee. Without its tension the knees would bend and the top of the pelvis would tip backward, thus pitching the trunk to a more horizontal position and accentuating the lumbar portion of the backward bend, which is not our aim in this beginning posture.

Lifting the hands overhead and pulling the arms back as much as possible protects the upper half of the body and provides lift and traction to the chest and abdomen. The rectus abdominis muscles resist this stretch in two ways: they provide further stability to the pelvis by way of their attachments to the pubic bones, and they support the role of the rectus femoris muscles in preventing acute bending in the lumbar region. What we want, and what our efforts give us from top to bottom, is a continuous arc of tension in muscles, tendons, and fasciae between the fingertips and toes. And that is why it can rightfully be called a whole-body backward bend.

It is important to keep your head upright in this posture and in an arc with the rest of the vertebral column because the head has more density than any other part of the body and is situated at the end of the most flexible part of the vertebral column. Allowing it to drop backward brings your attention to the neck and shifts your center of gravity to the rear so that you have to bend more in the lumbar region to keep your balance. Just draw the arms backward and keep the head between them if you can; otherwise, keep the head in a natural line with the rest of the spine. This precaution will also help prevent lightheadedness and fainting in those with low blood pressure.

The effects of gravity in this bend depend on the individual. Those who are strong and flexible will arc their bodies back and be aware of how gravity is affecting them from head to toe, but those who have a hard time standing up straight will be working against gravity just trying to pull themselves upright from a bent-forward position. Even so, it will still be useful for them to make the effort because it will improve their flexibility and help them balance their weight more efficiently in day-to-day life.

A RELAXED STANDING BACKBEND

When you are thoroughly confident with whole-body backward bends, you are ready to deepen the bend by exploring the nuances of relaxation in a standing backward bend. There are many poses that illustrate this principle, but for starters place your hands overhead without lifting them fully, and then bend backward without tensing the muscles of the thighs (fig. 4.20). Notice that this posture is entirely different from the last one. Relaxing the upper extremities allows more bending because the rectus abdominis muscles are under less tension, and relaxing the lower extremities takes tension off the iliopsoas and quadriceps femoris muscles, which in turn takes tension off the pelvis and the knees. The main place where stress and tension are felt is the lower back. For this reason, it is apt to call this a backbend rather than a backward bend.

A relaxed standing backbend such as this, with slightly bent knees and an arched-back lumbar region, places you more in the grip of gravity than a whole-body backward bend. But this gravitational assist is unwelcome unless you have a healthy lumber region as well as strong abdominal muscles and hip flexors. Even advanced students may find they are not comfortable holding this posture for a long time. So keep a watch on the lower back, and do not bend beyond your capacity to recover gracefully.

THE ROLE OF BREATHING IN STANDING BACKWARD BENDS

One of the secrets of successful standing backward bends is to maintain an even breathing pattern, but this cannot be taken for granted among beginners. Those who are not comfortable will tend to hold their breath after an inhalation, and this will obviously limit the time they can hold the posture. Teachers can spot at least three breathing patterns that should be avoided.

accentuated
lumbar lordosis

slightly bent knees

Figure 4.20. Relaxed standing backbend. This posture is advisable only for intermediate and advanced students because of the stress placed selectively on the lumbar region.

Some students will close the glottis entirely but come out of the posture with an "aaagh." Others will keep the glottis partially closed but make a soft grunting sound when they exhale and come up. Yet others will resist exhaling but without closing the glottis. Teachers may not notice this last pattern unless they are watching for it because it doesn't make noise. But in any case, students who hold their breath or breathe aberrantly in standing backward bends speak their discomfort and anxiety clearly, at least to those who have educated eyes and ears. If students cannot inhale and exhale smoothly in the posture, it is better for them to limit the bend—breathing constantly, keeping the lungs open to the atmosphere, and consciously depending for security on strong and healthy pelvic and respiratory diaphragms, abdominal muscles, and hip flexors.

A simple experiment will show two distinct ways that breathing can work in standing backbends. Ask a class of beginning students to come into the whole-body backward bend with their hands clasped overhead and lifted. Then ask them to breathe gently and notice that their hands move forward during inhalation and to the rear during exhalation. After they have come in touch with this, ask them to notice when they feel the most discomfort, or if not discomfort, wariness. Most of them will say it is at the end of exhalation, when their hands drop to the rear. It so happens that this is the moment in the breathing cycle that corresponds to the least tension in the abdominal muscles and in the respiratory and pelvic diaphragms, which furnish the most important support for the posture in everyone who is keeping the airway open. Because the spine is most vulnerable when tension is released in these supporting muscles, many teachers wisely suggest an alternative: that students consciously reverse their natural breathing patterns in relation to the bend. Instead of letting inhalation restrict the bend and exhalation accentuate it, they will suggest that students inhale their maximum inspiratory capacity during the deepest part of the bend and then purposely ease forward during exhalation.

FORWARD BENDING

There are three big issues in forward bending: gravity, the site where the body is bending, and breathing. Gravity plays only a minor role in whole-body standing backward bends, but it becomes somewhat more important in relaxed standing backbends, and it becomes an overwhelmingly important issue in forward bending. The reason for the latter is obvious: the upper part of the body is tipped far off axis. With this in mind, our first concern is whether to bend forward from the waist or from the hips, and our second concern is how to use breathing to purposely further our aims.

If you watch people bending forward in daily life, you will notice that they nearly always bend from the waist. This is the more natural movement.

You would look very odd indeed if you kept your back straight and bent forward from the hips to pick up an object from a coffee table. It is also easier to bend from the waist because there is less upper body weight above the waistline than above the hip joints. In hatha yoga we use both options. Bending forward from the hips is nearly always considered more desirable, but it is also more difficult, not only because there is more weight to control but also because by definition it requires a reasonable measure of hip flexibility, and this can't be taken for granted.

Our next concern is how to support the bend. Do we support it within the torso itself, which we'll call internal support, or do we support it with the help of the upper extremities, which we'll call external support. If you brace a forward bend externally with your hands on your thighs, it's natural to relax the torso, but if you slowly bend forward while allowing your hands to hang freely, your torso has to support itself internally all the way down, and under those circumstances it will be anything but relaxed. Every standing backbend, forward bend, and side bend should be considered with respect to these matters. As in backward bending, breathing is an important related issue: the more a posture is supported internally, the more it will have to be assisted by adjustments in the way you breathe.

FORWARD BENDS FROM THE WAIST

We'll begin with two simple and easy exercises in which you can experience the difference between externally and internally supported bends. Lean forward, bend your knees slightly, and, bending forward from the waist, slide your hands down the front of the thighs, gripping all the way. Settle your hands in place just above the knees, lower your head forward, and observe your posture. Your back is slightly rounded and relaxed, which is possible because you are supporting the torso with the upper extremities rather than with the back muscles (fig. 4.21). Your breathing is also relaxed because your respiratory diaphragm, pelvic diaphragm, and abdominal muscles are not having to contribute much to the posture.

Next, carry this process one step further. Drop your hands just below the knees and grip your legs firmly. Now we'll start to see big differences among students. If you are strong and flexible you will still be relaxed, but if you are stiffer you will start to feel some pulling in your erector spinae muscles. If that pulling is uncomfortable you will not be eager to go further down. So try to find a position that is just right, one that creates a little stretch but that minimizes discomfort. Next, sense the level of relaxation in your back, in your abdomen, and in your breathing, and then slowly release your hands without dropping further forward. The moment you start to release your hands, observe carefully and you will notice that you have automatically tightened up your lower back, your abdomen, and your respiratory and pelvic diaphragms (fig. 4.22). The difference between bracing the posture with your hands and releasing your hands without dropping down is the difference between an externally and an internally supported forward bend. If you are explaining and at the same time demonstrating this exercise to a class with the hands settled in place just above or just below the knees, your voice will be at ease before you release your hands, but the second you release them and yet hold the posture, your breathing and voice will become noticeably more labored.

Now that you can feel the differences between the two kinds of bends, stand in a relaxed position and roll forward slowly from the head, neck, upper back, lower back, and last of all from the hips. The hands can simply dangle. As you gradually pitch your upper body forward, you will sense tension gathering in your back and abdomen as well as in the respiratory and pelvic diaphragms. Work with this tension rather than struggling to resist it. Notice that the deep back muscles lengthen eccentrically and control your descent, but that purposely tightening the abdominal muscles and the respiratory and pelvic diaphragms provides the all-important increased intra-abdominal pressure that makes the movement safe. Come back up in reverse order, that is, beginning with the hips, and without too much delay. Muscles throughout the torso are already in a state of stretch and tension from supporting the posture internally, and they will lift you up naturally. Be aware of concentric shortening of muscles in the back as you lift up, but at the same time focus on the abdominopelvic unit as a whole, which is bounded by the abdominal muscles, the respiratory diaphragm, and the pelvic diaphragm. Activating this region will protect the lower back by spreading the vertebrae apart hydraulically, as described in chapter 3. Being attentive to the abdominopelvic unit will give you the sense of controlling the posture rather than the posture controlling you.

Figure 4.21. Externally supported forward bend. With the posture supported by the upper extremities, the back, abdomen, lower extremities, and respiratory and pelvic diaphragms can all remain relaxed.

It is better not to explore your limits at this stage. Just repeat the down and up movement several times without a lot of concern about stretching. Finally, roll down to wherever gravity carries you and explore the feelings—stiffness and discomfort if you are not accustomed to these stretches, a deep pull and comfortable tension if you are in good condition, or more complete relaxation if you are in excellent condition. Again, come up naturally.

FORWARD BENDS FROM THE HIPS

As you improve your hip flexibility, you will soon want to accomplish the more elegant forward bend from the hips. Let's start with an internally supported bend for beginners. Stand in the mountain posture with your feet either together or 6–10 inches apart and parallel. Establish a strong base in the lower extremities, and be aware that the lumbar region of the back is convex anteriorly. You are going to try to keep that arch intact as you bend forward.

Now bend forward from the hips. Average beginners will be able to bend about 10–30°, but advanced students with excellent hip flexibility may be able to bend up to 90° or even more. As with forward bending from the waist, if you are not bracing the posture with your upper extremities, you will have to support it with your back muscles, abdominal muscles, and respiratory and pelvic diaphragms. Even more than before, pay attention to pressing in gently but purposely with the abdominal muscles. Be aware of when you reach your limit of hip flexibility, and bend from the waist from that point on. Let the arms hang or interlock the hands with the opposite forearms. Do not tug against the ankles or bounce your torso up and down.

Relax as much as you can and still maintain your posture. Even though you started with a strong foundation and kept it while you were bending from the hips, you had to relax to some extent when you started bending

Figure 4.22. Internally supported forward bend. With the upper extremities dangling, abdominal muscles, back muscles, and the respiratory and pelvic diaphragms must support the posture along with the muscles, bones, and joints of the lower extremities.

at the waist. Your kneecaps are no longer lifted, and the hamstrings are in a state of relative relaxation, although your nervous system is keeping them in a holding pattern of activity. Depending on your flexibility and conditioning, the deep back muscles may be fairly relaxed or they may be active, eccentrically lengthening as gravity slowly eases you down. If you are in excellent condition you will be relaxed as soon as you are settled; everyone else will still be resisting gravity with the back, hamstring muscles, and the triumvirate of abdominal muscles, pelvic diaphragm, and respiratory diaphragm.

Stay in this posture, breathing evenly. The lumbar area is flexed forward as part of an arc of tension extending from the upper back to the heels. The tensions shift as you breathe. Since the crus of the diaphragm takes origin from the relatively stable lumbar vertebrae, and since the contents of the abdominal cavity are slightly compressed from the forward bend, each inhalation lifts the base of the thoracic cage, producing a slight lifting effect in the upper body and a slight increase in intra-abdominal pressure. Each exhalation then lowers you down slightly and relieves pressure from the diaphragm against the abdominal organs. Just feel that happening for about 30 seconds, and then gently roll up out of the posture while pressing in mildly with the abdominal muscles. Do not do any kind of intense maneuver such as lifting the head, then the upper back, and then straightening up from the hips; just roll up naturally.

This version of forward bending assumes that you are able to remain relatively relaxed in the posture. If you are struggling, all you'll notice is marked intra-abdominal pressure, difficulty breathing, and a tense back, abdomen, and respiratory diaphragm. The posture is for healthy beginners, and is contraindicated for anyone with acute lower back pain. If you go into this posture with pain in the lower back, you are likely to come out with more.

After you are comfortable in the beginning internally supported forward bend from the hips, you are ready for the intermediate version. Start with the feet together. This time develop a firm base and keep it, with the feet solidly on the floor, the kneecaps lifted, and the hips strong. Hold tension in the thighs, not only in the quadriceps femori but in the hamstring muscles and adductors as well. Bend from the hips as before. Now, however, when you reach your limits of hip flexibility, keep a strong base as you continue to bend at the waist. Hold tension purposely in the hips, thighs, and abdominal muscles as you bend down. Let gravity pull you down, and notice that you are aware of more subtleties of the posture, especially around the pelvis, thighs, and knees, than in the previous version.

After 15–30 seconds, experiment with assisting gravity by pulling yourself down actively toward the end of exhalation using the hip flexors (iliopsoas muscles). The abdominal muscles will now operate above and beyond a general effort for pressing in; they will assist the hip flexors synergistically in drawing you forward. Hold this pose, breathing as evenly as possible (fig. 4.23). As in the earlier exercise, each inhalation will lift you up and each exhalation will lower you down. To come out of the posture ease yourself up a little bit at the hips and in the lower back, press in purposefully with the lower abdominal muscles, and then come up in reverse order, first lifting the head, then the upper back to create a good lumbar arch, and finally extending the trunk back up from the hips. It is important to keep tension on the hamstring muscles as you come up to prepare for the extra tension on them when you raise the upper part of the back.

This is an impressive posture. It places so much tension on the muscles of the abdomen, pelvis, and thighs that you will hardly notice the accompanying increase in intra-abdominal pressure and increased tension in the deep back muscles. Nevertheless, it's all there—an experience of entering and exiting a forward bend from the hips that envelops you from head to toe.

Advanced students can take this posture one step further. Go into the forward bend exactly as in the intermediate posture, bending first from the hips and coming down with gravity, but then grasping the ankles or feet and assisting gravity with a combination of the abdominal muscles, hip flexors, and upper extremities (fig. 4.24). Once settled, this posture affects your breathing differently from the beginning and intermediate variations because now you are holding the trunk in place with the hands during inhalation. You'll sense little or no movement, only an increase in tension during exhalation and a decrease in tension during inhalation. Finally, if you wish to come out of this posture with an arched back, you should ease off the stretch slightly before lifting the head and upper back, exactly as in the intermediate version.

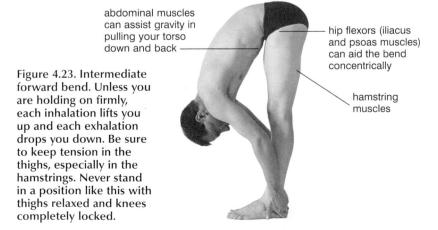

abdominal muscles can assist gravity in pulling your torso down and back

hip flexors (iliacus and psoas muscles) can aid the bend concentrically

hamstring muscles

Figure 4.23. Intermediate forward bend. Unless you are holding on firmly, each inhalation lifts you up and each exhalation drops you down. Be sure to keep tension in the thighs, especially in the hamstrings. Never stand in a position like this with thighs relaxed and knees completely locked.

Standing forward bends separate everyone roughly into two groups. Those with good hip flexibility have a gratifying experience: if their hips are flexible enough to press the chest against the thighs with the knees straight, the torso is inverted and the back is only mildly bent. The posture is rewarding and relaxing for this group of students because the full inversion of the torso, which is now hanging passively from the hips, allows the spine to stretch, much as it would if you were hanging upside down. In most people, however, short hamstring muscles and poor hip and back flexibility prevent this, and the torso arcs out from the lower extremities in a big semicircle. Teachers who have always been flexible enough to bend fully at their hips often find their students' situation incomprehensible.

MEDIAL AND LATERAL ROTATORS OF THE HIP

Up to this point we have dwelt only on the fundamentals of bending from the waist and from the hips. We'll now turn to the more subtle aspects of how shifts in foot position affect forward bending. We saw earlier that rotating the feet out (toes out, heels in) rotates the thighs laterally, and that rotating the feet in (toes in, heels out) rotates the thighs medially. What is more, rotating the thighs laterally stretches their medial rotators, and rotating the thighs medially stretches their lateral rotators. And these shifts are all important to us in forward bending. We'll soon see that both the medial and lateral rotators of the hips resist deep forward bending, at least in those who are not very flexible, so anything that stretches them even mildly at the start will limit performance of these postures.

The medial rotators of the hip joint are the gluteus medius and gluteus minimus, which lie beneath the gluteus maximus, take origin from the back of the ilium, and insert on the greater trochanter (figs. 3.8b, 3.10b,

Figure 4.24. Advanced forward bend. With the chest pressed tightly against the thighs, inhalation cannot lift you up significantly; it merely increases tension in the torso.

8.9–10, 8.12, and 8.14) of the femur, which is located laterally, posteriorly, and inferiorly to the bulk of the muscle. This architectural arrangement enables these muscles to both rotate the thigh medially and lift it out to the side for abduction. Three experiments will clarify their roles. First, stand with your heels together (for adduction of the thighs), toes out (for lateral rotation of the thighs), and knees straight (so that our experience with the medial rotators will be superimposed on stretched hamstrings). Now bend forward and notice how far you can come down and where you feel the tension. It's mostly in the hips, whose medial rotators will be rock-solid and resisting the bend from start to finish. Second, bring your feet parallel to one another, and you will immediately come further forward. Third, swing your heels out and toes in, and down you will go even more. What has happened is simple: when you rotated your thighs medially, you took tension off the gluteus medius and gluteus minimus, which were being stretched in the beginning by adduction and lateral rotation, and this permitted you to come further forward.

The main antagonists to the gluteus medius and minimus muscles are two of the lateral rotators of the thigh: the *adductor longus* and *adductor magnus* (see figs. 3.8–9 and 8.13–14 for general treatment of adductors). These muscles take origin from the inferior pubic rami and insert posteriorly enough on the back of the femur to rotate the thigh laterally as well as pull it in for adduction. To test their actions, we need to start with them in a stretched position, so stand with your feet about 3–4 feet apart for abduction of the thighs, and with your toes in and heels out for medial rotation. Then bend forward and pinpoint the site where you feel the most tension, which will be in your inner thighs. Next, swing your toes out enough to make your feet parallel, and notice that you can come further forward. Finally, swing your toes even further out to create lateral rotation of the thighs, and this will lower you down even more. Again, what has happened is straightforward—the exact counterpart to the experiments with the gluteus medius and gluteus minimus: when you take tension off the adductor longus and adductor magnus muscles by rotating the thighs laterally, it permits you to come more deeply into the forward bend.

THE ANGLE POSTURE

An excellent elementary posture, the angle pose, is a forward bend from one hip that further illustrates how foot position can affect forward bending. With the feet a comfortable distance apart, rotate the right foot 90° to the right and the left foot slightly to the right (about 30°). Keep them both firmly planted. Swivel the hips around so you are facing directly over the right foot. Grasping the right wrist behind your back with the left hand to help pull the torso around, bend backward slightly, and then bend forward first from the hips (fig. 4.25) and then at the waist (fig. 4.26). Keep the forearms flattened against the back for this variation. As you come into the posture you will be bending primarily at the right hip joint, so this is where it is most natural to place your concentration. Unless you are unusually flexible, you will notice that there will be a slight twist in the spine. Come up and bend back again moderately before swiveling around and repeating on the other side. Because you do the posture in both directions it is excellent for working with right-to-left imbalances, first and again last on the less flexible side. If your forearms are flattened against the back, this pose will be especially easy and you will be able to relax more into it. The alternative posture (figs. 6.26a–b) of lifting the arms away from the back is more demanding and will create a more intense experience.

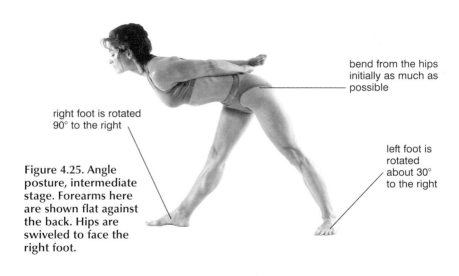

bend from the hips
initially as much as
possible

right foot is rotated
90° to the right

left foot is
rotated
about 30°
to the right

Figure 4.25. Angle
posture, intermediate
stage. Forearms here
are shown flat against
the back. Hips are
swiveled to face the
right foot.

Figure 4.26. Angle
posture completed,
with lumbar flexion
added to flexion of
the right hip joint.

Again returning to the importance of foot position, the angle posture is easiest if your rear foot faces straight out to the side, but when you rotate it somewhat medially, as suggested, or even more (up to 60°), it places increasing stretch on the lateral rotators of the thigh (the adductor longus and adductor magnus), and this starts to limit your forward bend. This is surprising: we might at first think that only the thigh which you are facing would limit a forward bend, since that's what's getting the hamstring stretch, but experimenting with different angles for the rear foot will make it plain that the rear lateral rotators can easily become the limiting elements to the bend.

The rear foot brings attention to the roles of the lateral rotators, but the front foot, the one you are facing, brings awareness to the medial rotators of the thigh (the gluteus medius and minimus) in addition to the obvious stretch in the hamstrings. The medial rotators are in a neutral position and are only moderately stretched any time the foot is pointed straight ahead, as is usual for this posture, and they will be relieved even of that tension if the foot is rotated medially 10–15°. For something really different, try rotating the front foot laterally, and you'll quickly sense the role of these muscles as medial rotators, because they instantly tighten up and limit the bend. These experiments reveal that analyzing the angle posture only with respect to the hamstrings of the front thigh is fine for an introduction to the posture, but just the beginning for anyone who is interested in serious study.

SIDE BENDING

In day-to-day life you bend forward to pick up objects, bend back moderately to stretch, and twist to look and reach. But except for cocking your head sideways, you do not often bend your spine to the side. We call this movement *lateral flexion*. It is unnatural in daily life and uncommon in hatha yoga, at least in comparison with other bends, but is usually just slipped in here and there during the course of more complex postures. Here we'll look at it in its pure form so we can recognize it within more intricate poses.

[Technical note: The idea of a "pure" side bend is an oxymoron. There is no such thing because the vertebral column, in adapting to lateral flexion, actually rotates slightly both above and below the main region of the bend rather than generating a strictly lateral movement. The same thing can be seen in a mechanical model in which metal cylinders represent vertebral bodies, and interposed short cork cylinders represent intervertebral disks. In accommodating to a bend, the combination of metal and cork cylinders (which together model the anterior functional unit of the spine) reveals mirror image rotations on either side of the middle section of the bend.]

THE STANDING SIDE BEND WITH FEET TOGETHER

For a whole-body side bend, stand with your feet together, including both the heels and big toes. Lift your hands overhead as in the whole-body backward bend, with the fingers interlocked and the palms tightly together. Lock the forearms and then pull the arms backward until they are even with the ears.

Create a firm base with the lower extremities by tensing all the thigh muscles and squeezing the hips together as tightly as you can. Then lift and stretch the upper extremities, and at the same time bend to the side in a minimal arc (fig. 4.27). You should lift so strongly with both arms and hands that when you bend to the right, the right arm does not feel weak in comparison with the left, and when you bend to the left, the left arm does not feel weak in comparison with the right. You will feel a whole-body bend from your ankles to your fingertips, with stretch throughout the side opposite to which you are bending. Don't relax anywhere; if you ease up on your effort even slightly, the emphasis changes from a whole-body bend to a side bend that is felt mostly in the lower thoracic region of the vertebral column.

Figure 4.27. Whole-body side bend, with priorities set by concentrating on lifting the hands as high as possible as in the overhead stretch (fig. 4.17) and maintaining equal strength in the two upper extremities. Assuming that both feet are planted firmly and that the lower extremities are the same length, no bending can occur at the hips. It's all in the spine.

Notice as well that the posture changes character if you place your feet even a few inches apart.

STANDING SIDE BENDS WITH FEET APART

Now we'll turn to side bends that are comparable to externally and internally supported forward bends. The movements for side bending are more limited than for forward bending, but some of the principles are similar. To illustrate: with your feet a comfortable distance apart (about two feet) and parallel, bend straight to your right while grasping your right thigh. Keep the body true to a frontal plane: bending straight to the side implies that you are keeping the hips facing the front. By contrast, if you let the right hip drop to the rear as you bend, you will shift from a relatively pure side bend to a combination of a side bend, twist, and forward bend. Watching carefully, notice that part of the bend is taking place at the hip joint on the side to which you are bending, and that the rest is taking place in the spine. As with supported forward bends, settle your right hand in a position on the thigh that allows you to remain relaxed

Figure 4.28. Side bend supported by the right upper extremity. As with externally supported forward bends, this is a comparatively relaxed posture. With each thigh abducted about 15°, only that much bending (by definition) is occurring at the hip; the rest is taking place in the spine.

(fig. 4.28). You'll not reach past your knee unless you are especially flexible. Then do the same experiments as for the forward bend—sequentially releasing and re-establishing your grip several times, noticing the effects on your abdominopelvic region in each case, and finally releasing your grip and coming up. Repeat on the other side.

Next, again starting from an upright position, make blades with your hands, palms facing the front, and bend to your right while keeping your right little finger near but not touching your right thigh. Now you will be supporting yourself internally all the way down, and because of that you will not need to make special preparations for coming up—your internal support system remains primed from start to finish. Come up and repeat on the other side.

Side bending is more complex than forward bending. For one thing side bending is asymmetric by definition, so to keep in balance you have to bend first to one side and then to the other. For another, forward bending involves flexion of both the spine and the hips, but side bending involves lateral flexion of the spine and abduction of the hip on the side toward which you are bending. Think this through: In a whole-body side bend with the feet together (in other words, with the thighs adducted), the pelvis can only remain perfectly in line with the thighs (fig. 4.27)—there is no such

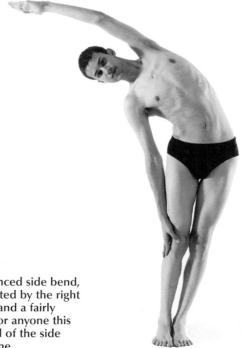

Figure 4.29. Advanced side bend, minimally supported by the right upper extremity, and a fairly relaxed posture for anyone this flexible. Nearly all of the side bend is in the spine.

thing as a side bend at the hip except in proportion to how much the thighs are abducted (fig. 4.28). Even in the case of an extreme 70° side bend in someone who is quite flexible (fig. 4.29), no appreciable bending takes place at the hips when the feet are kept together.

To come in touch with the "side bending" movement permitted at the hip joint, locate the greater trochanters (figs. 3.5–7 and 8.13–14) with your fingers and press your thumbs in the grooves just above them on each side. Then, keeping your spine straight and standing on one foot, swing your opposite foot to the side, noticing that the groove just above the greater trochanter is the site where the bend hinges. Abduction of the thigh at this joint varies from as little as 20° to as much as 90° in a full side-to-side splits position.

SIDE BEND WITH ONE KNEE ON THE FLOOR

Here is an alternative side bend that curves the entire spine to its maximum in lateral flexion (about 20° each in the lumbar and thoracic regions, and 35–45° in the cervical region), and that also produces maximum abduction at the hip joint. Although this is not strictly a standing posture, it allows you to bend further than can be accomplished standing up. Assume a kneeling position. Keeping the left knee on the floor, stretch the right foot out to the side and slide the right hand down the right thigh and leg. At the same time stretch the left arm up and over the head, with the palm down, and bend to the right (fig. 4.30). Keep your right foot flat on the floor in the

Figure 4.30. Kneeling side bend. Keep the hips in line with a frontal plane through the body, but allow the right foot to rotate to the most comfortable position.

most comfortable position (probably rotated out about 45°), but don't permit the hips to rotate out of the frontal plane. Come back up and repeat on the opposite side.

The overall sense of this posture is one of relaxation, in contrast to the standing side bend with the feet together or to externally and internally supported standing options. For this pose you maintain only enough tension to keep the body grounded in its frontal plane. In most students the posture is restricted in the lumbar region, mainly by the rib cage (in this case, the eleventh rib) butting up against the ilium. You can feel this for yourself if you insert your fingers between the rib cage and the ilium as you bend. Laterally, you can feel the tip of the eleventh rib, and just in front of that you can feel the inferior edge of the tenth rib, but the twelfth rib ends posteriorly beneath the erector spinae muscle, and you will probably not be able to locate this one unless you are slender and lightly muscled.

WHAT MAKES POSTURES DIFFICULT?

The standing postures we have discussed so far illustrate the principles that underlie standing stretches, bends, and twists, but they are not generally demanding. As we have seen so far in this chapter, the simplest and safest thing anyone can do standing is to stretch without either twisting or bending, and this is why we started with the mountain posture, the side-to-side stretch, and the overhead stretch. Then came stretching and twisting at the same time, but still without bending.

Next in difficulty are the three bending postures (backward bending, forward bending, and side bending) in which we have to be concerned with two more complications: how and to what extent we counteract gravity, and how bending affects our breathing. First we discussed backward bending, the most elementary of the three, in which the hip joints don't permit you to go back very far, in which gravity is not a major issue, and in which breathing helps you regulate your capacity for stretch. Next came forward bending, in which hip flexibility and the pull of gravity become all-important, and in which perhaps half the middle-aged students in a beginning class start asking themselves serious questions about the safety of their backs. Last came side bending, the least common movement in ordinary life and one in which the anatomical restrictions that protect you are not as well-honed and dependable as those for forward bending and backward bending.

We have focused on simple movements so far to illustrate fundamentals, but more demanding standing postures are also widely practiced. Apart from generally requiring more strength, aerobic capacity, and flexibility than the poses so far discussed, what specifically makes them challenging is that they usually involve complications—combining twisting with bending, doing twists and bends when some particular joint is in an unnatural position of stress, coming into a difficult internally supported stretch in which strength becomes a primary issue, and going from the final position of one difficult posture directly into another difficult posture. The triangle postures that follow illustrate all of these principles.

THE TRIANGLE POSTURES

Imagine two sticks, their bottom ends planted into the ground about three feet apart and their upper ends inserted loosely into sockets on either side of a ball, sockets which will allow the ball to rotate and swivel from side to side. Then imagine a flexible rod fixed to the top of the ball that can twist and bend to the front, back, or side in all possible combinations. Last, imagine a bamboo pole that runs perpendicularly across the back of the rod near its upper end, forming a cross. The sticks are the lower extremities, and the ball with which they articulate is the pelvis. The sockets that permit swiveling on either side of the ball are the sockets of the hip joints. The flexible rod is the spine, with the head on top and the pelvis on the bottom, and the pole is a combination of the two upper extremities that will remain lifted straight out to each side in one line in the triangle postures. In the advanced classic triangle, the rod-and-ball combination is bent straight to the side so that one end of the bamboo pole touches the bottom end of the stick on the same side. In the revolving triangle, the rod-and-ball combination is bent forward and then twisted 180°, so that the end of the pole touches the bottom end of the opposite stick.

THE PRELIMINARY STANCE

In all of the triangle postures you will abduct the thighs, rotate the feet asymmetrically, and then complete the specific pose. And because the triangle postures place unusual asymmetric tensions on the hip joints and on the muscles of the thighs, we'll look first at how we create internal resistance to simple standing twists with the thighs abducted. Start with the feet parallel and about two feet apart. As usual, tighten the thigh muscles all around. Then twist. This should be easy; nearly everyone can create a comfortable resistance to the twist if the feet are this close together. Now increase the distance to three feet, and a few people will start to get mildly uncomfortable in their hips and lower back. Finally, if we increase the distance to four feet, even good athletes may not find it reasonable to remain twisted for more than half a minute or so. The idea is to settle on your own personal stance, one that will allow you to keep the lower extremities comfortable and yet firm while twisting right and then left for a half minute each.

Next, with a stance established that allows you to keep the thigh muscles firm, turn the right foot fully to the right and the left foot about 30° to the right, that is, lacking 60° of being in line with the right foot.

Assume this position, however, without swiveling the pelvis and permitting the right hip to move markedly to the rear. The right thigh will be rotated laterally and the left thigh will be rotated medially. The medial rotators (the gluteal medius and minimus) of the right hip will now be in a state of stretch and resisting lateral rotation. If you are not convinced of this, allow the right hip to rotate back medially by swinging your right toes in, and you will notice that this instantly relieves tension in the right gluteus medius and minimus. On the left side, the lateral rotators—the adductor longus and adductor magnus—will be in a state of stretch and poised for resisting more medial rotation. And if you are not convinced of that, just allow the left hip to rotate back laterally to a more neutral position and you will notice that this immediately relieves tension in the left inner thigh. Instructors, of course, never describe the preliminary stance in this much detail. They usually just say "try to keep your hips in line with your chest and facing the front."

Notice that you can't hold the preliminary stance properly unless the left gluteal muscles remain tensed. If you are flexible this tension will not be too noticeable, but if you are stiff and relatively inflexible, you will need to contract the gluteals strongly or the posture deteriorates. You should also firm up the quadriceps femoris muscles, first on the right and then on the left, which lifts the kneecaps. Then tighten the hamstring muscles, again first on the right and then on the left, to make sure that the knee joints do not come into an undesirable state of hyperextension. Don't tighten the hamstring muscles enough to flex the knees, but take the effort up to that point. The only thing remaining is to check the placement of the toes, making sure you feel all of them solidly against the floor.

Finally, maintaining your stance, raise the arms and forearms until they and the shoulders are in a single line from right fingertips to left fingertips. The shoulders, arms, chest, and hips should all be facing the front. Don't bend; just stretch the fingertips laterally. Repeat everything on the other side. This is the basic preliminary stance. We'll now examine several different ways to complete the postures.

THE INTERNALLY SUPPORTED TRIANGLE FOR BEGINNERS

Stand with the feet parallel and 2–3 feet apart. Turn the right foot to the right 90°, and for this particular variation of the triangle keep the left foot facing straight to the front. Establish a solid base. Stretch the hands out to the sides and hold them at shoulder height with the palms facing forward. Bring the fingers together to make blades of the hands. To prepare for flexion of the trunk to the right, stretch out, reaching toward the right side and at the same time creating a feeling of lift in the torso. This will automatically create a priority for emphasizing the existing bend at the right hip in preference to the spine, and that is what we want.

Slowly flex the trunk to the right, being careful not to allow the right hip to swing to the rear any more than you have to. Letting that happen will enable you to bend more, but only at the expense of changing the side bend to a mixed side-and-forward bend. This is hard to avoid: just realize that any additional bending that you experience, at least over and above the initial sidebend that is defined by abduction of the right thigh, is in reality a forward bend. To complete this stage of the posture, bring the right hand toward the right thigh, knee, leg, ankle, or foot—whatever you can reach easily—keeping the palm facing forward and not making contact with the little finger (fig. 4.31). Move slowly; this will be an internally supported posture all the way down and up. Point the left hand straight up and twist your head to look up at it, but go only as far into the posture as is reasonable. The idea is to construct a personal posture that explores your comfortable limits while you are breathing evenly without pauses or jerks. If you are straining or feel you need to hold your breath, you have gone too far. Slowly come up and repeat on the other side.

Figure 4.31. Internally supported triangle for beginners. Don't drop down so far in this posture that it is no longer primarily a side bend. The pelvis should remain in line with a frontal plane through the body.

There are many things to notice about this posture. As in the preliminary stance, to keep from swiveling the hips while flexing to the right, beginning students have to keep the right quadriceps femoris muscle and the left gluteal muscles strongly engaged. Test this by allowing the hips to rotate slightly into more neutral positions, and then tighten these muscles, especially the left gluteals, to make the correction.

You initiate the move to the right with muscular activity on the right side of the body, and this stretches the skin, fascia, and muscles on the left. Then, as gravity carries you further to the right, the left side of the body resists that force and keeps you from tipping to the right too suddenly. To feel this, it is important to come into the posture supported internally by the abdominal muscles and the pelvic and respiratory diaphragms, and without resting your right hand on the right lower extremity. (If you do that, the left side of the torso will no longer need to restrain the movement to the right, and you will be doing the next posture—the externally supported triangle for beginners.)

The muscles and fasciae that are stretched on the side opposite the bend define your progress and stability in the internally supported triangle. If you are bending to the right, the left abdominal muscles, deep back muscles, and latissimus dorsi all lengthen eccentrically, keeping the beginner from lowering too fast into an unfamiliar position. With more experience these muscles yield to gravity, and the emphasis is now felt in the stretch of their connective tissues. The left adductor longus and magnus muscles are not stretched in this posture because the left foot is pointing straight to the front and the left thigh is not rotated medially. This is easily proven. If you come up momentarily and rotate the toes of the left foot in, and then return to the posture, you'll feel a substantial increase in tension in those muscles.

In the internally supported triangle the diaphragm tends to lift the trunk during inhalation and lower it down during exhalation, and hence you can drop further into the posture at the end of each exhalation. Then, during each ensuing inhalation you will feel additional tension from gravity, which will be resisting the diaphragm's tendency to lift you up. These effects will not be obvious unless you refrain from inhaling for a moment or two at the end of exhalation.

One of the lessons this posture teaches is the same one we explored with internally supported forward bending and side bending: you do not have to make any preparations to come out of it. Facing the palms forward and not touching the lower extremity forces you to evaluate your capacity constantly as you are flexing to the side. And since you are supporting yourself internally the whole time, you can come out of the posture without readjusting muscular tension, intra-abdominal pressure, or breathing.

Beginners will probably be consciously resisting their flexion reflexes on the side of the body that is being stretched, but experts move smoothly and unerringly into the finished posture with their flexion reflexes remaining well in the background. The muscles being stretched resist gravity without stress or strain, the connective tissues of the body limit the posture to a safe, predictable, and comfortable stance, and the action of the diaphragm works on those tissues when they are in a state of balance and equilibrium.

THE EXTERNALLY SUPPORTED TRIANGLE FOR BEGINNERS

To do an externally supported beginner's triangle, keep the same stance, with the right foot still 90° to the right and the left foot straight forward. Stretch the arms and forearms out, face the palms down, and lower the right hand to the right thigh. Grip it firmly. Then bend, sliding the right hand down the thigh and leg, but gripping and re-gripping all the way. Raise the left hand and look up at it. It is easy to come into this posture because you are supporting the weight of the torso with the right upper extremity, and the natural reaction of the torso under these circumstances is to remain relaxed (fig. 4.32). As a result you are more confident in the

Figure 4.32. Externally supported triangle for beginners. The right hip will usually have to move slightly to the rear, but adjust the hand position high enough on the leg to keep the hips facing the front as much as possible.

posture and can move more deeply into it, increasing the commitment to side bending at the right hip joint. It should be mentioned that many modern schools of yoga recommend resting the lowermost hand on one or more blocks of wood, which accomplishes the same end.

One important lesson of this posture, like that of externally supported forward and side bends, is learned coming out of it. Because you are supporting the posture all the way down with the arm, forearm, and hand, when you decide to come out of it you may be in trouble, like the cat that is able to climb a tree but unable to come back down. To come up you'll have to release your hand, and when you do that you will have to support the weight of the upper half of your body using only the internal muscles of the torso, which may not be prepared for the effort. The solution is twofold: first, don't go down very far until experience tells you that you can come back up gracefully; second, before you even attempt to come up, prepare yourself by tightening the muscles of the legs, thighs, and hips. This effort will naturally recruit more muscles in the pelvis and torso, and you will soon feel enough confidence to release your hand and come out of the posture.

THE CLASSIC TRIANGLE

As defined here, the classic, or standard, triangle is more of a balancing posture than the previous two because the left foot is turned 30° to the right (as in the preliminary stance) rather than pointing straight ahead. Stretch the arms out at shoulder height and face the palms down instead of to the front. Keep a sense of firmness in the lower extremities and a sense of stretch in the upper. Reach to the right and then bend, keeping a lifting sensation in the upper half of the body. As you go into the posture, twist your head around so that you are looking up at the outstretched thumb, and contact the floor (fig. 4.33), foot, or ankle with the right hand.

Keep a strong stance. Be aware that the gluteus medius and minimus remain under tension on the right side because the lateral rotation of the right thigh stretches the gluteals on that side. Tension should also be held willfully in the left inner thigh, the quadriceps femori on both sides, and the hamstrings, especially on the right side. The focus of the posture should be on the thighs, the hip joints, and the vertebral column. Just to experience what not to do in this regard, relax as much as you can without falling down and notice that this alters the dynamic of the posture completely—the hips will tend to swivel, the knee to which your bend is directed may become hyperextended and stressed, and the focus shifts away from the foundation of the body.

As you bend, keep the arms and forearms stretched out to form a single line, with the forearm of the upraised limb supinated and the fingers and thumb together in the same plane. If you cannot go very far, that's fine, but try to find some position in which you can be relatively comfortable for 5–10 seconds. Remember, in the beginning you are not going to try to accomplish any semblance of a finished posture. Your primary objective is to learn to move into and out of the pose without hurting yourself and without being in a state of pain or anxiety. If you bend so far that you are not at peace, you have gone too far.

Come out of the posture as you came into it, but make sure your stance has not deteriorated before you do so. If the pose has become slovenly, the movement back to the upright position presents new hazards, and a different pattern of sensations and nerve impulses will emerge from the muscles and joints of the lower extremities. If you have allowed the arm to act as a strut to support the pose you have two choices: either relax and ease yourself back up as gently as possible, or prepare to come out of the posture more confidently by first firming up the muscles of your extremities and torso.

Intermediate students can work with the classic triangle with the feet wider apart, so that the lower extremities approach a 90° angle from one another. But if the feet are so wide apart that you cannot maintain good control of the muscles of the thighs, back off. The triangle is the worst posture in the world for getting compulsive about increasing abduction of the hip joints: with the left adductors stressed from medial rotation, increased abduction, and gravity, it's too easy to pull a muscle in the groin.

Figure 4.33. Classic triangle. Bringing the right hand all the way to the floor will necessitate allowing the right hip to drop even further to the rear than in the case of the externally supported triangle for beginners, but at least try to keep in the spirit of a side bend rather than letting the hips swivel freely.

As you become stronger and more comfortable in the posture, you can intensify your efforts to bend, still being careful not to allow your base to deteriorate. By now you should have learned to use the postural muscles of the torso to control moving into the posture, using the hand as a guide, not as a stick for resting. An intermediate student may stay in the posture for 30–60 seconds before repeating on the other side.

In this posture the effects of breathing will be a little different from what we saw in the internally supported triangle for beginners. Since you will be maintaining your hand position, inhalation will no longer lift you up and exhalation will no longer drop you down. Nevertheless, as you bend toward your limit you will notice that each inhalation brings an increase in tension in the torso. Each exhalation then brings a release of tension, which gives you the option of slipping your hand down a little further. Gravity draws your head and chest downward and is sufficient to hold the upper half of the body in the lower position during the next inhalation. The tissues of the body gradually adjust to the increased tension and come into a new steady state. The complex unilateral dynamics of the classic triangle, as well as the manner in which this posture is affected by gravity and respiration, are what make it a more advanced posture than the internally supported triangle for beginners, in which there is no chance to get into a position from which you cannot gracefully rise.

After some practice it will be possible to come into the pose without compromising the dynamic aspects of maintaining it, and yet find 30–60 seconds of quiet and silence in the posture. An intermediate student should now take a little time to simply enjoy. You can come into a provisional stance, thinking mostly about maintaining your base. Then you can use breathing to ease your body more completely into the posture, observe it one-pointedly, and then slowly come up.

The manner in which you come out of the posture is important. If you have established the habit of maintaining your base even while you are in repose, you can just keep everything solid and come out of the posture as you came in. But if you have not maintained your base, coming out of the pose should be approached with care, especially if you have pulled yourself more fully into it with deep exhalations. If you have inadvertently relaxed the thigh and pelvic musculature, it may be hazardous to suddenly tighten everything up before rising because you will be stressing some of the deepest and least accessible muscles of the body when it is in an awkward position. So if you have relaxed those critical muscles, it will be better to simply rise out of the posture as gracefully as you can, or at the least, back off first into a position in which you are more comfortable, re-establish a firmer base, and then come out dynamically. Alternatively, if you are feeling too stressed, another possibility for exiting the posture safely is to bend the outstretched knee before coming up.

The feel of the advanced student's triangle is yet more refined from that of the intermediate student's triangle. It is the same posture, but you are now flexible and strong enough to go gently into it without much effort, placing your palm on the floor and looking directly up at the outstretched thumb. Because your thighs are more fully abducted, a greater proportion of the bend takes place in the hip joint. Now, as you breathe, your adjustments are almost all internal, and outside observers would not be mistaken if they thought that you looked comfortable. Each inhalation still brings a slight increase, and each exhalation a slight decrease in internal tension, but little of this results in changes in the position of the trunk or extremities.

THE REVOLVING TRIANGLE

The revolving triangle differs from the classic triangle in so many ways we could argue that it is completely unrelated. Ideally, the classic triangle is a modified side bend combined with hip abduction and torque-like stresses which act to minimize forward flexion and twisting. By contrast, the revolving triangle is an extremely complex posture which involves forward bending with the lower extremities in a partial splits position plus an additional twist of the trunk to the rear.

The first difference between the two triangles is in the placement of the feet. If you start with a bend to the right side (feet 2–3 feet apart for the beginner), the right foot is turned fully to the right (in this case laterally), and the left foot is turned to the right (in this case medially) about 60°, thus lacking only 30° of being parallel with the right foot. For beginners the 60° rotation of the left foot makes this posture even more wobbly than the classic triangle, in which the foot is rotated only 30°. And more than that, the extreme medial rotation of the left foot will increase tension in the adductors of the left thigh (recall: these are lateral rotators) even beyond that found in the classic triangle.

The next step is to twist to the right so that you are facing over your right thigh, much as you did for the angle posture. Everyone says "twist," but for those with excellent mobility in their hips this is not so much of a twist as it is a swivel, one in which the pelvis rotates 90° (fig. 4.34), causing the thighs to move from a position of abduction to one in which the right thigh is flexed with respect to the torso and the left thigh is hyperextended. In this sense the revolving triangle is diametrically opposed to the classic triangle, in which you maintain the abducted hips more or less in line with the frontal plane of the trunk.

For the next stage of the revolving triangle, draw an imaginary line through the pelvis from side to side, that is, perpendicular to the body's midsagittal plane, and slowly bend forward around that axis. Now we begin to see fundamental differences between beginning and advanced students. For advanced students it is easy to bend forward around that ideal axis (fig. 4.35), but those who are less flexible can come into the forward bend only by twisting their spines at the same time. Be watchful: either forward bending or twisting may be safe by themselves, but combining the two with your right thigh already flexed nearly to its maximum may not be within your reasonable capacity.

If you feel confident in continuing, you can move into the revolving triangle in one of two ways. In the first, bend forward from the hips with the arms stretched out to the sides. Proceed to your limit of forward bending, and as soon as that is reached, initiate a spinal twist so that you are facing the rear. Bring the left hand to the right leg or ankle, and swing the right hand upward in the same line with the left. Look up at the outstretched hand (fig. 4.36). The second and more common way of coming into the posture is to simply take the left hand to the right ankle in the most direct course possible, thus combining a forward bend with a spinal twist in one movement. This is more customary and natural, but the piecemeal method is more useful for precise exploration and analysis. After coming out of the posture in reverse order, repeat on the other side.

Flexible and not-so-flexible students will have completely different experiences working toward this final stage of the revolving triangle. Those

Figure 4.34. Revolving triangle, first position. This posture is identical to the beginning position for the angle posture except that the left foot is rotated about 60° to the right instead of 30°.

who are especially flexible can twist enough so that their chest will be facing to the rear and their upper extremities will lie in one line, just like the bamboo pole in our introductory description of the triangles. But those who are stiff will not be able to go nearly that far. First, it will be impossible for them to bend forward all the way down to the right thigh; second, their spines will already be twisted maximally just to come moderately forward, so they will not have much chance of furthering the twist in order to face their chest to the rear; and third, because their chest is still more or less facing the floor, they will have no chance of bringing their upper extremities into one line when they reach the left hand to the right foot and extend the right hand up in the air.

When we bend forward in the revolving triangle, almost every right-left member of paired muscles and ligaments on the two sides of the thigh and hip is stressed differently. As in any forward bend, the hamstrings are resisting stretch with a constant influx of nerve impulses, but much more so on the right side than the left. Likewise, the rectus femoris and iliopsoas muscles will be more active on the right than on the left, especially if you use them to pull yourself forward actively. The adductor longus and magnus will be placed under extreme tension on the left but not so much on the right.

The stretches for the calf muscles on the two sides are also reversed. With the left foot pulled around to a 60° angle and the left ankle flexed accordingly, the *gastrocnemius* and *soleus* muscles (the muscles that make up the bulk of the calf; figs. 3.10a–b, 7.6, 8.9–10, and 8.12) on the left side are stretched maximally, but on the right side, in which the ankle is extended, the calf muscles are not stretched in the least. Slight changes in the angle of the left foot alter the stretch on the calf muscles markedly; if the left foot is only at a 45° angle, the stretch on the left calf muscles is greatly moderated.

Figure 4.35. Revolving triangle, intermediate position. Again, this stage of the revolving triangle is similar to the angle pose except for reaching forward with the upper extremities.

As you bend forward, the ligaments of the hip joint are also affected differently on each side. As discussed in chapter 3, the ischiofemoral, iliofemoral, and pubofemoral ligaments form a taut spiral when the thigh is extended; when it is flexed, the spiral unwinds. So during the course of swiveling into the revolving triangle facing the right, which leaves the right thigh flexed and the left thigh extended, the spiral of these three ligaments is loosened on the right and tightened on the left. The resulting slack in the right hip joint is not cause for alarm—indeed, it would not be possible to come fully into the revolving triangle without the release of these ligaments—but it does caution us to watch the flexed hip joint carefully because that is the site where injuries are most likely to occur.

The all-important issue of hip flexibility is the main reason this posture is difficult. Beginning students who have gone to their limits in twisting just to get into a forward bend position over their front thigh can't be expected to twist an additional 90° to swing their chest, shoulders, and arms into one line. This is not a posture for them to stay in for more than a few seconds. The pose is so complicated that just attempting to approximate it requires full concentration.

Intermediate students will be more comfortable. Their feet can and should be further apart. Experience will allow them to maintain strength

Figure 4.36. Revolving triangle. The additional twist of the torso needed to complete the final position makes this a challenging pose for every-one who is not strong and flexible.

in the hips and thighs, and the posture will feel energizing instead of frustrating. This will start to happen when it becomes possible to hold the upper extremities in line with the shoulders and with one another. As the lower hand reaches further down and the upper hand reaches further up, intermediate students are empowered to work harder and with more directed concentration—reaching, stretching, breathing evenly, and exploring the postural base. Finally, advanced students are able to find repose in this posture without struggle or stress.

POSTURAL SEQUENCES AND TRIANGLES

Coming into a single posture, supporting it internally, and returning the same way you went in is the safest way to explore standing postures, but many instructors teach sequences of triangles and related poses. These are fine for intermediate and advanced students who have had a lot of experience, but they present particular dangers to beginners. Going from one strenuous position directly to another will require you to revise your musculoskeletal priorities at the same time you are being challenged, sometimes maximally, and this may require too much simultaneous feeling, analysis, and decision-making for novices.

We'll consider two examples, one easy and one hard. If you start in the classic triangle, you can confidently come into the forward bend position and work yourself into the revolving triangle. All you have to do is release the torque that is holding your hips facing the front, allow them to swivel around, and twist your torso to the rear, letting your arms follow naturally. It's easy. Coming from the revolving back to the classic triangle is another matter. You can come out of the twisted position of the revolving triangle into a forward bend without difficulty, but pulling your hips back into a straight position from the swiveled position, and at the same time reversing your arm positions, will create extraordinary strains in the hips and spine—strains for which the beginner is ill-prepared. Beginning students should try this kind of exercise only by moving back and forth between partial positions. If you go only halfway into a revolving triangle, and then go back halfway into the regular triangle, you will gradually develop the skill and confidence to do more demanding sequences.

TWO BALANCING POSTURES

All standing postures are balancing postures—it's just a matter of how much emphasis is placed on this property. But usually when we think of balancing postures in a standing position, we think of standing on one foot. Two such postures are the eagle and the tree.

THE EAGLE

In the eagle posture you stand on one foot as a sentinel, with one thigh, leg, and foot intertwined around the other. It's easy for those who are strong, slender, and flexible, but most students find it difficult to wrap their extremities around one another even when they are not balancing on one foot. To give it a try, stand first on your left foot, bend your left knee and hip slightly, and if you are a man, place your genitals either forward or to the rear. Then swing your right thigh forward and pull it tightly around your left thigh. Last, wrap your right ankle tightly behind your left leg and interlock your right foot even further around to the medial side of your left leg. Sometimes this is referred to as "double-locking" the legs. To complete the beginner's posture, swing your right elbow across to the near side of your left elbow, pull the forearms together, and interlock your wrists and hands (fig. 4.37). Fix your gaze, breathe evenly, and hold as long as is comfortable. Come out of the posture and repeat on the other side.

This is as much as most people will want to do. But to continue into a more advanced posture, bend your left knee and hip joints as much as possible consistent with keeping your back straight and upright (fig. 4.38). You miss the point of the posture if you bend your back and head forward.

Figure 4.37. Eagle posture. Interlocking the upper and lower extremities are the main challenges of this balancing pose.

This version of the eagle should be approached with caution. If you hold it for more than a few seconds, it may create cardiovascular effects that can cause you to faint when you come out of the posture, especially if you do it after other strenuous postures or exercises. Develop your capacity to lower your weight and increase your time in the posture over a long period of time.

If you consistently have trouble with balancing in the eagle, try the posture only at the end of leisurely hatha yoga sessions. Students who have difficulty at the start of a class may be able to do the posture easily at the end. And the eagle will also be more difficult if you have just had a strenuous musculoskeletal workout which has left you in a momentarily weakened condition.

THE TREE

After the complexity and contrived nature of the triangles and the eagle, we'll end this chapter with the most popular of all balancing postures—the tree—which is simply standing quietly on one foot. This pose speaks volumes not only for your state of physical balance, but also for your emotional and mental balance. Standing quietly on one foot is not as easy as it sounds.

Figure 4.38. Advanced eagle. Until you are certain of yourself, be wary of cardiovascular responses as you come out of this posture, especially if you remain in it more than 15–30 seconds.

Begin by standing on both feet, breathing evenly. As soon as you are calm and centered, lift one foot and place it on the opposite extremity. You can use one of several foot positions. In the beginning it is best to place one foot on the opposite ankle, but after a little experience, lifting the foot to the inner thigh (fig. 4.39) or into the half-bound lotus position (fig. 4.40) is more stable. At first the hands can be in the prayer position in the center of the sternum (fig. 4.39), but as you develop confidence they can be raised overhead, with the palms together and the fingers pointing away from the body. You can keep your eyes open and focused on a spot on the floor about six feet in front of you, or you can focus on some point in the distance. The room should be well lighted—for a real challenge, go to the opposite extreme and close your eyes.

You have to keep the knee extended in the tree posture, and you automatically create a solid pelvic base and root lock. And since both the hip and knee are extended, only the ankle is in a state of uncertainty. After you are able to stand quietly and confidently, you can start to examine how the muscles of your supporting toes, foot, and ankle balance the posture. You'll notice that as your weight shifts you compensate with muscular effort to avoid losing your balance; with experience the shifts in position become less obvious.

Figure 4.39. Tree posture. A plumb line of gravity in this balancing posture has an axial center in a slightly tilted off-sagittal plane that runs through the navel.

Yogis tell us that the tree posture is both grounding and centering, and that it will generate a sense of deep calm and endless patience. The lower extremity is like the trunk of a tree; the arms overhead are like its branches; and if you stand in the posture for ten minutes or so daily, you will feel as if your toes are rooted in the earth.

BENEFITS

A few hundred years ago, when everyone in agrarian societies walked and worked for hours at a time on their farms, there was little need to begin or end a day with standing yoga postures. But today most of us have sedentary jobs and are sorely in need of more exercise than we can get sitting at a desk. Standing postures, coupled with a moderate amount of aerobic exercise such as walking, running, or swimming, can fill the gap.

Figure 4.40. Tree posture, in a half-bound lotus. A plumb line of gravity comes closer to being in the mid-sagittal plane of the body than in the case of the previous posture because the right knee is not thrust out so far to the side.

Standing postures develop overall strength and flexibility, mildly stimulate the cardiovascular and respiratory systems, and accustom the nervous system to a range of body positions that are otherwise ignored. They flood the nervous system with information from all over the body, and they integrate the energy of the body from top to bottom and from inside out. They integrate the limbs with the torso, and they bring awareness to the respiratory and pelvic diaphragms as well as to the deep muscles of the abdomen, pelvis, and back. You can satisfy yourself of their special value by practicing them in the early morning for a few weeks, and then skipping them for a week, substituting sitting postures, lying-down postures, and inverted postures. When you reinstate them into your routine you'll know what you were missing.

"Our distant ancestors departed from this time-honored mode of locomotion (quadrupedal) and converted themselves into walking towers, with a high centre of gravity and a narrow base . . . . Nature should still be displaying the sign: 'Reconstruction in Progress, The Management regrets any inconvenience that may be caused.' "

— Elaine Morgan, in *The Scars of Evolution*, pp. 26–27.

CHAPTER FIVE
BACKBENDING POSTURES

"Disease, Debility, Doubt, Inadvertence, Sloth, Sensuality, Wrong Understanding, Non-attainment of the plane and Instability, these mental distractions are the impediments— Pain, dejection, unsteadiness of limbs, inspiration and expiration are the companions of the distractions— For their Prevention, the practice of one truth— The transparency of the mind comes from the development of friendship, compassion, joy, and neutrality regarding the spheres of pleasure, pain, virtue, and vice respectively—"

— Bengali Baba, in *The Yogasutra of Patanjali*, pp. 15–17.

The next three chapters bring us to the heart of hatha yoga—to postures that involve backbending, forward bending, and twisting. Of these three, backbending is the logical place to begin our discussion because it is relatively simple. But the two categories of backward bending and forward bending postures form a pair: the muscles that resist the bend in one category are the same muscles that pull us into the bend in the other category, and we need to see and understand them in reference to one another. To keep the comparisons in perspective this chapter will be about 90% backbending and 10% forward bending, and the next chapter will be about 90% forward bending and 10% backbending.

The plan here will be to first sum up the possibilities for forward and backward bending in the standing position, concentrating on limitations in the hip joints and lower back, and then to build on our discussion of the vertebral column by examining the spinal limitations to backbending in more detail. Next, we'll look at the relationships between breathing and backbending, and finally we'll turn to the myriad forms of backward bending in hatha yoga, beginning with the famous prone backbending postures— the cobra, locust, boat, and bow—and continuing with more specialized postures such as the fish, the wheel, and the camel. Two more backbending postures, the arch and the bridge, are traditionally part of the shoulder-stand series and will be deferred to chapter 9.

Some of the postures discussed in this chapter are only for those who are in excellent musculoskeletal health. A few guidelines are given in the course of the descriptions, but if in doubt about whether or not to proceed, take note of the specific contraindications at the end of the chapter.

THE ANATOMY OF FLEXION AND EXTENSION

To understand any function, envision being without it. For example, we can see at a glance how vertebral bending, both forward and backward, contributes to whole-body bending by examining how someone would bend if their spine were fused from the pelvis to the cranium. This is not an academic hypothesis. One who has had such surgery for severe osteoarthritis will bend forward only at the hip joints, just like a hinged stick—arms dangling, head, neck, and torso stuck out straight and stiff as a board. And yet this person may be comfortable and relaxed enough to practically take a nap. He may be able to bend forward up to 90° and hyperextend to the rear about 10°—entirely from the hips.

THE HIPS AND BACK IN COMBINATION

Hip flexibility in isolation is only of theoretical interest to us, at least for the moment, but the question of how hips and backs operate together for backward bending is eminently practical. Beginning with extremes, occasional circus performers—always women in images I've located—are able to extend their spines backward 180°, plastering their hips squarely against their upper backs. Images of 180° backbending can be seen in fig. 7.3 of Alter's *Science of Flexibility*, as well as in a beautiful sequence of video frames 7–8 minutes into the tape of Cirque du Soleil's *Nouvelle Experience*.

Maximum hip hyperextension appears to be about 45°, which is seen in occasional women who can drop down into the wheel posture and

Figure 5.1. 120° flexion at the hips is a common maximum, and is enough for laying the chest down easily against the thighs when accompanied by moderate flexion of the spine.

then scamper around on their hands and feet looking like daddy-long-leg spiders in a hurry. Both extremes—180° of back extension and 45° of hip hyperextension—are anomalous, and it is not advisable for anyone to attempt extending either the hips or the spine this much unless one's profession requires it. Even highly flexible dancers, gymnasts, and hatha yogis rarely try to bend backward more than 90°, ordinarily combining 20° of extension at the hips with an additional 70° of extension in the lumbar region. In this case the right angle between their thighs and their chests is more than enough to permit them to touch their feet to their heads in advanced hatha yoga postures (fig. 5.12).

Unlike the outermost limits for backbending, the outside limits for forward bending are all within a normal range for anyone with excellent general flexibility. Dancers, gymnasts, and hatha yogis, including both men and women, can often bend forward up to 120° at the hips with the lumbar lordosis arched and the knees straight (fig. 5.1). These same people may also be able to bend an additional 90° in the lumbar region, making a total of 210°. Since 180° is all that is required to lay the torso down against the thighs in a sitting forward bend (fig. 6.12), their full capacity for forward bending can't be tested except by measuring hip and spinal flexibility separately.

Figure 5.2. A relaxed standing backbend with straight knees includes about 20° of lumbar bending and 10° of hip hyperextension for a total of 30°of "backbending."

Next, let's consider the limits of backward and forward bending in some-one more typical, say a thirty-year-old male beginning hatha yoga student who has always been athletic and in good musculoskeletal health, but who has never shown an interest in any kind of stretching. He won't be able to bend backward very far, and what he can do is hard to appraise because he will invariably bend his knees and exaggerate his lumbar lordosis in preference to hyperextending the hips. Without a sharp eye it is difficult to differentiate among these three components.

To assess this young man's capacity for hip versus spinal backward bending as best we can, first have him warm up with an hour of vigorous hatha so we can see him at his best. Then, to make sure his knees do not contribute to the bend, ask him to do a relaxed standing backbend with his knees extended. His chest will probably be off vertical by only 30°, which suggests a combination of about 20° of backbending in the lumbar region with about 10° of hyperextension in the hips (fig. 5.2).

Forward bending in this same student is easier to evaluate. You can ask him to bend forward from the hips while keeping his lower back maximally arched. Then, just before his lumbar lordosis begins to flatten you can estimate the angle between his torso and thighs, which is likely to be about 30°. This represents forward bending at the hips (fig. 5.3). Then ask him to relax

Figure 5.3. A moderately flexible young athlete can typically bend only about 30° at the hips while keeping a sharply defined lumbar lordosis (simulation).

down and forward to his capacity. If he can bend forward a total of 90°, which allows him to reach down a little more than halfway between his knees and his ankles, it suggests that has achieved the additional 60° of forward bending in his lumbar spine (fig. 5.4).

CERVICAL, THORACIC, AND LUMBAR FLEXIBILITY

The spine's flexibility varies from region to region. Starting with the neck, the cervical region is especially mobile. If you have normal flexibility, your head can extend backward on the first cervical vertebra about 20°, and you can flex forward at this same site about 10°. The rest of the cervical region can bend backward 60° and forward another 80°, in this case touching the chin to the sternum. We'll look at these movements in detail, along with rotation, in chapter 7. Just below the seven cervical vertebrae, the thoracic region permits little forward or backward movement because the rib cage is too rigid. This means that most of the backbending in the torso takes place in the lumbar region between the bottom of the rib cage and the sacrum, that is, between T12 and S1. That's where we'll concentrate our attention.

Young people with good flexibility who have been practicing hatha yoga regularly (Table 5.1) might reveal a total of 60° of lumbar flexion for forward bending and 45° of lumbar extension for backbending. This is in addition to hip flexibility, which we'll say is 90° for forward flexion and 15° for backward bending. The knees, of course, must be kept straight to get accurate measurements. Someone with hip flexibility this good will be able to do an intermediate level posterior stretch (fig. 6.15) and will be able to arch back comfortably to touch their feet in the camel (fig. 5.35). If we break this down we can estimate the approximate mobility for flexion and extension that would be permitted between individual pairs of adjacent vertebrae (table 5.1).

Figure 5.4. Limited hip flexibility prevents this young man from reaching further down. This simulation reveals a combination of about 30° of hip flexion with 60° of spinal flexion for a total of 90° of forward bending. Such a student can and should bend the knees slightly to find a more rewarding pose, one that at least allows him to grasp his ankles.

WHAT LIMITS BACKWARD BENDING?

In chapters 3 and 4 we discussed the muscles and ligaments that limit backward bending in the hips. These include the psoas and iliacus muscles (figs. 2.8, 3.7, and 8.13); the quadriceps femoris muscles (figs. 1.2, 3.9, 8.8, and 8.11), especially the rectus femoris component (figs. 3.9 and 8.8–9); the abdominal muscles (figs. 2.7, 2.9, 3.11–13, 8.8, 8.11, and 8.13), especially the rectus abdominis (figs. 3.11–13 and 8.11); and the spiraled ischiofemoral, iliofemoral, and pubofemoral ligaments (fig. 3.6).

Turning to the torso, the main structural limitations to backbending in the thoracic region are the rib cage (figs. 4.3–4) and the spinous processes (figs. 4.6b and 4.7b), which extend so far inferiorly in the thoracic region that they quickly butt up against one another during extension. And in the critical lumbar region the first line of resistance to backbending is muscular—intra-abdominal pressure generated by a combination of the respiratory diaphragm (figs. 2.6–9), the pelvic diaphragm (figs. 3.24–29), and the abdominal muscles (figs. 3.11–13, 8.8, 8.11, and 8.13). As far as major skeletal and ligamentous restrictions to lumbar bending is concerned, there are four: the physical limitations of the vertebral arches (figs. 4.5a, 4.6a, and 4.12–13), the anterior longitudinal ligament that runs along the front sides of the anterior functional unit (fig. 4.13), the intervertebral disks, whose nuclei pulposi are driven anteriorly within the intervertebral disks (fig. 4.11), and finally, the superior and inferior articular processes (figs. 4.5–6 and 4.13b), which become tightly interlocked during extension.

It's anyone's guess as to which of these structures yield to permit 180° backbends in circus performers. It is possible, although I have not personally checked this out in anyone, that unusually mobile sacroiliac joints (chapter 6) might account for some of the capability of laying the hips down against the shoulders. In any event, after the bend is an accomplished fact the heavy spinous processes characteristic of the lumbar region (figs. 4.5b, 4.10a, and 4.13b) are probably butting up against one another.

sites where flexion and extension are permitted	bet. T12 and L1	bet. L1 and L2	bet. L2 and L3	bet. L3 and L4	bet. L4 and L5	bet. L5 and S1	bet. T12 and S1	at the hip joints	total: spine and hips
degrees flexion	5	6	8	9	14	18	60	90	150
degrees extension	4	4	4	9	14	10	45	15	60

Table 5.1. This chart estimates the degrees of flexion and extension permitted between individual vertebrae between T12 and the sacrum in someone who is moderately flexible. With 90° of additional flexion at the hip joints, the total forward bending permitted between T12 and the thighs is about 150°, which amounts to about 2.5 times as much forward bending as backward bending.

WHAT FACILITATES BACKBENDING?

What limits backbending is generally straightforward, but if you were to ask what assists backbending, the answer would have to be, "It depends." In standing backbends, as well as in passive supine backbends, which we'll cover later in this chapter, the answer is gravity. Standing, you simply lean your head and upper body to the rear, thrust your pelvis and abdomen forward, and let gravity carry you into the backbend. If you bend backward naturally, you also bend your knees, and that is why you have to extend the knees fully to evaluate backward bending in the hips and spine.

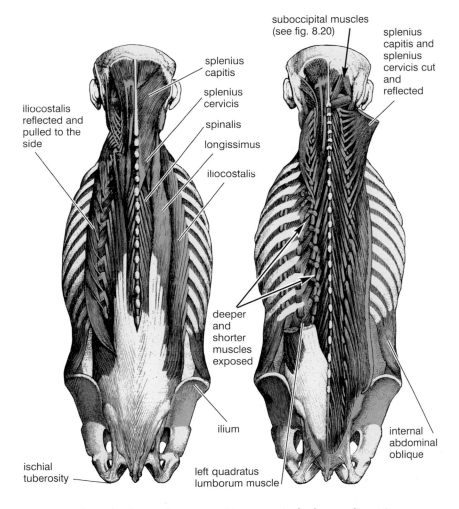

Figure 5.5. Deep back muscles exposed in successively deeper dissections following removal of the upper and lower extremities. The erector spinae are by definition the combination of the spinalis, longissimus, and iliocostalis (Sappey).

In prone backbending we are lifting one or more segments of the body against gravity. In this case the bend is accomplished by the deep back muscles (which lift you actively by extending the spine) or by the arms and shoulders (which can support a semi-relaxed prone backbend in any one of several ways). As we saw in chapter 4, the erector spinae muscles are responsible for extending the spine. They run from the pelvis to the cervical region and are composed of a complex of three muscles on each side—the *iliocostalis*, the *longissimus*, and the *spinalis* (fig. 5.5). Continuing into the neck are the *splenius cervicis* muscles, and to the back of the head, the *splenius capitis* (fig. 5.5). These latter muscles, which are also known as *strap muscles* from their characteristic appearance, run longitudinally. Deep to them are small, short muscles that run more or less obliquely between the spinous processes and the transverse processes—the *semispinalis* (fig. 8.14), *multifidus*, and *rotatores* muscles. Deeper yet are the *interspinales* muscles between adjacent spinous processes, and the *intertransversarii* muscles between adjacent transverse processes (shown in fig. 5.5 but not specifically labeled).

BREATHING AND BACKBENDING

It is obvious that muscles and gravity play major roles in creating hatha yoga postures, but anyone who has stood, lain, or sat quietly in a pose for a few minutes knows that something else is superimposed upon its equipoise. That something is breathing. A formal statement of the matter might run as follows: Under most circumstances of normal breathing, inhalations will either lift you more fully into a posture or create more tension in the body, and exhalations will either relax you further into the posture or reduce tension. From the perspective of the first four chapters of this book, we can now examine this statement with respect to backbending.

In whole-body standing backbending (fig. 4.19) we saw that it was natural to allow exhalation to lower you maximally to the rear (relaxing you further into the posture). Inhalation then lifts you up and forward (creating more tension in the body). We also saw that you could reverse this pattern on purpose by pulling backward more vigorously into the posture during inhalation (taking you more fully into the posture), and then relaxing and easing off the posture during exhalation (thereby reducing tension). We'll see variations on these principles in all the remaining postures in this chapter. In every case the diaphragm either restricts backbending, which we'll call diaphragm-restricted backbending, or assists it, which we'll call diaphragm-assisted backbending.

DIAPHRAGM-RESTRICTED BACKBENDING

When you are breathing naturally, inhalation restricts a standing backbend and exhalation assists it (fig. 4.19) because the diaphragm increases intra-abdominal pressure as it presses the abdominal organs inferiorly during inhalation. As we saw in chapter 3, it does this in cooperation with the abdominal muscles and the pelvic diaphragm: the increased intra-abdominal pressure restricts the bend by making the torso a taut, solid unit, thus protecting the critical lumbar region by spreading the vertebrae apart hydraulically and easing compression on all the intervertebral disks between the rib cage and the sacrum.

A simple experiment demonstrates diaphragm-restricted backbending. Stand up straight and interlock your hands behind your head, take a deep inhalation, and hold your breath while keeping the airway open. Then pull to the rear as hard as you can by tightening the muscles on the back of the body from head to heel (but don't pull your head back beyond the natural arc created by the torso). If you have taken a deep breath and your respiratory diaphragm is healthy, you'll feel the diaphragm stop the backward bend almost before it gets underway. Next, let just a little air out, and notice that your efforts to pull to the rear will be met with less resistance. Keep the airway open to make sure it is the respiratory diaphragm, acting in combination with the abdominal muscles, pelvic diaphragm, and the hydraulic nature of the abdominopelvic cavity, that is restricting and then easing the backbend. Increasing pneumatic pressure in the chest by closing the glottis would eliminate tension in the diaphragm and invalidate the experiment.

A propped backbending stretch will illustrate the same phenomenon. Stand with your back about two feet from a wall, and swing the hands overhead and backward to make contact. Adjust the distance so that you are in a comfortable backbend. Work the hands down the wall until you are just short of your comfortable limit of extension, but keep the knees straight. After you have relaxed and made yourself comfortable in the posture for 10–20 seconds, notice that each inhalation will diminish the bend in the lumbar region and straighten the body as a whole, and that each exhalation will allow your back to become more fully arched. More specifically, inhalation does two things: it pulls the thighs and legs slightly to the rear, and it lifts the chest and shoulders up and forward (fig. 5.6). It is easy to prejudice these results, however, and a fair test of diaphragm-restricted backbending requires that you keep constant tension in the upper extremities at all times and search out your most natural inclinations for bending and breathing. As we saw in chapter 4, you can easily reverse the results by purposely pressing more deeply into the bend during inhalation and then consciously easing off during exhalation.

DIAPHRAGM-ASSISTED BACKBENDING

You can feel the opposite phenomenon if you face the wall, bend forward from the hips, and support the bent-from-the-hips posture with your forearms. Keep the lumbar region arched forward, making the posture a backbend from the hip joints up (fig. 5.7). Breathe in the posture and watch carefully. The most natural result is that inhalation deepens the backbend and that exhalation eases it. Keep the abdominal muscles relaxed during inhalation. If you tense them the dome of the diaphragm can't descend very far, and you will lose the sense of being able to assist the backbend with inhalation. If there is any question about the results, aid exhalation with the abdominal muscles, and this will immediately and markedly diminish the lumbar lordosis (chapters 2 and 3). As with the propped backbending stretch facing away from the wall (fig. 5.6), you have to keep all conditions constant except for respiration. You can always pull yourself down consciously during exhalation to increase the lumbar bend, but that misses the point. Here we want to examine the effects of inhalation and exhalation in relative isolation.

The mechanisms underlying diaphragm-assisted backbending are straightforward. First of all, the depth of the backbend has been defined initially by the position of the feet on the floor and the forearms against the wall. This means that the abdominal muscles do not have to be tensed to keep you from falling, and it also means that intra-abdominal pressure will not be as pronounced as in a free-standing, internally supported backbend

Figure 5.6. Diaphragm-restricted backbend. Inhalation pulls the legs and thighs slightly to the rear, lifts the chest and shoulders up and forward, and increases intra-abdominal pressure, all of which combined straighten the body as a whole and diminish the backbend.

(fig. 4.19) or in the propped standing backbend (fig. 5.6). What is more, since the rib cage is relatively immobilized by the arm position, the crura of the diaphragm can only pull forward on the lumbar lordosis during inhalation to deepen the backbend. It's a three-part situation: the static body position defines the extent of the backbend in the first place, decreased abdominal tension and pressure allows the diaphragm to deepen it, and an immobilized rib cage requires the diaphragm to deepen it.

One more variable determines whether the diaphragm assists or restricts backbending in general, and that is how much the lumbar region is arched forward on the start. If students are stiff and wary, the arch will not be pronounced, the abdominal muscles will be tensed, and inhalation will create more stability, thus restricting the bend. That is what we usually see in beginning classes. With stronger and more flexible students, inhalations are more likely to increase the backbend, deepening it in proportion to how readily they can release intra-abdominal pressure and allow the iliopsoas, rectus femoris, and rectus abdominis muscles to lengthen. As we experiment with the prone backbending postures that follow, keep all possibilities in mind: notice that the descending dome of the diaphragm will either maintain restrictions in the torso and create diaphragm-restricted backbending, or, depending on the student's flexibility and on the positions of the upper and lower extremities, the diaphragm will force the front of the body forward during inhalation and deepen the bend.

Figure 5.7. Diaphragm-assisted backbend. In this forward bend from the hips, inhalation increases the depth of the lumbar lordosis provided the abdominal muscles are kept relaxed. The relaxed position against the wall permits the abdominal muscles to come forward and allows inhalation to occur freely, in contrast to the case of diaphragm-restricted backbending that is illustrated in fig. 5.6.

THE COBRA POSTURES

Of the many forms of backbending postures, the prone backbends—the cobra, locust, boat, and bow—are the most widely practiced in hatha yoga. In contrast to standing backbends, in which gravity assists your movement into the postures and resists your efforts to come up, in the prone backbends you lift parts of the body away from the floor against the pull of gravity and then return to your starting position with the aid of gravity. Prone backbends are harder than standing backbends because you are trying to overcome the resistance of connective tissues and skeletal muscles on the front side of the body while simultaneously opposing the force of gravity, but they are easier because coming out of the postures does not have to involve anything more than dropping slowly to the floor.

The cobra posture is named for the manner in which the magnificent king cobra lifts its head and flattens out its hood in preparation for striking its prey. It is probably the most well-known prone backbending pose in hatha yoga. Considered along with all its variations, it is worth the attention of everyone from beginners to the most advanced students.

Every variation of the cobra and its close relatives begins from a prone position and ends with the neck and back extended. In contrast to the caution we exercised in extending the head and neck in standing backbends, we have no reason to restrict that movement here, and we can work confidently with the cervical region without being concerned about losing our concentration on the rest of the posture or getting lightheaded from cardiovascular responses. Other characteristics of the postures vary. Depending on the specific exercise, from the top down, you can start either with the forehead or the chin on the floor, the deep back muscles either active or relaxed, the hip and thigh muscles firm or relaxed, the knees extended, relaxed, or flexed, and the feet together or apart.

THE CLASSIC BEGINNING COBRA

To introduce the series of cobra postures, we'll begin with the classic beginning pose even though it is not the easiest one. Start with the hands alongside the chest, with the palms down, the fingertips in line with the nipples, the heels and toes together, and the elbows close to the body. In this posture arch the neck enough to the rear to place the forehead against the floor, thus creating a reverse cervical curvature (fig. 5.8). If that position is uncomfortable, you can start with the nose or chin against the floor. On an inhalation, slowly lift the forehead, brush the nose and chin against the floor, and lift the head, neck, and chest slowly, vertebra by vertebra. Lift mainly with the back and neck muscles, using the hands only as guides. As you extend the spine, try to create a lengthening feeling. You should be at ease; if you are straining you have come up too far. Try to remain in the posture, breathing evenly, for 10–20 seconds, and then come back down in reverse order, ending with the forehead against the floor. This is the classic beginning cobra (fig. 5.9). If you are a novice, you may do little more than raise your head, and it might appear as if this involves only the neck muscles. But even this slight movement will engage the deep back muscles from the head to the pelvis, and over time you will slowly develop the strength and flexibility to come up further.

To relax you can turn your head to one side and rest. If that is too stressful for your neck, you can place a large soft pillow under your chest and head, which permits your head to be twisted more moderately. Or another alternative is to twist a little more insistently and at the same time press the side of the head firmly against the floor with isometric contraction of the neck muscles, which will stimulate the Golgi tendon organs and cause reflex relaxation of the associated muscles. In any case, turn in the other direction after the next variation, and then alternate sides each time you do another.

Teachers often tell students to place their heels and toes together but to stay relaxed while coming up into the pose. That can be confusing because anyone who tries to do this will find that holding the heels and toes together

Figure 5.8. Classic cobra, starting position.

Figure 5.9. In the classic cobra posture, the upper extremities serve mainly as guides, and the head and shoulders are lifted using the deep back muscles as prime movers. The muscles of the hips and thighs then act as synergists for moderately bracing the pelvis.

in itself requires muscular activity. It is best to explain at the outset that the classic cobra should be done with no more than moderate tension in the lower extremities, and that the act of holding the heels and toes together serves this purpose. Students should also pay close attention to their breathing; they will notice that each inhalation lifts their upper body and creates more pull in their lower back and hips.

THE COBRA WITH TIGHT LOWER EXTREMITIES

Now try a variation of the cobra with the lower extremities fixed solidly. With the chin against the floor, place the hands in the standard position. Then, keeping the feet together, tighten all the muscles from the hips to the ankles and lift up as high as you can with the back muscles, leading with the head and looking up. With the sacrum and pelvis stabilized by the tension in the lower extremities, you will be using only the erector spinae muscles to create the initial lift. As soon as you are up, each inhalation will lift the torso even more, and each exhalation will lower it down; both movements result from the action of the diaphragm. Inhale and exhale maximally if you are confused. The respiratory motion is more apparent here than with the classic cobra because the lower extremities are held more firmly in position. Except for the hand position, this posture is identical to the cobra variation we did in chapter 2 (fig. 2.10). There the movements of the upper body were discussed in terms of lifting the base of the rib cage in diaphragmatic breathing. The same thing happens here except that now we're calling it a diaphragm-assisted backbend.

A RELAXED COBRA WITH MILD TRACTION IN THE BACK

If your lower back is tender the classic cobra will be uncomfortable, and the most natural way to protect and strengthen the region will be to tighten everything from the waist down as in the previous variation. But there is an alternative—you can push up mildly into the cobra with the arms in a modified crocodile position. Instead of using the deep back muscles to extend the spine, which pulls the vertebrae closer together and compresses the intervertebral disks, we'll push up with the arms to lift the shoulders, place traction on the lumbar region, and remove tension on the intervertebral disks.

Start this posture with the hands on top of one another just underneath the forehead and with the elbows spraddled out to the side, or place the hands flat on the floor with the thumbs and index fingers making a diamond-shaped figure, the tips of the thumbs under the chin. Then, keeping the elbows, forearms, and hands planted against the floor, lift the head actively, push up with your arms, and create a gentle isometric pull with the arms as though you were wanting to pull yourself forward. At the same time observe that the leg, thigh, hip, and back muscles all remain relaxed. You are not going anywhere with the isometric lift and pull with the arms; you are only creating a mild traction in the back that encourages relaxation. This exercise protects the lower back just as effectively as keeping the hips and thighs firm because the back and lower extremities automatically stay relaxed as you lift up. The oddity of protecting yourself with both mechanisms at the same time will be obvious if you come into the posture and then tighten muscles generally from the waist down.

Notice how this posture affects your breathing. The tension on the chest from the arms and shoulders keeps it immobilized, in contrast to the diaphragm-assisted lift in the classic cobra and the previous variation. Most of the respiratory movement is felt as abdominal breathing exactly as in the stretched crocodile posture (fig. 2.23): the lower back lifts with each inhalation as the dome of the diaphragm descends, and the lower back drops toward the floor with each exhalation as the dome of the diaphragm rises.

CREATING TRACTION WITH THE HANDS AND ARMS

For this variation start with the hands alongside the chest, the heels and toes together, and the chin rather than the forehead on the floor. Then, instead of creating traction with the elbows and forearms, create it by pressing the heels of the hands toward the feet isometrically. This is similar to creating traction by pulling from the elbows, but the action is more difficult to control. You started this posture with just enough tension in the lower extremities to hold the feet together. Now try to let that melt away. Also try to minimize your tendency to push the torso up with the hands, even though that is hard to avoid while you are creating tension for pulling forward. This is a demanding whole-body concentration exercise. Notice how your breathing differs from that in the previous variation. The chest is not restricted—the diaphragm both flares the chest wall from its lower border and lifts the upper body, creating diaphragmatic rather than abdominal breathing.

RAISING UP AND DOWN WITH BREATHING

For this variation, start with your chin on the floor, the hands in the standard position alongside the chest, the heels and toes together, and the hips squeezed together. Then inhale while lifting your head and shoulders, and exhale back down until your chin touches the floor, breathing at the rate of about four breaths every ten seconds. You can experiment with keeping the hips somewhat relaxed, but it is more natural to keep them firm so that inhalations lift you higher. This exercise differs from the classic cobra in that it involves constant movement. You come all the way up and all the way down using a combination of the diaphragm and the back muscles, while in the cobra you hold the position as much as possible with the back muscles alone and allow the diaphragm to bob you up and down from there. A nice variation on this exercise is to turn your head to one side or the other with each inhalation: inhale, up (right); exhale, down (center); inhale, up (left); exhale, down (center); and continuing with a natural cadence for 10–20 breaths. During each successive inhalation, twist more insistently, lift more insistently, and expand your inspiratory capacity as much as possible with empowered thoracic breathing. This is a powerful and yet natural and comfortable exercise.

THE DIAPHRAGMATIC REAR LIFT

The next several variations of the cobra depend on reviewing the diaphragmatic rear lift (fig. 2.11). Summarizing from chapter 2, come into the standard preparatory posture for the classic cobra except that now the chin instead of the forehead is against the floor. Relax the entire body, especially below the chest, and then breathe deeply while keeping the chest and chin against the floor. Provided the deep back muscles, hips, and thighs are all free of tension, inhalation will arch the back and lift the hips, and exhalation will allow the lumbar region to flatten and drop the hips back down. At first, breathe quickly, almost as in the bellows exercise, and then slow your respiration down to observe the finer changes in tension and movement. In chapter 2 this exercise illustrated the connections of the diaphragm. Here it illustrates how lifting the hips (instead of the base of the rib cage) with the diaphragm creates another variation of diaphragm-assisted backbending.

ANOTHER COBRA WITH RELAXED LOWER EXTREMITIES

This next variation involves doing the cobra with the lower extremities completely relaxed, which is a posture that will challenge the concentration of even advanced students. Start with your chin on the floor and let your feet fall slightly apart into the position in which you will be most relaxed (heels in and toes out, or vice versa). Then slowly lift the head and chest, monitoring muscles in the lower half of the body to make sure they do not contribute to the lifting effort. This is easy enough at first, but it starts to feel unnatural as you rise more fully into the posture. Come up as far as you can, hold, and then come down slowly. Keep checking to make sure you do not feel a wave of relaxation on your way down, indicating that you tensed up as you lifted into the posture.

It's the gluteal muscles that are the most difficult to hold back in this pose. When you lift up into the cobra, you ordinarily support the effort of the deep back muscles by bracing the pelvis with the gluteus maximus muscles, and this insures that the erector spinae and other deep back muscles will lift only the upper part of the body. But if you relax from the waist down, the erector spinae muscles have two roles instead of one. They still lift the upper half of the body by way of their insertions on the chest, but now they also pull on the ilium and sacrum from above, deepening the lumbar lordosis and rotating the coccyx to the rear for an anterior pelvic tilt. If you have ever had back problems, this exercise will at once make you aware of your vulnerability, so do it only if you remain comfortable from start to finish.

Not surprisingly, the diaphragm contributes importantly to this posture; it acts in perfect cooperation with the erector spinae muscles by lifting both ends of the torso at the same time, thus assisting the backbend both from above and below. As we saw in chapter 2, this happens because the costal portion of the diaphragm lifts the rib cage and because the right and left crura lift the relaxed hips.

THE COBRA WITH REVERSE BREATHING

For an even more difficult concentration exercise, lie prone with your hands alongside your chest in the standard position, and come in and out of the cobra posture while reversing the natural coordination of diaphragmatic inhalations with the concentric shortening of the back muscles. Do this as follows: First, keep the chin on the floor while inhaling. Then exhale while tightening the hips and thighs (which holds them against the floor), and at the same time raise the head and shoulders. Next, inhale and relax the hips completely (which causes the hips to rise) as you lower the head and shoulders. Putting it differently: raise the head and chest during exhalation, and lower them during inhalation; tighten the hips and keep them down during exhalation, and relax them and permit them to rise freely during inhalation. This is disorienting until you master it. The exercise will work best if you take about one breath every four seconds.

After several exhalations up and inhalations down, try breathing more slowly. Relax completely during a deep but leisurely inhalation, and allow the crus of the diaphragm to deepen the lumbar lordosis from below. As exhalation begins, tighten the gluteals so the back muscles can lift the upper part of the body concentrically from a strong base without any help from the diaphragm. Then, during the next inhalation the chest will drop slowly to the floor, and the diaphragm will again lift the lower spine and hips as the back muscles and lower extremities relax. This is a difficult exercise, but after you have mastered it, along with the other variations of the cobra, you will have experienced all the possible combinations of breathing in relation to lifting up and down in the cobra. This posture helps place all the more natural possibilities in perspective. And apart from its value as a training tool, once you have succeeded in learning to do the sequence smoothly and rhythmically, the exercise is very soothing.

THE SUPPORTED INTERMEDIATE COBRA

Here is a good way to prepare for the advanced cobra. Lying prone with the chin on the floor, stretch the hands overhead with the arms and forearms parallel and the palms down. Keeping the elbows extended and the heels and big toes together, lift the head as high as possible, and pull one hand and then the other back toward the head in small increments. This will lift the upper half of the body. The back is passive; it is not doing the work of lifting you. One arm braces while the other pushes the body up, inch by inch. When you are up, find a relatively relaxed position with your weight resting on a combination of the hands, the lower border of the rib cage, and the pelvis. Or you can suspend the weight of your chest and abdomen between the hands and pelvis if that feels comfortable (fig. 5.10). Keeping the elbows extended is a feature of this posture alone. We'll dispense with doing that when we come to the full expression of the advanced cobra.

Most beginners make two mistakes in this exercise. One is to hang passively between their arms. Don't do that. Lift the chest and pull the scapulae down and laterally. With experience, you can find a position in which you are keeping the pelvis, or possibly the pelvis and the rib cage in combination, against the floor without hanging passively. The other common error is to let your attention stray from the forearm extensors, which permits the elbows to become slightly flexed.

Everyone will have a different limit to how far they can lift up and at the same time keep the pelvis on the floor. It will depend, obviously, on how much passive extension their lumbar spines can accommodate. Some will end up with their shoulders lifted up off the floor only a few inches; others may have enough flexibility to face the ceiling (fig. 5.11). If you are inflexible, notice that you feel vulnerable with the back muscles relaxed. Find a position

Figure 5.10. Supported intermediate cobra. In this pose the hands are pulled back incrementally (always keeping the elbows extended) until the pelvis is almost lifted off the floor. The head is pulled backward and the scapulae are pulled down and laterally (see chapter 8 for details of scapular movements), being careful not to hang the chest passively between the shoulders.

in which you can feel some of that vulnerability and yet remain still without pain or anxiety. As in other variations of the cobra, take deep empowered thoracic inhalations.

As soon as you are accustomed to the relaxed posture, pull the head back further and engage the back muscles isometrically. Now the inner feeling of the posture changes. The engaged back muscles make it feel safer even though your position has not shifted. Then, as soon as you have accommodated to this new feeling, bend the knees carefully, drawing the heels toward the head while keeping the elbows fully extended. Don't hurt your knees. This is an unusual position for them. It is fairly safe for the back because you have engaged the erector spinae muscles, but lifting the feet increases the intensity of the posture, so be watchful. You also have to concentrate on keeping the elbows straight because neurologic interconnections between motor neurons for flexors and extensors are such that the act of flexing the knees reflexly inhibits the motor neurons that innervate the extensors of the forearm (chapter 1), especially if the movement causes the slightest pain in the back or knee joints. Hold the posture for 10–20 seconds if you can do so confidently. Then slowly lower the feet back to the floor and slide the hands forward to the beginning position.

If you are not very flexible the lumbar region is bent to its maximum, especially when the feet are raised, and you will notice that breathing does not create marked external effects on the posture. Inhalation increases (and exhalation decreases) internal tension, but we do not see much accentuation or flattening of the lumbar region. In company with this, it may feel appropriate to breathe cautiously if you are at your limit. And even if you are flexible and comfortable, breathing deeply in this posture is not easy because the rib cage is constrained by the arm position. Nevertheless, if you would like to expand your inspiratory capacity (chapter 2), inhale thoracically as much as the posture permits.

Figure 5.11.
Supported cobra
facing the ceiling.
Flexible students
can bend their
spines enough
(about 70° in this
case) to face the
ceiling and yet
keep their thighs
on the floor.

THE ADVANCED COBRA

To do the advanced cobra, start from the same beginning position as the classic cobra, with the forehead on the floor and the fingertips in line with the nipples. Next, brush your nose and chin along the floor and slowly start lifting the head and chest with the back muscles. Then, keeping the back muscles engaged, slowly start to straighten the elbows until you have extended the back and neck to their limits. The extent to which the elbows are straightened will be a reflection of how much the spine is extended, as well as a reflection of the lengths of the arms and forearms. Beginners will not be able to come up very far, and it will be rare for even advanced students to straighten their elbows completely. It's not necessary anyway. The idea of this posture is to keep everything active. The deep back muscles, specifically, should be monitored constantly to make sure they are supporting the lift and not relaxing as the forearms extensors start contributing to the posture.

Keeping the back muscles active sounds like it ought to be easy, but for those whose spines are inflexible these muscles will be working against the antagonistic actions of the iliacus and psoas muscles, which maintain the first line of protection for restricting the bend, as well as the abdominal muscles, which stay tight to maintain the intra-abdominal pressure that is so important for minimizing strain on the intervertebral disks. It is a natural temptation to simply relax and support the posture entirely with the upper extremities. Don't do it. That's more like the next posture, the upward-facing dog.

As you progress in your practice of the advanced cobra, you will gradually become confident and flexible enough to allow the iliopsoas muscles and the abdominal muscles to lengthen eccentrically and even relax without releasing tension in the back muscles, and when that happens the back muscles will contribute to extension more effectively. The last step, after acclimating to the posture in its essential form, is to draw the feet toward the head (fig. 5.12).

Figure 5.12. In the advanced cobra, highly flexible students can bend their lumbar spines 90° and touch their feet to their head. For most students spinal and hip inflexibility (along with resistant hip flex- ors and abdominal muscles) limit coming fully into this pose.

As with the supported intermediate cobra, it is most important to keep the chest lifted and the shoulders pulled down and back. Nothing will violate this posture as certainly as allowing the chest to hang passively between the arms. And if you take the option of bending the knees and pulling the feet toward the head, be careful of stressing the ligaments that surround the knee joint.

Breathing issues in the advanced cobra are similar to those for the supported posture. The diaphragm will contribute to keeping the pose stable and restrict the bend for those who are less flexible, and it will deepen the backbend for those who find themselves flexible enough to come convincingly into the posture. In general, the advanced pose will not be very rewarding for anyone who is not flexible enough to sense that the diaphragm is either deepening the bend or creating tension for doing so, as well as getting out of the way of empowered thoracic breathing.

THE UPWARD-FACING DOG POSTURES

The upward-facing dog is not a cobra posture, but it begins in the same way and then goes one or two steps beyond. It is like a suspension bridge. The arms and forearms support the posture from above, the knees or feet support it from below, and the chest, abdomen, pelvis, and thighs are suspended between. Four variations are presented here, and in each one you support your weight differently.

To prepare for the upward-facing dog, start with the chin on the floor, the hands alongside the chest a little lower than for the cobra, the feet together, and the toes extended. Slowly lift the head and then the shoulders, keeping the muscles of the lower extremities engaged. As soon as you reach your limit of lifting with the back muscles, extend the elbows slowly, lifting your body even higher until your weight is supported by the arms, knees, and the tops of the feet. It is important to do this without relaxing the back muscles (fig. 5.13). The pose should be active front and back. Those who are especially flexible will have to keep the abdominal muscles engaged to avoid dropping the pelvis to the floor; those who are not flexible will not have this difficulty because their abdominal muscles are already tense. As in the advanced cobra, lift the head, neck, and chest. Don't allow the chest to hang passively between the shoulders. Come down in reverse order, taking a long time to merge the releasing of forearm extension into supporting the posture entirely with the deep back muscles.

Now try the same exercise with the toes flexed instead of extended. Keeping your knees on the floor and supporting yourself on the balls of the feet at the same time makes this a tighter posture because now the gastrocnemius muscles in the back of the calf are stretched. This places additional tension on the quadriceps femoris muscles, which (among their other roles) are antagonists to the gastrocnemius muscles; the tension in the quadriceps femoris is in turn translated to the front of the pelvis by way of the rectus femoris. The tension from the rectus femoris then restricts how far the pelvis can drop toward the floor. It's easy to prove. If you go back and forth between the two postures you'll feel immediately how the two alternative toe positions affect the pelvis—toes flexed and curled under, the pelvis is lifted; toes extended back, the pelvis drops.

In the full upward-facing dog, the knees are extended and you are supporting yourself between the hands and feet instead of between the hands and knees. This is a whole-body commitment requiring a lot more muscular tension in the quadriceps femoris muscles than the simpler posture. You can support your weight on your feet either with the toes flexed (fig. 5.14) or with the toes extended to the rear. Try both positions. Neither one is stressful if your feet are comfortable.

Breathing mechanics in the upward-facing dog are different from any other posture because the body is suspended in mid-air. You can easily rock back and forth or move from side to side like a suspension bridge in the wind, and this freedom of movement allows for deep thoracic inhalations and yet permits the diaphragm to deepen the backbend even in students who are not very flexible.

THE OPEN-AIR COBRA

This exercise requires good strength and athletic ability, healthy knees, and a prop made up of two cushioned planks; one (cushioned on top) is several feet off the floor, and another (cushioned on its underneath side) is slightly higher and situated to the rear of the first. The front plank will support the body at the level of the mid-thighs, and the rear plank will prevent the knees from flexing and the feet from flying up. Such a contraption is often found in health clubs. You will climb into the apparatus and lie in a prone position. The thighs will be supported from below by the front plank and the calves will be anchored in place from above by the rear plank. First, you allow the torso to hang down, flexed forward from the hips. From this position, you straighten the body and raise your head and shoulders as high as you can. The body from the thighs up will be suspended in mid-air as soon as you lift away from the floor.

This exercise is an excellent example of the manner in which gravity operates in relation to muscular activity. Little effort is required to initiate the movement for swinging the torso up the first 45°. Then, as the body comes toward the horizontal position you start getting more exercise. This feels similar to the classic cobra posture, except that it is more difficult because you are lifting the body from the fulcrum of the thighs instead of the pelvis. Then, as you arch up from that site you can begin to look right and left like a real cobra appraising its environment. Coming yet higher, the iliopsoas and abdominal muscles finally become the main line of resistance to the concentric activity of the back and neck muscles.

COBRAS FOR THOSE WITH RESTRICTED MOBILITY

The vertebral columns in older people sometimes become bent forward structurally, reverting to the fetal state of a single posterior curvature. The main problems with this, apart from not being able to stand up straight, are that the intervertebral disks have lost their fluidity, the joint capsules have become restricted, extraneous and movement-restricting deposits of bone have accumulated near joints, and muscles have become rigid. Those who have this condition are rarely able to lie comfortably on the floor in a prone position. But if they lie on cushions that support the body in a slightly flexed position and if the height of the cushions is adjusted carefully, all the simple variations of the cobra are feasible and will have beneficial effects throughout the body.

Figure 5.13. Upward-facing dog with knees down and toes extended. Come into this first of four dog postures systematically, and never hang between relaxed shoulders.

Figure 5.14. Upward-facing dog with knees up and toes flexed. Whole-body tension is required except in those who are so inflexible that their body structure keeps their thighs off the floor. The pose is like a suspension bridge.

ASHWINI MUDRA AND MULA BANDHA IN THE COBRAS

Ashwini mudra (chapter 3) is more natural in the upward-facing dog than in any other hatha yoga posture. The urogenital triangle is exposed, the genitals are isolated from the floor, the muscles of the urogenital triangle are relaxed, the gluteal muscles are engaged, and the pelvic diaphragm is automatically pulled in. Mula bandha (chapter 3), on the other hand, is natural in all of the postures in which the pelvis is resting against the floor, which means all the postures just covered with the exception of the upward-facing dog. Anyone who is confused about distinguishing between ashwini mudra and mula bandha can go back and forth between the upward-facing dog and an easy-does-it version of the classic cobra (in this case with the heels and toes together but with the gluteal muscles relaxed), and their confusion will vanish.

THE LOCUST POSTURES

The locust posture is named for the manner in which grasshoppers (locusts) move their rear ends up and down. The locust postures complement the cobras, lifting the lower part of the body rather than the upper, but they are more difficult because it is less natural and more strenuous to lift the lower extremities from a prone position than it is to lift the head and shoulders.

We can test the relative difficulty of one of the locust postures with a simple experiment. Lie prone with the chin on the floor and the backs of the relaxed hands against the floor alongside the thighs. To imitate the cobra, lift the head and shoulders. Look around. Breathe. Enjoy. This exploratory gesture could hardly be more natural. Notice that it doesn't take much effort to lift up, that it is easy to breathe evenly, that the upper extremities are not involved, and that the movement doesn't threaten the lower back. By contrast, to imitate the locust we'll need three times as many directions and cautions. Starting in the same position, point the toes, extend the knees by tightening the quadriceps femoris muscles, and exhale. Keeping the pelvis braced, lift the thighs without bending the knees. Don't hold your breath, and be careful not to strain the lower back. What a difference! While almost anyone new to hatha yoga can do the first exercise with aplomb, the second is so difficult and unfamiliar that new students have to be guided from beginning to end.

THE HALF LOCUST

The easiest locust posture involves lifting only one thigh at a time instead of both of them simultaneously. This is only about a tenth rather than half as hard as the full locust because one extremity stabilizes the pelvis while the other one is lifted, and this has the effect of eliminating most of the tension in the lower back. To begin, lie prone with the chin on the floor, the arms alongside the chest, the elbows fully extended, and the backs of the fists against the floor near the thighs. Point the toes of one foot, extend the knee, and lift the thigh as high as possible, but do this without strongly pressing the opposite thigh against the floor (fig. 5.15). Breathe evenly for ten seconds, come down slowly, and repeat on the other side.

The half locust is a good road map for the full posture because we see similar patterns of muscular activity, but it is easier to isolate and analyze the various sensations when one thigh is braced against the floor. To create the lift, the gluteus maximus and the hamstring muscles hyperextend the thigh against the resistance of the abdominal muscles, the iliopsoas muscle, and the quadriceps femoris. The hamstring muscles don't insert directly on the thigh. They insert on the back of the tibia, but in this case they act only on the pelvis because their tendency to flex the leg at the knee joint is prevented by strong isometric contraction of the quadriceps femoris muscle, which keeps the knee joint extended. It is as if you have attached a rope to the most distant of two boards that are hinged end-to-end; the far board is the tibia, and the near board is the femur. You want to lift the two boards as a unit to keep them aligned, but the rope goes only to the distant board. So another set of supporting lines has to run on the front (locking) side of the hinge to prevent the boards from folding up. The hamstring muscles are the ropes; the quadriceps femoris muscles are the supporting lines.

The half locust posture is worth more attention than it usually gets. In a slightly different form it is commonly prescribed by physiatrists and physical therapists for the recovery period that follows acute lower back pain. If you are on the mend from such a condition, and if you are able to lie in the prone position without pain, you can rapidly alternate what might be called thigh lifts—extending the knees and raising them (one at a time) an inch off the floor at the rate of about four lifts per second. If you repeat the exercise twenty to thirty times several times a day, it will strengthen the back muscles from a position that does not strain the lower back.

Figure 5.15. Half locust. This posture, which should be done without pressing the upper extremities and the opposite thigh strongly against the floor, is excellent for leisurely analysis of complex muscular and joint actions.

THE SUPPORTED HALF LOCUST

A more athletic posture supports the lifted thigh with the opposite leg and foot. Lie in the same prone position with the chin on the floor. Place the right fist alongside the right thigh, with the back side of the hand against the floor, and place the left hand, palm down, near the chest. Twist your head to the left and bend the right knee, flexing the right leg 90°. Then, using an any-which-way-you-can attitude—in other words, the easiest way possible—swing the left thigh up and support it on the right foot just above the left knee. Nearly everyone will have to lift their pelvis off the floor to get the left thigh high enough, and that is the purpose of twisting the head to the left and of having the left hand near the chest to help you balance. Try not to end up with the entire body angled too far off to the right, however. Use your breath naturally to support coming into the posture, taking a sharp inhalation on the lift, and then breathing cautiously but evenly while the foot is supporting the thigh. Even though you came into the posture with a swinging movement, try to come down slowly by sliding the right foot down the left leg. Repeat on the other side.

You can refine this exercise to make it both more difficult and more elegant if you come up slowly instead of with a swinging movement. Concentrate on breathing evenly throughout the effort and on keeping the pelvis square with the floor. Settle into the posture by slowly relaxing the abdominal muscles and hip flexors, which increases extension of the back. Finally, if you are flexible enough, deepen the backbend with your breathing, supporting the full posture both with the diaphragm and with deep thoracic inhalations (fig. 5.16).

THE SIMPLE FULL LOCUST

As soon as your are comfortable with the half locust you can begin to practice the full locust. The basic posture, which we'll call the simple full locust, is a difficult pose, but we place it first to give an idea of the posture in its pure form. The last three variations form a logical sequence which we'll call the beginning, intermediate, and advanced locusts.

To do the simple full locust, place the chin on the floor, the arms alongside the thighs, the forearms pronated, and the backs of the fists against the floor. If you want to make the posture more difficult, supinate the forearms and face the backs of the fists up. In either case point the toes to the rear, tighten the gluteal muscles, and last, keeping the knees extended, hyperextend both thighs, allowing them to become comfortably abducted at the same time. Do not try to aid the effort for hyperextension of the thighs with the arms at this stage. That will come later. If you are a beginner you may not have enough strength to make any external movement at all, or you may barely be able to take some of the weight off the thighs, but you will feel the effects in the lumbar, lumbosacral, sacroiliac, and hip joints, and you will still benefit from the effort.

When you raise the thighs in the simple full locust, you are trying to hyperextend them with the gluteus maximus muscles acting as prime movers, and doing this with both thighs at the same time makes this posture a great deal more difficult than the half locust: you are lifting twice as much weight, the pelvis is reacting to the muscular tension instead of stabilizing the posture for lifting just one side, and the lumbar lordosis is accentuated in one of the most unnatural positions imaginable. To make this seem a little easier you can take the option of allowing the knees to bend slightly, which will have two effects: it will permit the hamstring muscles to be more effective in aiding extension of the thighs, and it will facilitate their roles as antagonists to the quadriceps femoris muscles. The reason for allowing the thighs to become abducted brings us back to the hips; the gluteus medius and gluteus minimus (figs. 3.8b, 3.10a–b, 8.9, 8.12, and 8.14) are abductors, and holding the thighs adducted keeps these muscles in a stretched position and generally impedes hip hyperextension. The simple full locust is a challenging posture if your measure of success is external movement, but if you practice it daily you will soon be able to lift up more convincingly.

THE BEGINNER'S FULL LOCUST

The next variation, the beginner's full locust, is the easiest in the series. Keeping the elbows straight, place the fists under the thighs, pronating the forearms so that the backs of the fists are against the floor, and pull the arms and forearms under the chest and abdomen. Again, keeping the heels, toes, and knees together, try to lift the thighs while holding the knees fairly straight (fig. 5.17). This variation will affect a higher position in the lumbar region than the first posture, and your attention will be drawn to the genitals

Figure 5.16.
Supported half
locust. Beginning
students can learn
this posture by first
turning the head in
the same direction
as the thigh that
will be raised, then
swinging the thigh
up and catching it
with the opposite
foot; refinements
to provide for
more grace and
elegance can come
later.

rather than the anus, favoring activation of mula bandha over ashwini mudra.

The beginner's full locust is easier than the simple version because the fists provide a fulcrum that allows you to lift the thighs into extension from a partially flexed position. In the simple version of the full locust in which you are trying to lift your thighs from an extended to a hyperextended position, most of your effort goes into the isometric effort of pressing the pelvis more firmly against the floor. For unathletic beginners this is the end of the posture. But they should still experience both—the simple full locust to feel the essence of the basic pose, and the beginner's full locust to feel a sense of accomplishment.

THE INTERMEDIATE FULL LOCUST

You need to develop more strength in your shoulders, arms, and forearms for the intermediate variation of the locust. It is exactly like the previous posture except that you use the arms, forearms, and interlocked hands to press against the floor, and this helps you lift up much further. It requires a whole-body effort involving all the muscles on the anterior sides of the arms and shoulders, plus the deep back muscles, the gluteal muscles, and the hamstrings. The intensity of the commitment needed to raise the knees just a few inches off the floor is likely to surprise even a good athlete. But lift as high as possible and hold (fig. 5.18). Many benefits are gained just by increasing the isometric tension in your personal end position.

Even though this posture requires a whole-body effort, you feel it most significantly in the lower back. You can check this in someone else by placing your hands on either side of their vertebral column as they initiate the lift. In everyone, you will feel the muscles in the lower half of the back bulge strongly to the rear, and in those who are able to lift their knees six inches or more off the floor, you will notice the bulge spreading throughout the back as more and more of the erector spinae is recruited into the effort.

Figure 5.17. This beginner's full locust is easier than the simple full locust (not illustrated) because the position of the fists under the upper portions of the thighs permits them to act as a fulcrum for lifting the thighs. This posture also favors holding mula bandha over ashwini mudra (which is more in character for the simple full locust).

One of several unique characteristics of the locust posture is the extent to which the pelvis is braced. All of your efforts to lift are countered by numerous muscles acting as antagonists from the anterior side of the body: the rectus femoris pulls on the anterior inferior spine of the ilium; the psoas pulls on the lumbar spine; and the iliacus pulls on the pelvis. All of these muscles and their synergists act together from underneath to brace the body between the knees and the lumbar spine. And with this foundation stabilized, the gluteus maximus muscles, hamstrings, and erector spinae operate together to lift the pelvis and lower spine as a unit. The gluteus maximus muscles will first shorten concentrically and then act isometrically to place tension on the iliotibial tracts, which run between the ilia and the proximal portions of the tibias and fibulas (figs. 3.8–9 and 8.12). The actions of the gluteus maximus muscles are supported synergistically by the hamstrings, which, like the gluteus maximi, pull between the pelvis (in this case the ischial tuberosities) and the legs.

By themselves the gluteus maximus muscles and their synergists would not take you far, as in the case of the simple and beginner's full locust, but when the arms and forearms are strong enough to help drive you up, the muscles on the back side of the body are able to act more efficiently. This is a powerful posture but one of the most unnatural postures in hatha yoga, and since much of the tension for raising the thighs is brought to bear on the lower back, it is for intermediate and advanced students only.

The whole-body muscular efforts needed to maintain the intermediate full locust are intense, and since the abdominal muscles and the respiratory and pelvic diaphragms have to support the effort from beginning to end, inhalations will not be very deep, and the externally visible effects of breathing will be negligible.

Figure 5.18. This intermediate full locust is manageable only by intermediate and advanced students; those who are not both strong and flexible will not be able to lift this far off the floor. In any case, what differentiates the posture from the beginner's full locust is the way the pose is supported using the upper extremities.

THE ADVANCED FULL LOCUST

The advanced full locust is one of the most demanding postures in hatha yoga. To do it, those who are able to lift themselves up moderately in the intermediate practice now roll all the way up in one dynamic movement, balancing their weight overhead so the posture can be maintained without much muscular effort. This is only for athletes who are confident of their strength, flexibility, and the soundness of their spines. Those who can do it always seem to be at a loss for words when they try to explain what they do—speaking vaguely about concentration, breath, flexibility, and intention. It's a whole-body effort. If any link is weak the posture cannot be done.

To press up into this posture, nearly everyone has to lock the elbows and then interlock the hands underneath the body in some way. You can interlock the little fingers and keep the rest of the fingers and palms against the floor. Or you can clasp the fingers together as in fig. 5.18, starting the position with the hands cupped around the genitals. In either case, lift up into the intermediate position using the arms and back muscles. And then, without hesitating, inhale, bend the knees, press the arms and forearms against floor more forcefully, and in one fluid movement lift into the final posture with the feet straight up. Ideally, this is a balancing position. Once you are in it you will need to keep only moderate isometric tension in the back muscles, you will not have to keep pressing so vigorously with the arms, and you can flatten the backs of the hands against the floor. Flexibility for backbending really pays off here, the more the better, and the easier it will be to balance without holding a lot of tension in the deep back muscles. You can either keep your feet pointing straight up or lower them toward the head (fig. 5.19), which makes it even easier to balance.

Breathing is one of the most important elements of the advanced locust, and most students will find it necessary to take a deep inhalation to assist the action of coming up into the pose. After that there are two schools of thought. One is to exhale as you come up and keep the airway open according to the general rule for hatha yoga postures. This is the best approach because the pose is executed and supported by a combination of the upper extremities, the deep back muscles, and intra-abdominal pressure—not by intrathoracic pressure (chapter 3). But if you can't quite do that you can close the glottis to lift up and then breathe freely once you are balanced.

If you do not have enough flexibility in the back and neck to remain comfortably balanced after you are in the posture, you can place the palms flat against the floor, bend the elbows slightly, and support the lower part of the chest (and thus the whole body) with the arms and elbows. This will enable you to build up time in the pose.

The advanced locust places the neck in more extreme, and forced, hyper-extension than any other posture, and to prepare for this students will find it desirable not only to work with backbending postures in general, but also with special postures that extend the neck to its maximum. The cobra postures (figs. 5.9–12), the upward-facing dogs (figs. 5.13–14), and the scorpion (fig. 8.31b) are all excellent for this purpose.

Although the advanced locust has to be treated as a dynamic whole, try to do it slowly. Many students have been hurt by falling out of the posture when they have tried to toss themselves up into the full pose before they have developed sufficient strength and control. If you are almost able to do the advanced locust, you will soon be able to master it by developing just a little more lumbar flexibility in combination with more strength in the arms and back.

THE PRONE BOAT POSTURES

The prone boats curve up at each end like gondolas, lifting both ends of the body at the same time, and in this manner they combine elements of both the cobra and the locust. We discussed a simple prone boat in chapter 1 to illustrate movements of the body in a gravitational field. And since these postures work against the pull of gravity as well as against muscular and connective tissue restraints, they can be difficult and discouraging to many beginning students, who often cannot even begin to lift up into them. To add insult to injury, instructors may tell students to come up into the postures and relax! We'll remedy this by starting with an easy version. After a few weeks of regular practice, all of them become easy.

In the first version of the prone boat, lie with your chin against the floor, the heels and toes together, the arms along the sides of the body, and the back of the hands against the floor. Point the toes to the rear, tighten the quadriceps femoris muscles to extend the knee joints, and simultaneously lift the head, neck, chest, shoulders, thighs, and hips (fig. 5.20). If you have plenty of strength, lift the hands as well (fig. 1.15). Breathe evenly, and notice how the respiratory diaphragm lifts both the chest and the lower

Figure 5.19. The advanced full locust is easier in certain respects than the intermediate version because here one is balancing rather than support-ing the pose with muscular effort. No one should try this posture, however, who is not confident in their athletic prowess and 90° flexibility for backbending.

extremities with each inhalation and lowers them down slightly with each exhalation.

In the second version, which is a little harder than the first, again lie with your chin against the floor, with the heels and toes still together, but now stretch the arms out perpendicular to the body, palms down. Make blades of the hands, stretch the fingers, point the toes, extend the knees, and simultaneously lift all four extremities.

The third and most difficult version of the boat (fig. 5.21) is the same as the second except that the hands are stretched overhead with the elbows extended. Beginners usually resist this if you ask them to do it first, but if you start by leading them through the first and second versions they may be surprised to find that they have generated enough energy and enthusiasm to try the third. And rather than telling students to relax in the posture, instructors can suggest that they imagine they are lengthening the body in addition to lifting at both ends. This somehow enables them to raise up more efficiently and to feel more at ease.

Figure 5.20. In this easiest version of the prone boat (with oars coasting in the water), muscles are held firm on the back side of the body. Superimposed on that effort, the respiratory diaphragm lifts both the upper body and the lower extremities concentrically during each inhalation and drops them eccentrically during each exhalation, making this is a whole-body, diaphragm-assisted backward bend.

Figure 5.21. This third version of the prone boat is the most difficult one, and few students will be able to lift up this far, but if students start with the first two versions as preparatory poses, this one becomes easier.

As with the cobra postures, many older people who are chronically bent forward can benefit from a propped-up prone boat by placing supporting cushions under the torso. If the cushions are chosen and adjusted perfectly, their heads, arms, and feet will be in a comfortable beginning position. They can then lift up at both ends and be rewarded with a beneficial exercise.

THE BOW POSTURES

When an advanced student has come into the bow posture, it resembles a drawn bow: the torso and thighs are the bow, and the taut upper extremities and the legs are the string, which is drawn toward the ceiling at the junction of the hands and the ankles. The beginner's posture is not so elegant. Most of its length is flattened down against the floor, and the pose is acutely hinged at the knee joints.

THE BEGINNER'S BOW

To begin, lie prone, flex the knees, and grasp the ankles. Try to lift yourself into a bow, not by backbending with the deep back muscles, which remain relaxed in the beginning posture, but by attempting to extend the legs at the knee joints with the quadriceps femoris muscles. An attempt to straighten the knee joints is little more than an attempt, since you are holding onto the ankles, but it does lift the thighs and extend the knees moderately (fig. 5.22).

It seems extraordinary that the quadriceps femoris muscles (figs. 3.9, 8.8, and 8.11) are the foremost actors for creating this posture, but that is the case. At first they are in a state of mild stretch because the knees are flexed. Then, shortening these muscles concentrically against the resistance of the arms and forearms creates tension that begins to pull the body into an arc. The quadriceps muscles are performing three roles simultaneously: extending the knee joints from a flexed position, lifting the thighs, and creating tension that draws the bow. The tensions on the knee joints can be daunting for a beginner, and many students do well just to grab their ankles. If that is the case, they may not lift up at all but merely contract the quadriceps muscles isometrically. That's fine. Just doing this every day will gradually strengthen the muscles and toughen the connective tissue capsules of the knee joints enough to eventually permit coming up further into the posture.

THE INTERMEDIATE AND ADVANCED BOWS

Intermediate students will approach the bow with a different emphasis. They may use the quadriceps femoris muscles to lift the knees in the beginning, but they aid this movement by engaging the deep back muscles (the erector spinae) to create an internal arch. Then, as they come higher they will use the gluteal muscles in the hips to provide even more lift. And as the gluteus maximus muscles shorten concentrically for extending the hips, they lift the pelvis away from the floor and indirectly aid extension of the knee joint.

As advanced students lift into the posture, the lumbar region becomes fully extended and the hips become hyperextended. Such students will be dividing their attention among at least five tasks after they bend their knees and grasp their ankles: paying attention to stretch and tension of the quadriceps; maintaining a strong connection between the ankles and shoulders; watching the knee joints, which are receiving an unconventional stress; overseeing the complex muscular interactions between the quadriceps femoris muscles on one hand and the gluteal muscles that lift the thighs during extension of the hip joints on the other; and breathing, which is rocking the upper half of the body up with each inhalation and dropping it forward with each exhalation (fig. 5.23). Once these conditions are established, advanced students can take the final option of drawing the feet toward the head.

The nervous system orchestrates all of this complex musculoskeletal activity. In beginners, the quadriceps femoris muscles, deep back muscles, and gluteals do not receive clear messages to lift strongly into the posture. Instead, they are inhibited by numerous pain pathways that take origin from the joints (especially the knees), and from the front side of the body generally. Such reflex input should not be overridden by the power of will. Even advanced students who are able to lift up more strongly may find that discomfort in the sacroiliac, hip, and knee joints limits the posture. Experts honor these signals mindfully. Then, as soon as there is no longer any hint of pain, the only limitations to the posture are muscle strength and connective tissue constraints. Advanced students, having practiced the bow thousands of times over a period of years, know exactly how much tension can be placed safely on each joint and how to come down from the posture without harm.

Figure 5.22. In this beginner's version of the bow posture, the main concentration is on lifting the body using the quadriceps femoris muscles (knee extensors on the front of the thighs).

THE KNEE JOINT

Studying the bow posture in detail leads us first and most obviously to the knees. When the knee joint is extended (fig. 5.24) it is almost invulnerable to injury because all of its component parts fit together perfectly and protect the joint from torques and impacts from all directions. The flexed knee joint is another matter. Its internal and external supporting ligaments have to become loose to permit flexion, and any tension superimposed on the flexed joint, whether from extension, twisting, or even additional flexion, can make the joint vulnerable to injury.

Standing postures, even with the knees flexed, are rarely a problem. The knee joint is the largest and strongest joint in the body, and toting our weight around from place to place is its forte. Under ordinary circumstances it can support tremendous muscular stresses—hopping up and down a flight of stairs on one foot, running down a mountain, and for those who are prepared, jumping to the ground from heights of five feet or more. All of these activities require only moderate flexion, and the knee joint is made to order for them. The problem comes with yoga backward bending postures that flex the knee under unnatural circumstances.

The bow posture is a case in point. The beginning pose must start with a flexed knee joint which is then forced into extension from an awkward and unfamiliar prone posture. Indeed, it is hard to imagine a more unnatural or demanding role for this joint than to use it as the primary tool for lifting into the bow. The intermediate and advanced postures are not so much of a problem because they make more efficient use of the erector spinae muscles, the hamstrings, and the respiratory and pelvic diaphragms for coming up into the posture. They use knee extension merely to aid the lift, and they do so from a less completely flexed position than in the beginner's bow.

The two advanced cobra poses stress the knees in an entirely different way. We come into these postures with the knees straight, which does not stress them at all, but once the rest of the body is settled we can take the additional option of bending the knees and pulling the feet toward the

Figure 5.23. In this advanced bow (the string of the bow is drawn to the rear at the junction of the hands and the ankles), the lumbar region is extended 90° and the hips are hyperextended. The quadriceps femoris muscles are not nearly so important as in the beginner's bow; they can help initiate coming up as an option, but they are not important as prime movers for the final pose.

head. This increases tension on the quadriceps femoris muscles at the same time the joint ligaments start to become lax. Unfamiliar and disquieting sensations warn us promptly to be wary, and only after much experience will students not have misgivings about completing this last refinement of the cobra. To understand how the strongest joint in the body can be so vulnerable to injury, we must examine the anatomical components of the knee that often create problems. We'll begin with the menisci.

THE MENISCI

The two bones that stand in end-to-end apposition to one another in the knee joints, the tibia and the femur, can withstand the repetitive shocks of walking and running because they are well cushioned. The lower ends of the femurs (*femoral condyles*) and the upper ends of the tibias (*tibial condyles*) are covered with thick layers of articular *hyaline cartilage* that

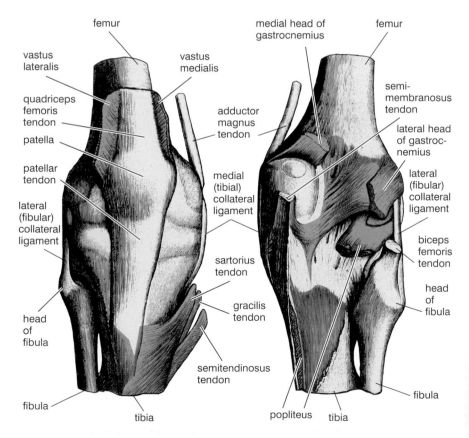

Figure 5.24. On the left is an anterior view of a superficial dissection of the right extended knee joint, and on the right is a posterior view (from Sappey).

have a slippery surface for permitting flexion and extension. In addition, donut-shaped wafers of fibrocartilage called the *medial and lateral menisci* (or *medial and lateral semilunar cartilages*), cushion the mating surfaces of the condyles (fig. 5.25). Torn menisci are difficult to treat because they have lost their blood supply by the time we reach our mid-20s, and if they are damaged after that time, usually in dance or athletics, they are essentially irreparable. This is why "torn cartilages" are greeted so apprehensively by adult athletes.

The menisci move freely during the course of flexion. This is not ordinarily a problem because we assume that they will come back into their home position when the knee joint is subsequently extended. But that does not always happen, and if it doesn't the menisci can get crushed by the opposing condyles. This might happen when you kick a ball or start to get up from a squatting position, and if you ever encounter unusual resistance to extension under such circumstances, carefully sit down and massage the knee before you try to straighten it. If you crush the menisci the only remedy may be trimming them surgically, if not removing them outright.

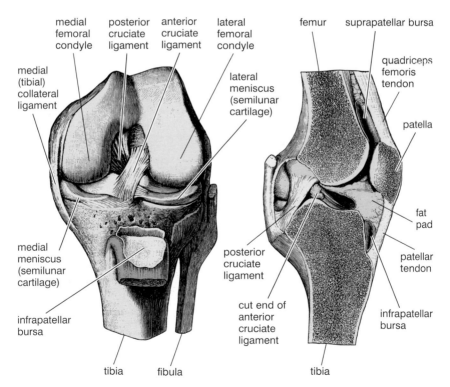

Figure 5.25. On the left is an anterior view of a deep dissection of the flexed left knee joint, and on the right is a sagittal cut through the extended knee joint (from Sappey).

THE CRUCIATE LIGAMENTS

The femur and tibia meet together in a hinge, and two internal ligaments, the *anterior and posterior cruciate ligaments* (fig. 5.25), keep the bones in alignment. Viewed from the side, the cruciate ligaments form an X. The anterior cruciate ligament runs from the back of the femur to the front of the tibia and constrains anterior displacement of the tibia in relation to the femur, while the posterior cruciate ligament runs from the front of the femur to the back of the tibia and constrains posterior displacement of the tibia.

The cruciate ligaments, especially the anterior cruciate ligament, are commonly stretched or torn in knee injuries, and even though they do not lose their blood supply entirely, as do the menisci, they are still so poorly vascularized that they do not easily heal. Many professional football careers have been brought to an abrupt end by an anterior cruciate injury. Envision a linebacker in a crouched position, clipped from behind by an opponent who throws his full weight against the top of the unsuspecting tibia. This will certainly tear the anterior cruciate ligament. It would be surprising, to say the least, to hear of anyone injuring a cruciate ligament doing advanced hatha yoga postures, but someone with an old injury might find that certain postures bear watching.

THE MEDIAL AND LATERAL COLLATERAL LIGAMENTS

The other two commonly injured ligaments in the knee are the *medial and lateral collateral ligaments* (fig. 5.24–25), which are important accessory ligaments on the medial and lateral aspects of the joint capsules. These are frequently injured by weekend athletes playing sports in which the body is wrenched and twisted in a direction at cross-purposes to foot position. It might be touch football, racquetball, or any sport in which your consciousness is directed away from your foundation. The collateral ligaments are often sprained under such conditions, meaning that some of their connective tissue fibers have been torn away from their bony attachments. Such injuries are slow to heal. If someone says they have a "trick knee," it is often after a long history of collateral ligament sprains. As with cruciate ligament injuries, yoga postures are less likely to create new injuries than to irritate old ones.

THE PATELLA AND BURSAE

The main role of the *patella*, or kneecap (figs. 5.24–25), is to make extension of the knee joint more efficient. It is a *sesamoid bone*, that is, a bone that is inserted in a tendon, in this case between the quadriceps femoris tendon and the patellar tendon (fig. 5.24). During flexion it slides down the *intercondylar notch*, a deep vertical gutter between the femoral condyles (the articular end of the femur); during extension it acts like a pulley on the front of the knee. The patella is important in any activity in which the knee is supporting the weight of the body in a partially flexed position. You can imagine how useful it is in a standing lunge (fig. 1.2), or when you are rising up from a half squat with a barbell on your shoulders. In these activities your entire body is supported by the quadriceps femoris muscles, quadriceps tendons, patellas, and patellar tendons.

The up and down movement of the patella is made possible by the *suprapatellar bursa*, a lubricant-filled sack that is situated between the patella and the underlying tissues (fig. 5.25). *Bursae* are common accessories to many joints whose attaching tendons move in relation to their underlying tissues. If adhesions resulting from disease, injury, or inactivity develop within a bursa, the tendon can no longer slip back and forth easily. And this, in the case of the suprapatellar bursa, is one of the commonest causes of a "stiff knee" following traumatic injury. A circular problem may develop—you can't flex the knee because of the injury, the bursa develops adhesions, and the adhesions further inhibit mobility. Again, the practical concern in hatha yoga is with old injuries. If they flare up, students so affected should be cautioned not to do any postures that stress the knees until the problem is resolved.

MINOR KNEE PROBLEMS

What can be done to prevent and resolve minor knee problems? The answer is simple—regular and prolonged muscular tension applied to the extended knee joint. Under those circumstances all parts of the joint fit together perfectly, allowing it to withstand intense isometric contraction of the surrounding muscles. If you have knee pain which is not due to serious internal injuries, the following series of standing postures can be highly therapeutic.

Keep the feet parallel and as widely separated as possible while holding the thigh muscles firmly, especially the adductors. Extending the knees fully is fine provided you do not hyperextend them beyond 180°, and provided you hold tension in the hamstrings as well as the quadriceps femoris muscles. Holding a firm base with isometric tension is the whole point of this series. Twist right, then left, then face the front; in each direction bend forward and backward, holding each position for 2–7 breaths (about 5–30 seconds). That's six combinations. Hold the arms in various positions—elbows grasped behind the back (fig. 7.17a–f); arms and forearms stretched laterally; hands in a prayer position behind the back; arms overhead with forearms behind the head and catching the elbows; hands interlocked behind the back and pulled to the rear; arms in a cow-face position first one way and then the other; and hands on the hips (fig. 7.18a–g). That makes six times eight which equals forty-eight poses.

Start modestly, doing the postures only five minutes a day, and then gradually increase your commitment. If you spend fifteen minutes a day on this series, you cannot help but strengthen the muscles that insert around the knee and place a healthy stress on the capsule of the knee joint, as well as on its associated tendons and ligaments. And this works wonders. After a few months of regular practice, the connective tissues of the joint will have gained enough strength and integrity, at least in the absence of serious medical problems, to withstand not only reasonable stresses on the knees in flexed positions, but also the stresses of sitting in cross-legged meditative postures.

SUPINE BACKBENDING POSTURES

The standing and prone backbends play an obvious role in hatha yoga. But what is a supine backbend? How can one do such a thing? Oddly enough, this is indeed possible, and several of the most important traditional back-bending postures in hatha yoga, including the fish, the wheel, and the bridge, either start or end in a supine or semi-supine position. In addition to that, in more recent times postures that support the back with a bolster or a ball have become regular fixtures in many practice plans, especially those with a therapeutic orientation.

Like the prone backbending postures, we can divide supine backbending postures into two categories, active and passive, depending on whether you lift yourself internally into a backbend or lie supine on a prop that allows you to relax. The simplest active supine backbending posture is to lie on your back and lift the abdomen toward the ceiling. We can call this a lumbar lift, and that's where we'll start.

THE LUMBAR LIFT AND LUMBAR PRESS

To do the lumbar lift, lie in the corpse posture, relax your abdomen, and arch the lower back forward by tightening the deep back muscles. The effect of this is to drag the rear of the pelvis and coccyx along the floor toward the shoulders, and thus increase the lumbar lordosis. This posture, however, is not entirely stable: you can't maintain it except by continuing to pull on the hips from above. To create a posture that you can observe at your leisure, heave your hips up away from the floor while holding your weight momentarily between your heels and shoulders, and at the same time tug the hips forcibly toward the head using the deep back muscles. Then maintaining that tension with the back muscles, drop the hips back to the floor. If you do this two or three times in succession, readjusting each time to a deeper backbend, you will create a stable lumbar arch.

Once you are acclimated to it, you will notice that this posture is a good illustration of how inhalations either take you further into a posture or increase internal tension, and how exhalations either relax you toward a neutral position or decrease internal tension. If you keep the abdominal wall relaxed after you have lifted up, a deep abdominal inhalation pushes the abdomen further out and merely increases intra-abdominal pressure; if you hold moderate tension in the abdominal wall after you have lifted up, a full diaphragmatic inhalation increases internal tension, accentuates the lumbar lordosis, and takes you further into the posture (fig. 5.26). Each exhalation, by contrast, either decreases intra-abdominal pressure in the case of abdominal breathing, or relaxes you down slightly toward a neutral position in the case of diaphragmatic breathing.

When you are breathing diaphragmatically the lumbar lift becomes another diaphragm-assisted backbend, and except that it is supine, the posture is comparable to a prone diaphragmatic rear lift. In both cases it is impossible to create the lifting effect with inhalation unless the gluteal muscles are relaxed, and that is why it helps to pull your hips forcefully into a higher settled position. After feeling the effects of breathing in the new posture, you can come out of it by thrusting your hips back inferiorly to flatten your back toward the floor. Then relax in the corpse posture and notice that the diaphragm is no longer affecting the spine.

The opposite and complementary posture to the lumbar lift might be called a lumbar press, and this is another challenging exercise in concentration and awareness. Again start in the corpse posture, but this time exhale while you flatten your back against the floor (fig. 5.27) using the abdominal muscles (figs. 3.11–13, 8.8, 8.11, and 8.13), as discussed in chapter 3. Then keep the back immobile as you inhale, and notice the adjustments you must make internally to accomplish that. Each inhalation tends to pull the lumbar lordosis forward, and that has to be prevented with extra intra-abdominal pressure and tension from the rectus abdominis muscles, even though this pressure and tension restrains inhalation. It's a balancing act. Exhalation is also a balancing act because it tends to create a slight release, and if you are not watchful the lower back will start to return to its slightly arched neutral position. So in the end both inhalation and exhalation have

Figure 5.26. Lumbar lift. A full diaphragmatic inhalation superimposed on having first dragged the pelvis toward the shoulders and having relaxed the gluteal muscles is illustrated here. The pose is a diaphragm-assisted backbend (arrow).

to be monitored constantly to keep the back pressed against the floor. Another complicating factor is that most students concentrate on the press so intently that they forget their extremities, and this will usually mean tensing them. Try not to let that happen. This is a breathing and concentration exercise for the lower back and abdomen.

The lumbar press is more delicate than, but otherwise similar to, the preparatory position for the basic supine double leglift (fig. 3.17). There we were preparing for a tremendous amount of tension on the lumbar lordosis from the hip flexors, and we had to make a no-holds-barred effort to keep the lumbar region flat against the floor. Here we are looking at something more refined. We are lying supine with the lower back pressed to the floor and exploring an exercise without movement, holding the lower back flat against the floor during both inhalation and exhalation.

The lumbar lift is obviously contraindicated for anyone with lower back problems, but the lumbar press is a different matter. Anyone who can master that posture will have a potent weapon for dealing with low back problems. The exercise will strengthen the abdominal muscles along with the pelvic and respiratory diaphragms, and one can gradually learn to use breathing to bring conscious awareness to the abdominopelvic region. Coupled with mild prone backbending postures that strengthen the back muscles, the lumbar press is a powerful means to lumbar health.

THE FISH

Most fish regulate their buoyancy with swim bladders into which oxygen is secreted from the blood (which helps them rise toward the surface), and from which oxygen is absorbed into the blood (which helps them descend). Human beings do something similar: if they barely have enough body fat to enable them to float, they can regulate their buoyancy in the water by the amount of air in their lungs. And so it is that one of the most effective ways to stay afloat is to settle into the classic yogic fish posture (fig. 5.28) with the feet folded into the lotus position, the lumbar region deeply

Figure 5.27. The lumbar press is a demanding breathing and concentration exercise for the back and lower abdomen in which the tendency of the diaphragm to pull the lumbar region forward is fully countermanded by intra-abdominal pressure that strongly presses the lower back against the floor (arrow) and that restrains inhalation.

arched, the head and neck extended, and the anterior surface of the chest just above the surface of the water. Under these circumstances the body will rise and fall in the water with each inhalation and exhalation.

Although the fish is a backbending posture, we also placed it in chapter 3 because an advanced abdominal exercise—the superfish leglift (fig. 3.19b)—can be derived from the pose. We'll analyze the simplest beginning fish here, and cover other variations of the posture with the shoulderstand, with which it is traditionally paired as a counterstretch (fig. 9.19). To start, lie supine with the hands under the hips (palms up or down as you choose), knees extended, and feet together. Lift up on the forearms, arching your back, head, and neck, and then place a little of your weight on the back of the head, just enough to touch the floor. Extend the spine as much as possible as you support the posture with the forearms (fig. 3.19a).

Respiratory responses in the fish posture depend on the extent of your lumbar arch and on how active you choose to be in the posture. In students who are not very flexible, the costal insertion of the diaphragm cannot move the base of the rib cage because its bottom rim is already open, and the crural insertion cannot lift the lumbar lordosis because that region is already arched to its maximum. The only thing that can obviously move is the abdominal wall, which moves anteriorly with each inhalation and posteriorly with each exhalation. Advanced students who are more flexible and able to arch up more convincingly will have an altogether different experience. They will be doing diaphragm-assisted backbending, and their relaxed posture in combination with a favorable arm position also permits them to enlarge their upper chests for empowered thoracic breathing more convincingly here than in any other posture.

THE WHEEL

The wheel is one of the most dynamic whole-body postures in hatha yoga. Its Sanskrit name, *chakrasana*, implies that it activates the body from head to toe. In its fullest circular expression, with the hands touching the heels, it is the backbend of backbends.

The conventional way to come into the wheel is to lie supine with the feet flat on the floor and as close to the hips as possible. Place the palms

Figure 5.28. This classic fish posture with the feet in the lotus position is rewarding only for advanced students. Others will enjoy the more elementary posture with the feet outstretched (fig. 3.19a).

near the shoulders with the fingers pointing down. Then lift the pelvis and push the hands against the floor, straightening the elbows to complete the posture. Let the head hang passively. If you have limited strength but plenty of flexibility (fig. 5.29) you can easily come up, and if you have lots of strength but limited flexibility you can push yourself up into some semblance of the posture with brute force. If you are limited in both realms the posture will be challenging, and all you may be able to do is push against the floor isometrically.

Given average spinal flexibility, the key to the posture is efficient use of the triceps brachii muscles. But the problem is that when the shoulders are against the floor, the extreme position of the upper extremities (with the wrists extended 90° and the elbows fully flexed) challenges the ability of the triceps brachii muscles to extend the forearms. If students find themselves struggling just to lift their hips, they will probably find that extending their elbows to complete the posture is impossible. And even if they can push themselves partway up, the iliopsoas and the rectus femoris muscles, acting as extensile ligaments, will start to limit them and prevent the requisite hyperextension of the hips. These problems are not insurmountable, and the usual remedy is to keep trying every day, gradually increasing strength and improving the backward bending capacity of the spine and hips. You can also lift up on your fingertips and the balls of the feet to make it easier. Sooner or later everything comes together and it becomes possible to partially extend the elbows, and when that happens the triceps brachii muscles will be more efficient in lifting you up. How the posture appears at that point will depend on the amount of backward bending in the spine and hips. With moderate flexibility the posture resembles an arch bridge, and with excellent flexibility it becomes chakrasana.

Figure 5.29. The wheel posture requires a combination of sacroiliac nutation (see chapter six), moderate hip hyperextension, 90° of lumbar extension, and enough strength in the arms (especially the triceps brachii) to push up into the posture. Alternatively, those who are especially flexible can drop into the pose from a standing position.

We know that the diaphragm generally limits backbending in those who are not flexible, and this is especially true of the wheel, where the need for muscular effort overwhelms the ability to deepen the backbend with inhalations. Advanced students who are lifted up in the posture don't have that problem. They can easily adjust their breathing so that inhalations arch them up more powerfully.

If you are an advanced student or gymnast who can drop backward into the wheel from a standing posture, you will find this an excellent study in re-ordering priorities. Instead of establishing a solid base and moving your attention and efforts from distal to proximal (chapter 4), you lean backward from a standing position, exhale, and drop into a provisional backbend with your focus on the lower back. Then you alter the positions of the upper and lower extremities, moving them closer together to bring the core of the body under tension.

As you lower yourself into the posture, the quadriceps femori come into their own, resisting all the way. The pectoral muscles also come under increased tension and impart that tension to the chest. All of this happens fast, but you are able to do it because you have prepared for a long time and are confident that the abdominal muscles, the respiratory and pelvic diaphragms, and the psoas, iliacus and rectus femoris muscles will have enough strength to protect the lumbar region.

Advanced students will always exhale as they drop into the wheel from a standing position. Intermediate students often do not have that much confidence, and if they are uncertain of themselves they tend to hold their breath as they drop back, knowing consciously or unconsciously that this will protect the back. As they stiffly lower their hands to the floor, often with a plop, it is obvious that they are going beyond their reasonable capacity. Such students should concentrate on coming into the wheel from the supine posture until they are more flexible and confident.

Both intermediate and advanced students will find that they need to be watchful of their knees, especially if they wish to lift from the wheel back into a standing posture. The quadriceps femoris muscles are already exerting a lot of tension on the knee joints in a flexed position, and that tension will increase during the forward-thrusting maneuver to stand up.

PASSIVE SUPINE BACKBENDING

The deep muscles that ordinarily hold the spine erect also keep it compressed, and it is only when you are lying down that they can relax and allow the vertebral bodies to spread apart (chapter 4). Any reclining position can accomplish this aim to some extent, but it is especially useful to relax in a passive supine backbend in which you lie on a bolster that permits your spine to extend without muscular involvement.

In hatha yoga classes we can accomplish this by lying passively on a ball placed underneath the spine anywhere between the pelvis and the head. This will open and release the facet joints of the superior and inferior articulating processes, and you can relax and let gravity do the work of stretching the spine, with the ball acting as a fulcrum all the way up and down the vertebral column. The idea is to relax muscles from the knees to the head on both the front and back sides of the body. It is a good idea to be wary of doing these exercises too much in the beginning. They are deceptively simple. They increase back and neck mobility, and they stretch you out more than you realize. It is easy to do too much and not know it until the next day.

You can experiment with different types and sizes of balls, but standard 8 1/2 inch playground balls (sometimes called foursquare balls or action balls) that contain 1–2 pounds of air pressure per square inch work well. This size and type of ball is excellent if you are in good to average condition. It is relatively soft, it does not press too sharply against the spinous processes and back muscles in any one place, and it's big enough to give most people a good stretch. Basketballs, volleyballs, and soccer balls are too hard for the average person. Softer plastic balls can be found for those who find the standard playground ball too hard. But don't try these exercises at all if you have acute back pain or a recent history of chronic back problems.

First place the ball under the pelvis with the knees extended, and roll from side to side to mobilize and stimulate the sacroiliac joints, applying pressure to different regions of the pelvic bowl and noticing whether or not you can relax (fig. 5.30). If you can't, it means that you are having to protect yourself for some reason, and you may want to use a smaller and softer ball or place some pillows under the hips and shoulders to lessen the stretch. Then, after working with the pelvis and sacroiliac joints, roll the body down an inch at a time so the ball makes contact with L5, then L4, and on toward the mid-lumbar region, continuing to roll from side to side in each new position. Take your time and explore the sensations. You may feel some discomfort posteriorly as well as muscular resistance in the thighs, muscular resistance that pulls downward on the front of the pelvis. This tension (on both right and left sides) comes from the rectus femoris muscle, which pulls inferiorly on the *anterior inferior iliac spine* (of the pelvis), and from the *sartorius muscle*, which, as the longest muscle in the body, pulls inferiorly on the *anterior superior iliac spine* all the way from the medial border of the tibia. Notice that breathing has little impact on the posture: your lumbar region is already arched so deeply that the crura of the diaphragm cannot further increase the lordosis.

You will have different sensations when the ball is under the mid-lumbar region (in comparison with the pelvis) because this segment of the spine permits a lot of backbending. If you have any lower back tenderness, your weight on the ball at this site will expose it quickly, but if you are entirely comfortable you will soon feel relaxation spreading down the body. First the shoulders relax and the upper chest sinks toward the floor; then the abdominal muscles start to relax, which allows your breathing to become slow and smooth; then the hip flexors (the psoas and iliacus muscles) begin to relax, along with the quadriceps femori, the adductors, and the hamstring muscles. As that happens the relaxation deepens and the back conforms more and more to the curvature of the ball (fig. 5.31).

When you roll the ball to the lumbar-thoracic junction, the lumbar lordosis becomes less pronounced but is capable of more mobility, and deep inhalations will accentuate its arch. Now it becomes more difficult to roll from side to side, but stretching the arms overhead adds to the stretch in the chest and opens the rib cage. This creates a perfect moment to take deep thoracic inhalations and expand your inspiratory capacity. The work in this region will now start to affect the neck because the head drops back passively, and the neck will have to be strong and flexible to support its dead weight. If you are uncomfortable you can place one or more pillows under your head to temper the stretch and tension.

As you roll the ball under the middle and upper thoracic region you can gradually lower the hips to the floor and slowly feel resistance melt away (fig. 5.32). As soon as that happens the lumbar region will become passive and will not be affected by your breathing, and you will notice that the exercise mostly affects the neck and rib cage. As you relax with the ball under the chest, you may feel small adjustments taking place at tiny *facet joints* located between the ribs and the vertebral bodies, and between the ribs and the transverse processes (figs. 4.6a, 4.7a–b, and 5.33). This can be helpful if a rib has become dislocated at the site of these joints by a sudden unusual movement when the spine is in a bent position. Someone who understands the anatomy of the joints between the ribs and the spine can work the head of the rib back into its correct position and bring relief, but passive supine backbending on a ball can also alleviate minor dislocations.

Figure 5.30. Passive supine backbend with an 8 1/2 inch playground ball under the pelvis and rolled slightly to one side. Here we are stimulating and mobilizing the sacroiliac joints.

With the ball under the neck you should try to relax from the lower chest down. At first you may experience some pulling sensations in the upper chest that are difficult to release, but after you have worked with the various postures in the shoulderstand series (chapter 9) and your neck has become strong and flexible, this posture becomes so comfortable that you can easily fall asleep in it. Do be watchful of this. It is easy to develop a stiff neck if you stop paying attention to what you are doing.

After you have gotten used to placing your weight straight back symmetrically, you can roll the head and neck from side to side against the ball, mashing down against it actively, working your way down the transverse processes of the vertebrae between C1 and C7. This is both stimulating and relaxing, and it is an intense but effective remedy for stiffness. It can be done softly against a spongy ball by someone who is delicate, more vigorously against a playground ball by the average person, and even more aggressively against a soccer ball, volley ball, or basketball by bodybuilders.

A KNEELING BACKBEND—THE CAMEL

A camel's back is about seven feet off the ground. Fortunately for its rider, a camel can kneel, so one does not have to embark and disembark to and from that height, and that is where this kneeling hatha yoga posture gets its name. Only intermediate and advanced students should try the full version, and even they should do the following test exercises if they have not done the posture before.

Figure 5.31. A ball placed directly under the lumbar region is supremely relaxing if the lower back is healthy, but if the subject is unable to relax the abdomen and thighs, this is evidence for current or incipient lower back problems.

Figure 5.32. A ball placed under the mid-chest region permits the pelvis to drop to the floor, and mobilizes the facet joints between the ribs and thoracic vertebral bodies, and between the ribs and the thoracic transverse processes (see figs. 4.6–7 and 5.33).

Kneel with your thighs at right angles to the floor with your weight placed on your knees and the bottom surfaces of your flexed toes. Next, place the hands on the hips, fingers down. Stretch the pelvis forward. Keeping enough tension in the quadriceps femoris and iliopsoas muscles to keep you from falling backward, lower your head to the rear and extend the spine backward until you are at the limit of the quadriceps and iliopsoas muscles' ability to hold the position comfortably. Watch your breath: if you have to restrain it at the glottis to protect your spine, you have gone too far.

At this point look and reach back with only one hand, swinging the other arm forward for balance. Do this first on one side and then on the other (fig. 5.34), making a mental note of how difficult it would be to place both hands against the heels or ankles at the same time. Now try the same exercise with your toes pointed to the rear. It is quite a bit harder. Because the heels are closer to the floor you have to reach back further than you do with the toes flexed, and reaching back further will require more stretch and strength of the quadriceps femoris muscles, which are already being worked and perhaps overworked just to support the posture. Finally,

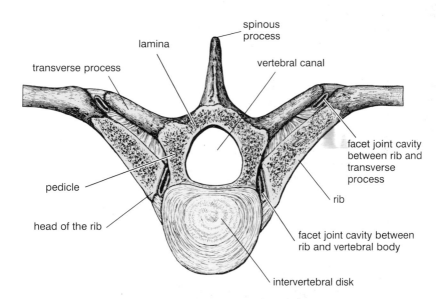

Figure 5.33. A cross-section through a mid-thoracic vertebra and two ribs. The viewpoint is from above, and it portrays an intervertebral disk along with the vertebral arch, pedicles, laminae, and vertebral canal. Four synovial facet joint cavities are also apparent, two between the heads of the ribs and the vertebral body, and two more between the ribs and the transverse processes (from Morris). See also figs. 4.6a and 4.7a–b.

if you are comfortable in the test exercises and confident that you can do the full posture and come out of it gracefully, flex your toes again, reach back, and either touch your heels momentarily with your middle fingers or rest your hands on your ankles. You can easily come out of either position by pushing up asymmetrically from one side. Come up, then come back down into the posture and push up from the other side first. It will be harder on the thighs to come up symmetrically, lifting up and away from both ankles at the same time. When you can finally come in and out of this posture gracefully, either one side at a time, or from both side symmetrically, start all over and work with the pose with the toes extended (fig. 5.35).

The camel posture might be considered a modification of a standing backbend, except that it is more advanced because you can't come into it halfway. You're either in it or you're not. In a standing backbend you can arch back to within your reasonable capacity, stay there for a while, and then come up, but the only way you can do the camel is to reach back for your heels or ankles, and keep on reaching until you make contact. There is another less demanding alternative if you are a beginner, and that is to support the lower back with your hands and arch back as far as your capacity permits. Over time, you will develop enough confidence to reach for one heel.

The camel generally places the body in an unusual combination of joint positions—hyperextension of the hips, stress on the knees from tension in the quadriceps femori, extension of the ankles, and extension of the spine. You have to focus your attention from head to toe, and at the same time you

Figure 5.34. Beginner's training pose for the camel. Reaching back toward the heel (toes flexed) with one hand while the other is reaching forward for balance is easy for healthy beginners.

have to intuit how far to reach back in order to touch your ankles. Just as important, you should come out of the posture the same way you came in, but now you are pulling up and forward with the already-stretched quadriceps femoris muscles, which are now at a mechanical disadvantage. Even though it may be difficult to pull yourself up and forward with those muscles, it's not a good idea to simply collapse down because that places even more stress on the knees.

CONTRAINDICATIONS

By most standards of hatha yoga, the cobra exercises given through and including the section on raising up and down with breathing are mild, meaning that anyone in average condition can do them. The rest of the cobras, however, as well as many of the other backbending postures that follow, including most of the locusts, boats, bows, the lumbar lift, the fish, the wheel, passive supine backbending, and the camel, may be overly stressful for anyone with a history of chronic back pain, or even worse, for anyone with acute back pain. Therefore be conservative, and consider those postures contraindicated for such conditions.

Any posture that increases intra-abdominal but not intrathoracic pressure may exacerbate hiatal hernia, and many of the more challenging backbending postures create these exact pressure differentials. If you sometimes have pain in the upper abdomen on the left side, especially after eating, re-read the discussion on hiatal hernia in chapter 3 so you can form your own judgment about doing the backward bending postures. And if you are still uncertain, seek medical advice.

Figure 5.35. Camel posture. Build up to it cautiously and systematically from the half-camel, first with the feet and toes flexed and finally with them extended, always making sure you can come out of the posture gracefully.

BENEFITS

Backbending, forward bending, twisting, inverted, and standing postures each have individual and group personalities. Among these, backbending postures are the most exhilarating and invigorating. As yoga postures they stimulate the sympathetic nervous system and prepare you for activity. In ordinary life activities, mild backbending relieves the tedium of sitting hunched over a desk, the stress from manipulating objects in the environment, and the habit of too much resting and reclining. In backbending you extend yourself, literally and figuratively, by lifting your posture, raising your spirits, and preparing yourself for action.

"Attachment follows the experience of pleasure— Aversion follows the experience of pain— Clinging-to-life is the sentiment which causes its own potency to flow equally even in the wise— Their operations are to be got rid of by meditation— "

— Bengali Baba, in *The Yogasutra of Patanjali*, pp. 34–35.

CHAPTER SIX
FORWARD BENDING
POSTURES

"As a result of practicing asana, you begin to understand your own body language. The body develops its own sensitivity and knows whether the food you eat is "right" or not. Your internal clock regulates your schedule precisely, and your body lets you know if you're exercising too much, if you're sleeping too much, and so forth."
— Pandit Rajmani Tigunait, in *Inner Quest*, p. 68.

"Can you touch your toes?" Flexibility is equated first and foremost with "flex-ability," and flexing forward—reaching for your toes—is its universal standard. It is what a hatha yoga teacher first sees, what a high school track coach on the alert for hamstring injuries is aware of, and it is the first thing personal trainers check when they measure your flexibility in health clubs. For testing purposes it means forward bending with the knees straight; if the knees are bent most people can fold forward almost completely.

Whether we are bending our knees or keeping them straight, we make vastly more use of forward bending than backbending. It is common enough to face an object, reach forward, and pick it up, but few can reach backward from an upright posture into the wheel, and even those who are able to do so would never reach over backward to retrieve something from the floor. We also spend hours every day in sitting forward bends—in front of computer screens, in car seats, in movie theaters, or on couches—generally with our backs rounded to the rear and our hips flexed. Everything considered, forward bending is more deeply ingrained in our bodies and nervous systems than any other posture.

In chapter 5 we saw that backward bending takes place mostly in the vertebral column, and that forward bending includes hip and ankle flexion as well as spinal bending. And because we can bend forward generously at the hips, we see many more forward bending than backbending postures in hatha yoga. They are included in one form or another in practically every chapter of this book.

We'll start by looking at the anatomy of all the sites where forward bending can take place, from head to toe. Then we'll focus on several forward bending postures in detail: the posterior stretch and its variations, the down-facing dog, and the child's pose. And finally, because forward bending relies so heavily on sacroiliac and hip flexibility, we'll turn to the various postures and exercises that encourage movement in the sacroiliac joints and that open the hip joints and make them more accommodating.

FORWARD BENDING: HEAD, NECK, AND CHEST

We saw in chapter 5 that the seven cervical vertebrae permit about 90° of forward flexion. In an upright posture, you can initiate a forward bend of the head with the tiniest nudge from the *sternocleidomastoid* muscles (fig. 8.11), which have a dual origin on the sternum and the clavicle and run from there up and back to the heavy bony protuberance (the *mastoid process*) just underneath the ear. After initiating the bend with these muscles, gravity carries the head further forward, controlled (as should by now be familiar) by eccentric lengthening of extensor muscles on the back side of the neck, as well as by tension in the elastic tissue in the ligamenta flava and the ligamentum nuchae (chapter 4).

As soon as the head comes forward, you can rest in that position and explore how the tissues respond. That is enough. It would be unnatural to make an extra effort to flex the neck while you are focused primarily on bending in the lower back and hip joints. Our main concern with the neck is that it be comfortable. Pain in one part of the body often has effects elsewhere, and neck pain doesn't have to be very serious before it stops your desire to do anything else.

The articulations of the ribs with the twelve thoracic vertebrae posteriorly (chapter 4) and with the costal cartilages and sternum anteriorly (figs. 2.5 and 4.3–4) creates a hollow basket, a fixed unit that cannot easily accommodate either forward, side, or backward bending. Forcing the thoracic vertebrae and the rib cage into such curves would only break ribs, disarticulate costal cartilages, and collapse the chest. Only about 10° total of forward to backward bending can take place within the normal thorax of an average twenty-year-old, which means that the front-to-back mobility of the upper part of the torso depends almost entirely on spinal flexibility in the lumbar region, sacroiliac mobility, and hip flexibility.

LUMBAR AND LUMBOSACRAL FORWARD BENDING

In chapter 5 we saw that most forward bending in the spine from T1 and below takes place in the six intervertebral disks in the lumbar area between the twelfth thoracic vertebra and the sacrum. Up to 90° of forward bending is occasionally seen here, but 30–80° is more common. A moderately flexible young athlete might reveal 40° of flexibility in the lumbar region plus 75° of flexibility in the hips, and just be able to grasp his toes with his outstretched fingers (fig. 6.1).

Students who have good spinal flexibility might be able to bend forward more than 50° in the lumbar area, but they will not ordinarily choose to do so in hatha yoga postures. It is more convenient for them to bend forward at the hip joints, where cartilage slips against cartilage in synovial fluid, than to perturb the six intervertebral disks between T12 and S1. For the average person hip flexibility does not permit this, and such people usually try to compensate by trying to bend forward more in the lower back. Yoga teachers know that this is a formula for trouble, and that is why they always say "bend from the hips."

THE DEEP BACK MUSCLES

It takes only one glance at a class of novices attempting sitting forward bends to see why it is not a good idea to force such postures: the resistance of the deep back muscles compresses the vertebral column, stresses the intervertebral disks, and strains ligaments from the sacrum to the head. So if this is the case, what do we want from the spine and deep back muscles? The answer is strength and adaptability: deep back muscles that can relax completely in the corpse posture, play their rightful role as extensile ligaments for maintaining an upright posture, and still act strongly as agents for bending and twisting.

Figure 6.1. This forward bend illustrates 75° of hip flexion and 40° of lumbar flexion in a moderately flexible young athlete (simulation).

SACROILIAC NUTATION AND COUNTERNUTATION

Even though the sacroiliac joints are synovial joints, their opposing surfaces usually fit together tightly enough for every movement of the pelvis to affect the sacrum (and therefore the spine as a whole), and for every movement of the sacrum to affect the pelvis. This view—that the sacroiliac joints are essentially immobile—has practical value, and it was in fact the only view until the 1930s, but it is an oversimplification: the synovial structure of the healthy sacroiliac joint is now known to provide its groove-and-rail architecture (figs. 3.3 and 6.2b) with the capacity for a small amount of slippage—movements that have been called *nutation* and *counternutation* by the French orthopedist I.A. Kapandji.

Even though nutation and counternutation are often minimal, these movements are important for doing backward bending, forward bending, and seated meditation postures with the niceties and refinements that expert hatha yogis take for granted. The difficulty is that few hatha yogis (or for that matter exercise specialists of any variety) have encountered discussion of how the movements function in posture, athletics, and day-to-day life. To understand how and why nutation and counternutation are important to us, we must look at their scope and nature in detail.

NUTATION AND COUNTERNUTATION

Nutation and counternutation are not complicated concepts as long as one understands three points: The first is that the sacrum rotates roughly in a sagittal (front-to-back) plane *within* the pelvic bowl. Nutation rotates the *promontory* (the top front border) of the sacrum anteriorly (toward the front of the body) and it rotates the coccyx posteriorly (toward the back of the body); counternutation rotates the promontory of the sacrum posteriorly and the coccyx anteriorly. It is important not to confuse these specialized sacroiliac rotations with anterior and posterior pelvic tilts, which are movements of the pelvis as a whole. Nutation and counternutation are sacroiliac rail-and-groove slippages of the sacrum *between* the pelvic bones (fig. 6.2a), not tilts of the entire pelvic bowl.

The second point to understand is that the planes in which the sacroiliac joints lie are not parallel to one another. If they were—if the joints were situated in exact parasagittal planes (parallel to the midsagittal plane of the body)—the sacrum could rotate without disturbing the configuration of the pelvis as whole. That doesn't and indeed can't happen because the sacrum has a broad wedge-like shape with the leading face of the wedge pointing to the rear. And since the mating surfaces of the sacroiliac joints always remain in close apposition, nutation pulls the ilia closer together (that is, toward the midline) as the promontory of the sacrum rotates forward, and counternutation forces the ilia laterally (that is, away from the midline) as the promontory of the sacrum rotates to the rear (fig 6.2a).

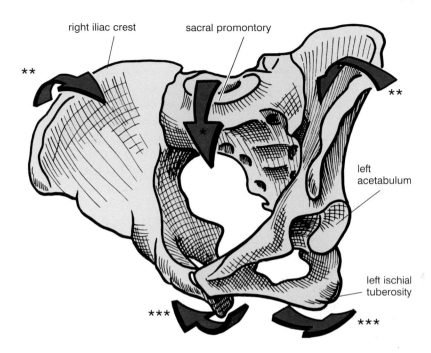

Figure 6.2a. The gross movements of the pelvis and sacrum that are involved in sacroiliac nutation are indicated by arrows. The promontory of the sacrum is thrust forward (*), the iliac crests are shifted medially (**), and the ischial tuberosities are spread apart (***). For counternutation, the shifts are in the opposite direction: the iliac crests move laterally, the sacral promontory moves posteriorly, and the ischia move medially (from Kapandji, with permission).

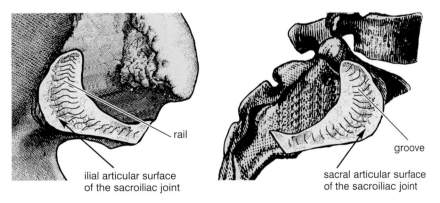

Figure 6.2b. Enlargement of fig. 3.3 showing the matching rail (on the left) and groove (on the right) architecture of an idealized sacroiliac joint. Such a joint might permit up to 10° of slippage (essentially a rotation) between full nutation and full counternutation. The pelvic bone (left) has been disarticulated from the sacrum (right) and flipped horizontally to reveal the articular surfaces and to suggest how a healthy sacroiliac joint could permit this much movement (Sappey).

Finally, remember from chapter 3 that each pelvic bone (one on each side) is composed of the fused ilium, ischium, and pubis, with the ilia on top at the waistline, the pubic components in front above the genitals, and the ischia with their tuberosities below and behind (figs. 3.2–4). When the ilia are pulled toward the midline during nutation, the ischia and ischial tuberosities have to swing laterally, and when the ilia are spread apart by counternutation, the ischial tuberosities rock back toward one another. We should note that even though nutation and counternutation were named in recognition of the movements of the sacrum alone (nutation means nodding forward, in this case nodding forward of the promontory of the sacrum), the accompanying movements of the pelvic bones are equally important (fig. 6.2a).

The way these movements are orchestrated during childbirth further clarifies their function. Again, according to Kapandji, during the early stages of labor counternutation draws the promontory of the sacrum to the rear and spreads the ilia, making more space for the baby's head as it approaches the birth canal. Then, as the head enters the vagina, nutation draws the coccyx to the rear and spreads the ischia, opening the base of the body and easing delivery.

NUTATION AND COUNTERNUTATION IN HATHA YOGA

Four fundamental movements illustrate how the concepts of nutation and counternutation apply to hatha yoga. First, to create maximum nutation, stand with the thighs abducted (feet perhaps 3 feet apart) and come into a forward bend purely from the hips. To avoid forward bending in the spine, most people should reach out to a desktop or wall (fig. 5.7, except with the thighs abducted). Get settled into a comfortable position in which you can monitor your pelvis and vertebral column, and then try to thrust your coccyx even more posteriorly and the promontory of the sacrum even more anteriorly. If your sacroiliac joints permit, you will feel some peculiar shifts within the pelvis, shifts which flexible students report as "spreading the sitting bones" or "feeling the thighs come apart." Such descriptions are not metaphors; these movements actually happen during the course of nutation. You may be able to feel them in yourself, and you can certainly appraise them in a highly flexible partner, either by placing your thumbs against the insides of their ischial tuberosities from behind, or by lying down on the floor and placing your hands against the inside of their upper thighs. You may not be able to feel the movements of your partner's sacrum, but the movements of the ischial tuberosities and of the femurs are unforgettable. Easing up on the posture will produce counternutation and pull the ischia and upper thighs back together.

Next, consider relaxed standing lumbar backbends (figs. 4.20 and 5.2), which provide yet another example of nutation. As you bend lackadaisically to the rear you can easily imagine that the ilia are pulled backward and medially in relation to the upper part of the sacrum. The top of the sacrum moves backward as well, but its promontory is squeezed *forward* relative to the ilia. The coccyx, by contrast, moves to the rear in relation to the pelvic bowl, and the ischial tuberosities spread apart. These conditions all define nutation—the nodding forward of the promontory of the sacrum. It may seem counter-intuitive that nutation accompanies a relaxed standing backbend, but that is what happens.

Third, in contrast to forward bending from the hips and relaxed lumbar backbends, easy standing forward bends from the waist result in counter-nutation. Here the ilia are first pulled forward and laterally in relation to the upper part of the sacrum, which means that the sacral promontory moves backward (again relative to the pelvic bowl). And as the ilia are spread apart, the coccyx moves forward and the ischial tuberosities come closer together.

Fourth, recall from chapter 5 that most hatha yoga teachers ask students to squeeze their hips together and create whole-body backward bends rather than relaxed lumbar bends. Squeezing the hips together, along with keeping plenty of eccentric tension in the iliacus and rectus femoris muscles, maintains the attitude of counternutation by holding the ischia close together and by keeping the promontory of the sacrum pulled to the rear rather than allowing it to be squeezed forward. Yoga teachers intuitively recognize that it's a sharper, safer posture, especially for beginners.

Reiterating: to avoid getting confused, the reader must constantly remember that nutation and counternutation describe movements of the sacrum and pelvic bones purely in relation to one another, and that these movements may or may not be the opposite of pelvic bowl movements during the course of backbending and forward bending. For backbends in general as well as forward bends in general, movements of the pelvic bowl as an entity are exactly what one would figure out logically: the upper rim of the pelvis tilts backward during backbending (a posterior pelvic tilt) and forward during forward bending (an anterior pelvic tilt). Just don't forget that such pelvic tilts are entirely separate from the shifts between nutation and counternutation.

It is difficult to feel and measure the movements of the ilia during nutation and counternutation, but the medial and lateral excursions of the ischial tuberosities and thighs provide us with windows that allow us to evaluate the other components of these specialized movements. Assuming that you have some sacroiliac mobility, place your middle fingers tightly against the medial borders of your ischial tuberosities, and ask someone to measure the distance between your fingernails during a relaxed standing backbend. What they'll find is that this span measures about 2 inches in the male and 3 1/2 inches in the female, which represents maximum nutation (ilia in, ischia out; fig. 6.2a). It is important to stay relaxed for the measurement, however, because if you tighten your hips, the gluteal muscles will pull your fingers closer together and skew the measurement to the low side.

Next come all the way forward (bending from the waist), keeping your fingers tightly in position. Now a flexible young man might show 1 1/2 inches across the same span between the fingernails (a decrease of 1/2 inch), and a flexible young woman might show 2 1/2 inches across the same span (a decrease of 1 inch). This represents maximum counternutation (ilia out, ischia in; fig. 6.2a).

Now you know, or can at least imagine, why yoga teachers ask you to bend forward from the hips (fig. 6.10–11). In contrast to what happens in a forward bend from the waist, the first thing you will accomplish, or at least try to accomplish, is nutation. If you have sufficient sacroiliac mobility, this takes place automatically even before you bend at the hips. And the opposite is true of whole-body backbends. For such bends, the safest attitude, especially for beginners, is full counternutation (fig. 4.19). The nutation that is established as a priority in relaxed lumbar backbends (fig. 4.20) is best reserved for experts who are fully confident in their lower backs and who need full nutation for coming into extreme backbending positions such as the advanced cobra (fig. 5.12) or the wheel (fig.5.29).

FORWARD BENDING AT THE HIP JOINTS

If your ability to bend forward were to be tested casually or even objectively with devices that measure how far forward you can reach in a sitting forward bend, little or no distinction would be made between how much of that flexibility is in the hips and how much is in the lower back.

[Technical note: Under these circumstances, no consideration would be given to the possible role of sacroiliac movements. That's just as well, because whether nutation or counternutation is more significant will depend on the person: nutation will be more prominent in those who have good hip flexibility, and counternutation will be more characteristic of those who have to bend mostly from the waist. Sacroiliac movements take care of themselves naturally and will therefore be ignored in the following discussion.]

If you are unusually flexible and can lay your chest down against your thighs while keeping the knees straight, we can estimate that the first stage of the bend will be about 120° of flexion at the hip joints and that the second stage will be about 60° of flexion in the back. The fact that someone who is this flexible can bend twice as much in the hips as in the lumbar spine gives some idea of how important hip flexibility is to forward bending in general, and it also accounts for why yoga teachers place so much emphasis on it: they know how useful hip flexibility is to them personally. Here we'll focus on its main impediments—hamstrings and adductors—for those who are not so flexible.

THE HAMSTRING MUSCLES

The hamstring muscles include the *biceps femoris*, the *semimembranosus*, and the *semitendinosus muscles*, and except for the short head of the biceps femoris, which takes origin from the femur, they all take origin from the ischial tuberosities and run from there along the rear of the thighs all the way to the bones of the leg—the tibia and fibula (figs. 3.10b, 8.9–10, and 8.12).

Because the hamstring muscles bypass the distal end of the femur, flexing the knee joint brings their insertions closer to their origins and releases tension throughout the backs of the thighs during any forward bend. Extending the knees in a forward bend, on the other hand, stretches the hamstrings and generates tension that pulls on the underside of the pelvis. This in turn makes it difficult to rotate the top of the pelvis forward (in an anterior pelvic tilt) and remove tension from the flexed-forward lumbar region.

A simulation of these effects in a flexible young man will clarify the role of the hamstrings in forward bending. First he should lie supine with one knee straight and the other knee flexed, and then he should draw the bent knee toward the chest without pulling it so far that the pelvis is lifted away from the floor. Keeping the knee flexed relieves hamstring tension so effectively that the hip can now be flexed through a range of about 150° from its supine starting position, creating a 30° angle with the floor (fig. 6.3). Then as he extends the knee, the hamstring muscles will first be pulled taut and then force a partial extension of the thigh to a less acute angle, perhaps to a nearly straight up 80° (fig. 6.4). This means that when his knees are extended the hamstrings limit hip flexion by 50°.

THE ADDUCTOR MUSCLES

Some of the adductor muscles of the thigh (figs. 2.8, 3.8–9, and 8.13–14) have a hamstring character. That is, in addition to drawing the thighs together they pull them posteriorly. And since some segments of the adductors (chapter 4) take origin on the inferior pubic rami near the ischial tuberosities (chapter 3), it is obvious that they will limit forward bending exactly as the hamstrings do—by exerting tension on the underside of the pelvis and creating a posterior pelvic tilt. It is also plain that the adductors will be stretched the least when the thighs are together and the most when the feet are wide apart. Therefore, with the thighs adducted, the limitations to hip flexibility are mostly from the hamstrings (figs. 3.10b, 8.9–10, and 8.12).

Abducted, any additional restriction to forward bending can only be from the adductors. How this operates depends on the person. As shown here, the adductors limit hip flexion the most in those who start with less hip flexibility (figs. 6.5–6).

FORWARD BENDING AT THE ANKLES AND IN THE FEET

If you have average flexibility in the ankles, you will be able to *flex* (*dorsiflex*) the foot no more than 20–30° and *extend* (*plantarflex*) the foot no more than 30–50°. But you need to flex the ankle 45° to drop the heels to the floor in hatha yoga postures such as the down-facing dog (figs. 6.17 and 8.26), as well as to sit in a squat (especially with the feet parallel) without lifting the heels (fig. 6.7). Mobility for extending the ankle is also needed for postures such as the upward-facing dog (fig. 5.13) and for sitting comfortably on the heels with the toes pointed to the rear. The pained facial expressions in a

Figure 6.3. Hip flexion of up to about 150° is permitted when tension on the hamstring muscles is released by flexing the knee.

Figure 6.4. When the knee is held fully extended, the hamstrings limit most people to less than 90° of hip flexion, in this case a simulated 80°.

room filled with beginning hatha students trying to sit in such a posture reflects how rarely it is used in Western societies.

Flexion and extension at the ankle takes place at the *talocrural joint*, which is located between the distal ends of the tibia and fibula on one hand and one of the tarsal bones of the foot, the *talus*, on the other (fig. 6.8). The distal ends of the tibia and fibula form a hemi-cylinder (a cavity shaped roughly like half a cylinder) which articulates with the pulley-shaped upper surface of the talus. The rest of the bones of the foot permit little movement except for the metatarsals and phalanges, which become important when the feet and toes are flexed (that is, when the feet are dorsiflexed and the toes are curled toward the head) and when they are extended (that is, when the feet are plantarflexed and the toes are pointed away from the body).

As in all joints, movements of the ankle are limited by muscular tension, ligaments, and bone, and it is tension in the soleus and gastrocnemius muscles (figs. 3.10, 7.6, 8.9–10, and 8.12) on the back of the leg that first prevents you from pressing the heels to the floor in an attitude of flexion (that is,

Figure 6.5. With the thighs adducted, the hamstring muscles permit about 80° of hip flexion.

Figure 6.6. With the thighs abducted, the increased tension on the adductors (some of which have a hamstring character) can further limit hip flexion, in this case to about 60°.

dorsiflexion) in the down-facing dog. Extension (that is, plantarflexion) is more likely to be checked by ligaments and bone, although tension in the flexors, which are situated on the anterior sides of the bones in the legs, will curb extension in everyone whose ankle flexibility is severely limited. Over time, extension of the ankle joint takes care of itself with the practice of postures such as the upward-facing dog with the toes extended (fig. 5.13). Developing the capacity for flexion is more difficult because the gastrocnemius and soleus muscles, as well as the ankle joints, have developed lifelong habits for functioning within limited lengths and ranges of motion. Besides the down-facing dog, possibly the best practice for correcting an incapacity for flexion is simply to sit in a squat for 2–3 minutes several times a day. At first you can squat with your feet well apart and your toes angled out widely enough to permit your heels to reach the floor (fig. 6.9), and after you have acclimated to that you can gradually bring your feet closer together and parallel.

Figure 6.7. Excellent ankle flexibility for flexion is required for squatting flat on the floor with the feet together and parallel.

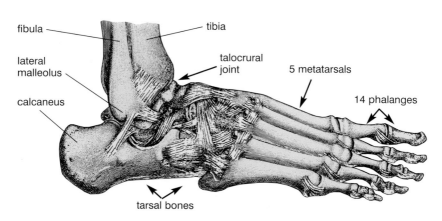

Figure 6.8. The 28 bones and 25 joints associated with each foot and ankle support the weight of the body and accomodate to uneven surfaces for walking and running. The talocrural joint is the one which we are constantly aware of in hatha yoga because that is where we experience the flexion and extension of the ankle needed (and frequently lacking) for so many postures (Sappey).

CLINICAL MATTERS AND CAUTIONS

Contraindications to forward bending are obvious. Don't do any forward bending postures if you have acute back pain; instead, get yourself under the care of a professional, who by tradition will probably tell you to go to bed and rest. But there are many grey areas that indicate caution rather than contraindication. Muscular tension sometimes edges over into mild discomfort or pain, and even if you have come into a posture carefully you may still experience sensations that you do not know how to interpret. Explore these carefully and try to analyze their nature and source. Try to discover if they include pain, stretch, or a combination of the two, and then try to localize the specific sites of the sensations. The idea is to learn where the forward bend is being limited so you can attend more specifically to that site. You may decide to limit yourself to being watchful, or you may decide to plan a program for working with the region more actively.

LOW BACK PAIN IN GENERAL

The causes of low back pain are legion, and attempting to consider them in detail is beyond the scope of this book. Nevertheless, a few comments on locating the pertinent anatomical hot spots are appropriate. Looking from above down, lumbar and lumbosacral pain appear just above the pelvis and close to the midline posteriorly. If pain just lateral to this region is found only on one side, it may be due to stress and weakness in one of the *quadratus lumborum* muscles, which are interposed between the psoas muscles on one hand and the erector spinae and abdominal muscles on the other, and which strengthen the all-important connections between the ilia and the rib cage (figs. 2.7, 3.7, 5.5, and 8.14). Alternatively, and possibly a little lower and more precisely localized, tension on the *iliolumbar ligaments*, which run between the fourth and fifth lumbar vertebrae to the pelvis (fig. 3.4), may manifest as a slight pulling which extends from the

Figure 6.9. Squatting with the feet well apart and angled out is one of the best methods to gradually develop ankle flexibility for flexion.

lowest two lumbar vertebrae laterally to the crests of the ilia on both sides. These important ligaments stabilize the lumbar region in relation to the pelvis and augment stabilization of the sacroiliac joint. One or more of the iliolumbar ligaments is occasionally stressed, generally on one side of the body, and is a common source of low back pain. If you experience some relief from digging your thumbs into the grooves just medial to the crest of the ilium on each side, especially while you are doing a standing backbend, you are probably alleviating strain on these ligaments.

Foreshortened hamstrings, adductors, and hip flexors are often contributing elements to low back pain. Because our ordinary activities do not keep these muscles stretched and flexible, they can gradually become shorter, occasioning chronic tension, injury, and subsequent muscle spasm in the deep back muscles and hips, not to mention a general resistance to forward bending. A long-term course of stretching and strengthening, with emphasis on the latter, is indicated.

SACROILIAC SPRAIN

Continuing inferiorly, malfunctional sacroiliac joints are also frequently associated with low back pain. Their architecture varies, not only from person to person, but from one side of the body to the other. Only one characteristic is constant—reciprocal mating surfaces that match one another (figs. 3.3 and 6.2b). An indentation on one surface of the joint always matches a tubercle (bulge) on the other, and a ridge on one side always matches a groove on the other. If the matching surfaces are smooth, movement will be free, but if they are irregular, with many peaks and valleys, movement will be limited, and over a long period of time we would not be surprised to see that fibrotic connections have formed that bind the opposing surfaces together. If that process is not interrupted with daily exercise, stretching, and hatha yoga postures that are designed to encourage nutation and counternutation, *ankylosis* (that is, partial or complete fusion) of the joints may develop—and if that happens, as mentioned earlier, any dislocation can produce extreme pain and trauma.

Pain from dislocations of the sacroiliac joint appears lower and slightly more lateral to sites that are exhibiting lumbar, lumbosacral, and iliolumbar ligament pain, usually more on one side than the other. If the surrounding muscles do not keep the joint protected when it is under stress, the fibrous tissue that binds the sacrum to one or the other side of the ilium can be torn, with repercussions that can last for many years if the injury is not diagnosed and properly treated. This condition, called *sacroiliac sprain*, may be so painful that the surrounding muscles immobilize the joint. Then, as such a sprain heals, the offended joint often becomes locked while the one on the opposite side compensates with too much mobility.

If that happens you can try to restrain movement on the side that is overly mobile and promote movement on the side that is locked, and if you get to work on the condition before the joint becomes completely fused, you may be able to gradually regain mobility and balance. Asymmetric standing postures are ideal, as is the pigeon, which we'll cover later in this chapter. You can work with these postures three times, favoring the tight side first and last, so long as you keep in mind that any program trying to deal with anything more than minor problems should be approved by a professional.

SCIATICA

There is one more thing to consider—the *sciatic nerve* (fig. 3.10a)—that is loosely associated (at least by the general public) with lower back problems. Most nerves course alongside and among the flexor muscles of joints, so that folding the joint releases tension on the nerve. But the sciatic nerve, which passes posteriorly through the hip on the extensor side, is a glaring exception: hip flexion places it under more rather than less tension. This is ordinarily not a problem, but when muscles deep in the hip are injured, scar tissue forms during the healing process and frequently restricts movement of the sciatic nerve somewhere along its course through the muscles supporting the hip joint. This usually happens in the region of the *piriformis muscle* (figs. 3.8b, 3.10a, and 8.12). The result is *sciatica*—pain that radiates down the back of the thigh. If, after a seemingly minor injury, you get a dull persistent pain in one hip and thigh when you are forward bending, or even just walking or sitting, it is probably sciatica. It can last for days, weeks, or even years, but assuming the source of the problem is in the hips, it can usually be treated successfully by manual medicine and bodywork, often in combination with a program of stretching.

LOW BACKS: LIVING AND LEARNING

Working with all lower back problems that do not go away with a few days of rest and recuperation is likely to require commitment to a long-term program. To be on the safe side, drop false expectations and think in terms of 5–20 years of consistent, patient effort. Here are your specific aims: relief from and healing of low back pain in the short run, strengthening the back and abdominal muscles in the intermediate span of time when you are likely to be afflicted with chronic stiffness, gradually increasing flexibility in the long run, and above all, professional care from beginning to end. Surgery? Maybe, but do your own research and get a second and even third opinion. And no matter what, get attention from someone who is truly interested in working with these problems in a three-way partnership: the professional, the patient's interest and enthusiasm, and the aggravated back.

THE POSTERIOR STRETCH

If a traditional hatha yoga instructor were to teach a student only five postures, they would probably be the headstand, the shoulderstand, the sitting half spinal twist, the cobra, and the posterior stretch—the definitive forward bend. The posterior stretch can be initiated properly only after full nutation of the sacroiliac joints; after accomplishing that, its essence can be known only if the pose is hinged primarily at the hip joints and completed with a minimum amount of spinal flexion. Although we'll discuss posterior stretches for students with a wide range of skills, the plain truth is that the full experience of this posture is denied everyone who lacks good sacroiliac and hip flexibility.

To do the posterior stretch, you start in a sitting position with the back straight, the knees extended, and the heels and toes together (fig. 6.10). Stretch the hands overhead and thrust the sacral promontory forward through the action of the psoas muscles in order to emphasize nutation. Then, keeping the promontory of the sacrum forward, the ilia closer together, and ischia apart, fold forward slowly from the hips. As soon as you have reached your limits of hip flexion (fig. 6.11), bend forward in the lumbar region with the aid of gravity. This movement should not be passive; it should be accompanied by actively lengthening and stretching the torso. Finally, as you flatten your chest down against your thighs (fig. 6.12), the sacroiliac joints slip back into a neutral position between nutation and counternutation. In the final posture, the knees are straight, and the ankles, feet, and toes are all flexed, completing a literal posterior stretch from head to toe.

Figure 6.10. Beginning position for posterior stretch for advanced students. Notice that the model starts with his hips flexed about 100° (10° beyond vertical) merely by reaching straight up with his hands.

BEGINNING FORWARD BENDING

The above instructions are fine for a select few, but preposterous for the average person. In the first place, when they are sitting on the floor in the starting position, inflexible beginning students begin the pose with their sacroiliac joints in full counternutation instead of nutation. They probably won't be aware of this, but they will be acutely aware and frustrated that they are already bent forward at their hips to their limit just trying to sit up straight. They also won't be very appreciative of the instructor who softly intones "let gravity gently carry you forward" when gravity is pulling them nowhere but backward.

Still, an alternative is needed for those who are not flexible enough to roll forward into the classic posture, and the best is a natural sequel to one of the sit-up exercises in chapter 3. Sit flat on the floor with the knees extended, the head forward, and the back rounded. For now, at least, forget about sacroiliac movements. Place each fist in the opposite armpit, and slowly pull forward using the hip flexors (the iliacus and psoas muscles).

Figure 6.11. An intermediate position for the posterior stretch in which the hips are flexed about 130°. The model has now "bent forward from the hips" about 30° (from 100° to 130°). Even so, the lumbar lordosis has already started to flatten in comparison with the full lumbar arch shown in fig. 6.10, which is evidence for having already begun the process of bending forward "at the waist."

Figure 6.12. The completed posterior stretch, with knees straight and the chest flattened down against the thighs.

Locate the pulling sensation from these muscles deep in the pelvis. This is easy since they are doing nearly all the work. It is also important to keep the knees fully extended and the thighs in place by tensing the quadriceps femori. We want those muscles to stabilize the thighs and knees, thus enabling the iliopsoas muscles to act only on the pelvis to pull the torso forward. (If the quadriceps femoris and iliopsoas muscles were to act in the other direction to lift the legs and thighs, we would end up in one of the boat postures, as discussed in chapter 3, rather than a sitting forward bend.) When you have flexed forward at the hips as far as you can, hold the bend to your comfortable capacity (fig. 6.13). This posture is excellent for beginners, not only because everyone can do it, but also because it brings attention to the places where the posture is being limited: the lower back, the pelvis, and the hip joints. It is safe as well as rewarding.

The hip flexors pull you forward concentrically in this exercise, but they also eccentrically resist dropping to the rear if you slowly roll part way back while keeping the head forward and the lumbar region flexed (fig. 6.14). Do not lower yourself so far to the rear that you don't have enough strength to lift up gracefully. Go back and forth between the two positions (all the way forward and moderately to the rear) 5–10 times to become familiar with the sensations. If you wish you can always roll all the way down and

Figure 6.13. The hip flexors (iliacus and psoas muscles) and the rectus femoris muscles are the prime movers for pulling forward concentrically, while the quadriceps femoris muscles as a whole keep the knees extended.

exhalation permits down and forward movement

inhalation lifts the posture up and back

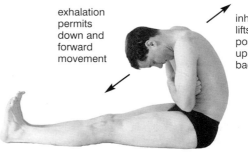

Figure 6.14. Leaning slowly to the rear with the fists still in the armpits, the hip flexors and rectus femoris muscles lengthen eccentrically to restrain the movement. Then, as you come back up and again pull forward, the same muscles shorten concentrically.

combine the exercise with yoga sit-ups. Just make sure that you can come back up without having to hold your breath and throw yourself up from the supine position.

The next stage might be called a posterior stretch modified for beginners. To do it, rest your hands on your thighs and draw the toes toward the head. Reach forward with the hands, flexing the neck, and bend forward by flexing the lumbar region and the hips at the same time, concentrating on using the hip flexors exactly as in the exercise in which the fists were in the armpits. Create moderate tension in the quadriceps femori to brace the knees against the floor. Then, after sliding the hands down the lower extremities, settle into a comfortable position holding on to the legs, ankles, or whatever you can reach without pulling or tugging. Remain in the posture for 1–5 minutes. If you are not comfortable it means you have ensconced yourself into a pose that is beyond your reasonable capacity. In any case, to come out of this beginning posture, slowly roll up, releasing first in the more vulnerable hips and lumbar spine, and then in the rest of the body. As in the previous posture, don't be overly concerned with sacroiliac movements.

THE INTERMEDIATE POSTERIOR STRETCH

Once you have achieved a little hip flexibility and are familiar with the operation of the hip flexors, you can try a real posterior stretch. Sitting on the floor with the knees extended, the feet together, and the hands stretched overhead, try to bend forward leading with the promontory of the sacrum so as to start, at least, with your full capacity for sacroiliac mutation. What you are doing is trying to keep the lumbar lordosis intact as long as possible. This may be difficult because if your hamstrings and adductors are too short, they will keep the top of the pelvic bowl tilted to the rear in a posterior pelvic tilt. The hamstrings and adductors are not an impediment to the previous exercises for beginners, in which you were purposely rounding the back to the rear, but here they prevent those who are inflexible from even sitting up straight, much less flexing forward at the hips. There is no quick remedy for this state of affairs, and you will probably have to work with the hamstrings and adductors for a long time before you see much difference in your forward bending.

Stretching your hands overhead as high as possible in the starting position will bring you an immediate understanding of your limits in this posture. If you have good hip flexibility, you will feel the back arch forward in the lumbar region and create an anterior pelvic tilt merely by stretching up. In this case, your torso might be at a right angle from the floor, defining a 90° forward bend from the hips. It is more likely, however, that you do not have that much hip flexibility, and just trying to come into the initial stretched-up posture will stretch your hamstrings to their limits. In any case, to complete the posture as gracefully as you can, maintain as much lift as possible with the upper extremities, keeping the knees extended and the feet together, and bend forward slowly, allowing gravity to help the back become rounded.

As soon as you reach your limit of bending with the upper extremities stretched out and forward, drop the hands to the thighs, knees, ankles, or feet, and remain in that position for a minute or two. Relax. Most students will try to tug on their feet or legs with their hands to pull themselves down, but this should be discouraged. Only when you have achieved good hip flexibility is it advisable to pull on the feet with the hands (fig. 6.15), and even then the object should be lengthening and stretching the back, not forcibly flexing the lumbar spine. Sitting on a cushion will also help because it will take some of the tension off the hamstrings and permit the pelvis to rotate forward at the hip joints without depending so much on spinal flexion.

To come out of the posture intermediate students should think first of lengthening the back, firming up the lower half of the body, and then stretching forward from a feeling of strength in the abdomen and lower back. From there they can lift the arms, raise the head, and unfold at the hips. Coming up this way adds to the tension in the lumbar region and hips; this is inadvisable for beginners but should work for experienced students. If there is any doubt, forego the elegance and release the pose initially in the hip joints and lower back as recommended for beginners.

THE LATERAL ROTATORS OF THE THIGH

After the intermediate levels of the posterior stretch have been experienced, it will be helpful to explore the muscles that impede your progress more systematically. We'll start with certain of the lateral rotators of the thigh that also act as abductors in sitting positions. Coming into the posterior stretch with your heels and toes together, you may notice a sense of lateral pulling deep in the hip. This comes from stretching four paired muscles: the *piriformis*, the *obturator internus*, and the *superior and inferior gemelli* (figs. 3.8b, 3.10b, and 8.12), all of which are lateral rotators of the thighs.

Figure 6.15. The intermediate student working on the posterior stretch should still concentrate on internal work with the hip flexors and rectus abdominis muscles rather than trying to pull forceably into the posture with the upper extremities.

To pin down for certain the extent to which these four paired muscles are limiting your forward bending, sit on the floor with your knees extended and your feet far enough apart for the big toes to barely touch when the thighs are rotated medially as much as possible. Then bend forward, taking note of exactly how far you can come into the posture and of how much lateral pulling this produces in the hip. Then come up a little, swing the toes out (which rotates the thighs laterally), and again bend forward. This relieves the tension, allows you to come further forward, and demonstrates how much the lateral rotators were limiting your bend.

The piriformis, obturator internus, and gemelli muscles are located beneath the superficial gluteal muscles. Even though they are deep, it is easy to envision their roles if you understand that their origins are located on the anterior side of the sacrum, and that they pass across the hip to insert on the lateral side of the femur from behind. In an upright posture they function to rotate the thigh laterally, but in a sitting position with the thighs flexed 90° they become abductors as well. Therefore, they are stretched the most when the thighs are *adducted*, rotated medially, and flexed. Don't confuse the actions of these muscles with the other lateral rotators of the thighs, the adductor longus and adductor magnus (see figs. 3.8–9 and 8.13–14 for general treatment of adductors). As we saw in chapter 4 those muscles are stretched most effectively when the thighs are *abducted* (rather than adducted), rotated medially, and flexed. To understand how the two sets of lateral rotators differ in limiting forward bending, go back and forth between the standing test with thighs abducted (chapter 4) and the sitting test just described with thighs adducted.

Because most of us are not perfectly symmetrical, the piriformis, obturator internus, and gemellis are often under more stress on one side than the other, and sooner or later this results in shorter muscles on the stressed side. If right and left footprints are asymmetrical, this is suggestive of such imbalances: left footprints that angle out to the side more than right footprints indicates tighter left lateral rotators. The posterior stretch will also reveal muscular imbalances because the disparity will leave the foot on the easy side upright and the foot on the problematic side dropped to the side. For example, if the right foot is rotated out and the left foot is upright in the posterior stretch, it means that the lateral rotators of the right thigh are under more tension than those on the left. The diagnosis is clear, the remedy obvious: asymmetric postures of all kinds that attend more insistently to the tighter side.

THE HAMSTRINGS

If the lateral rotators of the thigh are not under too much stress, the hamstrings are the muscles most likely to limit the posterior stretch since they extend all the way from the ischial tuberosities to the tibia. If you feel stretch in the tendons on either side of the knee joint and in the muscles on the back sides of the thighs when you bend forward, you are feeling the hamstrings (figs. 3.10b, 8.9–10, and 8.12). To test this, bend the knees slightly; what releases can only be the hamstring muscles and the adductor muscles that have a hamstring-like character.

THE PSOAS, ILIACUS, AND RECTUS FEMORIS

The psoas and iliacus muscles are classically thought of as hip flexors, as when they flex the thighs in leglifts from a stabilized pelvic bowl (chapter 3). But when the thighs are fixed against the floor in the posterior stretch, the origins and insertions of these muscles are reversed, with origins on the femurs and insertions on the ilia (in the case of the iliacus) and the lumbar spine (in the case of the psoas). Those muscles are responsible, along with the rectus femori (figs. 3.9 and 8.8–9), for pulling the torso forward at the hips in the sitting forward bend.

The iliacus, psoas, and rectus femoris muscles all act synergistically to create forward bending at the hips, but they act at different sites on the torso. The action of the iliacus is the most straightforward. From the stabilized femur it pulls on the inside of the ilium, tilting the top of the pelvic bowl to flex the pelvis forward on the thighs at the hip joint. We can think of the iliacus muscles as pure pelvic bone flexors in this posture. And since they pull forward exclusively on the ilia, they also create a tendency for counternutation (fig. 6.2a, movements opposite to the arrows). Ultimately, however, since the spine is attached to the pelvis at the sacroiliac joints, the vertebral column (and thus the rest of the body) follows the forward movement of the pelvic bones.

The action of the psoas muscles is more complicated. Because they attach to the lumbar vertebral bodies they pull forward only on the spine, and since the spine includes the sacrum they pull its promontory forward. In contrast to the actions of the iliacus muscles, the psoas muscles pull the sacrum forward between the two ilia (independently of the pelvis as a whole), and in so doing they create nutation, leaving the ilia behind where they will be drawn medially, spreading the ischial tuberosities apart, and opening the base of the pelvis (fig. 6.2a, arrows). (Recall that this situation is opposite from what we saw in a standing forward bend from the waist, which created counternutation instead of nutation.) The movements of the ilia are difficult to feel, but the psoas-induced lateral movement of the ischial tuberosities in those who have good sacroiliac flexibility are easily discernable. All you have to do is place one finger under each sitting bone and bend forward from the hips. If you place the strongest possible emphasis on keeping the lumbar region arched forward as you initiate the bend, you will feel the ischial tuberosities spread apart laterally.

The action of the rectus femoris is similar to that of the iliacus. When the knees are fully extended and the thighs are fixed against the floor, the only thing this muscle can do is pull forward on the anterior inferior iliac spine, making it, like the iliacus, another pelvic bone flexor. The main difference is that the iliacus acts from the upper part of the femur and the rectus femoris acts from the extended knee joints.

THE ABDOMINAL MUSCLES

Even if you can bend forward 100–120° at the hips, you will still have to bend forward in the lumbar region between T12 and S1 (chapter 4) to complete the posterior stretch. Gravity can assist flexion of this region if you are limber enough, but the main aids to completing the posterior stretch are the abdominal muscles, especially the two rectus abdominis muscles (figs. 3.11–13 and 8.11). These run between the chest and the pelvic bones, not between the chest and the thighs, so they act purely between the pelvis and the rib cage to flex the lumbar region forward (fig. 6.15). Therefore, they complete the forward bend without having the slightest effect on hip flexion.

To translate this discussion into experience, assume the preparatory position for the posterior stretch with the spine perpendicular to the floor and the knees extended, but with the upper extremities relaxed and the hands resting in a neutral position. Slowly come forward, bending as much as possible from the hips. Starting from below, notice three patterns of muscular tension on each side of the body: first, a line of tension in the rectus femoris muscle running along the front of the thigh between the patella and the pelvic bone; second, another line of tension within the iliacus muscle running from deep within the upper thigh to deep within the pelvis; and third, a line of tension within the psoas muscle running between the upper part of the thigh and the lumbar vertebrae.

Now try two more experiments that are designed to show you exactly what is happening in the thighs. First try to bend further, but now keep the thighs relaxed and notice that pulling forward will tend to bend the knees ever so slightly. This is because the stretched hamstring muscles are pulling between the ischial tuberosities and the far side of the knee joints, causing the knees to buckle. To see the contrast, again re-establish tension in the quadriceps femoris muscles on the front side of the thigh. The four heads of the quadriceps, operating together, keep the knees fully extended while the rectus femoris head of the quadriceps again pulls on the pelvis from the front.

With the pelvis and knees once more stabilized from below, most of the remaining impetus for pulling down and forward is provided by the abdominal muscles. You can feel how these operate by lifting up slightly and poking your fingers deep into the abdomen, first in the region of the vertical bands of the rectus abdominis muscles, and then on the sides of the abdomen. As you make the specific effort to pull forward the rectus abdominis muscles push the fingers out strongly, while the other three layers of abdominal muscles (which are located more laterally and posteriorly) push the fingers out less vigorously.

COMPLETING THE POSTERIOR STRETCH

Let's say you are an advanced intermediate student who is flexible enough to come most of the way into the posterior stretch using the iliopsoas, rectus femoris, and abdominal muscles. You have had enough experience to know your weaknesses and vulnerabilities, and you have analyzed the posture enough to know approximately how much you are bending at each site. For example, if you bend 90° at the hips and 70° in the lower back, you will have 20° to go to make a 180° bend. That's not so much, and you should now be able to complete the posture by pulling yourself down the rest of the way with your upper extremities—interlocking the fingers around the feet and flexing the hips an additional 20° to pull yourself the rest of the way down. Hold for as long as you are comfortable.

When the most flexible students complete the posterior stretch with the upper extremities, they will gain an additional 30° of bending at the hips, or 120° total at that site. But this would fold them a total of 190°, which is neither necessary or even possible. What actually happens is that as they come forward, their backs become less rather than more rounded to the rear, and they settle into a final 180° bend by easing off 10° in the lumbar region (fig. 6.12). Pulling forward with the hands is not advised for those who do not have good hip flexibility because when they try to complete the posture with their upper extremities, they will increase rather than decrease the lumbar portion of the forward bend, and this can strain the back.

Coming out of the posture, beginners should exit by gently unrolling from the top down; intermediate students can begin to explore coming out with a little more authority; and advanced students will feel confident enough to reach forward from their most extreme bend and lift up from the hips.

THE PLANK POSTURE

After coming out of the posterior stretch, students at all levels can counterbalance that pose with the plank posture. Sit with your feet outstretched in front of you, bring the heels and toes together, place the hands on the floor behind you with the fingers pointed toward the feet (not illustrated), and lift the pelvis until your body comes into one plane from head to foot. Alternatively, for increased stretch in the forearm flexors, you can point your fingers to the rear (fig. 6.16). The plank posture is accomplished with the same muscles that were stretched in the posterior stretch—the deep back muscles, the gluteals, and the hamstrings—and holding them in an isometric position, as you must in this posture, brings a welcome relief from the posterior stretch.

THE DOWN-FACING DOG

After the posterior stretch, the down-facing dog is possibly the most fundamental and widely taught forward bending posture in hatha yoga, but it is completely different from the posterior stretch in its structure and mood. It's a pure nutation and hip flexion stretch, or at least it should be, not a lumbar bend, and it's a more active posture because it requires use of both upper and lower extremities.

In its ideal form the down-facing dog assumes the shape of an upside-down V, with only the hands and feet touching the floor. The hips are flexed sharply, the ankles are flexed 45°, and the lumbar lordosis is kept intact. The posture is related to a standing forward bend initiated from the hips, but it is safer and more effective because leverage from gravity will not strain the back, as can happen in the standing position.

To do the down-facing dog assume a hands-and-knees position with the arms, forearms, and thighs perpendicular to the floor, the fingers pointing forward, and the toes flexed. Lift the hips into a V (often called a piked position) while keeping the lower back arched and attending to nutation of the sacroiliac joints as a first priority. The upper two-thirds of the body from the pelvis up contains one plane of the V, and the thighs and legs

Figure 6.16. The plank posture complements forward bending poses because it holds the deep back muscles, gluteals, and hamstrings (all just previously stretched by forward bending) in a state of isometric contraction in a straight, safe position.

contain the other, joined to one another at a 60–90° angle at the hips (fig. 6.17). This represents 90–120° of hip flexion. The heels should be on or reaching for the floor. It may not be possible for beginners to assume this position, and instead of looking like an upside-down V, the posture will more likely resemble a croquet hoop, with the arms and forearms constrained to an obtuse angle from the torso, the lumbar region rounded to the rear, the hips flexed only 45–60° instead of 90–120°, and the heels lifted off the floor. Such students can make the pose more attractive by bending their knees to take tension off the hamstring muscles. That's fine. Doing this will permit more hip flexion, keep the lumbar region flatter, and anchor the pelvis more effectively in sacroiliac nutation (fig. 8.27).

[Technical note: Recall that the amount of hip flexion by definition refers not to the angle *between* the pelvis and the thighs (which is a measure of the angle displayed by the V in the advanced student's down-facing dog), but to the total *excursion* of the thighs (from the anatomical position) relative to the pelvis. That's why 110° of hip flexion reveals a 70° angle between the torso and the thighs in the expert's down-facing dog (fig. 8.26).]

In intermediate students the lower back is probably flat rather than arched forward, and the arms come more in line with the shoulders. The heels may still be off the floor but the piked position begins to appear, with perhaps a 100° angle between the two planes, which represents 80° of hip flexion.

In an ordinary standing or sitting forward bend, beginners usually round the back at the expense of the hip joints. In the down-facing dog, however, they can lift their heels, which takes tension off the calf muscles (the gastrocnemius and soleus) and allows them to arch the back. They can then focus on trying to achieve more hip flexion.

Figure 6.17. This simulation of an intermediate level down-facing dog (the heels are still slightly off the floor) should be taken as something to work toward by the beginner, who will probably have to be content with a hoop-shaped dog posture until developing better hip and ankle flexibility (see fig. 8.26 for the advanced pose).

To understand why lifting the heels helps you flex the hips, we have to look at the design of the lower extremities as a whole. First, because the gastrocnemius muscles take origin from the femur just above the knee joint and insert (along with the soleus muscles) on the heels (figs. 7.6 and 8.10), they have two actions: extension of the ankles and flexion of the knees. Second, the hamstrings, which are the primary limiting elements to hip flexion, also have two actions: extension of the thighs and flexion of the knees. And since the gastrocnemius and hamstring muscles share one of these functions—flexion of the knees—it follows that if you lift the heels and bring the insertions of the gastrocnemius muscles closer to their origins (thus reducing tension on them), this will allow you to stretch the hamstrings in relative isolation. And that is exactly what happens when you lift your heels in the down-facing dog. If you try it you can instantly feel the lumbar lordosis become more pronounced and allow increased flexion at the hips. Then as you lower the heels to the floor you can feel tension increase both in the gastrocnemius muscles and in the hamstrings, which in turn causes the lumbar region to flatten or even become rounded to the rear.

To put these principles into practice, students should lift up on the balls of the feet as they come into the piked position, arch the lower back forward to establish both nutation of the sacroiliac joints and a convincing lumbar lordosis, and then try to press the heels toward the floor while keeping the back arched. If this is difficult they can bend one knee and press the opposite heel to the floor, making an asymmetrical posture in which they stretch one side at a time and incidentally take pressure off the lumbar region and the opposite hip.

The down-facing dog is supported equally by the upper and lower extremities, but a common beginners' error is to compromise the spirit of this principle and leave the arms and shoulders relaxed. The extended knees automatically keep the lower half of the body in one plane, but the scapulae, which connect the arms to the torso, are held in place almost entirely with muscular attachments (chapter 8), and for the down-facing dog to be properly supported, these muscles have to remain engaged at all times. As you extend the arms and actively flex your hip joints into the piked position, the scapulae should be pulled down and laterally; otherwise, they will be drawn in and up, causing the posture to degenerate. Instructors correct this error by telling students not to let their chests hang passively between their arms.

In the down-facing dog, tension should also be maintained in the hands. The fingers should be spread out with the middle fingers parallel, and pressure should be exerted against the floor with the entire hand. This activates flexors of the wrists and hands when they are in a moderately stretched and extended position. Special attention should be paid to tension in the thumbs, index fingers, and the medial aspects of the arms and forearms. Holding the arms and shoulders correctly will also create a more substantial stretch in the pectoral muscles on the anterior surface of the chest.

Students who are having difficulty with the down-facing dog because of stiff hips and ankles can try some preparatory stretches. They can stand with the feet 2–3 feet apart, slide the hands down the thighs and legs, bend the knees, plant the hands on the floor, and walk the hands forward until the body is in the shape of a hoop. From that point they can explore in any one of several directions to create stretches that prepare for the full posture: they can walk the feet further apart to create an adductor stretch; they can walk the feet closer together to focus on the hamstrings; and they can bend one knee at a time while pressing the opposite foot toward the floor to stretch the gastrocnemius and soleus muscles.

THE CHILD'S POSE

We'll end this section with the simple child's pose—simple that is, if you are as flexible as a child, for the posture requires the entire body to be folded in on itself in the fetal position. You can come into the pose from a hands-and-knees position, with the feet and toes extended, by first sitting back on your heels, thus lowering the thighs tightly down against the calf muscles, and then flexing the torso down against the thighs, resting your forehead on the floor in front of your knees. Ordinarily, you will lay your upper extremities alongside the legs, with the palms up and the fingers lightly flexed (fig. 6.18), but if you have need of a milder posture that does not fold you so completely into flexion, you can place your hands alongside your head. The posture is relaxing and refreshing, and so long as you do not fall asleep, you may hold it for as long as you like. The child's pose is often done between other forward bending postures because it stretches the spine from end to end in a non-threatening manner.

Figure 6.18. The full expression of the child's pose is sometimes a challenge, but one that is easily remedied by placing the hands in a more neutral position alongside the head and by the use of props such as a small pillow under the ankles, another one between the legs and the thighs, one or two thick pillows between the thighs and the torso (especially helpful for stiff backs), and yet another small pillow under the forehead.

Those who have good flexibility in the spine and healthy hip and knee joints will not have trouble with the child's pose, but some people will be in discomfort. The remedies are simple—one or more pillows between the torso and the thighs for a stiff back, another pillow between the thighs and the legs for tight knee joints, a small pillow just underneath the ankles for feet that resist full extension, and a cushion for the forehead. With one or more of these props the child's pose can be adapted to almost anyone and still yield its benefits. If it is done with care the posture can also be a welcome palliative for those with chronic back stiffness because it places the lumbar region under mild traction.

BREATHING AND FORWARD BENDING

Breathing in forward bending postures will be experienced differently by those who are relatively stiff than by those who are stronger and more flexible. Advanced students have many options, but those who are inflexible in the hip joints have to tense the abdominal muscles just to maintain the posture, and this creates many repercussions.

BREATHING IN THE POSTERIOR STRETCH

In the posterior stretch for beginners, the abdominal organs are compressed by the forward bend, and if you are holding lightly onto your thighs, legs, or ankles with your elbows slightly flexed, each inhalation will lift your torso as the dome of the diaphragm descends. Each exhalation then lowers your torso back forward and down.

If you are an intermediate student and are committed earnestly to the posture, you will probably be holding tightly onto the lower extremities with your hands to maintain a deep forward stretch that is close to your limits, and under those circumstances it is harder for the diaphragm to lift you up during inhalation. There is increased tension during inhalation as the diaphragm presses down against the abdominal organs and a release of that tension during exhalation, but you are still not in the full posterior stretch, and you are still at the mercy of a pneumatic system that tends to lift you up and down. Under those conditions this bent-forward posture with incomplete hip flexion can never be perfectly stable or satisfying.

Finally, if you are able to complete the posture by flattening your torso down against your thighs and holding it firmly in place with your upper extremities, you will have yet a third experience. The fixed and flattened torso prevents the posture from lifting and lowering during inhalation and exhalation, and the resulting sense of stability and silence is the reward for your efforts.

As soon as you are coming close to this third experience, you can use breathing to increase your capacity. Come into the forward bend with the knees extended. Relax the shoulders and arms, and rest the hands on the ankles or feet. Breathe in and out normally, and confirm that inhalation is lifting you up and creating more tension in the trunk. Now, in one easy sequence, breathe out to your full capacity (as in agni sara), and at the end of exhalation pull your torso forward with the rectus femoris, iliacus, psoas, and rectus abdominis muscles. Then grip a lower site firmly and hold the posture, inhaling and exhaling several times until your body has adjusted to the new stretch. You are not pulling yourself into position with the upper extremities. Instead, you are using the hands only to hold yourself in a position established by the muscles of the torso and the proximal muscles of the lower extremities. Repeat this sequence several times. If you start with reasonably good hip flexibility, you will be amazed at how much you can draw yourself forward.

No matter what your flexibility, if you pull yourself firmly enough into a forward bend with your hands so that your breathing cannot lift you out of the posture, you will notice that tension from inhalations, especially from deep inhalations, spreads throughout the body and is redirected into stretching the calf muscles, hamstrings, and the deep back muscles, as well as joints and muscles throughout the upper extremities. During exhalation the diaphragm permits a release that lulls the stretched muscles into submission. Hatha yoga teachers are speaking literally when they tell you to let the breath stretch the body.

BREATHING IN THE DOWN-FACING DOG

Breathing in the down-facing dog is different from that observed in any of the other forward bending postures. Since this pose is semi-inverted, the diaphragm presses the abdominal organs toward the ceiling during each inhalation in addition to drawing air into the lungs, and during each exhalation the diaphragm eccentrically resists the fall of the abdominal organs toward the floor. And finally, the weight of the abdominal organs against the underside of the diaphragm causes you to exhale more completely. This illustrates a pattern of breathing that we'll see in a more extreme form in the headstand and shoulderstand.

Since beginners will be forming a relaxed hoop in this posture rather than a V with a taut abdomen, their bellies will remain relaxed and they will be breathing abdominally. Respiration simply pooches the abdomen out during inhalation and relaxes it during exhalation, so this mode of breathing has little effect on the posture.

Intermediate and advanced students who do the down-facing dog more elegantly have a different experience. They press enthusiastically into an upside-down V, concentrating on maintaining the deepest possible lumbar lordosis. The arch in the spine creates a backbending posture superimposed on 90–120° of hip flexion, and when these students are at the same time working consciously with the breath, the result is diaphragm-assisted back-bending (chapter 5). To experience this, come into the posture keeping the lumbar lordosis intact, lifting high up on your toes if that is necessary, and take long, deep inhalations while at the same time committing yourself to coming more completely into the posture—that is, accentuating the acuteness of the angle between the pelvis and the thighs. You will immediately sense that the diaphragm is a powerful influence for assisting this effort.

BREATHING IN THE CHILD'S POSE

This is an easy one. Because the body is folded upon itself in the child's pose, inhalation increases tension throughout the torso, and exhalation decreases it. Both inhalation and exhalation are active. In addition to drawing air into the lungs, inhalation has to press against the abdominal organs, which are incompressible (chapter 3), and that is why you feel a sense of increased tension. Exhalation is also active, or it should be, because you are breathing evenly, and even breathing requires that you not exhale with a whoosh. The point is easily proven if you take a deep inhalation and then suddenly relax your respiration; the air rushes out, and you realize that you normally resist this.

SACROILIAC FLEXIBILITY

Sacroiliac flexibility has until now been overlooked by those who write manuals on exercise and flexibility, and the terms nutation and counternutation are rarely encountered. This is not surprising since the sacroiliac movements are limited to only 5–10° (except during the end stages of pregnancy), and these are overshadowed by the grosser movements of the spine and pelvis as a whole.

Even though the range of sacroiliac movements is narrow, however, healthy and mobile sacroiliac joints make for safer, sharper postures. Indeed, the proper execution and full expression of backward bending, forward bending, and seated meditation postures presupposes the ability to establish nutation and counternutation at will. And because the concepts are unfamiliar and complex, some reiteration and review is in order. First recall where the movements take place. They're not spinal movements (as happen at intervertebral disks and other joints in the spine), and they're not movements at the hip joints (as happen at the acetabula between the pelvic bones and the femurs). Rather, they are literally the *only movements* permitted between the axial skeleton and the appendicular skeleton for the lower extremities (fig. 6.2a, arrows for nutation, and imagine their opposites for counternutation). And they are subtle: think of movements within the pelvis itself. If you want to understand the concepts, you will have to both think the movements through intellectually and appraise them experientially, and you also have to do this while envisioning them not only in isolation but within larger bending gestures that involve the spine and the hip joints. These are not minor challenges.

NUTATION IN FORWARD BENDS

For both intermediate and advanced students, establishing nutation as a first priority in forward bends can be summed up easily: while maintaining the arched-forward lordosis in the lumbar region and while keeping the iliacus components of the iliopsoas complex relaxed, create a selective pull in the psoas muscles. You will sense little external movement, but the psoas muscles pull sharply forward on the lumbar region, and this in turn pulls the promontory of the sacrum forward, which favors nutation. The ilia are left behind and pulled medially by default as a result of keeping the iliacus muscles relaxed. Although not ordinarily verbalized in this way, this is what hatha yoga teachers want you to do. It is the preferred beginning step for forward bending, whether standing or sitting. Only after this subtle maneuver is accomplished should you bend forward at the hips and then in the spine. As you bend forward at the latter sites, the sacroiliac joints will readjust themselves, moving to a more neutral position between nutation and counternutation.

The down-facing dog works especially well for evaluating and sensing sacroiliac movements in advanced students (fig. 6.17, and even more in fig. 8.26), because experts have enough hip flexibility to settle into the posture with an arched-forward back. From this position, they can go back and forth between counternutation (pulling the ischia together, tightening the abdominal muscles, and pressing the promontory of the sacrum to the rear in relation to the ilia) and nutation (sharply pulling the lumbar lordosis and sacral promontory forward with the psoas muscles, relaxing the abdominal muscles, and allowing the ischia to be drawn apart). It is useful for the advanced student to keep the thighs moderately abducted for the posture, because as described earlier, an observer can monitor the movements of the upper thighs by feel: they shift medially during counternutation and laterally during nutation. Keep in mind, however, that the down-facing dog does not work well for those who are not flexible because the hoop-shaped dog posture favors pulling the ilia laterally and forward, thus creating a priority for counternutation. The remedy is simple: place the hands on a chair or table so there is plenty of leeway to keep a prominent lumbar lordosis.

One of the most useful forward bending postures for stressing nutation, as well as a posture that is accessible to beginning and intermediate students, is the one presented to illustrate diaphragm-assisted backbending (fig. 5.7). This pose, which was already mentioned in the section on anatomy, combines three elements: a forward bend at the hips, a backbend in the upper half of the body, and full nutation of the sacroiliac joints. (As in the case of the down-facing dog, it is helpful to come into the pose with abducted thighs.) Be careful to keep the abdominal muscles relaxed; if you don't, they will drive the lumbar lordosis to the rear and compromise your effort. With that caveat, this is one of the best postures for getting into the most extreme nutation you can manage. You can select a hand position on the wall that permits full relaxation of the abdominal muscles, your most expressive lumbar lordosis, and a specific effort with the psoas muscles that rotates the promontory of the sacrum forward, the coccyx up and back, and the ischia apart—all with minimal alteration to the appearance of the posture.

We also see good potential for nutation in many other postures that contain elements of forward bending from the hips. For those who are flexible enough to keep a deep lumbar lordosis during the course of forward bends, such postures include the superfish leglift (fig. 3.19b) and the straight-backed boat (fig. 3.22b). And for those who are less flexible, simple and useful postures include cat stretches with maximum lumbar lordotic curvatures (figs. 3.30, 3.34a–b with the dashed lines, and 3.36), sitting on your heels or on a bench in the adamantine posture (fig. 10.9), and any seated meditation posture in which you can demonstrate a deep lumbar lordosis, whether you accomplish this by virtue of excellent native hip flexibility or a supporting cushion (chapter 10). In the cat stretches and sitting postures, even beginners can learn to relax the abdominal muscles, pitch the lumbar region and the promontory of the sacrum forward with a selective contraction of the psoas muscles, permit the ilia to come closer together, and spread the ischia. And one more added benefit is that these simple postures permit you to alternate full nutation with full counternutation: pushing the lumbar region maximally forward favors nutation, and pushing it maximally to the rear favors counternutation.

NUTATION IN BACKBENDS

If you have a healthy back, you can do relaxed symmetrical backbending postures to encourage nutation. These include all the gravitationally-aided backbending poses, beginning with the relaxed standing lumbar bend (just mentioned in the section on anatomy as well as in chapter 4), in which nutation accentuates the lumbar lordosis, squeezes the promontory of the sacrum forward between the ilia, and spreads the ischial tuberosities (fig. 4.20). Or try this: stand with the thighs comfortably abducted and place your hands astride the ilia with the thumbs against and directly behind the top of the sacrum. Relax and bend backward to produce maximum nutation. You may not be altogether certain of feeling the top of the sacrum moving forward in relation to the ilia as you bend backward, but as you slowly shift forward from the extremity of the backbend and move into counternutation, you'll feel a dramatic shift of the ilia as they move forward and laterally on either side of the sacrum. It almost feels like a gear shifting in the manual transmission of an automobile.

Another excellent posture favoring nutation is the propped, diaphragm-restricted backbend leaning against a wall (fig. 5.6), except that here you modify the posture by aiming for a diaphragm-assisted backbend. You do this by bending the knees, working your hands somewhat further down the wall, and relaxing the abdomen so as to permit the diaphragm to accentuate the bend. This creates full nutation by squeezing the promontory of the sacrum forward in relation to the ilia.

Next, try the variation of the upward-facing dog in which the feet and toes are extended (the tops of the feet facing down) and the knees are left on the floor (fig. 5.13). In this posture gravity does the work of dropping the pelvis, with the promontory of the sacrum leading the way and creating nutation. If you move slowly, you can also get the same feeling with the toes flexed, resting on the balls of the feet, knees, and hands.

Next, try lying supine with an 8 1/2 inch playground ball under the lumbar region (fig. 5.31). If you can relax the abdominal muscles and allow gravity to lower the upper back and pelvis toward the floor, this posture will encourage nutation; otherwise you will protect your back with an attitude of counternutation (resistant abdominal and iliacus muscles, straighter body, squeezed-together hips, and spread-apart ilia).

Finally, for those who are flexible enough, push up into the wheel posture from a supine position (fig. 5.29), and allow nutation to take place as a first priority, with the promontory of the sacrum squeezed forward and the ischia pulled apart. The abdomen and hip flexors, especially the iliacus muscles, must be relaxed, for only under those circumstances will full nutation complement maximum spinal and hip extension. And as it happens, the preference for counternutation, or even sacroiliac joints that are frozen in that attitude, is a common impediment to pushing up into the wheel for many students.

[Technical note: The most advanced students, such as dancers and gymnasts who are extraordinarily flexible, may be able to do this posture one better—keeping the sacroiliac joints in an attitude of partial counternutation. The most flexible students, in fact, may feel this is desirable for protecting themselves, given that full nutation may take place too readily for their comfort. We can see a continuum of possibilities for the wheel posture: inflexible beginning students who show little or no sacroiliac movement; intermediate students who can come partially into the wheel by pushing to their limits of nutation; advanced students with excellent sacroiliac mobility who feel comfortable in the posture with full nutation; and last, those who have more sacroiliac flexibility for nutation than they feel comfortable using.]

COUNTERNUTATION IN VARIOUS POSTURES

Nutation is natural in upward-facing dogs in which you support the posture between the knees and the hands, but counternutation is more natural when the upward-facing dog is supported between the feet and the hands. For the latter, squeeze the hips together while keeping the toes either flexed (fig. 5.14) or extended. The main object here is to engage the abdominal, gluteal, and deep back muscles strongly enough to initiate coming into the posture with a relatively straight body and the fullest possible counternutation. You can feel it: the ilia are pulled forward in relation to the sacrum by the iliacus muscles, and that movement is supported by squeezing the hips together along with the ischia. Once this posture is established, lower the pelvis carefully so as not to release the counternutation. The abdominal muscles (along with the respiratory and pelvic diaphragms) will act synergistically with the iliacus muscles to support counternutation: they will resist lowering of the pelvis eccentrically but powerfully; and because in combination they maintain a high intra-abdominal pressure throughout the breathing cycle, they will also assist in keeping the lumbar spine straight and keeping the promontory of the sacrum well to the rear in relation to the ilia. Again, you can feel all of these tendencies if you have a clear concept of the anatomy.

The most common postures that support counternutation are standing and sitting forward bends from the waist. All you have to do is flex the spine forward (as opposed to flexing the hips), and this will encourage counternutation. Health-club crunch exercises (fig. 3.1), the fire exercise (fig. 3.16), yoga sit-ups (figs. 3.21a–b), the round-bottom boat (fig. 3.22a), the phase of standard cat stretches that push the lower back toward the ceiling with the abdominal muscles (figs. 3.34a–b, halftones), a relaxed and externally supported standing forward bend for beginners (fig. 4.21), and the beginner's forward bend with the fists in the armpits (fig. 6.13) all foster counternutation—keeping the ischia together, the ilia apart, and the promontory of the sacrum to the rear. And these postures are all safe and easy.

The other supremely important standing postures that support counternutation were mentioned earlier: standing whole-body backbends (as contrasted to lumbar backbends) in which the hips and ischia are squeezed together and the main priority is keeping the promontory of the sacrum to the rear and the ilia spread apart (fig. 4.19). It's another posture for those who require maximum lumbar protection, particularly when the maximum bend is accompanied by deep empowered thoracic inhalations.

In general, counternutation is preferred by those who are uncertain of themselves. They keep the hips squeezed together, the pelvis tucked under in a posterior pelvic tilt, and maintain tense abdominal muscles, all of which are classic postural adjustments for everyone who has a stiff back. If this describes you, don't fight the reality: go with it. This is the work you need to do. After a year or so of conditioning, you may feel inclined to pursue more postures that release counternutation and favor nutation.

ASYMMETRIC POSTURES AND THERAPEUTIC APPROACHES

If one sacroiliac joint is more restricted than the other, you can use asymmetric postures to free up the joint on the tight side. But you need to be careful, because it is easy to make a mistake and work selectively on the wrong side. So to be certain of your diagnosis, first go back and forth for 20–30 minutes between postures that favor extreme nutation and others that favor extreme counternutation. Then watch and wait for 24 hours. If you have sacroiliac discomfort only on one side in the form of a vague ache in the region of the sacroiliac joint, it probably means that the sacroiliac joint on that side is more restricted than it is on the other. Do make sure, however, that you are not feeling symptoms discussed earlier in this chapter—unilateral iliolumbar ligament strain, lower back pain on one side, or sciatica.

As soon as you know which sacroiliac joint is more restricted, keep working mostly on symmetric postures, but think of adding some that are asymmetric. The preliminary pigeon, as well as folding forward from that posture (figs. 6.25a–b), are excellent, and will tend to open up the sacroiliac joint associated with the front knee. Do them three times, first and last for the tight side. The best and simplest asymmetric standing posture that selectively affects one sacroiliac joint is the first stage of the angle posture in which you are initiating a bend from one hip (figs. 4.25 and 6.26a). If the right side is tight, come forward facing the right foot only to the extent that you can maintain a full lumbar lordosis, and then pull selectively and insistently with the right psoas muscle to encourage full nutation in the right sacroiliac joint. Don't come any further forward, as this is likely to release the nutation. As usual, face the right foot, then the left, then the right once more. It is best to work with simple postures that can be analyzed without doubt. Asymmetric standing postures such as triangles, side bends, and lunges, as well as asymmetric sitting postures and twists, are all so complex that it is better to work with them in each direction equally. Unless you are certain of what you are doing, you might end up favoring the wrong side.

HIP FLEXIBILITY

Good hip flexibility is the most important single requirement for at least half the postures in hatha yoga—sitting and standing forward bends, lunges, triangles, sitting spinal twists, many variations of the inverted postures, and meditative sitting postures. So it is not surprising that we treat this topic over and over, and that we pick this chapter on forward bending to do so definitively. We'll look at it here mostly in supine, sitting, and standing postures, with the thighs both abducted and adducted and with the knees both flexed and extended. Later, we'll discuss the topic in twisting postures (chapter 7), inverted postures (chapters 8 and 9), and meditative sitting postures (chapter 10).

The problem with talking about hip flexibility is that most people do not ordinarily trouble themselves to define it precisely. If students can't bend forward in the posterior stretch because of tight hamstrings, or if they cannot abduct their thighs very far because of tight adductors, or if they cannot extend their hips because of tight hip flexors, or if their sacroiliac joints are frozen, is it appropriate for hatha teachers to call these problems of hip flexibility? They usually do. But in one sense poor hip flexibility is the result, not the cause of these situations, just as hip inflexibility can be the result of excess weight in the abdomen in those who are obese.

To see hip flexibility in its purest form, we would have to look at someone who is both slender and devoid of functional hamstrings, adductors, hip flexors, and hip extensors—in short, someone whose range of motion at the hips is limited only by ligaments and bony constraints within the hip joint itself. But even if we could find such a model, it would not help us plan a useful hatha yoga practice because we are mainly interested in hip joint mobility in those whose extremities are intact. We'll therefore discuss hip flexibility in the broadest possible terms, concerning ourselves with the final result and considering all possible limitations to mobility. Our objectives are twofold—working guardedly to improve mobility of the hip joint itself, and doing stretches to increase the lengths of the muscles that form the first line of resistance.

Any time we work with the hip joint, as well as with most other synovial joints, we must be sensitive to when limitations in movement are caused by muscle, when they are caused by ligaments, and when they are caused by bony stops. If movements are being abnormally restricted by muscles, we can work consistently to lengthen them. When a normal range of motion is restricted by ligaments or bony stops, we should be wary of attacking these restrictions aggressively, realizing that overstretching ligaments can cause their associated joints to become destabilized, and understanding that bony stops are built into our body plan. When we practice also affects these matters. If you wake up after eight hours in bed, muscles will create the most restrictions, but after an hour of hatha yoga, especially in the evening, the muscles are not as assertive, and you may have more awareness of bony constraints and of ligaments that now require more tender treatment.

In the following discussion we'll work from safe and simple to challenging and complex. Supine hip-opening stretches are first because in that position the lower back is stabilized against the floor. Inverted postures are next (although these are not covered in detail until chapters 8 and 9), because we can explore hip flexibility more delicately when the hip joints are bearing only the weight of the inverted lower extremities. Then come the sitting postures in which we have to divide our attention among several tasks— stabilizing the lower back and pelvis, stretching the hamstrings and adductors, and maintaining awareness of ligamentous and bony constraints within and immediately surrounding the hip joint. Postures such as the pigeon are yet more challenging because gravity places the weight of the body directly on the muscles and ligaments of the hip when they are in already-stretched positions. Standing postures are the most challenging for three reasons: first, they require the hip joints to be held in set positions defined by the position of the feet; second, the weight of the head, neck, upper extremities, and torso is brought to bear on the hip joints, often when they are in extreme positions; and finally, tension on the hips from turning, twisting, and lunging adds to the stresses imposed by gravity.

SIMPLE SUPINE HIP OPENING

The simplest and most fundamental hip opening postures are those that work in moderation to free up the ligaments, joint capsules, and synovial surfaces of the hip joint while muscular restrictions are minimized. And since it is obviously not feasible to minimize those restrictions by detaching our hip and thigh muscles from their insertions, we do the next best thing and flex the knees. The following six stretches and movements can be done in the early morning even while you are still lying in bed, and if you try them after an hour's session of hatha yoga they have an even deeper action.

First, lie on your back and draw the knees toward the chest with the hands, keeping the thighs adducted (fig. 6.19a). This is the first and easiest position: The hamstrings are not stretched because the flexed knees bring their insertions on the tibia and fibula closer to their origins on the pelvis (figs. 3.10b and 6.3); the adductor muscles on the insides of the thighs (figs. 2.8, 3.8–9, and 8.13–14) are not stretched because the thighs are together; and finally, if you are slender the abdomen will not get in your way. As soon as you are satisfied that muscle, fat, and other tissues of the thigh and groin are not limiting the stretch, you can be certain that you are working with limitations in and around the hip joint itself.

The first thing you notice about this stretch is that pulling your knees tightly against the chest can go only so far without prying the pelvis up and away from the floor. This is a lever action, accomplished by the two femurs in combination, whose necks, at least in students who are not restricted by soft tissues, pry against the thin cartilaginous rims of the acetabula (the sockets of the hip joints). You can see this even more clearly if you lift one knee at a time and pull it diagonally across the body—the upper rim of the acetabulum is not horizontal but lies at an angle, and pulling the femur diagonally accesses this cartilaginous rim directly and pries the same side of the pelvis away from the floor.

You can make the first exercise more effective by resisting the lifting effect on the pelvis isometrically. You do this by trying to roll the pelvis back down against the floor, pressing the sacrum toward the floor using the deep back muscles at the same time you are pulling on the long end of the lever (the knees) with the hands.

For the second position, pull the knees slightly apart, with the hands still grasping them from the outside. The thighs are slightly abducted, although not enough in most people to stretch the adductors. Now the neck of the femur will be in contact with the rim of the acetabulum in a slightly different region, lateral to the first point of contact, and the shaft of the femur may also be butting up against the anterior superior iliac spine on the front of the ilium (figs. 3.2–6 and 6.19b). Again, you can intensify the stretch by pulling the pelvis toward the floor with the deep back muscles.

Figure 6.19a. First of six supine hip-opening poses: With the knees together, pull them toward the chest with inter-locked hands and fingers, at the same time prying the pelvis away from the floor.

Figure 6.19b. Second: With the knees apart, pull on them from the sides with the hands, providing a slight stretch for the adductors.

Third, grasp the knees from the inside, abducting the thighs even further to the side. Depending on your body type and flexibility, the shafts of the femurs may now be lateral to the anterior superior iliac spines, enabling you to pull the knees closer to the floor (fig. 6.19c).

Fourth, grab the ankles and pull them toward you, and at the same time dig the elbows into the thigh muscles, pressing them out and increasing abduction. The soles of the feet will probably be together in this position (fig. 6.19d). If you are feeling stretch in the inner thighs, you are feeling the adductors, and if you are not, the primary limitation to the movement is still the hip joint.

Fifth, catch the lateral sides of the feet and pull them closer to the chest, and at the same time dig the elbows into the calf muscles, pressing the knees even further to the sides (fig. 6.19e). This stretches the adductors maximally and aligns the femur in such a way that its shaft has the potential for dropping down laterally to the lateral border of the ilium. If you are flexible enough, someone may even be able to push your knees all the way to the floor.

Sixth, hold onto the soles of the feet from the inside so that the legs are perpendicular to the floor, and cautiously pull the knees straight down (fig. 6.19f). The thighs will not be abducted as much as in the previous position, but the necks of the femurs can still clear the ilia. Here again, this is easier if you have a partner to help you. In any of the last two or three positions in which the neck of the femur is not prying directly against the rim of the

Figure 6.19c. Third: Grasp the knees from the inside and pull them down and laterally for more stretch of the adductors.

Figure 6.19d. Fourth: Grasp the ankles and pull the feet toward the head, pressing the elbows sharply against the thigh muscles. This tends to lift the shoulders, and you may want to have a pillow to support your head.

hip socket or the front of the ilium, you may still get stopped by tenderness in the groin. Don't force the issue if that is the case, because numerous delicate tissues run through this region.

CIRCUMDUCTION

The hip joint is a ball-and-socket joint that accommodates "rotation" during the course of any combination of six movements—flexion and extension, abduction and adduction, and lateral and medial rotation. Even though in a literal sense all of these motions rotate the head of the femur in the acetabulum, by convention only the last two are termed anatomical rotation. These of course can be superimposed on any of the others. For example, if you sit down and spread your thighs apart keeping the knees straight, and then turn your toes out, you will be superimposing lateral rotation on flexed and adducted thighs; turning the toes in from the same position is medial rotation. "Hip-opening" in hatha yoga means developing a full range of motion for all of these movements plus one more—*circumduction*—that sequentially combines *flexion, abduction, extension,* and *adduction*.

You can circumduct the thigh in any position in which the floor or some other object is not in the way. We'll look at it in a standing position to explore the principle and then in supine postures to see how various muscles restrict the movement. If you balance on your left foot, extend the

Figure 6.19e. Fifth: Grasp the lateral aspects of the feet and pull them toward the head, pressing the elbows sharply against the calf muscles. This provides the fullest stretch for the adductors in this series.

Figure 6.19f. Sixth: Grasp the soles of the feet from their medial borders and pull the knees toward the floor on either side of the chest. Be careful, because the arms can pull more powerfully than is temperate for the hip joints.

right knee, and swing the right foot around in a circular motion, you will be circumducting the thigh. You can start with adduction, continue forward for flexion, swing the foot out for abduction, to the rear for extension, and back into home position with adduction. If you project an imaginary tracing on the floor with your foot, you'll notice that the movement is kidney-shaped rather than circular. There are two reasons for this: the leg you are using to support your weight gets in the way of the one that you are swinging around, and you can flex the thigh forward further than you can extend it to the rear. Try this exercise with both lower extremities and notice if the excursion is different on the two sides.

As you swing your right thigh in front of the left thigh and leg, you can first feel the right hip joint and the left thigh limiting how far you can pull the right thigh to the left and forward. Then as you flex the thigh straight forward the right hamstrings limit the movement. Swinging on around to the side, the right adductors (or the right hip joint in those who are especially flexible) start resisting. Continuing to the rear, the right hip flexors limit extension, and finally, just before you bump into the left thigh, the right hip joint again stops you.

We have already taken note of the straightforward effects of knee extension and flexion on hip flexion (figs. 6.3–4), and we have seen how important this is to forward bending with the thighs adducted, as in the posterior stretch (fig. 6.12). Now we'll look at how flexing the knee assists circumduction of the thigh in general and abduction of the thigh in particular. You can do this only when you are lying supine with the hips near the edge of a table or firm bed, because you want to be able to hyperextend the thighs beyond the lower edge. The first thing to do is repeat in the supine position what you just did standing, extending the knee and projecting the tracing for circum-duction on the opposite wall rather than on the floor. Then, with this as a basis for comparison, bend one knee and project the same kind of tracing on the wall from an imaginary line running down your thigh.

You will notice immediately that you get a much bigger projection when your knee is bent. Swinging the thigh around in the same direction as before, it does not make much difference at first whether the knee is bent or extended. But as soon as you have the thigh flexed straight to the front, knee flexion enlarges that segment of the projection considerably over what is apparent with the knee extended. Moving on around, as you abduct the thigh straight to the side, you get only marginally more thigh abduction with the knee flexed than extended. Finally, everything is reversed as you try to bring the thigh to the rear. Knee flexion at that point stops thigh hyperextension cold because the rectus femoris muscle comes under extreme tension and keeps the thigh lifted. Be careful at that stage not to hurt your knee.

Next, to examine how this works in a dynamic movement, improvise freely, circumducting the thigh with varying combinations of knee flexion and extension, always sweeping the thigh in as wide a "circle" as possible. These are all valuable exercises for hip opening in their own right even though they are not practical for a class. They work best at the end of a firm bed.

As you play with the different options and combinations for knee flexion and thigh abduction, you can begin to understand how the hamstrings, adductors, and hip flexors in combination affect circumduction of the thigh, and you will be encouraged to see that lengthening these muscles even a little can improve hip flexibility. Circumduction of the thigh also clarifies for us how hip flexibility is limited by the inherent structure of the joint itself, and once you become aware of the limits imposed by the individual muscles, you can begin deeper work. As an experiment, do a series of hip-opening postures, both the simple ones outlined at the beginning of this section and some of the more demanding ones described later, then try a balanced hatha yoga practice for an hour, and come back to the same hip-opening exercises at the end. After you are warmed up, you will not only be more flexible, you will be conscious of more bony and ligamentous constraints.

A SUPINE HALF LOTUS HIP-OPENER

The next exercise—a supine half lotus hip-opener—does two things: it improves hip flexibility, and it stretches the adductors and deep back muscles. And because the back is stabilized against the floor, it is safe as well as effective. It is less rewarding early in the morning than after you are thoroughly warmed up, but once you get acclimated to it the pose will become a favorite. Lie supine and draw the heels toward the hips, keeping the feet on the floor. Then bring the left ankle to the near side of the right knee, resting the ankle against the thigh and pressing the left knee away from the body. Pass the left hand into the triangle formed by the two thighs and the left leg, lift the right foot off the floor, and grasp the right shin with the left hand just below the knee. Pull the left foot down closer to the pelvis with the right hand, place the right forearm above the left ankle, and interlock the fingers around the right shin just below the knee (fig. 6.20).

Figure 6.20. To do this supine half-lotus hip-opener, proceed as directed in text, or modify it as needed, for example by permitting the right ankle to rest above the right wrist instead of below, or by grasping the back side of the right thigh instead of the right shin.

If you are not flexible enough to get into this position, you can keep the right wrist below the left ankle, or you can grasp the thigh instead of interlocking the fingers around the knee. Do it any way you can. Rock from side to side as far as possible without falling. Then draw imaginary circles on the ceiling with the right knee. Go as far to the left as possible without toppling over, and pull on your right knee, deeply stretching the adductor muscles on the left side that attach posteriorly along the inferior pubic rami. Then go as far to the right as possible, again without toppling over, and pull, feeling the stretch higher in the back. In this position, the right thigh is flexed straight toward the chest, so the right adductors are not being placed under much tension, but the left adductors are stretched by the modified half lotus combination of flexion, abduction, and lateral rotation. Repeat on the other side.

GOLGI TENDON ORGAN STIMULATION

The following exercise lengthens the hamstrings and reaffirms the principles of working with feedback circuits between tendons and their muscles (chapter 1). It is safe for beginners because the back and pelvis are stabilized against the floor. Locate a length of cloth or a belt that can be thrown over one foot and grabbed with the hands. Then do a standing forward bend to test the initial length of the hamstrings. Next, lie down with the buttocks against a wall, the feet facing the ceiling, and the thighs flexed 90° from the trunk. Keep the knees extended and the feet together, and toss the cloth or belt across the sole of one foot.

Keeping one leg against the wall, draw the other foot away from the wall by tugging on the belt with the opposite hand. Keep both knees extended. With the other hand, first locate the ischial tuberosity on the side you are working with, and feel the hamstring tendons that lead distally, up and away (toward the ceiling) from that point of origin. Second, locate the cordlike hamstring tendons that connect the bellies of the hamstring muscles to their insertions on the tibia and fibula. These tendons can be felt just proximal to the medial and lateral sides of the knee joints. Once the tendons are all located, hold the cloth or belt firmly and press the thigh toward the wall isometrically, bringing the hamstring muscles into a strong state of contraction. Then, keeping that tension on the hamstring muscles, deeply palpate their musculotendinous junctions with your free hand, first near their origins and then near their insertions. Repeat the exercise on the other side. To complete the experiment, stand up, again try the forward bend, and notice how much further you can come down.

Vigorously palpating Golgi tendon organs of the hamstring muscles when the hamstrings are in a state of isometric contraction relaxes the muscles, and we can see the evidence moments later when they accommodate to a greater length under conditions of passive stretch. For example, if we guess that the hamstring muscle fibers had been receiving 30 nerve impulses per second before the isometric endeavor and massage, they might receive only 20 nerve impulses per second in the same stretch after the treatment, releasing some of their tension and enabling us to bend forward more gracefully. Even more to the point, the diminution in motor neuronal input seems to last for as long as a day or two, supporting the usual advice to do hatha yoga postures every day.

HIP OPENING IN INVERTED POSTURES

When the body is inverted, hip-opening is both safe and effective because the hip joints are not bearing the weight of the body as a whole. In either the headstand (chapter 8) or the shoulderstand (chapter 9), you can stretch the hamstrings on one side at a time by pulling one foot overhead, and you can stretch the adductors by allowing gravity to abduct the thighs. From the shoulderstand you can come into the plow or half plow to stretch the hamstrings on both sides, and you can do that while abducting the thighs maximally to stretch the adductors. You can also fold one foot into a half lotus position and lower the other foot overhead toward the floor to stretch the hamstrings on one side. Finally, you can work within the hip joint by folding up the knees and hips in any number of ways that reduce muscular tension.

FORWARD BENDING WITH ONE FOOT TUCKED IN

Sitting forward bends with one foot tucked in are among the most useful hip flexibility stretches for beginners and intermediate students. They do not place as much stress on the lower back and sacroiliac joints as the posterior stretch, and they are asymmetrical postures that are helpful for working with right-left imbalances.

To begin, sit on the floor, stretch one leg out in front of you, and pull the other foot in toward the perineum. The thighs will be at about a 90° angle from one another, and you will be facing about halfway between the two. Next, to work with this posture conventionally, twist the spine 45° to face the outstretched leg and come forward without lifting your hands overhead or making an attempt to bend initially from the hips. Let the hands rest, depending on your flexibility, on the outstretched thigh, leg, or foot (fig. 6.21a). Remain in the posture for about half a minute and then slowly unroll, first at the hips and then in the lumbar region. Finally, lift the head and neck. Repeat on the other side.

This forward bend toward one foot is useful for several reasons. First, with one knee flexed, it is stretching the hamstrings on only one side. Second, even though it is stretching the adductors to some extent on both sides, it creates more stretch on the side to which you are reaching because the knee on that side is extended. This is one of the best possible postures for working with the adductors on one side at a time. Finally, the forward bend with the pelvis angled 45° creates different and generally fewer stresses on the hip joint and lower back than the posterior stretch, making this posture safer and less discouraging for beginners.

After getting accustomed to this pose, try a variation. Again come into the preliminary position with the right knee extended, but instead of reaching out directly toward the right foot, press your left forearm against your left knee. In most students, this will pry the right hip off the floor, and that's fine. Now reach out and slide the right hand against the floor halfway between the two thighs, approximating a 30–45° angle from the outstretched leg. Keep reaching toward your limit even though it lifts your right hip even more. The idea is to stretch the adductors on the right more than the previous posture. Then try reaching out closer to the right foot, perhaps 10–20° off axis from the right leg, in order to increase tension even more in the right adductors (fig. 6.21b). These variations are mainly for beginners and intermediate students. Those who are already flexible will not find them very interesting because their adductors can easily accommodate to all of the stretches.

Finally, after exploring the poses in which you are reaching out at an angle, come back to the original posture and reach straight toward the right foot. You will find that you are able to come further forward (fig. 6.21c). The hamstrings are still resisting almost as much as before, but the stretches off axis from the extended knee have relaxed and lengthened the adductors on that side, and the increase in how far you can now reach is a rough measure of how much they were contributing to your limitations— over and above restrictions from the hamstrings—in the initial posture.

These are all elementary stretches, and in keeping with the spirit of meeting the needs of novices, they should all be explored by simply rolling forward naturally, working from distal to proximal, coming into the stretch first with the upper extremities, head, chest, lumbar region, and hips, and then releasing in the opposite direction one step at a time from the hips to the upper extremities. Since these are asymmetric postures, you should repeat the series on the other side. After a warm-up, you can take the option of moving briskly back and forth from one side to the other to determine if one side is tighter, and then concentrate your attention accordingly.

When students are comfortable with these postures, they can start thinking about re-ordering their priorities by reaching up first with the hands and bending from the hips, then the spine, and then the head and neck. Only advanced students with good flexibility should try the final step of catching the outstretched foot to pull themselves fully into the pose, however, because we see the same problem here as in the posterior stretch: advanced students come into the posture by releasing rather than increasing tension on their spines, and beginners who pull forward with their hands may strain their lumbar region. Coming out of the pose, beginners should roll up and out as always (even if they came into the posture bending from the hips), intermediate students should release slightly in the hips and lower back before reaching forward, out, and up, and advanced students can do whatever they want, including reaching out and then up as a first priority.

Figure 6.21a. With the left foot placed against the right thigh, come forward in an initial trial to feel and evaluate hamstring tightness.

Figure 6.21b. Pressing the left forearm against the left knee, reach out at various angles (in this case 20°) to stretch the adductors on the right side.

Figure 6.21c. Notice the improvement. This is due to having lengthened the adductors, some of which have a hamstring character and which limit the forward bend for the same reasons as the true hamstrings.

Another variation of this series of postures that may be of more interest to intermediate and advanced students is to place the pulled-in foot high up on the opposite thigh before undertaking the forward bends. This variation is not recommended for those who have poor hip flexibility or for anyone with chronic low back pain, since it places peculiar and unanticipated stresses on the lower back.

FORWARD BENDING WITH THIGHS ABDUCTED

Forward bending with the thighs abducted stretches both hamstrings and adductors. In its usual form it is a symmetrical posture for intermediate and advanced students who have already achieved good hip flexibility. Start with the knees straight and the thighs abducted. Then initiate a forward bend leading first with the promontory of the sacrum to achieve your personal maximum for nutation. Then bend from the hips, or try to do so, before bending additionally in the lumbar spine. Those who have good sacroiliac flexibility will feel their ischial tuberosities spread apart and will be able to flex the pelvis forward while keeping a prominent lumbar arch (fig. 6.22), but as in the case of the posterior stretch, those who are inflexible are likely to be at their limits of sacroiliac nutation and hip flexibility merely trying to sit up straight.

In this posture tight adductors add to the problem of tight hamstrings for two reasons. First, spreading the thighs apart places the adductors under tension even before you start to bend forward. Second, because some of the adductors take origin posteriorly along the inferior pubic rami, they will pull forward on the underside of the pelvis just as surely as the hamstrings. Compared with the posterior stretch, the additional difficulty you have coming forward is due to the adductors, and if you are not very flexible you are likely to be struggling. If you can't separate the thighs more than 90°, it means that both the hamstrings and the adductors are limiting the stretch. The simplicity of the problem makes this perhaps one of the most maddening postures in hatha yoga for stiff novices.

Those who have good sacroiliac and hip flexibility have a completely different experience. They may even be able to bend all the way to the floor, keeping the back straight and even keeping the lumbar region arched forward.

Figure 6.22. In this advanced forward bend with hip joints abducted, the thighs literally get out of the way of the pelvis, which can drop all the way forward to the floor in the most flexible students.

This 180° bend from the hips can happen only because full sacroiliac nutation in combination with extreme abduction gets the thighs out of the way quite literally, and permits the front of the pelvis to drop down between them. This, in fact, has to happen in the most extreme cases of abduction, in which the thighs are spread straight out to the sides. What happens at the level of bones and joints under those circumstances is that the pelvis rotates forward and allows the anterior borders of the ilia to drop down between the necks of the femurs just as we saw with some of the hip-opening stretches (figs. 6.19e–f). The only difference is that here, those who are especially flexible are able to both abduct and to flex their hips fully when their knees are extended.

The extreme abduction in this posture, in combination with the pitched-forward pelvis, reveals one more feature. In comparison with the posterior stretch, it actually takes tension off the hamstrings because the sacroiliac joints slip to the extreme of nutation, spreading the ischial tuberosities laterally and positioning them closer to the insertions of the hamstrings, which are located out to the side in this posture. This means that those who have a lot of sacroiliac mobility will not be limited by either the hamstrings or the adductors in this stretch and will be working on limitations within the hip joint itself.

It is even more important to be attentive to foot position in this posture than in the posterior stretch, because with the feet spread so far apart, you may not notice that one foot is angling out more than the other. The cause of this is right-left imbalances in the medial and lateral rotators of the thighs, and these cannot be corrected except by paying attention to detail over a long period of time. If you are relatively inflexible, you should find the foot position that interferes the least with your attempts to bend, so long as you keep both feet at the same angle, but if you are more flexible you should analyze carefully which foot position gives you the most useful stretch. Many instructors who are watchful of such matters will suggest that you try to keep both feet perfectly upright.

CHURNING

An alternative for those who cannot come very far forward in the previous posture is to work with it asymmetrically and dynamically, combining the pose with a mild spinal twist. From the starting position reach with your right hand toward your left foot (or thigh, knee, or leg) while at the same time swinging your left hand to the floor in back of you and giving yourself a little push from behind to aid the forward bend. Then come up partially and reverse the position, reaching with your left hand toward your right foot and pushing yourself forward from behind with your right hand. Exhale each time you come forward, and inhale each time you come up.

This dynamic churning exercise is more rewarding for beginners than the previous pose because the emphasis on movement allows them to feel as if they are accomplishing something. It's also helpful because the asymmetry of the movement allows them to work with extra concentration on the side that is showing more restrictions.

THREE MORE VARIATIONS WITH ABDUCTED THIGHS

Three more variations of sitting forward bends with the thighs abducted are commonly taught. For the first one simply face the left foot and bend forward. If the thighs are spread to a 90° angle in the first place, this will require a 45° twist in the spine in comparison with bending straight forward toward the floor. Repeat on the other side.

For the second and more demanding variation (fig. 6.23a), tuck the left foot in toward the perineum, and while remaining upright, twist left. Follow this with a side bend to the right, at the same time reaching overhead with the left hand toward the right foot. This superimposes lateral flexion to the right onto a spinal twist to the left. Again, repeat on the other side.

The last variation is similar to the second one except that both knees are extended. With the thighs abducted, sit straight, twist enough to the right—45° if the thighs are at a 90° angle from one another—to face the right foot, and do a side bend toward the left foot. Repeat on the other side. Like the previous pose, this posture is more fitting for advanced students who can reach overhead with the free hand, grasp the toes, and pull further into the side bend (fig. 6.23b). This should done carefully, especially if you do the posture after you are warmed up, because so much of your attention is placed on whole-body twisting and bending followed by pulling with the free hand that you may not notice you have just dislocated your hip. One time a friend of mine got carried away doing a demonstration and did exactly that. So be watchful. If you feel something give, come back delicately to a neutral position and stop doing all postures for a day or so to evaluate what has happened.

Figure 6.23a. To come into this twisted side bend, tuck the left foot in, twist left, bend to the right, and reach overhead with the left hand toward the right foot. This superimposes lateral flexion to the right onto a spinal twist to the left.

These last two postures are among the few side bends that are possible from a sitting position. Even so, they are not pure side bends because they are superimposed on sitting spinal twists. This makes them less natural and more complex than standing side bends (chapter 4), in which the thighs are extended and in line with the torso, and in which a relatively simple lateral flexion of the spine is possible.

SITTING TWISTS

Sitting spinal twists, which we'll discuss in the next chapter, create stretches and stresses in the hip joint not found in any other type of posture. Every sitting spinal twist in which you lift one knee and pull it toward the opposite side of the body opens the hip joint, and does so without many encumbrances from muscular attachments, and, as we have seen repeatedly, when the muscles are not limiting hip flexibility we are down to bare bones, joints, and ligaments.

THE KNEELING ADDUCTOR STRETCH

To create the purest passive stretch of the adductors, especially for students who are not very flexible, warm up for at least half an hour, and then kneel down on a well-padded carpet with each knee on a small sheet of cardboard. Drop the head and chest, and settle onto the forearms with the forehead resting against the crossed hands. Slowly slide the knees apart, letting the feet move to whatever position is most comfortable as the body weight abducts the flexed thighs, possibly to an angle of about 120° from one another. Stay for a while in a position that permits the most abduction.

Next, slowly move the torso forward. This may permit you to slide the knees further apart (fig. 6.24). Then slide the knees somewhat closer together and lower the hips to the rear for more flexion. Try to relax. As you flex the hips and take your weight to the rear, you will be stretching different parts of the adductors than when you bring your weight forward, and anyone who is not very flexible will find that moving to the rear is an

Figure 6.23b. For this twisted side bend, abduct the thighs, twist right, bend left, and again reach overhead with the free hand toward the opposite foot.

intense stretch that should be approached with caution. Last, you can try coming all the way forward, but be careful of this if you have lower back problems because the pose creates an acute backbend that places a lot of stress on this region.

These exercises are among the most effective stretches available for the adductors, and working with them five minutes a day as a part of a balanced practice will soon show results in all postures that depend on adductor flexibility, including all forward bends, all standing postures with the feet wide apart, and all cross-legged sitting postures.

THE PIGEON POSTURE

The pigeon has a large, puffed-out chest, which we mimic in the pigeon posture. It is a superb advanced backbending pose in its completed form, but a preliminary version can be done by nearly everyone. We place it here because both versions work with hip flexibility on the right and left sides simultaneously, producing extension of the hip joint for the thigh directed to the rear and flexion of the thigh facing the front.

To come into the preliminary pigeon, start in a hands-and-knees position, pull the right knee forward to flex the right hip joint, and thrust the left foot back to extend the left hip joint. Allow the right foot to end up wherever it naturally falls, which will usually be near the genitals. Keep the elbows extended and the shoulders pressing the hands against the floor. Pull the head up and back, thrust the chest forward, and feel the essence of the pose (fig. 6.25a). If you feel any discomfort in the right hip joint or groin, stop right there. This is as far as you should go. Later on, after you have gotten accustomed to the basic position, you can lower your weight all the way down so that you are lying on the right thigh, which flexes the right hip joint completely and lessens hyperextension of the left hip joint (fig. 6.25b). Neither the hamstrings nor the adductors are being stretched, but you may feel the effects in the hip joint itself. Or, because all of your body weight is pressing against the forward thigh, you may feel tightness in the groin on that side.

When you are comfortable with the preliminary pigeon, you can try the advanced posture. Gradually swing the front foot out so the leg is perpendicular to the long axis of the thigh and torso. Don't rush. As you bring your leg around to about a 45° angle, you may start to notice tension in

Figure 6.24. This kneeling adductor stretch should be done with caution, especially by those who are not very flexible.

Figure 6.25a. This preliminary pigeon, allowing the front foot to come to a neutral position, is comfortable for everyone who has healthy knees. The right thigh is flexed, the left thigh is hyperextended, and the back and neck are fully extended.

Figure 6.25b. Folding into a forward bend from the previous posture is a comfortable and richly rewarding posture for many students. The pose increases flexion of the right thigh and lessens hyperextension of the left thigh.

Figure 6.25c. The advanced pigeon, with the front leg approaching a 90° angle from the front thigh, and featuring a 90° lumbar backbend, should be approached with respect and caution.

your knee joint and a deep pulling sensation in your hip joint on that same side. Inexperienced students should never lower their weight down against the front thigh in this position because it places too much stress on the knee. Intermediate students can moderate the tension on the front knee by working with the classic pigeon posture—lifting up on the fingertips, pulling the head and shoulders up and back, thrusting out the chest as much as possible, and taking empowered thoracic inhalations to increase their inspiratory capacity. Advanced students complete this posture by grasping the sole of the rear foot and placing it against the top of the head (fig. 6.25c).

HIP FLEXIBILITY STANDING

Improving hip flexibility in standing forward bends requires a different mind-set from working with hip flexibility in supine, inverted, or sitting positions. Standing postures present special problems because we place the feet in positions that often depart radically from the norms of ankle, knee, and hip movements needed for walking and running. Any time we plant the feet and then come into a standing posture with a twist, bend, lunge, or some combination of these three, we place stress on the hip joints and their supporting muscles. The problem is that we do not ordinarily readjust the feet to make the posture easier, which indeed would miss the point; instead, we work with the posture by increasing the tension in muscles around the hips and knees. To this end, many instructors recommend standing on non-slip rubber mats to keep the feet firmly in place and immobile. Beginners are well-advised to begin with moderate foot positions to minimize stress, which allows them to explore standing postures in their mildest form before working with more demanding versions. To illustrate we'll do two experiments, one for the hamstrings and one for the adductors.

We'll begin with an adaptation of the angle posture (figs. 4.25–26 and 6.26a–b) to stretch the hamstrings. Stand with the feet about three feet apart. Then rotate the right foot 90° to the right and the left foot 30° to the right. Given this foot position, the torso will most naturally face about 45° to the right, and the hip joints and the muscles that restrain them will be relatively comfortable. To continue, swivel to face the right foot as square-ly as possible, and notice that this movement alone creates intense tension in the left quadriceps femoris. Counter that by tightening the left gluteals. The left thigh is now hyperextended, causing the head of the femur to be driven into tight apposition with the acetabulum. If you keep standing up straight, hyperextension of the left thigh forces the lumbar region into a deeper lordosis. There is little or no tension on the right hamstrings because the right lower extremity is still in a neutral position.

Then, facing the right foot, establish full nutation of the right sacroiliac joint by tightening the right psoas muscle, and then slowly fold forward, first from the hips (fig. 6.26a) and then in the lumbar region. As you come down, the right hamstrings come under more tension (fig. 6.26b). Keep the muscles of both thighs active to make the posture feel more secure, and notice that this also makes you feel more confident in coming forward. Come up and repeat on the other side.

This posture involves a swivel and a forward bend at one hip joint which is intensified by the weight of the upper body. The resulting stretch is more demanding than hamstring and hip stretches in supine, inverted and sitting postures because now you have to be attentive to the body as a whole—and this is one of the easiest and least complex of the standing postures. Most others, especially the triangles and lunges, place even more stress and tension on the hip joints.

Figure 6.26a. Coming into this intermediate stage of the angle posture provides one with a golden opportunity for the study of muscle and joint mechanics. First, swiveling to face the right foot before starting to bend forward at the right hip creates intense tension in the left quadriceps femoris muscle, which should be countered by tightening the left gluteals. Second, hyperextension of the left thigh (still standing straight) forces the lumbar region into a deep lordosis, which should be maintained even as you start to bend forward. It is at this point (third) that a sharp asymmetric effort is made with the right psoas muscle (the above posture) to maintain the lumbar lordosis, maximize nutation of the right sacroiliac joint, and bend at the right hip. This is a magnificent asymmetric pose in its own right and is worth extended study. Repeat on the other side.

To stretch the adductors, try standing on a non-slip surface with the thighs abducted almost as far apart as possible, still, however, being somewhat conservative. Keep the feet parallel, and make sure the back and pelvis are in a comfortable upright position. Then bend forward slightly, keeping the lower back arched, and notice that you may be able to abduct the thighs even further. The idea is to refine this posture to find a revised maximum but fairly comfortable limit for abduction. Then bend in tiny increments, first coming forward from the hips to stretch the hamstring muscles and adductors that originate from the rear, and then straightening up and bending backward to stretch those segments of the adductors that originate more anteriorly on the inferior pubic rami. Notice that the stretches become intense in both directions. It is important to be careful even with these small shifts in position for two reasons: because the exercise was started with near-maximum abduction, and because the feet are planted and the weight of the entire upper body is brought to bear on the hip joints and adductors. The directions are telling: bend forward slightly; move in tiny increments; stretches become intense in both directions. Such comments would not be necessary if you were working with hip flexibility and adductors in supine, inverted, or sitting postures.

Figure 6.26b. Folding forward in the angle pose places intense stretch on the hamstring muscles of the front thigh, stretch which should be countered by holding those muscles in an isometric state of contraction along with moderate tension in the rest of the muscles of both lower extremities. This pose provides an excellent example of the inadvisability of holding an intense posture in which the knee joint is fully extended (chapter 4) without keeping tension in the muscles that support the joint antagonistically.

NONATTACHMENT

Those who are flexible in the hips and those who are not are like ships passing in the night. Anyone with good hip flexibility can press the chest against the thighs in a standing forward bend (with the knees straight) and allow the lumbar region to be stretched passively by gravity. They are relaxed and comfortable, and they feel a sense of inversion. In sitting postures it is easy for them to press the chest to the thighs (again with the knees straight) in the posterior stretch or to lower the abdomen, chest, and head all the way to the floor with the thighs abducted. By contrast, those whose hips are not flexible are practically a different species. As they try to lower into a standing forward bend at the hips, their backs are crunched over rather than stretched passively by gravity, and they get little or no sense of inversion. And as far as doing a credible sitting forward bend with the thighs abducted is concerned, they might as well try flying.

Even if it is not possible for you to flatten your chest against your thighs while keeping your knees straight, be comforted. This is of little consequence, at least by itself. Hatha yoga is a science of mind as well as body, and beneficent changes in the mind-body continuum are available even to those who are stiff and past their youth. For achievement of health and peace of mind, consistent practice is more important than accomplishing some arbitrary standard of flexibility, and if you work consistently with a balanced set of postures, progress will be realized on many fronts, some of them unexpected. All you need is commitment plus a playful sense of observation and experimentation. Such an attitude also cultivates *vairagya*—nonattachment—and if that accompanies your quest, you will, by definition, be successful.

BENEFITS

Forward bending postures are generally more subdued than back-bending, twisting, and inverted postures. They tend to quiet rather than stimulate the somatic nervous system and the sympathetic limb of the autonomic nervous system (chapters 2 and 10). On the other hand, as soon as students are able to flex forward enough in the spine and the hips to compress the abdomen, forward bending postures seem to have mildly invigorating effects on the abdominal organs, possibly stimulating the enteric nervous system (chapter 10), and thereby enhancing digestion and assimilation of foodstuffs from the bowel, as well as relieving constipation. Finally, because forward bending postures are so important for hip flexibility, they are among the most important training postures for meditation.

"If you want to learn about meditation, you first need to know something about concentration. According to yoga, concentration means focusing the mind on one object. An undisciplined mind—the kind most of us have—tends to shift continually from one object to another. Steadying the mind by focusing it on one object helps you to gradually overcome this ever-wandering habit of the mind. After prolonged practice, the mind is able to focus on one object for longer and longer intervals. When the mind remains concentrated on one object for a period of twelve breaths, this is called meditation. Thus, meditation can be defined as the uninterrupted flow of concentration."

— Pandit Rajmani Tigunait, in *Inner Quest*, p. 71.

CHAPTER SEVEN
TWISTING POSTURES

"This sadhana is not something esoteric or mysterious. It is simply a way of awakening knowledge about ourselves. And when we have learned how to know ourselves, we will also know the world around us and the supreme reality as well. . . . Through Vedanta you can gain the knowledge about yourself and your relationship with the universe; through these practices you can gain experiential knowledge about these same things."

— Tapasvi Baba, from a lecture given at the Himalayan Institute in Honesdale, PA on July 17, 2000 (translated by D. C. Rao).

Imagine what it would be like if we couldn't twist. We couldn't swing a bat or a golf club. We couldn't greet someone standing next to us without turning our entire body. We couldn't even twist the lid from a jar. We would walk like marionettes and dance like robots. Twisting is needed for every activity that involves moving to the side and front simultaneously. Lean diagonally across a table—you twist. Throw a ball—you twist. Scratch your left foot with your right hand—you twist. Not only that, every inquiry about anything not directly in your line of sight requires twisting. It might be something as simple as looking around a room to determine where to go and what to do next, or it could be something as complex as twisting the head, neck, and trunk around 135° in an automobile seat to confirm that no one is alongside and slightly behind you before you change lanes.

Anatomically, all asymmetric whole-body activities, even those that principally involve flexion and extension, contain elements of twisting. Take walking. As you stride forward with one foot the opposite hand comes forward—right foot left hand, left foot right hand, right foot left hand, left foot right hand. This creates a moderate whole-body twist, and a slight twist in the gait not only balances the body, it calms the nervous system. If you are skeptical, try it the other way—right foot and right hand forward at the same time, clunk; then left foot and left hand forward at the same time, clunk. It's jarring.

In hatha yoga we have many twisting postures to pick from, including all asymmetrical postures that are not pure sidebends. In this chapter we'll examine a few of them in detail, including basic spinal twists, various combinations of flexion, extension, lateral flexion, and twisting of the head and neck, and certain standing postures that combine twisting with backbending and forward bending. We'll start with an analysis of the fundamentals of twisting, then look at the anatomy of twisting from head to toe, and complete the chapter by looking at supine, standing, inverted, and sitting twists.

THE FUNDAMENTALS OF TWISTING

Forward and backward bending postures, as well as flexion and extension in general, always take place in relation to earth's gravitational field, but twisting is fundamentally different because you can twist the body or some part of the body without altering its relationship to gravity. For example, you can twist your head as far as possible right and left, but unless you combine this with flexion, extension, or lateral flexion, the relationship of the head to earth's gravity is unchanged.

Bends and twists not only differ in nature, they differ in how they come about. Any movement that involves bending—whether whole-body bending, or flexion and extension of a limb—gets its impetus either from interactions with gravity or from a force like that created when children in swing sets kick their feet forward and then backward to get themselves going, or like that created when you push off from the end of a swimming pool. All the twisting motions with which we are concerned here, however, are initiated by *torque*. It is torque that starts someone spinning around on a rotating chair, and it is torque that the rotators of the hip use for rotating the long axis of the thigh with respect to the pelvis.

[Technical note: To be accurate, it has to be admitted that nearly all movements at joints use torque, whether bending your elbow, kicking a football, or grasping a pencil. That is why I was careful not to say that *linear* movements are used for pushing off from the end of a swimming pool. That would have been incorrect; torque is used there as well to extend the hips, knees, and ankles. The difference is that for those movements the axial center of rotation is in each specific joint, whereas in this chapter our main concern is with forces that operate to create twist on the long axis of the body or limb from the perimeter of an imaginary circle that surrounds the axis of the affected body part.]

Bending and twisting differ in at least three more ways. First, forward and backward bending are often symmetrical, but twisting can never be: it always pulls structures on the right and left sides of the body in opposite directions. Second, forward and backward bending need not increase axial tension in the body, but twisting, unless it is utterly unresisted, always compresses structures that lie in the axis of the twist. Last, while forward and backward bending are comparatively simple expressions of flexion and extension, there are several different kinds of twists: rotations of synovial joints, more constrained spinal rotations, and whole-body swivels that combine both of the above.

TORQUE

Simplistically, and for purely practical purposes, we can state that torque is any mechanical force that can produce a rotation. It is initiated by muscular effort, but like any other force, that effort does not have to produce a visible result. It is like a push. You can push against a sapling and bend it, or you can push against a tree trunk to no avail. And so it is that a torque can either actuate a twist or it can be an isometric effort that attempts a twist but fails—trying to twist a locked doorknob, trying unsuccessfully to escape having both shoulders pinned to the floor in a wrestling match, or tugging on the rope of a frozen lawnmower engine. In hatha yoga the muscular effort to come into a spinal twist creates torque throughout the body, actuated in some regions, in others not.

SYNOVIAL ROTATIONS IN THE EXTREMITIES

The simplest kind of twisting involves free and easy rotation at synovial joints, in which the slippery cartilaginous mating surfaces of bones offer little or no resistance to movement. We see this when we "twist" a screw into a board, alternately pronating and supinating the forearm. As shown earlier (figs. 2.8 and 4.3), these rotations take place because two pivot joints, the proximal radio-ulnar joint at the elbow and the distal radio-ulnar joint at the wrist, permit the ulna and radius to come into an X configuration for pronation and into a parallel configuration for supination. Other familiar examples of synovial rotations are the rotary movements of the femur in the ball-and-socket hip joint and the rotary movements of the humerus in the glenoid cavity of the shoulder joint (fig. 1.13).

SPINAL ROTATION

The second kind of rotation, and the one that the hatha yogi first thinks of as a twist, is rotation of the spine. Whether standing, sitting, supine, or inverted, spinal twists involve the entire torso, but they all start with axial rotations between adjacent vertebrae. Taken together, such rotations—24 of them in all between C1 and the sacrum—add up to a lot of movement, even though this takes place against the resistance of intervertebral disks, facet joints between the vertebral arches, the rib cage, and muscles and ligaments from the head to the pelvis. We rarely do complete spinal twists in everyday life, but they are one of the five fundamental gestures in hatha yoga.

STANDING SWIVELS

Another kind of twist might better be called a swivel. It involves rotating the pelvis and thighs around so that the torso faces to one side, and it invariably begins in a standing position with the thighs at least partially abducted. The swivel can be limited to the lower extremities, but more commonly it is a whole-body twist in which the hips, shoulders, and torso are all rotated in the same direction, usually at the same time. Swiveling is a combination of spinal and synovial twisting—a spinal twist superimposed on synovial rotation of the hip joints. In sports we often see these movements taken to their extremes, as when skiers negotiate a steep drop with short, side-to-side excursions while keeping their shoulders perpendicular to the fall line of the downhill slope. And in everyday life, any time you face the torso in a direction other than straight ahead but do not shift the feet, you swivel at the hips. In hatha yoga practically any standing posture that involves planting the feet and then turning the rest of the body is a swivel—a spinal twist combined with medial rotation of one thigh and lateral rotation of the other.

For practical purposes, how a posture feels to us is our major concern, so from this point on, if it feels like a twist we'll call it a twist, whether it is a free and easy synovial rotation, a constrained spinal rotation, a swivel, a torque that goes nowhere, or a torque that produces a movement everyone recognizes as some kind—any kind—of a twist.

STABILITY IN TWISTING

Try to imagine how difficult it would be architecturally to design a joint that is stable enough to twist as well as permit flexion and extension. You would probably limit flexion by allowing the joint to fold in on itself completely, but you would need to include both ligaments and bony stops to limit extension. That's easy enough. But to permit that same joint to twist as well, you would have to include a complex of muscles to activate the twisting, and you would have to superimpose any number of specialized ligaments and muscles on the joint to keep the twisting within reasonable bounds.

This is not a small order, and as a general rule, wherever extensive flexion and extension take place, we see that twisting is limited. In the lumbar region of the spine, which is the site of most spinal forward and backward bending, little twisting is permitted, but in the chest, where backbending and forward bending are limited, we see excellent potential for twisting. And in the extremities, the fingers and toes permit flexion and extension but little twisting. Only in the cervical region of the spine, in the hips and shoulders, and in the flexed knees will we see extensive flexion and extension as well as the additional potential for rotation, and those regions are as stable as they are only because of robust muscular support and numerous restraining ligaments. Even so they are all hot spots for dislocations and other injuries.

ASYMMETRY AND TWISTING

Whole-body twisting is always accomplished by pairs of obliquely oriented muscles, one on the right side of the body and one on the left. The muscle on one side shortens concentrically, creating the twist, while the muscle on the other side lengthens, resisting the twist. The external and internal abdominal oblique muscles in the abdominal wall are a case in point. As you twist to the right, the right external and the left internal abdominal obliques shorten concentrically, and the right internal and left external abdominal obliques lengthen against resistance. On the other hand, vertical muscles such as the right and left rectus abdominis, as well as horizontal muscles such as portions of the right and left transversus abdominis, remain the same length during a twist and simply come under isometric tension equally on both sides.

All ligaments that run obliquely on the two sides of the body are also brought under asymmetrical tensions by twisting. If you bend your knees and twist the torso to the right, as you do when you ski to the left while your shoulders are facing downhill, excessive lateral rotation of the left leg will be checked by the collateral ligaments of the left knee, and excessive medial rotation of the right leg will be checked by the cruciate ligaments of the right knee. Skiing to the right, of course, mirrors these tensions. By contrast, if you stand in a symmetrical knock-kneed position with the toes facing in and heels out, the ligaments are stressed symmetrically rather than asymmetrically. Excessive medial rotation of both legs is checked equally by the cruciate ligaments on both sides. Or if you stand with the feet wide apart, heels in and toes out, excessive lateral rotation of both legs is checked equally by the collateral ligaments of both sides.

Ordinary activities such as walking keeps the two sides balanced, but when you twist you usually favor one side—always holding a book to one side and twisting the neck in the same direction, or always coming up on the same side for air when you are swimming freestyle. Or if you consistently hold the top of the handle of a snow shovel with the right hand and throw the snow to the left, you will develop more strength, stamina, and flexibility for twisting to the left.

Twisting habitually to the same side during the course of daily activities distorts the body's bilateral symmetry, and such biases sooner or later produce asymmetries in its structure. For this reason twisting postures should always be done in both directions, and to correct imbalances they should be done three times—twice on the less flexible side.

COMPRESSION

To get the most water out of a washcloth you don't roll it up and bend it—you roll it up and twist it, and this squeezes and compresses the water out of the cloth from top to bottom. By the same token, in twisting postures all structures that lie in the axis of the twist (an imaginary line around which the rotation takes place) are squeezed and compressed. We have seen that throughout the body, oblique muscles create twists, while vertical and horizontal muscles resist them isometrically. Every time you do a spinal twist or even establish the torque to create one, the obliquely and vertically oriented muscles of the back and abdomen compress the spine and torso axially, and when that happens the compression is transferred to the spine and torso as a whole. We also see this phenomenon in the extremities. Every rotation of the arm, forearm, thigh, or ankle creates axial compression in the long axis of the extremity, the amount of which is in direct proportion to the intensity of the effort.

You can feel this compression from head to toe in a whole-body twist and swivel. Stand with your feet 2–3 feet apart and combine a twist of the head, neck and torso with a swivel in the hips and resistance in the knees, ankles, and feet. The muscular effort tightens joints and ligaments everywhere from head to toe. The body feels like a tightly wound spring. Twist in the other direction for balance, and each time you release the posture, notice the release of axial compression.

THE SKULL, THE ATLAS, AND THE AXIS

The cervical region between the head and the chest is one of the few in which flexion, extension, and twisting can all occur at the same time. Flexion, extension, and lateral flexion take place between the head and the first cervical vertebra (C1) at bilateral synovial joints (no intervertebral disk is present here); rotation takes place at a synovial joint between C1 and C2 (again, there is no intervertebral disk at this site); and all four movements, singly or in combination, take place in the rest of the cervical region from C2 to T1, the segment of the cervical spine which contains joints with typical intervertebral disks and synovial articular processes (figs. 4.8, 4.10, and 4.13a). We'll start our discussion at the top and work down.

Of the seven cervical vertebrae, the top two have been given special names that reflect their anatomical and functional relationships. The top one (C1) is the *atlas* (figs. 7.1–2), named for Atlas, the Greek god whose shoulders supported the world. In this case the atlas is a ring of bone which supports the skull, and its relationship with the cranium permits side-to-side and back-and-forth movements but little rotation. The second cervical vertebra (C2), the *axis*, has a tooth-like protuberance called the *odontoid process*, or the *dens*, which extends up from the vertebral body and provides the axis around which the atlas together with the cranium can rotate (figs. 4.8, 4.10, and 7.2).

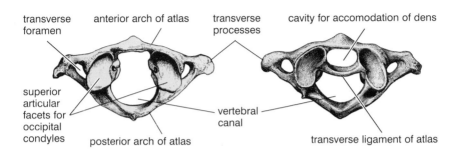

transverse foramen   anterior arch of atlas   transverse processes   cavity for accomodation of dens

superior articular facets for occipital condyles   posterior arch of atlas   vertebral canal   transverse ligament of atlas

Figure 7.1. Superior views of the atlas (left) and of the atlas with its transverse ligament (right). The transverse ligament helps delimit the region within which the dens of the axis rotates. The foramina in the transverse processes accomodate the vertebral arteries. The combination of the two superior articular facets for the occipital condyles can be likened to the concavity of a spoon subserving flexion, extension, and lateral flexion of the cranium on the atlas, but not rotation (Sappey).

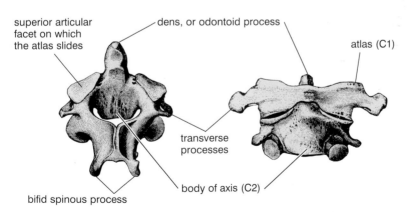

superior articular facet on which the atlas slides   dens, or odontoid process   atlas (C1)

bifid spinous process   transverse processes   body of axis (C2)

Figure 7.2. On the left is a posterosuperior view (from behind and above) of the axis (C2), and on the right is an anterior view (from the front) of the atlas (C1) and axis in combination. The atlas in combination with the cranium rotates around the dens (odontoid process) of the axis. Notice how much wider the atlas is than the axis and compare this image with the drawings made from the roentgenograms in fig. 4.8 (Sappey).

Even though the cranium and the atlas together can rotate around the axis, the cranium cannot rotate on the atlas because of the shape of the matching articulating surfaces. If you can find a spoon whose converse mirror image matches that of an egg, you can slip the egg back and forth and from side to side, but you cannot turn it around and around. The cranium sits on the atlas in much the same way; it slips on the atlas in only two directions, from front to back and from side to side. The articulating surfaces at the base of the skull—the *occipital condyles*—lie on either side of the *foramen magnum*, which is the hole that accommodates the spinal cord where it emerges from the base of the brain. The cranial surfaces of the occipital condyles (the egg) slip back and forth (flexion and extension) and from side to side (lateral flexion) on matching articulating surfaces on the upper surface of the atlas (the spoon).

FLEXION AND EXTENSION

The articulating surfaces between the head and the atlas allow slippage in the form of 10° of flexion, 20° of extension, and 15° of right and left lateral flexion. To get a sense of the articulation between the skull and the atlas, nod your head back and forth without bringing the rest of the cervical vertebrae into the motion. Just be aware of a sliding feeling, with a little more motion permitted backward than forward. Envision the axis of rotation extending from ear to ear. As you move your forehead forward the chin is tucked, and as you move your forehead backward the chin is jutted forward. This is the feeling of flexion and extension between the cranium and the atlas. Then after exploring that motion, bend the neck backward and forward to your limits. The contrast is clear. You can feel the cranium sliding on top of the neck in the first motion, and you can feel the entire cervical region bending in the second.

LATERAL FLEXION

Next explore lateral flexion at the same articulations. Keeping your head balanced upright, slowly tilt your forehead to the right and your chin to the left, and then repeat in the opposite direction, envisioning the axis of rotation extending between the mouth and the back of the head. Again, don't bend the rest of the neck. To the extent that you can feel the chin and the forehead move in opposite directions, you will be feeling the side-to-side slipping of the cranium on the upper surface of the atlas. If you concentrate on carrying this particular movement to its maximum, it feels extensive— at least 15° in each direction. Again, for contrast, as soon as you have reached the limit of movement between the atlas and the head, flex the head and neck as far as possible laterally. This gives you a different sensation, like the movement of the head in a Jack-in-the-Box.

THE MOVEMENT OF THE ATLAS AROUND THE AXIS

Now we can return to our more immediate interest in twisting as we explore the rotation of the atlas (plus the skull) around the odontoid process of the axis (figs. 4.8 and 7.2). Do the following experiment. Sit quietly with the spine and head upright. Without dipping the head either forward, backward, or to the side, slowly turn it to one side and then the other through a total excursion of 60–90°. In other words turn 30–45° to the right and 30–45° to the left. Try to relax. Turn slowly, then a little more quickly, then again slowly. This is the movement of the atlas around the odontoid process. You can envision it if you make a circle with the thumb and index finger of one hand, insert the thumb of the other hand (representing the odontoid process) against the back of the circle, and rotate the circle through an excursion of 60°, in other words 30° each way.

Next, for contrast, turn the head as far as possible in each direction. If you have fairly good flexibility you can turn about 90° each way. The first portion of the twist is 30–45° of rotation between the atlas and the axis, and the remainder of the 45–60° represents twist in the rest of the neck.

To return to the delicate twisting that takes place only between the atlas and the axis, notice what happens as you move past the symmetric central balancing position with the head facing straight ahead. Something takes place—a change of speed, first a slowing down, then a speeding up as you pass center, a movement so subtle that you will not feel it unless you are relaxed. What is happening is that a cam action is lifting your head slightly as you pass center. It's as if a motorized toy car were approaching a little hill. You can feel your head rise as you approach the hill, and a gathering resistance that peaks at the top and then diminishes as you cross over. Find the little hill, park exactly on top, and then move slightly to one side and then the other. If you watch carefully you will notice that the high point is not perfectly on line in the midsagittal plane of the body but is usually somewhat to the left or right of center, and that you habitually adjust your body posture so that your head rests on the side that keeps you facing relatively straight to the front.

THE RELATIONSHIP BETWEEN THE AXIS AND THE ATLAS

To understand why this happens we need to look more carefully at the relationship between the axis and the atlas. As we discussed in chapter 4, each individual vertebra is composed of a vertebral body and a vertebral arch. The exception to this is the atlas, which is simply a ring of bone (figs. 7.1–2). In the embryo what had originally been the body of the atlas became incorporated into the axis, bestowing on the axis the equivalent of two fused-together vertebral bodies, the top portion of which is the odontoid process that stands up from the rest of the vertebra (fig. 7.2). It is around this process that the atlas rotates, and it is from this function that the axis was named (figs. 4.8 and 7.2).

The reason you feel the rising-on-center sensation as you twist is that the lateral joint surfaces between C1 and C2 are not in perfect apposition unless the head is turned slightly to one side or the other. Accordingly, rotation off that center is accompanied by a slight vertical descent of the head; keeping the head directly on the high point once you have located it is an interesting exercise in balancing and concentration. Enthusiasts for working with right-left balance might suggest that shifting the head to the side on which it does not usually rest will gradually bring the rising-on-center sensation into the midsagittal plane of the body and make the joint more symmetrical. It's an interesting idea, although speculative.

MOVEMENTS OF THE HEAD AND NECK

Below the axis, from C2 to T1, typical intervertebral disks separate the vertebral bodies of the spine. This limits movements between any two adjoining vertebrae (figs. 4.10 and 4.13a), but this segment of the spine still provides us with mobility far above and beyond what we saw between the skull and C2—a total of about 90° of additional rotation, 80° of additional flexion, 50° of additional extension, and 40° of additional lateral flexion on each side.

If you concentrate, it is possible to isolate the movements between C2 and T1 from those between the skull and C1. Thrust the chin forward and at the same time flex the entire neck forward and down. Then after coming back to a neutral position, pull the chin backward and at the same time extend the neck to the rear and down. Last, initiate lateral flexion to the right and left from the base of the neck rather than head first. In all of these cases you can avoid, at least to some extent, the smaller motions between the skull and C1. By contrast, twisting selectively between C2 and T1 without first rotating between C1 and C2 is more difficult. Only if you tighten the strap muscles of the neck generally (fig. 5.5), and do so with some determination, can you initiate a twist from the base of the neck and avoid the rotation between C1 and C2.

HATHA YOGA NECK EXERCISES

We ordinarily move the head and neck in highly stereotyped patterns, and because these movements are common and natural in day-to-day life they are also the safest. Hatha yoga neck exercises are another matter. They are done in isolation and are usually taken further than the moderate flexion, extension, and twisting found in our usual activities. In hatha yoga exercises we flex the chin all the way to the sternum, extend the head back as far as possible, flex the ear toward the shoulder laterally, and rotate as far as possible right and left (fig. 7.3). We must explore the cervical region in detail to understand these and other possible neck movements, and then we must augment our knowledge with practice, experimentation, and observation. Take your time and be conservative. No movement that causes neck pain should be repeated injudiciously, and in the event of injury, practice should be stopped until the problem is resolved.

FLEXION

Flexion is the most natural movement of the neck—it happens any time you look down. To do it as an exercise just lower the head forward as far as possible and let it rest in that position. After a few weeks of practice you may be able to pull the chin all the way to the sternum, hold it there isometrically, and be confident that the extensor muscles in combination with the restraining ligaments of the spine will maintain the vertebrae in a state of healthy alignment. Do not aim for a chin lock (chapter 3), in which you try to place the chin into the suprasternal notch. Just lay it down against the front of the sternum. If you flex forward a total of 90°, the first 10° of movement is between the cranium and the atlas, and the balance takes place between C1 and T1.

EXTENSION

Extension of the neck is also a natural movement—it happens any time you look up. To do it as an exercise extend your head to the rear and hold in that position for a few seconds. The first thing you'll notice is that extension is stopped by restraining ligaments and bony stops long before the back of the head can reach the upper back. Even so, you can reach the end of a relaxed excursion to the rear and then pull further back vigorously against the bony and ligamentous stops so long as you increase your capacity gradually. If you can bend the head backward a total of 60°, the first 20° will be from the slipping motion between the cranium and the atlas, and the last 40° will be from the incremental vertebra-to-vertebra shifts between C1 and T1. After you have explored extension in isolation, go slowly back and forth between full flexion and full extension.

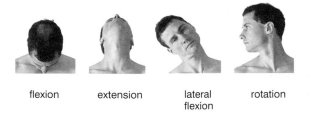

flexion   extension   lateral   rotation
                                flexion

Figure 7.3. Neck movements: flexion, extension, lateral flexion, and rotation.

LATERAL FLEXION

Most people both have and use a considerable range of motion for flexion and extension, but it is rare for anyone to cock their head to the side more than 30–40°. About 15° of lateral flexion occurs at the synovial joint between the cranium and the atlas, and the rest takes place between C1 and T1. Lateral flexion is rarely done in isolation but is commonly found in combination with rotation. Try it. Bend the head toward the left shoulder, and notice that it is natural to combine that movement with a slight twist either to the right or left that rotates the nose up or down. Notice how far you can safely take this movement, and compare it with plain lateral flexion. You will quickly become aware that combining the two is what you are accustomed to, as when you flex your head laterally to the right and look slightly down to scan titles on bookshelves. A pure lateral flexion is a curious movement, and when you do it with some insistence you will sense deep bone-to-bone restrictions on the side to which you are bending. Explore lateral flexion to the right and left separately, and then go back and forth between the two.

TWISTING

Twisting in the neck could hardly be more natural, and can be safely taken to its limit, at least by anyone who is practicing hatha yoga regularly. Possibilities for minor variations abound. If you twist as though you were looking for a pencil in a drawer at waist level and to your right, you will notice that this twisting movement includes lateral flexion to the right. And if you twist as though you were looking for an object on a shelf to your right at shoulder level, the movement includes lateral flexion to the left. Finally, if you twist as though you were looking for an object above you and to your right, the movement includes lateral flexion to the left, extension, and twisting.

Now, minimizing flexion, extension, and lateral flexion, try a pure twist of the head to the right and then to the left. In other words keep your head level and envision an axis of rotation that runs from the top of the head straight down through the spine. A moderate twist is a fine exercise, but if you continue until you come to your limit and then keep pulling, you will come in touch with a clear sense of axial compression. And if you move slowly you can also feel tension gather in the various structures of the neck that finally stop the twist—first muscles, then restraining ligaments, and finally skeletal stops. If you can twist a total of 90° in each direction, the first 45° of the movement is the rotation at the synovial joint between C1 and C2, and the rest is between C2 and T1. Pay attention to all the sensations that accompany the full twist, and explore new limits as you become confident that you will not hurt yourself. You will be startled to find how robust the neck is and how vigorously you can pull into a fuller twist if you increase your capacity gradually. And if you keep pulling isometrically when you reach the limit of the twist, you will increase both muscular strength and the hardiness of the ligamentous and skeletal structures. After working with each side separately, go sequentially from side to side.

OTHER MOVEMENTS IN THE NECK

Many other neck movements can be explored. One of the best is to twist the head 30°, 45°, or 60° to one side, and from that position, to swing it back and forth linearly—right front to left rear, and left front to right rear. A less natural movement, and one that should be approached more respectfully, is to take the head and neck through the same linear movements without first twisting. Going forward at an angle feels safe enough, but going backward you will encounter the same sort of unusual restrictions that you experienced going to the extremes of lateral flexion.

Neck rolls, in which you swing the head around slowly in a motion that is similar to circumduction of the thigh or arm, are questionable exercises, and this is easy to demonstrate. Let's say you are looking down in your lap and suddenly your attention is called to a bat in the upper right corner of the room. You don't have to think. Your head will move quickly and safely in a straight line to face the object of your concern, and muscles and restraining ligaments will protect you from going too far. By contrast, if you connect the two points with a fast neck roll instead of a linear motion, you will immediately see why such movements deserve to be treated with caution. Instead of moving linearly from a neutral position, you are circling your head around in a highly artificial movement. Even though most people are unconsciously wary of going anywhere close to their limit, these movements can still cause injuries in those who are doing them for the first time, and it is for this reason that many hatha yoga teachers say they should not be taught at all. In any event, if you are determined to do them, at least move slowly and well short of your limits.

THORACIC TWISTING

If you watch your back in a mirror while you stand and twist, you will see a surprising amount of twisting taking place in the thoracic region: 30–40° of rotation in each direction between T1 and L1 in an average young adult. The design of the anterior and posterior functional units of the vertebral column (chapter 4) permits these movements. To summarize: the anterior functional unit of the spine is a flexible rod composed of the stack of vertebral bodies and intervertebral disks, and the posterior functional unit is a tube (enclosing the vertebral canal) that is composed of the stack of verterbral arches, superior and inferior articulating processes, and various restraining ligaments. In general, the anterior functional unit permits twisting and bending, and the posterior functional unit restricts twisting and bending.

The thoracic region of the spine permits a lot of twisting for three reasons. First, it contains twelve vertebrae and intervertebral disks—half the vertebral column—and this means that there are twelve sites for rotation (fig. 4.7). Second, the axis of rotation for twisting in the chest runs approximately down the center of the anterior functional unit (fig. 4.6a, large dot), permitting the hydraulic system of each intervertebral disk (fig. 4.11) to work perfectly, compressing its nucleus pulposus axially, stretching the elastic fibers of its annulus fibrosus evenly all around its perimeter, and causing the disk to bulge moderately on all sides. The elastic fibers in the annulus fibrosus that are oriented obliquely will be either stretched or released depending on their orientation and the direction of the twist, and those that are oriented vertically and horizontally will be stretched evenly all around.

A third reason twisting takes place easily in the thoracic region is that the synovial joints for the superior and inferior articulating processes of each thoracic vertebra are oriented in a frontal plane, one roughly parallel to the back surface of the chest, and this orientation allows the joints to slip efficiently with respect to their neighbors above and below (figs. 4.6–7).

The large number of vertebrae and intervertebral disks, the ideal axis of rotation, and the ideal orientation of the articular processes—all of these facilitate thoracic twisting. And the numbers add up. With twelve possible sites for rotation, we need only an average of 3° between adjacent vertebrae to make a total of 36°, and this means that twisting here requires only slight adjustments between adjacent vertebrae and their joints. In fact, the articular processes in this region can be displaced from one another so readily that it is not the spine but the rib cage that is the main limit to twisting. Were it not for restrictions there, this part of the spine could probably rotate 120° in each direction.

Like the articular processes, the ribs are also roughly oriented in a transverse plane, and only a slight shearing effect between adjacent ribs is needed to allow a small amount of spinal rotation between adjacent vertebral bodies. And this is what accounts for the 30–40° of rotation that we actually experience in the chest. Beyond 40°, the rib cage becomes the main impediment to twisting because the ribs connect to the sternum in front by way of the costal cartilages, creating a stabilized cage that can rotate only so far.

LUMBAR TWISTING

Just as the design of the chest permits extensive twisting, so does the design of the lumbar region prevent it. There are two main obstacles. First, in this region the axis of rotation runs down the spine through the bases of the spinous processes in the posterior functional unit (fig. 4.5a, large dot) rather than through the center of the anterior functional unit. This by itself would make twisting almost impossible—it would require lateral displacement of one vertebral body in relation to its neighbors above and below. The same forces of torque that act to rotate one vertebral body in relation to its adjacent vertebral body in the thoracic region can act in the lumbar region only to sheer the lumbar disks from side to side. Second, the superior and inferior articulating processes on the vertebral arches start shifting from a frontal to a sagittal orientation in the lower thoracic region, and in the lumbar region, this shift has become an accomplished fact; it stops rotation between adjacent vertebrae almost completely (figs. 4.5 and 4.13b). According to the textbooks, these two mechanisms taken together limit twisting to an average of about 1° between adjacent vertebrae or a total of 5° between L1 and the sacrum.

TWISTING IN THE TORSO—THE BIG PICTURE

Flexible young students who are twisting in a standing posture with their thighs adducted, their feet parallel, and their knees extended might end up with their shoulders 70° off axis from their feet. The twist would consist of about 35° of rotation between the feet and pelvis, 5° between the pelvis and the chest, and 30° between L1 and the shoulders, and 45° of rotation between the shoulders and the head, for a total of 115° between the feet and the head (fig. 7.4). Come into such a posture and notice the sensations as you pull into as much of a twist as possible. If you are attentive, you can

Figure 7.4.This standing twist exhibits approximately: 35° of rotation between the feet and the pelvis, 5° of rotation between the pelvis and the chest, 30° of rotation between the chest and shoulders, and 45° of rotation between the shoulders and the head. This makes 115° total rotation between the feet and the head. The model might reflexly gain 10° more between the feet and the head by looking to her far right instead of back toward the camera.

feel the oblique muscles in the back and abdomen creating the torque that rotates the thoracic region and tries but fails to rotate the lumbar region.

In a whole-body standing twist it is the internal and external abdominal obliques and the deepest of the obliquely running back muscles (figs. 3.11–13, 5.5, 8.8, 8.11, and 8.13–14) that initiate the torque. Some of the latter promote twisting over as little as one or two segments of the spine and others create a twisting effect between three or more. By contrast, the rectus abdominis muscles and the long segments of the erector spinae resist the twist and increase the sense of axial compression that is the hallmark of spinal twisting.

THE LOWER EXTREMITIES

By definition, the torque for twisting in the lower extremities begins at the sacroiliac joints, but externally visible rotations are not observable except in the hip joints, the flexed knees, and the ankles. We'll begin our discussion at the source and work down.

TORQUE IN THE SACROILIAC JOINTS

Whole-body twists impart severe torque to the sacroiliac joints, and if we had to depend only on ligaments to keep them stable, we might be in trouble. Fortunately, the surrounding muscles provide additional protection, especially if they are kept in a state of moderate tension. The gluteus maximus muscle (figs. 3.8, 3.10, and 8.9–10) serves in such a role. This muscle takes origin from the sacrum, the coccyx, and the ilium, and it inserts into the iliotibial tract (figs. 3.8–9, 8.8, and 8.12), as well as directly on the femur (fig. 3.10b). You can feel how these muscles support the sacroiliac region during twisting postures if you stand with the feet parallel, 2–3 feet apart, and twist gently to the right keeping the gluteal muscles relaxed. Observe the sensations and dynamics carefully. Then tighten the gluteus maximus on the left side, and observe how its contraction resists the twist by tugging on the left rear side of the sacrum and ilium, keeping them both pulled slightly to the rear and protecting the pelvic bowl from excess stress. Even so, if you were to stand for 3–5 minutes in such a posture, you might come in touch with a vague ache in the left sacroiliac joint. Repeat the exercise on the other side for balance.

Now stand with the feet together rather than 2–3 feet apart and again twist to the right. When you resist the twist in this case, you will feel the left hip tighten up even more solidly than before because you are stretching and activating the left gluteus medius and gluteus minimus muscles as well as the gluteus maximus. These are medial rotators and abductors of the thigh (figs. 3.8, 3.10a–b, 8.9–10, 8.12, and 8.14, with details in chapter 4), and when you twist to the right with the thighs adducted, the left thigh is swiveled into a lateral rotation that automatically places these muscles under increased tension. Along with the gluteus maximus muscles they create a field of muscular activity that supports the entire pelvis, including the sacroiliac joints.

The sacroiliac joints can be overly stressed if good judgment is not used in some of the standing and sitting spinal twists. This is particularly true if you are thoroughly warmed up and the ligaments and muscles that normally inhibit sacroiliac twisting have become lax. If you are hurting yourself, the tip-off is pain at the lateral edge of the sacrum on one side, and if you continue twisting in the face of discomfort, the joint becomes vulnerable to more serious injury.

TWISTING AT THE HIP JOINTS

To examine twisting at the hip joints, stand with the knees extended and the medial edges of the feet parallel and about two feet apart. Then tighten all the muscles of the lower extremities and twist the body to the right from the waist down, keeping the abdomen, chest, and shoulders in the same plane with the pelvis. Below the hip joints you will feel a combination of torque and twist in the ankles, legs, knees, and thighs, especially on the left side. Keep the feet flat on the floor, and this will diminish their tendency to slip. Repeat on the other side.

When the muscles of the thighs and legs are tensed while keeping the legs extended, the ankles and knees permit only a small amount of rotation, but most people will be able to twist at least 45° at the hips, using a combination of medial rotation of one thigh and lateral rotation of the other. And a few rare students can rotate their hips almost 90° from their feet, with only a small proportion of this twist in their ankles and knees (fig. 7.5.) In either case, you will observe only moderate effects of the twist in the right hip joint, but on the left side, the hyperextension that results from swiveling to the right produces a pronounced pull on the tightly wound spiral of the pubofemoral, ischiofemoral, and iliofemoral ligaments (fig. 3.6).

Feel the gluteal muscles on the left side with your fingers as you twist to the right. As discussed in chapter 4, the more vigorously a healthy person twists, the more these muscles resist, and to give maximum support to twisting postures we need tension not only in the gluteal muscles on the side opposite the direction of the twist, but also in the quadriceps femoris on the same side and in the adductors and hamstrings on both sides. Beginners will find that their enthusiasm for working with challenging twisting postures will be in direct proportion to their ability to support them with muscular tension.

TWISTING AT THE KNEES

Until now, our discussion of the knee joints has focused on their actions as hinge joints that permit flexion and extension. In chapter 5, we saw that extension places the ligaments of the knee under tension and holds all components of the joint together, and we saw that flexion permits the ligaments to become lax. In this chapter we'll explore the one movement not yet mentioned—rotation of the flexed knees.

If you are sitting in a chair with the thighs fixed and parallel to the floor, and with the legs perpendicular to the floor, you will be flexing the knees 90°. If you have good flexibility, you can rotate your feet out laterally from this position about 40°, and you can rotate them medially about 30°. You see the movements of the foot, but almost all of the rotation is happening at the knees. If you try the same experiment sitting on a high bench with the knees flexed at a 30° rather than a 90° angle, you'll notice that the amount of knee rotation is diminished to about 30° of lateral rotation and about 20° of medial rotation. And of course, if you return to a standing position and extend the knees, rotation is stopped completely. To make these comparisons fairly, of course, you have to keep the thighs stabilized or you will add hip rotation to knee rotation and confuse the two.

Figure 7.5. This twist, which started with the thighs adducted and the feet parallel, reveals almost 90° of rotation of the pelvis relative to the feet. This is unusual; most people cannot swivel their hips much beyond 45° with their thighs abducted and feet parallel.

The muscles that rotate the flexed knees are the hamstrings (figs. 3.10b, 7.6, 8.9–10, and 8.12) and a small muscle on the back of the knee joint, the *popliteus*. Two of the hamstrings, the *semitendinosus* and *semimembranosus* muscles, insert on the medial side of the tibia, and are thus medial rotators of the flexed knee; the *biceps femoris* inserts laterally on the head of the fibula, and is thus a lateral rotator of the flexed knee. To experience this sit upright with your knees flexed 90°. Then grasp the tendons of the semimembranosus and semitendinosus muscles on the medial side of the knee joint, rotate the knee medially as strongly as possible, and feel the tendons get tighter. You can do the same thing with the biceps femoris if you rotate the leg laterally as strongly as possible.

The popliteus muscle is visible posteriorly (figs. 5.24, 7.6, and 8.14); it takes origin from the lateral surface of the femur, runs inferiorly and medially, and inserts on the lateral surface of the tibia. Its anatomical disposition therefore allows it to do double duty as a medial rotator of the tibia and a lateral rotator of the femur. The former is what you notice when you rotate your feet in while you are sitting on a chair, and the latter is what you notice when the muscle torques the thigh laterally from a fixed foot position. This is the more common situation in sports because you frequently rotate the thigh (and with the thigh the rest of the body) with one foot planted on the ground, as when you recover from serving a tennis ball.

Because the knee joints are among the few that permit flexion and extension as well as twisting, and because they are vulnerable to injury when they are flexed, everyone should use caution in approaching hatha yoga postures that involve a combination of flexion at the knees and whole-body twisting. And except for the simplest lunging postures, sun salutations, and squatting on the floor in a symmetrical pose, almost every posture in hatha yoga in which the knees are flexed involves either rotation or torque in the knee joint.

THE ANKLES AND FEET

An astounding number of bones (28) and joints (25) are associated with each foot and ankle (fig. 6.8), and in combination their architecture enables us to support the weight of the body, propel us forward, and accommodate to surface irregularities on the ground. And because most of the basic movements in the foot-ankle complex involve stresses from both torque and rotation, they are included in this chapter.

In chapter 4 we saw how foot position in standing postures affects the hip joints. With the knees extended, what we referred to as rotating the foot out stretches the medial rotators of the hip, and what we called rotating the foot in stretches the lateral rotators of the hip. Later, in chapter 6 we turned to the ankles and saw that 30–50° of extension (plantar flexion) takes place when we lift up on the balls of the feet, and that 45° of flexion (dorsiflexion) is needed for pressing the heels to the floor in the down-facing dog.

Twisting (that is, true axial rotation) at the ankle joint is so minimal that it is usually not even listed in elementary texts, but careful studies have shown that those with average flexibility at this site can rotate the

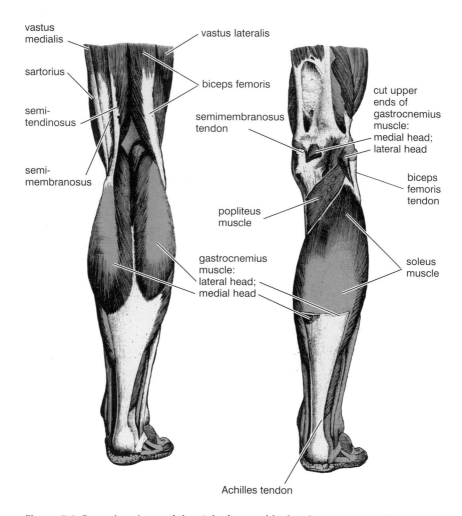

Figure 7.6. Posterior views of the right foot, ankle, leg, knee joint, and lower portion of the thigh. A superficial dissection is illustrated on the left, and a deeper dissection (following the removal of the bulk of the two heads of the gastrocnemius muscle and the hamstrings) is illustrated on the right. When the knee is bent and the thigh is stabilized, the popliteus muscle rotates the leg medially, but when the foot is stabilized, the popliteus muscle rotates the thigh laterally. In the instance of the right thigh shown above, the popliteus muscle has the effect of rotating the body as a whole around to the right (Sappey).

foot medially about 7° and laterally about 10°. To experience these rotations, stand with the knees extended and the heels and toes together. Place the hands just above the knees to brace them and hold them together, and tighten all the muscles around the thighs. Then twist to the right keeping the feet flat on the floor. This is critical: the slightest lifting of the heels or the edges of the feet brings other movements into the picture. Under these carefully controlled circumstances you have rotated the right foot medially and the left foot laterally, both at the ankle joints. These axial rotations at the ankles may be minimal, but they are seen in many standing postures, and are therefore an important practical concern.

[Technical note: This is a different use of the word rotation for the feet and ankles than we have used previously. The circumstances just above refer to axial rotation within the ankle joint, not swinging the feet in or out. To keep terminology within reach of lay audiences, "rotation of the feet in and out," or "rotation of the feet medially and laterally," which is the same thing, will always refer to the movements of the feet as a whole unless axial rotation within the ankle joint is specifically indicated.]

To explore rotation of the feet as a whole, stand with the knees extended, the heels together, and the medial borders of the feet at a 90° angle from one another. Under these circumstances each foot will be rotated out (laterally) 45°. Most people can go a little further, perhaps to 70° for each foot (fig. 7.7). Now bring the medial borders of the feet parallel and next to one another. The feet are now rotated to a neutral position. Next, bending the knees as necessary, bring the big toes together and spraddle the heels out 90° from one another to rotate the feet in (medially) 45° (fig. 7.8). And finally, abduct the thighs widely, bend the knees deeply, and try to bring the feet into a straight line. This is a moderately difficult balancing posture in which you have rotated each foot out (laterally) 90° (fig. 7.9).

[Technical note: Always keep in mind that "rotation of the feet," as defined in this book, is to a large extent reflective of rotation of the thighs when the knees are extended, or rotation of the legs when the knees are flexed. Don't get confused: the terminology is logical. If you are looking at and thinking of the thighs, say rotating the thighs. If you are looking at and thinking of the legs, say rotating the legs. And if you are looking at and thinking of the feet, say rotating the feet, even though you are aware that most of that we call rotating of the feet actually reflects rotation of the legs or thighs. Just say what you see, and everyone will know what you are talking about: it could hardly be more simple.]

SUPINE TWISTS

Supine twists, especially the easy ones, are welcome both at the beginning and end of a hatha yoga class because they are relaxing, energizing, and do not require much effort. Along with simple hip-opening postures (chapter 6), they are a good barometer both of stiffness at the start of a hatha yoga session and of improvements in flexibility at the end.

Figure 7.7. The feet here are rotated "out" (that is laterally) about 70°.

Figure 7.8. The feet here are rotated "in" (that is medially) about 45°.

SIMPLE SUPINE TWISTING

For the first of three simple supine twists, lie on your back, draw the heels toward the hips, and stretch your hands out to the sides. Then lower the knees about 45° to one side, keeping the feet, legs, and thighs together, and holding the soles of the feet relatively flat against the floor (fig. 7.10). The ilium on the side opposite the twist will be lifted slightly off the floor. This is a subtle concentration exercise that generates a twist in the hip joints and creates a moderate torque in the sacroiliac joints and lumbar region—all without creating a twist in the chest. Mild muscular tension in the lower extremities is required to keep the soles of the feet against the floor and to prevent the knees from dropping too far to the side. Repeat on the other side.

Next create a more obvious twist by lowering both knees all the way to the floor, this time facing the soles of the feet toward the opposite wall (fig. 7.11). Try to remain relaxed, but keep both knees together and both shoulders against the floor. This is a completely different posture from the first one because now the thighs are only slightly swiveled with respect to the pelvis. You will feel this twist in the sacroiliac joints, the lumbar region, and the chest—a torque in the sacroiliac joints and the lumbar region that may get your attention if you remain in the posture for more than a minute or two, and a twist in the chest that may yield little clicks and readjustments in the facet joints of the thoracic spine and ribs. If you feel the lower back with the palm of the hand, you will be able to confirm that it remains in nearly

Figure 7.9.
Here the
feet are
rotated out,
or laterally,
about 90°.

the same plane with the pelvis and that most of the twist takes place in the chest (fig. 7.11). If you can't settle into this posture without lifting the opposite shoulder off the floor, place one or more pillows under the knees.

Finally, twist to one side after drawing the knees closer to the chest. To keep the shoulders down in this posture, it is helpful to spread the arms and forearms straight out to the sides. You will also have to use muscular effort to keep the knees near the chest, which incidentally makes this pose a forward bend as well as a twist. Notice that the posture brings the pelvis to a full right angle with respect to the floor (fig. 7.12) and leaves the thighs in a neutral unswiveled position with respect to the pelvis. Tension is completely removed from the sacroiliac joints and largely gone from the lumbar region, and you will feel most of the twist high in the chest. If as a result it is difficult to keep the opposite shoulder against the floor, prop pillows under the knees to diminish the twist. It is best to keep the head in a neutral position for all three of these postures: twisting it in the opposite direction distracts your attention from analysis of more interesting effects in the rest of the body. Repeat on the other side.

Figure 7.10. This twist, keeping the feet flat on the floor, is a subtle concentration exercise that affects only the hip joints, sacroiliac joints, and lumbar spine.

Figure 7.11. Dropping the knees all the way to the floor and keeping the feet together creates effects in the sacroiliac joints, lumbar region, and chest.

Figure 7.12. Twisting with the knees kept close to the chest removes all tension from the hips and sacroiliac joints, and most of the tension from the lumbar region. This pose primarily twists the chest.

A RELAXED SUPINE TWIST

The three previous postures require at least some muscular activity. To contrast them with a relaxed supine twist, place the hands flat on the floor and straight out to the sides, draw the heels toward the hips, cross the right knee over the left, and twist the lower part of the body to the right so that the knees are lowered toward the floor (fig. 7.13). There is little or no tendency for the opposite shoulder to lift off the floor in this mild posture, but it still may be tricky to relax in it from head to toe, especially at the beginning of a hatha yoga session. Adjust the amount of flexion in the knees and hip joints (determined by how far away from the hips you place your feet) so you can relax as much as possible; the more flexion the greater the twist, but the greater the twist the more challenging it will be to relax.

The main characteristic of this pose is that you can analyze the sensations of passive stretch at your leisure. The posture pries the left thigh in the direction of the twist, pulls the head of the left femur slightly away from its socket, and lifts the left side of the pelvis off the floor. As a result, the pose places torque on the lumbar region and on the left sacroiliac joint. Repeat the posture, crossing the left knee over the right and twisting down to the left. If you try this relaxed twist at the beginning and then at the end of a session of hatha yoga postures, the experience will be markedly different. In the beginning you may feel slight pain in some of the ligaments of the hip and sacroiliac joints, making it difficult to relax from the waist down, but when you are warmed up the situation will have improved. And as your musculoskeletal health improves you will gradually find yourself able to relax more completely.

THE SUPINE TWIST, FULL RESISTANCE

This posture takes the above relaxed supine twist to the other extreme. Lie on your back and again cross the right knee over the left, but now interlock the ankles as well, the right under the left. And instead of twisting to the right as you did before, twist to the left. With the hands outstretched, and

Figure 7.13. This relaxed supine twist is very complex, and it may challenge the ability of beginners to relax, especially at the start of a session of hatha yoga.

with the upper extremities stabilizing the torso, strongly press the right shoulder to the floor and twist the knees as much to the left as possible (fig. 7.14). For this to feel comfortable you will have to create internal muscular resistance to the twist, just as we tightened the muscles of the lower extremities to create a healthy musculoskeletal framework for standing postures (chapter 4). The need for creating muscular resistance in this posture is even more obvious. It's like wringing out a washcloth. The muscles that are creating the twist have to be matched from the shoulders to the feet by their antagonists, all of them in a state of isometric activity, some creating the twist, others resisting. It's unthinkable to relax the antagonists in this posture. Slowly release the pose and repeat on the other side.

THE DOUBLE LEGLIFT SUPINE TWIST

The double leglift supine twist is an intense abdominal and back exercise in addition to a twist. To do it start from a supine position with the upper extremities extended straight out to the sides, palms down. From there exhale, press the lower back to the floor, and taking care to keep the knees extended fully, lift the thighs perpendicular to the floor in a double leglift (fig. 3.17). Then slowly allow gravity to carry the flexed thighs and extended legs to one side while you turn your head in the opposite direction. Try to keep the soles of the feet within an inch or so of the same plane, bracing the posture with the upper extremities and keeping both shoulders against the floor. As always, repeat on the other side.

This is an advanced posture, and you have to have excellent flexibility as well as abdominal and back strength to do it comfortably. Just coming into the initial position with the thighs perpendicular to the floor requires 90° of hip flexibility and that's before you even think of adding the twist. Those who can complete the posture will end up with their pelvis about 15° short of being perpendicular to the floor, so if the shoulders are kept flat, the posture will require 75° of twist between the shoulders and the pelvis. If we allot 5° of twist to the lumbar region, this means 70° of twist in the thoracic segment of the vertebral column, or 25° more than the average of 45°.

Figure 7.14. A double-locked-leg twist is best done against full muscular resistance.

If the arm and shoulder opposite the direction of the twist are lifted off the floor, or if you can't lower the legs to the floor while keeping them flexed 90°, there are several ways to moderate the posture. You can keep the knees extended and lower over only to the point at which you can comfortably come back up, exhaling as you go to the side and inhaling as you come back up (fig. 7.15). This requires strong muscular activity in the abdominal wall and creates intense sensations in the hip joints, sacroiliac joints, and lower back. Or you can keep the lower back straight and the knees extended, but flex the thighs less than 90° rather than trying to keep them perpendicular to the torso. As before, if you can't lower the feet all the way to the floor, lower them only part way down. Simplest of all, flex the hips 120° and the knees 90° (fig. 7.16), and then twist to one side with the knees together until they touch the floor. Either relax in the pose, or keep breathing and immediately raise back up in a continuous movement before slowly twisting to the other side.

Figure 7.15. The double leglift supine twist, done according to the specifications in the text, is a valuable study in the dynamics of the musculoskeletal architecture of the body, and will challenge even the strongest and most flexible athlete. Swinging the lower extremities over 45° as shown here is more reasonable for most students.

Figure 7.16. For those who do not have enough flexibility to do the full double leglift supine twist, dropping the lower extremities to the floor from this easy preparatory position is still useful. For the best exercise, go from side to side coming to within an inch or two of the floor in a continuous movement, and relax and rest in this upright position instead of resting with the legs all the way to the floor on one side.

If you can easily do the full posture with the knees straight and the hips flexed 90°, you can sharpen it further by keeping the feet together, which requires pushing the upper foot out so it remains in line with the foot that is closest to the floor. To the extent that the feet remain in line with one another, the pelvis will approach being perpendicular to the floor, and if you keep the shoulders against the floor, the final posture will require 90° of twist between the shoulders and the pelvis instead of the 75° mentioned earlier.

All of the double leglift supine twists can be approached in two ways— as movements or as postures. As far as building strength is concerned, you can get the most benefits by going back and forth from one side to the other continuously, without quite touching the floor and without pausing to relax. On the other hand, you can come all the way down to the floor in a relaxed or semi-relaxed hatha yoga posture. In any case, all of these exercises and postures are useful for exploring twisting in the thoracic spine because the hips are not swiveled when the thighs are against the floor and because the pelvis and shoulders are stabilized. As you come into the twist, you can flex the thighs and knees to suit yourself in order to place the lumbar and thoracic spine under as much tension as they can accommodate in the relaxed posture.

STANDING TWISTS

In chapter 4 we saw that standing in unfamiliar and unconventional postures can be challenging when the force of gravity comes to bear on the joints of the lower extremities. Such challenges are magnified when twisting is added to the equation. You may do fine with standing back-bends and forward bends, but add a twist to the posture and suddenly problems emerge front and center.

The extent to which a standing twist is challenging depends on a number of issues: on how the feet are positioned (rotating the feet in or out places torque on the ankles); on whether or not the knees are flexed (bending the knees places extra tension on its collateral and cruciate ligaments); on the extent to which the thighs are abducted (abducted thighs places extra tension on the adductors); on whether or not the thighs are rotated (rotating the thighs laterally or medially places extra tension on the medial or lateral rotators respectively); on whether or not the shoulders are in line with the pelvis (rotating the shoulders in either direction with respect to the pelvis will twist the thoracic spine and place torque on the lumbar region and sacroiliac joints); and on whether or not the head is in the same frontal plane as the torso (twisting the neck tends to take your attention away from the rest of the body).

Because standing twists present so many challenges, there are four basic rules that should always be honored: establish a solid foundation from the waist down; be aware of and make adjustments for aspects of the posture that are unnatural; appraise how easily you can get into the posture; and look at whether or not there are potential difficulties connected with your exit. Then act on your knowledge by practicing within your capacity. Soon you will be able to do difficult postures comfortably, and this in turn will improve your general strength and flexibility.

STANDING TWISTS AND BENDS

The simplest standing twists combined with bending are the ones discussed in chapter 5 in relation to strengthening the knee joints: you twist in one direction and bend forward and backward from that position; you twist in the other direction and again bend forward and backward; and facing the front you again bend forward and backward. These six bends, done repetitively with the upper extremities in different positions, are excellent whole-body twisting and toning exercises. Using them as a warm-up is also a good preparation for more formal standing postures such as the triangles and the warrior poses.

Here we'll look at the exercises in detail, using the simplest position for the upper extremities, which is grasping the elbows or forearms behind the back. Stand with the feet parallel and as far apart as is comfortable. Keep the kneecaps lifted, the hamstrings strong posteriorly, the adductor muscles firm on the medial aspects of the thighs, and the hips firm. This creates a solid foundation for the postures from which you can be aware of your limits. How far apart you adjust your feet is the single most important feature of the stance. They should be placed as far apart as you can manage and at the same time maintain strength and control in your foundation when you twist and bend. If the feet are too close together you will not feel as if you are getting much exercise from the waist down, and if they are too far apart you can't gather enough strength in the hips and thighs to stabilize the posture. The other general policy is to bend to please yourself. Intermediate and advanced students usually prefer to bend from the hips, but beginners and those who have poor hip and sacroiliac flexibility will find it more natural to bend from the waist. If you have any low back problems, work with extreme caution. Few movements are as hazardous for bad backs as bending the spine after having twisted it.

Twist right and slowly bend forward. Notice that gravity helps you in part, but that you have to make an effort to keep from dropping straight to the front. Try to find an easy point of equilibrium among muscles, ligaments, and joints that are under stress as gravity pulls you forward.

Observe that keeping yourself twisted requires that you hold the posture isometrically (fig. 7.17a). Take 3–7 breaths. Slowly come back up, stay twisted, and bend backward, keeping the head in line with the torso rather than thrown excessively to the rear. Be aware of the muscles that are holding you isometrically in this asymmetric twisted backbend (fig. 7.17b). Create a feeling of lift in the backbend, try to squeeze the hips together, and again take 3–7 breaths. Keep the knees straight. Slowly come to an upright position, twist in the other direction, and repeat the forward bend (fig. 7.17c) and backbend (fig. 7.17d) on the other side. Come up again, face straight to the front and again bend forward (fig. 7.17e) and backward (fig. 7.17f). Squeeze the hips together to create as much counternutation as possible in all the backbends. In the forward bends, those who have good hip and sacroiliac flexibility will bend from the hips and at least start their bends with nutation; everyone else—namely those who can't bend easily from the hips—will keep their sacroiliac joints in counternutation throughout the series.

Turning to the side in these postures will create a whole-body swivel that includes a pronounced torque in the feet, ankles, legs, knee joints, and thighs, as well as a torque plus a twist in the hip joints, torso, and neck. You have to keep the feet parallel to get these results; if you allow them to rotate in the direction of the twist, the entire body swivels around at the hips and you will get little twisting of the vertebral column. The purpose of these exercises is to effect the twist mostly in the hips and chest. Average students in good health may rotate a total of 80° between the feet and the shoulders, including about 5° of axial rotation in the ankles, 40° of swiveling in the hips, 5° of twist in the lumbar spine, and another 30° in the chest.

Beginning, intermediate, and advanced students will all have different reactions to the gravitational field. Beginners and those who are especially stiff will quickly butt up against their limits. They may come forward only a little, and their "backbends" may involve little more than standing up straight. Intermediate students, after a week or so of practice, may come into an uneasy equilibrium with gravity and still not be entirely comfortable in the full forward and backward bends. Advanced students can actively pull themselves down and back, or in the case of the twists, down and around, and back and around, beyond the point at which gravity alone would take them. As they do this they will protect themselves either by firming up their antagonistic muscles or by releasing tension in a manner that the body recognizes as safe.

As your musculoskeletal health improves, you will feel like pulling more deeply into each position. And as you intensify each movement, you will be strengthening all of the muscles involved in creating the postures, not only the agonists but all their synergists and antagonists. You will be aiding gravity with your musculoskeletal efforts, but at the same time you will be stimulating reflexes that relax the muscles that are antagonists to the prime movers, and this in turn allows you to come further into each position. In addition to catching the forearms or elbows behind your back, you can do these exercises with six more arm positions, all of them subtly different from one another in the way they affect the twists and bends. These are: with the arms and forearms stretched laterally (fig. 7.18a); with the arms overhead, forearms behind the head, and hands catching the elbows (fig. 7.18b); with the hands, palms together, in a prayer position behind the back (fig. 7.18c); with the hands interlocked behind the back and pulled to the rear (fig. 7.18d); with the upper extremities in the cow-face position with one hand behind the back reaching from below and the other catching it from overhead (this one should be done twice, once on each side; fig. 7.18e); and with the hands astride the ilia, bending from the hips instead of the waist for a change of pace (figs. 7.18f–g). Repeat the series of six bends for each hand position.

TWISTING IN THE TRIANGLE POSTURES

In chapter 4 we discussed the triangle postures in detail, but did not stress the fact that all of them include elements of twisting. Focusing on this aspect in the classic triangle, come into the preliminary stance with the thighs abducted and the feet parallel, and envision a frontal plane running through the body from ear to ear, shoulder to shoulder, hip to hip, head to toe. Then rotate the right foot 90° to the right and the left foot 30° to the right, come into a side bend to the right while keeping the spine relatively straight (fig. 4.33), and notice where the plane becomes distorted. It remains true through the shoulders, chest, and abdomen, but because the right gluteal region is pulled slightly to the rear, the former frontal plane reveals a moderate kinking through the pelvis. It also reveals a slight deviation in the left lower extremity, depending on how much the left foot is rotated to the right. The plane shifts almost 90° in the right lower extremity because the right foot is pointed straight to the right, and it shifts 90° in the head and neck because you are looking straight up at your left hand. All in all, this adds up to a lot of twisting.

We saw in chapter 4 that most of our efforts in the triangle involve resisting the tendency to twist the pelvis. But even though students are routinely cautioned to keep their hips facing the front, this is an unrealistic expectation in the triangle except perhaps for ballet dancers and gymnasts. Average intermediate students can accomplish the 90° rotation of the foot on the side toward which they are bending, but only by allowing the opposite side of the pelvis to rotate slightly forward. It's a mistake to make this too much of an issue, and the usual injunction—try to keep the hips facing the front as much as possible—is a reasonable middle ground.

a.

Figure 7.17a. First of six standing twists and bends with hands grasped behind the back. Starting with the feet parallel and comfortably apart (see discussion), twist right and bend forward to within your reasonable capacity, just a simple gesture toward a twisted bend if you are frail, bending with the aid of gravity for the average person, and supplementing gravity with the abdominal activity for those who are confident. Keep the thighs strongly engaged.

Figure 7.17b. For the second position, come up, stay twisted to the right, and bend backward, which may not mean much more than standing straight for some, pulling mildly to the rear for the average person, and pulling strongly to the rear for those who are confident. Keep the thighs willfully engaged and keep breathing. In all of the backbending positions, try to keep the hips squeezed together in an attitude of ashwini mudra (chapter 3).

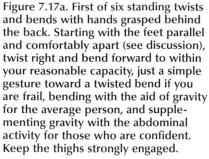

b.

c.

Figure 7.17c. For the third position, come up, twist to the left, and bend forward as in fig. 7.17a except in the opposite direction. (The order in which direction to go first can be switched on alternate days, as in going to the right first on even days of the month and going to the left first on odd days of the month.)

Figure 7.17d. Come up, stay twisted to the left, and bend backward, with muscles engaged from head to toe. It's fine to keep the knees straight so long as they are not hyperextended beyond 180° and so long as the thighs (especially the hamstring muscles) are kept in a state of strong isometric contraction. Experienced students can of course use their own judgement about how and where to relax in standing poses (see chapter 4).

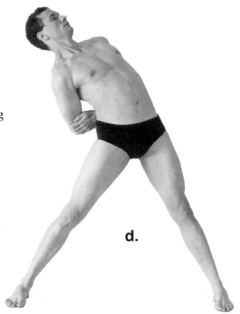

d.

Figure 7.17e. Come up, face the front, and bend forward, again keeping the muscles in the thighs engaged. Notice that gravity pulls you straight forward and that it is not necessary to keep pulling yourself around to the side as in the twisted poses. Strong students can still aid gravity with muscular activity.

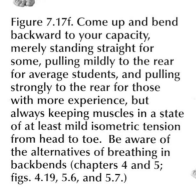

e.

Figure 7.17f. Come up and bend backward to your capacity, merely standing straight for some, pulling mildly to the rear for average students, and pulling strongly to the rear for those with more experience, but always keeping muscles in a state of at least mild isometric tension from head to toe. Be aware of the alternatives of breathing in backbends (chapters 4 and 5; figs. 4.19, 5.6, and 5.7.)

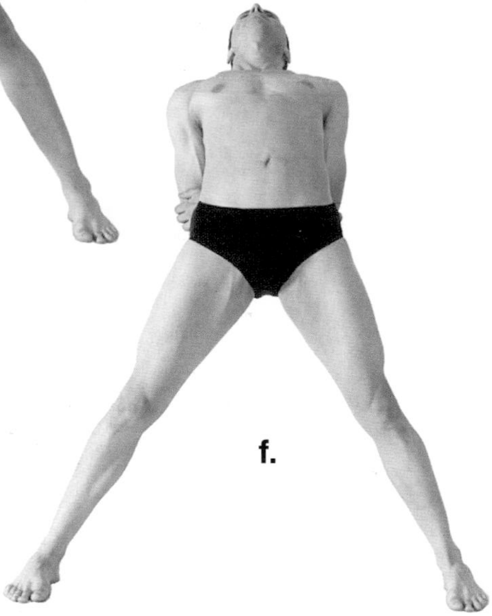

f.

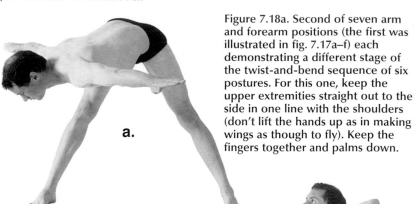

Figure 7.18a. Second of seven arm and forearm positions (the first was illustrated in fig. 7.17a–f) each demonstrating a different stage of the twist-and-bend sequence of six postures. For this one, keep the upper extremities straight out to the side in one line with the shoulders (don't lift the hands up as in making wings as though to fly). Keep the fingers together and palms down.

a.

Figure 7.18b. Grasp the forearms or shoulders behind the head for the third series of six bends. Watch the extra weight overhead that has to be managed in the three forward bending positions. With this series in particular, beginners and those who have less musculoskeletal confidence will want to start with six small bending gestures rather than immediately exploring their limits. With more experience, one can always ratchet up commitment to the full postures.

b.

c.

Figure 7.18c. For the fourth series of six bends, place the hands in a prayer position behind the back, or work toward that position by touching the third and fourth fingers of one hand to the third and fourth fingers of the opposite hand, and then gradually lifting the paired hands higher to approximate the palms as much as possible. Never hurry this process, as doing so can cause repetitive stress injuries.

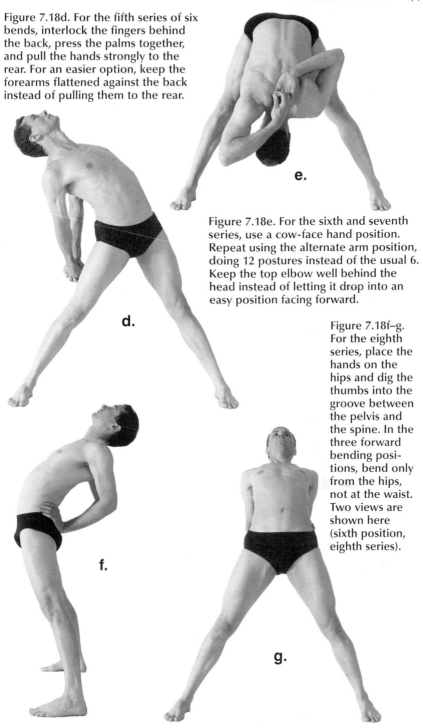

Figure 7.18d. For the fifth series of six bends, interlock the fingers behind the back, press the palms together, and pull the hands strongly to the rear. For an easier option, keep the forearms flattened against the back instead of pulling them to the rear.

e.

Figure 7.18e. For the sixth and seventh series, use a cow-face hand position. Repeat using the alternate arm position, doing 12 postures instead of the usual 6. Keep the top elbow well behind the head instead of letting it drop into an easy position facing forward.

Figure 7.18f–g. For the eighth series, place the hands on the hips and dig the thumbs into the groove between the pelvis and the spine. In the three forward bending positions, bend only from the hips, not at the waist. Two views are shown here (sixth position, eighth series).

d.

f.

g.

The revolving triangle (fig. 4.36) contains even more extensive rotations than the classic triangle. In the first stage, turn the right foot 90° to the right and the left foot 60° to the right, and then swivel the trunk around to face the right foot. Ideally, you are not twisting the spine very much in this position. Instead, the whole-body swivel causes the left thigh to end up hyperextended and the right thigh to end up flexed, both depending on how far the thighs were abducted initially. Beginners are properly advised to gradually accustom themselves to spreading their feet further apart in order not to end up in awkward positions that are beyond their capacity.

If you do not have enough flexibility to swivel the pelvis until it is perpendicular to the right lower extremity, you will have to accomplish the rest of the twist in the spine. Even so, whichever way you do it, you will still twist a frontal plane running through the shoulders 90° to the right thigh and leg. To complete the posture you add a 90° forward bend and an additional 90° twist of the spine so the shoulders end up facing the rear. To top it off you twist your neck an additional 90° to look up at your right hand. What had originally been a frontal plane through the body would twist 90° and then bend 90° to become horizontal in the pelvis. Then the plane would twist another 90° to be vertical through the shoulders, and finally it would twist the last 90° in the neck—making a 270° twist altogether.

TWISTING IN THE LUNGING POSTURES

When fencers lunge they thrust the body down and forward in one swift movement, and the lunging postures imitate this thrust. They all involve 90° of flexion of the thigh and knee on one side and full extension of the thigh and knee on the other. The torso can either be twisted or bent, and the upper extremities can be lifted overhead or stretched in other directions. As a group, athletes are attracted to these postures because they are helpful for developing muscular strength, moderate aerobic capacity, and a full range of motion for flexion and extension, which happen to be the basic movements needed for most limit-pushing athletic endeavors. Since these postures require some athletic prowess, it is wise to be conservative about plunging directly into them. They are relatively natural poses, however, in comparison with the standing twists and triangles, and once you are acclimated to them they are among the best all-around postures for energizing and strengthening the body as a whole. As asymmetric standing postures they are also ideal for working with right-left imbalances.

TWISTING IN THE STANDING WARRIORS

The standing warrior postures are named for their wide-bodied, powerful-looking stances. They speak for themselves. Although the two postures shown here are not ordinarily classified with twists, careful analysis reveals some twisting in both of them. In relation to a frontal plane in the anatomical position, the first posture (warrior II in Iyengar's classification) exhibits a 20–30° twist in the pelvis, a 20–30° countertwist in the upper body, and a 90° twist in the neck; the second posture (warrior I in Iyengar's classification) exhibits a 90° twist in the pelvis.

To come into the warrior II posture, stand with the feet 3–4 feet apart, and swivel the pelvis 20–30° to the right. This is accomplished automatically by turning the right foot 90° to the right and the left foot 10° to the right. The right side of the pelvis will move slightly posterior in relation to the left as you adjust the feet, so this will give you a different feeling from the triangle, in which you try to keep the hips in line and facing the front. Stretch the arms and forearms out so they are in line with the shoulders and the lower extremities. Since the pelvis is swiveled 20–30° to the right, the arms and shoulders will end up twisted 20–30° back to the left. Then lower your weight, flexing the right knee and hip while hyperextending the

Figure 7.19. The standing warrior II is a fairly natural pose in which several kinds of twists make themselves apparent: about 25° in the pelvis, a 25° countertwist in the upper body, and a 90° twist in the neck. Aim in the final posture to keep the front leg perpendicular to the floor as a first priority, so as not to thrust the knee beyond the ankle. Then widen your stance and drop your weight (as strength and flexibility permit) until the front thigh is parallel to the floor. Keep the upper body upright, leaned neither forward nor backward. Beginners should compromise by not dropping their weight so far and adjusting their foot position so the leg is still perpendicular to the floor and the thigh angles up toward the pelvis.

left hip and at the same time stretching the hands out in front and back, palms down, and making blades of the hands and fingers. Keep the arms, forearms, and shoulders in a single line, and look toward the right hand (fig. 7.19).

There are three musts in this posture: keep all the muscles of the extremities firm; keep the torso straight and perpendicular to the floor rather than leaned forward; and don't flex the front knee more than 90°. To complete the posture in its ideal form, the right leg should be perpendicular and the right thigh parallel to the floor. As your capacity for lowering your weight improves, you will have to widen your stance so that the right knee does not push out too far. Hold the feet as flat as possible, being careful not to raise the lateral edge of the left foot. Repeat in the opposite direction.

This is a good posture for beginners and those in poor physical condition because it can be modified to meet everyone's personal needs. All you have to do to make it easier is start with a narrower stance, swivel as in the full posture, and lower your weight until the leg is perpendicular to the floor. Go back and forth from one side to the other, and over a period of weeks or months gradually widen your stance. As soon as you reach the point at which your front thigh is parallel to the floor (and the front leg perpendicular, as usual), and as soon as you can do this in both directions while keeping the torso erect, you have arrived at the full posture.

As you come into the posture, hyperextension of the left hip joint tightens its spiral of iliofemoral, pubofemoral, and ischiofemoral ligaments (fig. 3.6) at the same time that flexion of the right hip unwinds its same three ligaments and allows the head of the femur the freedom needed to rotate in the acetabulum. Doing the posture in the opposite direction will reverse these situations, tightening the spiral in the right hip joint and loosening it on the left.

This is an elementary but at the same time complex pose. The twist is accomplished by swivelling at the hip joints in one direction and twisting the chest back in the other direction. With the shoulders facing the side, and with the head facing the outreaching hand, the neck will also have to be twisted 90°. Finally, because the pelvis is at a 20–30° angle from the shoulders, the hyperextended rear thigh forces it into a forward tilt that creates a moderate sidebending posture. To ease the tension with respect to the rear thigh, beginning students often minimize side bending by leaning forward instead of keeping their torso upright. This looks unseemly, however: it is better to compromise the posture with a narrower stance and less flexion of the front leg and thigh. When students do not try to lower their weight into the full posture prematurely, the demands on the pelvis for side bending will not be so urgent, they can remain upright without difficulty, and the posture becomes relatively easy.

The standing warrior II pose strengthens and stretches muscles throughout the lower extremities, especially the hip flexors and knee extensors. Facing the right, the right quadriceps femoris muscles and the left psoas and iliacus muscles are stretched to whatever limits you are exploring as they lengthen eccentrically, support the posture, and lower your weight. Everything is reversed when you face the left.

The next variation of the standing warrior (warrior I) is more demanding. Begin as in the previous posture, but in this one turn the rear foot in about 20° instead of 10°, lift the hands overhead with the palms together, and pull them to the rear. Swivel the hips around so the head, chest, and abdomen face the front thigh squarely. When this is accomplished, the pelvis will be rotated all the way around and will approach a 90° angle with respect to the thighs (figs. 1.2 and 7.20). Repeat on the other side.

Because the pelvis now faces the front, and because the rear leg and thigh are perpendicular to the torso, the rear thigh is more acutely hyperextended than in the previous posture. And since you are still keeping the

Figure 7.20. The standing warrior I is more demanding than the warrior II pose, because you swivel the pelvis (insofar as possible) 90° to face the front knee, and because hyperextension of the rear thigh produces the need for a sharp backbend. For an even more demanding posture, extend the head and neck to the rear and face the ceiling (fig. 1.2). As in the case of the the warrior II pose, beginners should compromise by keeping the leg perpendicular to the floor and by not dropping their weight so far.

torso upright, the only way you can adapt to the extra hip hyperextension is with a lumbar backbend. Therefore, as you come into the pose you should create a whole-body feeling by strongly lifting the hands overhead and pulling them to the rear, thus lifting the rib cage and chest, drawing the shoulders back and down, and taking excess tension off the lower back. To make the spirit of spinal extension complete, you can also extend the neck enough to look toward the ceiling (fig. 1.2). As with the previous posture, beginning students can also make this one easier by narrowing their stance and not lowering so far toward the ideal position. Everyone should breathe deeply, expanding the chest as much as possible and inhaling their full inspiratory capacity.

THE EXTENDED LATERAL ANGLE POSTURES

The extended lateral angle postures are more difficult than the warrior postures because they are lunges combined with both twisting and bending, and because it is hard to find comfortable intermediate positions that are less demanding than the full posture. Beginners can make triangles easier by simply coming part way down, and in the warrior postures they can stand with the front thigh and knee only partially flexed, but in the extended lateral angle postures they will either have to counteract gravity with internal muscular activity or come deeply enough into the postures to brace themselves with the upper extremities.

For the simplest posture, stand with the feet a comfortable distance apart, turn the left foot 90° to the left, and keep the right foot pointing straight ahead. Then keeping the pelvis and chest facing as much to the front as possible, flex the left knee and bend to the left, either bracing the left forearm against the left knee (fig. 7.21), or for the full posture, placing the left hand against the floor (fig. 7.22). Try to adjust the width of your stance to make the left leg perpendicular to the floor and the left thigh parallel to the floor. This is a lot easier if you are bracing the posture with your forearm. As in the triangle, the right side of the pelvis will probably come slightly forward. Bring the right upper extremity overhead to come into a straight line with the right thigh and leg. Repeat on the other side.

Except for the bent knee and the fact that the opposite upper extremity is pointing to the side instead of straight up, this posture is similar to the triangle. With the elbow braced against the knee the posture is easy; with the hand all the way to the floor it is demanding.

The revolving lateral angle posture is as unlike the lateral angle posture as the revolving triangle is unlike the triangle. It is another twisting posture, in which, like the revolving triangle, you will twist all the way around and face the rear for the final position. To come into the posture stand with the feet 3–4 feet apart, rotate the left foot 90° to the left, and

Figure 7.21. Supporting this easy extended lateral angle posture with the left arm makes it possible even for novices to spread their stance enough to bring the left leg perpendicular and the left thigh parallel to the floor.

Figure 7.22. The extended lateral angle posture requires good athletic strength and flexibility, and should be approached conservatively, gradually widening the stance and dropping the weight to bring the front thigh parallel to the floor.

point the right foot straight ahead. Next, instead of keeping the pelvis facing the front, rotate it around as much as possible, creating a 90° swivel, and bend forward while at the same time continuing the twist in the torso for another 90°. You will end up, as in the revolving triangle, with a 270° twist in the frontal plane of the body. As with the lateral angle posture, you can either brace the opposite elbow on the knee or place the hand on the floor to create a more demanding pose (fig. 7.23). Repeat on the other side.

INVERTED TWISTS

Twisting in variations of the headstand (chapter 8) and shoulderstand (chapter 9) complements twisting in standing postures because the lower extremities are free to move around in space rather than remaining fixed in positions defined by the placement of the feet on the floor. In the classic shoulderstand that is supported by the upper extremities, you can expedite the twist by pushing on the back more firmly with one hand than the other (fig. 9.7). A sharper posture, provided you are capable of remaining straight, is to twist the body in an internally supported shoulderstand (figs. 9.1 and 9.6b) while keeping the upper extremities alongside the torso, thus twisting the trunk between the hands. And to create intermediate level poses, you can accompany twisting in the shoulderstand with additional stretches of the lower extremities such as twisting in one direction in the shoulderstand and taking the opposite foot overhead toward the floor in

Figure 7.23. The revolving extended lateral angle pose is a complex posture that in the end produces a 270° twist in the frontal plane of the body. As in the case of the extended lateral angle pose, one can legitimately compromise this posture by supporting it with the elbow resting on the thigh rather than on the floor.

the same direction (leaving the other thigh and leg fully extended), coming into a twisted half or full plow after twisting in the shoulderstand, coming into a twisted half lotus one-legged plow (fig. 9.8), or coming into twisted knee-to-the-opposite ear poses (figs. 9.9–10). Other related postures such as variations of the threading-the-needle pose (figs. 9.17a–b) are especially helpful for working with the full spinal twist, which we'll discuss in the next section.

In the headstand, twisting the body with the lower extremities straight overhead is both comfortable and rewarding. The easiest way to do this is to abduct the thighs (fig. 8.33) and then swing them around into a position in which one is flexed and the other is extended. Then you continue in that same direction, pulling the feet as far as possible through an arc as you twist the body between the hip joints and the neck. Then come back to a neutral position with the thighs abducted and swing the feet around in the other direction.

SITTING SPINAL TWISTS

All sitting spinal twists, by definition, have two features: they are always upright, and the hips are always flexed. And because many of these postures take tension off the hamstrings and adductors, and yet flex the hips, they generally produce more intense stretches in the hip joints, pelvis, and spine than supine, standing, and inverted twists. This often makes it difficult to complete sitting spinal twists comfortably and attractively, but it also enables us to work deeply with native hip flexibility that is limited by that joint and its restraining ligaments rather than by muscles.

FOUR SIMPLE TWISTS

For the simplest and purest sitting twist, sit cross-legged on the floor, thrust the lumbar region forward, and establish full nutation of the sacroiliac joints. To aid those efforts pull on the knees frankly with the hands, which of course tenses the arms and shoulders. Next pull the shoulders down but at the same time create an axial extension of the torso. After getting settled, relax the upper extremities but without allowing the posture to deteriorate. Then, keeping your head level and your gaze parallel to the floor, twist the spine, including the neck, as far as possible.

Twisting left in this simple posture, you can place the left hand behind you on the floor and the right hand on the left knee to help pull the shoulders around (fig. 7.24), or you can place the hands in any comfortable position. What is important is that the simplicity of the twist allows you to hold the spine straight, retain a full lumbar lordosis, and keep the sacroiliac joints in full nutation. Repeat in the other direction, then reverse the positions of the feet, and twist in both directions again, making four possibilities.

Even though it is simple, this posture should not be overlooked, especially in a beginning class. It may be the only sitting twist in which many
students can remain both fully upright and comfortable. Folding in the
lower extremities in a simple cross-legged position, especially when lifted
up by a supporting cushion that raises the pelvis slightly away from the
floor, places the least tension on the adductors and hamstrings of any
sitting twist, permitting the pelvis to remain oriented perpendicular to the
floor and allowing students to twist while still remaining upright. If they
feel their back with their hands they can confirm that little or no twisting
is taking place in the lumbar region, and if they watch themselves in a
mirror they'll see that nearly all the twisting is in the chest and neck. They
will also notice that the twist is maintained internally: some of the muscles
in the torso are creating the twist while others are resisting it.

Another simple twist, the second one in this series, is to sit as nearly
upright as possible with the feet stretched straight out in front of you,
heels and toes together. From that position, twist right and place the right
hand on the floor behind you and the left hand to the right side of the right
thigh. Keep the torso pushed upright with the right hand while twisting, and
then pull yourself further into the twist with the left hand (fig. 7.25). Keep
both knees extended and the hamstrings relaxed. Repeat on the other side.

This posture is more challenging than the cross-legged twist because it
mixes a sitting spinal twist (with the hips flexed, by definition) with fully
extended knees. Accordingly, it favors sacroiliac counternutation, places
tension on the hamstring muscles, and tends to tuck the pelvis (creating a
posterior pelvic tilt), which of course rounds the lower back. If you prefer
a pure spinal twist without spinal flexion you can simply move your right
hand further to the rear, arch the lumbar region forward, and reestablish
the twist with the torso leaned to the rear. If you are keeping a convincing
lumbar lordosis, the hamstrings are still under tension, especially on the
left side (if you are twisting right) and that tension tends to buckle the left

Figure 7.24. This simple cross-
legged sitting twist, the first of
four, should never be overlooked,
because it is one of the few twists
that does not depend on relatively
flexible hips, and is one of the few
twists that nearly all beginners can
do easily. Pull with the hand on
the knee, and push the posture
straighter using the hand that is
bracing the posture from behind.

knee, so it is important to be particular about keeping it extended with the aid of tension in the quadriceps femoris. Keep the head on axis with the body rather than pulling it forward or dropping it backward so the cervical region receives a pure twist.

The third twist in this series is even more difficult, at least for those who have poor hip flexibility, and that is to remain upright for a sitting spinal twist while abducting the thighs with the knees extended. To try this posture, spread your feet apart, aiming for a 90° angle between the abducted thighs, and twist the head, neck, and shoulders to the right, again bracing your back with the right hand on the floor behind you and aiding the twist as best you can with the left (fig. 7.26). Repeat on the other side.

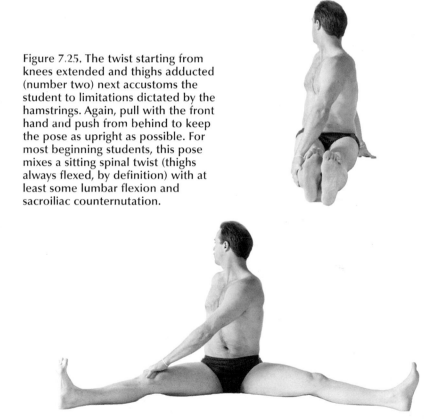

Figure 7.25. The twist starting from knees extended and thighs adducted (number two) next accustoms the student to limitations dictated by the hamstrings. Again, pull with the front hand and push from behind to keep the pose as upright as possible. For most beginning students, this pose mixes a sitting spinal twist (thighs always flexed, by definition) with at least some lumbar flexion and sacroiliac counternutation.

Figure 7.26. The twist with knees extended and thighs abducted (third in this series) next illustrates limitations dictated by the adductors as well as the hamstrings. Beginners should adjust the amount of abduction to make the pose easier or lean back to maintain a lumbar lordosis. The back should not be rounded excessively to the rear.

428   *ANATOMY OF HATHA YOGA*

If the thighs are abducted to their limits in this third posture, twisting in one direction creates intense stretch not only in the hamstrings, but also in the adductors on the opposite side. Both sets of muscles will be tugging on the base of the pelvis, making this the most demanding of the sitting twists described so far. To find the ideal amount of abduction for improving hip flexibility, mildly stretching both the adductors and hamstrings, and yet still working productively on the spinal twist, go back and forth between partial and full abduction several times, twisting to one side and then the other. The idea is to strike a reasonable compromise that does not degrade the lumbar curvature too much and give in totally to sacroiliac counternutation. Sitting on a support will also allow a more graceful twist.

The fourth and last set of twists in this series folds both legs in the same direction and is especially useful for relieving the stress of other sitting twists. Pull the left foot in toward the groin and swing the right foot around to the right. The left foot ends up against the right thigh. Twist to the left, placing your left hand behind you on the floor and your right hand on your left knee. Twist further, aiding the gesture with both hands. Notice that you naturally lean back as you come further into the twist, thus straightening the spine and permitting more sacroiliac nutation (fig. 7.27a). Then come up and twist to the right with the aid of the right hand on the floor and the left hand on the left knee. Twisting in the same direction in which the legs are pointing will require most novices to lean markedly forward, thus flexing the lumbar spine, although that requirement is not apparent in this illustration (fig. 7.27b).

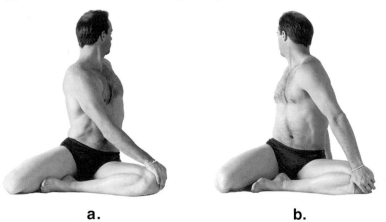

a.   **b.**

Figure 7.27. Twists with the feet both folded in the same direction (fourth in this series) provide a welcome relief to students frustrated with other sitting twists. Twisting in the opposite direction of the legs (a) is easy and results in leaning slightly to the rear, which straightens the spine. Twisting in the same direction (b) requires leaning forward and some spinal flexion in everyone who is not gifted with excellent hip flexibility (obviously that does not apply to this model).

Twist your neck in this fourth posture to please yourself, either rotating it in the same direction as the rest of the twist or keeping it in a neutral position perpendicular to the shoulders. If your hip flexibility is poor, you can moderate the posture by sitting up on a cushion, helping you to create a more pure spinal twist without flexing the spine and placing so much tension on the hip joints. The posture should be repeated, as always, on the other side, making a total of four possibilities, two with the feet facing to the right and two more with the feet facing the left.

HALF SPINAL TWISTS

The next six postures (figs. 7.28–32 for five of the six; the first one is not illustrated) are loosely called spinal twists, but technically they are half spinal twists. The simplest and easiest (not illustrated here) is a leaned-back twist for those who have serious problems with hip flexibility. To come into the posture sit flat on the floor with thighs adducted and knees extended. Raise the left knee and place the left foot on the floor on the left (medial) side of the right knee, or perhaps a little closer to the groin. Twist left moderately, hook the right forearm around the left knee, and brace the posture with the left hand far enough behind you on the floor to enable you to lean backward with your elbow extended and keep the lumbar spine arched forward. As soon as you are in a comfortable position, twist a little more to the left and readjust the placement of the left hand against the floor to increase the twist and yet sharpen the lumbar curve. Look as far to the left as you can and repeat the pose on the other side.

This posture has much to offer. Because the adductors are not being stretched at all, and because you are leaning well to the rear, the hamstring muscles are not creating much tension on the base of the pelvis, and you can concentrate all of your awareness on sacroiliac nutation and twisting the spine. It's also an excellent introduction to using the upper extremities for prying yourself around in the classic half spinal twist.

The classic half spinal twist (fig. 7.30) is a complex posture, but it is also one that can be approached in stages (figs. 7.28-30) that are useful for beginners. For the least demanding version, start from a cross-legged easy posture (fig. 10.10a) with the right foot drawn into position first (ending up with the left leg in front of the right), and raise the left knee, leaving the left foot flat on the floor in front of the right leg. Twist the body provisionally to the left, looking as far to the rear as you can while keeping the head and neck on axis with the thoracic spine. Place the left hand behind the hips and support the body with the forearm extended. To complete the easy version of this posture, hook your right forearm around the left knee and pull it in against the chest and abdomen (fig. 7.28), push more insistently with the left hand, and sit up as straight as possible. Both ischial tuberosities should be on the floor. Those who are inflexible can work with this posture for a month or so before going to the next stage. Repeat on the other side.

For the next stage, continuing from before, lift the left foot and place it flat on the floor on the right side of the right knee. This is now starting to be a more complex posture, especially for beginners, because they usually have to lean forward, losing their starting position of maximum nutation along with their lumbar lordosis. Even so, they should continue within their limitations, pulling the left knee more vigorously to the chest (still with the right forearm) and twisting to the left as much as possible (fig. 7.29). As in the case of the first stage (fig. 7.28), if this is as far as you can go without creating stress and strain, work with this pose for an additional month or two before proceeding. Repeat on the other side.

Assuming you wish to go on to the classic version of the half spinal twist, pry yourself around further into the twist by bringing the back side of the right arm to the left side of the left knee and then reach down to grasp the right foot (which is easiest), the right knee (fig. 7.30), or if you are flexible enough, the left foot. This twist is even more complex than the previous version (fig. 7.29). Only if you are unusually flexible will this be a respectable half spinal twist. For most people cranking the body around with the shoulder changes the posture from a spinal twist to something that combines hip prying and triceps brachii massage with three movements of the spine—forward flexion, lateral flexion, and twisting. This is not without virtue, but the posture is no longer a pure spinal twist. There is nothing wrong with it so long as everyone realizes what is happening, but it is important to understand that the pose will never look and feel complete until hip flexibility is sufficient to obviate the need to bend forward in the lumbar region. As with the other sitting twists, elevating the buttocks with a pillow eases the posture for those who are not flexible enough to feel comfortable flat on the floor. Repeat on the other side.

All sitting half spinal twists that begin from a cross-legged position can also be done with one leg extended. Most of these are more demanding than the cross-legged versions because the hamstring muscles that insert on the extended leg will now be pulled taut, tugging on the ischial tuberosity on that same side and making it even more difficult to sit straight than in the classic half spinal twist shown in fig. 7.30. The most common variation of this twist differs only slightly in knee, shoulder, arm, and hand position from the half spinal twist which introduced this series. For this one (number five in the series) sit straight with the right leg stretched forward and the left knee upright, but instead of placing the left foot on the floor on the medial side of the right thigh, lift the left foot and bring it all the way across to the right (lateral) side of the right thigh. Now instead of hooking the left knee with the right forearm, as in the case of the introductory pose,

Figure 7.28. To study the important half spinal twists, it is helpful for novices to proceed in stages. After pulling in the right heel in toward the upper left thigh, place the left foot in front of the right leg and pull the left knee toward the chest with the right hand and forearm. Except for the prying effect on the left hip joint, this is a pure sitting twist and is an excellent pose for beginners. They need not go to the next stage until they are ready. Repeat on the other side.

Figure 7.29. Next, lift the left foot and place it on the right side of the right thigh. Pull both feet in as close as possible, the right heel toward the groin and the left foot toward the lateral aspect of the right hip. Once the feet are in position, again pull the left knee closer to the chest with the fore-arm, and notice that this pose is more demanding of flexibility than the first one shown in fig. 7.28. Again, one need not go further. Repeat on the other side.

Figure 7.30. Last, for the standard half spinal twist, place the back of the right arm in front of the left knee to increase the prying effect on the left hip joint. Grasp the right foot, the right knee (as shown here), or the left foot with the right hand. For many students, this is a complex and difficult posture that combines a hip pry and forward bending with a spinal twist. Repeat on the other side.

twist more fully to the left and bring the back of the right arm against the left thigh and catch either the right knee or the left foot with the right hand to accentuate the twist. The left hand again braces the posture from behind (fig. 7.31). This posture is identical to the leaned-back twist (the one that introduced the half spinal twists) except that here you are sitting straighter and starting with the pulled-in foot on the lateral rather than the medial side of the opposite knee, which insures that the back of your arm will be more effective in pressing the upright thigh against your torso As before, sit on a pillow to moderate the posture if you are not flexible enough to feel graceful in it, and as always, repeat on the other side.

If you have sufficient flexibility in your hips, spine, and shoulders, you can pull yourself into more complete twists by catching your hands together behind the back. You can start cross-legged or with one leg outstretched, but in either case you will have to pull the upright thigh and leg in closely enough to place the foot flat on the floor on the lateral side of the opposite knee. Try it first cross-legged (number six in this series). Draw the left heel in and place the right ankle to the left (lateral) side of the left knee. Twist to the right and pull the right knee close to the chest. Then, twisting even more to the right, anchor the back of the left arm and shoulder against the right knee, reach with the left hand between the right thigh and leg (from the front) to catch the right hand, which is reaching around from behind. Clasp the wrists or fingers together (fig. 7.32). Unless you are quite flexible you will probably have to lean forward to some extent. In any case, interlocking the hands requires excellent flexibility in the right hip joint. Otherwise you will have to lift the right hip off the floor and try to balance on the left hip. That is not an unthinkable compromise in spinal twists in which you are able to brace the right hand against the floor, but it won't work here because the right hand is interlocked with the left and no longer available to keep your balance.

Figure 7.31. Another variety of the half spinal twist keeps one knee extended, which again brings the hamstring muscles of that side to your attention. This pose requires some spinal flexion of all but the most flexible students. You can push strongly with the rear hand to improve hamstring flexibility.

This posture is excellent for working with hip flexibility, but it should be approached with respect. If you have to bend forward excessively, lift the hip off the floor, and struggle to get your hands interlocked, you may let your attention lapse from the posture as a whole and overlook vulnerabilities that leave you exposed to injury.

BREATHING ISSUES

In chapter 2 we discussed the four modes of breathing: abdominal, diaphragmatic, thoracic, and paradoxical. Twisting the torso constricts abdominal breathing because it makes the lower abdominal wall taut and prevents its expansion. Under ordinary circumstances this facilitates diaphragmatic breathing, in which the descent of the dome of the diaphragm lifts the rib cage, but that also is limited because thoracic twisting keeps the upper abdomen taut as well, which in turn limits flaring of the base of the rib cage. The twist in the thorax even limits the ability to lift the chest for thoracic breathing. So where does this leave us? If you come into a simple cross-legged spinal twist, you can feel restrictions in breathing everywhere, but at the same time you will notice small respiratory movements throughout the torso: some in the lower abdomen (abdominal, or abdomino-diaphragmatic breathing), some flaring of the rib cage (diaphragmatic, or thoraco-diaphragmatic breathing), and some lifting of the rib cage (thoracic breathing). The one mode of breathing you'll not see is paradoxical breathing, and this makes spinal twists a special blessing for anyone trying to break that habit.

In the more complex twisting postures, the relationship of the thigh to the torso adds another complication. If your posture presses the thigh against the lower abdomen, abdominal breathing is restricted. And if the thigh is pressed strongly against both the abdomen and the chest, you can

Figure 7.32. The fullest expression of the half spinal twist requires interlocking your hands together behind your back. This posture can be attempted by intermediate level students, but it will not be very rewarding until you have enough hip flexibility to sit upright without having to push the posture straight with the rear hand. Like all versions of the half spinal twists, this one improves hip flexibility but appears awkward until good progress toward that end has been made.

only breathe by lifting the chest for thoracic breathing. Finally, this pressure on the abdominal wall frequently causes a slowing of inhalation and short bursts of exhalation. These adjustments in breathing cannot be helped; the postures demand them. They can only be watched and minimized.

THE FULL SPINAL TWIST

When you are able to do the half spinal twist, catching the hands gracefully and comfortably behind your back, and at the same time sitting straight and keeping the ischial tuberosities flat on the floor, you will be able to approach the quintessential sitting spinal twist—*matsyendrasana*, the famous spinal twist done from the full lotus position. Few people who grow up sitting in chairs will have enough flexibility to get into it, and even for them years of preparation may be needed. The more demanding variations of the half spinal twist are obvious requirements, as are the half lotus, the full lotus, and any or all of the dozens of exercises that increase hip flexibility (chapter 6).

If you can do the lotus posture (fig. 10.15) comfortably, and if you can sit straight in the half spinal twist without resorting to forward and lateral flexion, you can begin to work with another preparatory posture—the spinal twist in the half lotus. Start with the left foot pulled up to the top of the right thigh so the heel is pressed into the abdomen. Then raise the right knee and draw the right foot in toward the perineum (holding the

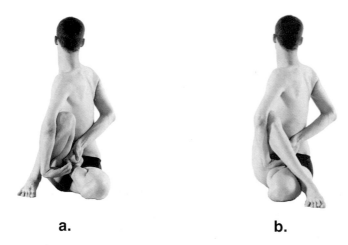

a.   b.

Figure 7.33. The full spinal twist (b) and its most immediate preparatory posture (a) are among the most difficult poses in hatha yoga. To do the preparatory pose, come into the half lotus (left foot pulled to the top of the right thigh) and then lift the right knee and pull the right foot in toward the perineum (a). To come into the full spinal twist, go from the preliminary pose and place the right foot on the floor on the left side of the left knee. Twist to the right, looking back, and catch the hands, interlocking the fingers behind the back (b).

foot flat on the floor in front of the pelvis), all the while keeping the back straight. Raising the right knee is probably what you will not be able to do. Here is where working with a half lotus modification of the trapezius stretch in chapter 9 (fig. 9.17b) will help.

When you are able to come into this position convincingly, twist right from the hips to the head, keeping the right thigh near the torso, bringing the left elbow against the lateral side of the right knee, and catching the hands together near the left foot (fig. 7.33a). Work equally on both sides, or favor the side which is more problematic. You should work with this posture until you are able to remain comfortable in it while you are keeping the spine straight.

Only when you are comfortable in the full lotus and can do the spinal twist in the half lotus is it time to try the full posture. Again place the left foot on top of the right thigh, but now swing the right foot over the left knee and plant it on the floor lateral to the left thigh and pulled in as much as possible. Twist to the right, looking back. You can use one of two hand positions depending on your flexibility: simply dropping them where they fall naturally, or reaching with the left hand between the right leg and thigh to catch the right hand behind the back (fig. 7.33b). As with all twisting postures, repeat on the other side.

BENEFITS

Twisting postures in hatha yoga complement forward and backward bending by exercising muscles in more complex ways than is accomplished by the symmetrical movements of flexion and extension alone, and in some cases they exercise muscles that are highly specialized for twisting. Beyond that, the axial compression of the spine and other structures of the torso improves nutrition to the intervertebral disks and squeezes blood out of the internal organs of the abdomen and pelvis, thus improving circulation in the great supportive systems of the body. For these reasons, twisting postures are essential for a complete practice and must always be included in any balanced program of hatha yoga.

"Vedanta is like a blueprint for a house. It is knowledge about the house, but you cannot live in the blueprint. You have to build the house using bricks and mortar. And that is tantra. So Vedanta gives you knowledge about the supreme reality; tantra enables you to experience that supreme reality in yourself."

— Tapasvi Baba, July 17, 2000 (translated by D. C. Rao).

CHAPTER EIGHT
THE HEADSTAND

"Sirshasana is really a blessing and a nectar. Words will fail to adequately describe its beneficial results and effects. In this Asana alone, the brain can draw plenty of Prana and blood. This acts against the force of gravity and draws an abundance of blood from the heart. Memory increases admirably. Lawyers, occultists, and thinkers will highly appreciate this Asana. This leads to natural Pranayama and Samadhi by itself. No other effort is necessary."
— Swami Sivananda, in *Yoga Asanas*, pp. 15–16.

To be "stood on your head" is to be surprised and shocked, and this is the essence of the headstand—turning the world topsy-turvey and adjusting to being upended. To launch this jolt to our spirit, we balance on the top of a spheroidal surface—the cranium—which can be likened to balancing the pointed end of an egg on a button. The headstand not only inverts our vision of the world, it inverts the pattern of blood pressure in the body—increasing it in the head and dropping it to practically nothing in the feet. And because the increase in blood pressure in the head may be the first deciding ingredient in whether or not it is prudent to try the headstand, we'll begin with a discussion of the cardiovascular system. Most of the rest of the chapter focuses on the musculoskeletal anatomy of the headstand and related postures: two techniques for doing the headstand and how each of them affects the neck and body; the anatomy of the upper extremities and methods for developing the strength needed for coming into the headstand safely; correction of front-to-back imbalances; breathing adaptations for inverted postures in general and the headstand in particular; and combining the headstand with backbending, forward bending, and twisting. Finally we'll examine the question of how long one can remain in the posture.

THE CARDIOVASCULAR SYSTEM

When you stand on your head the first thing you feel is pressure—pressure on top of the head, pressure in the arteries and veins, and pressure in the soft tissues of the head and neck. And along with these comes more subtle aspects of pressure—the demand for maintaining your balance and the psychological urge to come out of the posture. These physical and psychological pressures affect every system in the body in one way or another: musculoskeletal, nervous, endocrine, circulatory, respiratory, digestive, urinary, immune, and reproductive. We'll concentrate here on the most obvious one, which is circulation.

The heart pumps blood through two sequential circuits—pulmonary and systemic—from the right ventricle to the lungs and back to the left atrium in the *pulmonary circulation*, and then from the left ventricle to the body and back to the right atrium in the *systemic circulation*. In the pulmonary circulation blood picks up oxygen in the lungs and releases carbon dioxide; in the systemic circulation blood picks up carbon dioxide from the tissues of the body and releases oxygen. The flow of blood is unidirectional—from right atrium to right ventricle, to pulmonary artery, lungs, pulmonary vein, left atrium, left ventricle, aorta, body as a whole, veins, and back to the right atrium—around and around continuously from birth to death (fig. 2.1).

Each circuit contains arteries, capillaries, and veins. The pulmonary circulation to the lungs is a *low-pressure* (22/8 mm Hg), *low-resistance* circuit; the systemic circulation to the body is a *high-pressure* (120/80 mm Hg), *high-resistance* circuit. They are both affected by inverted postures, but we'll concentrate on the systemic circulation first because most of our interest is in how the headstand affects the body as a whole.

BLOOD PRESSURE AND FLOW IN THE SYSTEMIC CIRCULATION

Any time someone says they have a blood pressure problem, what they are talking about is blood pressure in the arteries of their systemic circulation. Like atmospheric pressure, alveolar gas pressures, and blood gases (chapter 2), blood pressure is measured in millimeters of mercury (mm Hg). When we say, "Normal blood pressure is 120 over 80 mm Hg," or more simply 120/80, we are referring to the pressure of the blood against the inner walls of medium-sized arteries of the systemic circulation, usually measured in the arm when we are either sitting quietly in a chair or lying down (fig. 8.1). As pressures go these are low—the equivalent to about 2 pounds per square inch of air pressure in the tires of your car.

The two figures are significant. The blood pressure in an artery of the arm rises to 120 mm Hg as the heart pumps blood from its contracting left ventricle; this is the *systolic pressure*, which is named after the Greek word meaning "contraction." Between contractions the pressure drops to 80 mm Hg as the left ventricle fills; this is the *diastolic pressure*, or the pressure between contractions. Blood pressure in an artery of the arm is only a small part of a bigger picture, however. From the left ventricle, blood is propelled successively through the aorta to large arteries, medium-sized arteries, arterioles, capillaries, venules, and veins, and blood pressure decreases from segment to segment. Within the heart itself—in the left ventricle—systolic blood pressure is 120 mm Hg and diastolic pressure is a mere 10 mm Hg, because the latter drops almost to nothing while the ventricle is filling with blood. The textbook standard of blood pressure is 120/80 mm Hg between the aorta and small arteries, and beyond the arterioles in the capillary bed it drops to about 15 mm Hg. On the venous side of the systemic circulation, blood pressure continues to drop in the venules and veins, and it is essentially 0 in the vena cava where that vessel opens into the right atrium of the heart (fig. 8.1).

Blood pressure in medium-sized arteries depends both on the heart acting as a pump and on *peripheral resistance*. The importance of the pump is obvious: a harder-working heart creates more pressure in the system. But the resistance to flow in the arterioles is just as important: as peripheral resistance increases, blood pressure in the arteries also increases. There are many neurological, hormonal, and other physiological factors that influence the heartbeat and peripheral resistance, but they are beyond the scope of this book; here we'll note only that any time you become especially active or anxious, the sympathetic nervous system and hormones from the adrenal gland increase blood pressure by increasing both peripheral resistance and the strength and rate of the heartbeat.

Figure 8.1. This graph shows blood pressure in different parts of the systemic circulation at heart level. The continuous curves in the portions of the graph for large arteries, small arteries, and arterioles represent variations in systolic (top) and diastolic (bottom) blood pressure, and the dashed curve in the same regions represents averages (for example, about 100 mm Hg in large arteries). Systolic and diastolic pressures are no longer detected separately in capillaries and veins, and blood pressure drops essentially to 0 mm Hg where the vena cava empties into the right atrium (Dodd).

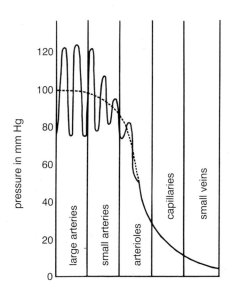

Blood pressure also varies in different parts of the body. It increases below the heart and decreases above the heart because the weight of the column of blood in an artery adds to (or subtracts from) the pressure generated by the heart and by peripheral resistance. In a standing position with blood pressure in medium-sized arteries at 120/80 mm Hg at heart level, blood pressure will be about 210/170 mm Hg in the arteries of the feet and about 100/60 mm Hg in the brain (fig. 8.2a). The only circumstances under which we'll see blood pressure equalized throughout the body at 120/80 mm Hg is if we neutralize the effect of gravity by lying prone or supine, by submerging ourselves in water, or by taking up residence in a space capsule that is orbiting earth.

Turning upside down in the headstand reverses the figures seen standing in a straightforward fashion. Blood pressure will remain at 120/80 at heart level, at least if you are not under too much stress, but the pressure in the arm will rise to about 140/100 mm Hg because the arm is alongside the head and below the heart instead of level with it. We can calculate that blood

100/60 mm Hg

120/80 mm Hg
(average of
100 mm Hg)

210/170 mm Hg

40/0 mm Hg

120/80 mm Hg

140/100 mm Hg

150/110 mm Hg
(average of 130 mm Hg)

Figure 8.2a. Arterial blood pressure in a standing posture in muscular arteries in different parts of the body.

Figure 8.2b. Calculated arterial blood pressure in the headstand in muscular arteries in different parts of the body.

pressure will only be about 40/0 mm Hg in the feet, with diastolic blood pressure dropping to zero, and with the systolic blood pressure of 40 mm Hg barely sufficing to perfuse the capillaries. Even then it's marginal, which is why your feet may "go to sleep" and get pins-and-needles sensations if you remain in the headstand for a long time.

We can calculate that blood pressure at the top of the head increases from 100/60 mm Hg in a standing position to 150/110 mm Hg in the headstand (fig. 8.2b), or even higher if you are not confident of the posture. The headstand is therefore contraindicated for anyone who has abnormally high blood pressure for the simplest of reasons: the posture can increase blood pressure in the brain to dangerous levels—perhaps well above 150/110 mm Hg. Conservative medical opinion also recommends that you avoid the headstand even if high blood pressure is brought to a normal level with medication.

As important as blood pressure is, we can't understand the cardiovascular responses to inverted postures without also considering the flow of blood through the system—both bulk flow through the major segments of the system and the rate of flow though specific vessels. Since it's a one-way circuit, the same volume of blood per unit of time (about 5 liters per minute at rest) has to flow through each segment of the cardiovascular system. And there is also the question of rate of flow through individual arteries, capillaries, and veins. Just as a river carries water sluggishly where the river is wide, and briskly where it is constrained by tubes and by turbines that generate electricity, so does the rate of flow vary in the vascular system. The flow is speediest through arterioles, where it is choked off the most as well. It is slowest in the capillaries, and it flows at an intermediate rate through the veins, which carry blood back to the heart.

THE VENOUS RETURN

If the capillary beds and veins were static tubes with fixed diameters, blood would stream from the arterioles into the capillaries, pour from the capillaries into the veins, and be pushed all the way back to the heart by arterial pressure. But this is not the way the system operates. The capillaries and veins are expandable: they could easily accommodate all the blood in the body. And this can create a serious problem because within certain limits the amount of blood brought to the heart per minute (the *venous return*) regulates the volume of blood pumped by the heart per minute (the *cardiac output*). Here is what happens: As venous return increases, the additional blood stretches the walls of the ventricles, and when that happens, the stretched muscle fibers in the ventricles automatically pump more strongly, thus increasing cardiac output; as venous return decreases, the ventricles pump less vigorously, thus decreasing cardiac output. Therefore, the mechanisms for moving blood from capillaries, venules, and veins back to the heart are critical. If too much blood stagnates in those parts of the system, which can happen for many reasons, cardiac output decreases and the heart may not receive enough blood to pump to the brain and other vital organs.

When we are in a normal upright posture, the venous return from veins located above the heart is unimpeded, and blood drops like a waterfall to the right atrium. And at heart level (in the middle segment of the arm, for example) venous pressure is about 15 mm Hg. Since this is more than the 0 mm Hg where the blood enters the right atrium, it is still pushed easily back into the heart. The lower extremities are another story, however, and to get blood back to the heart from the feet, at least in an upright posture, the venous return has to overcome a pressure of about 140 mm Hg, which reflects the height of a static column of venous blood below the heart.

The mechanism for getting blood back to the heart from the lower extremities is beautiful in its simplicity and elegance. The veins below the heart contain one-way valves, and contraction of the skeletal muscles surrounding these valves acts as a "*muscle pump*" to squeeze blood through them and back toward the heart. When the muscles relax, the valves close to prevent backflow, insuring that the flow is unidirectional

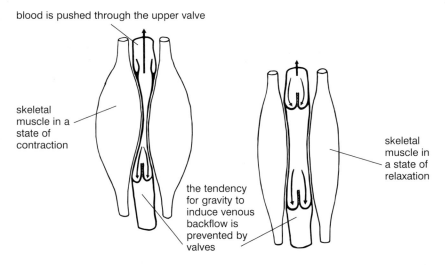

Figure 8.3. Skeletal muscle pump for venous return. On the left the muscle contracts, and blood can escape this segment of the vein only by being pushed through the upper valve. On the right, the muscle relaxes, and the upper valve is closed by venous back-pressure. As skeletal muscles become active throughout the body, alternately contracting and then relaxing, blood is pushed back to the heart mechanically (Dodd).

(fig. 8.3). Drill instructors in the military may not be aware of this mechanism, but they know that skeletal muscular activity is needed to get blood back to the heart, and that is why they instruct new recruits who are standing at attention on a hot day to isometrically contract and then relax the postural muscles of their lower extremities, which keeps them from fainting.

There are no valves in the head and neck: our upright posture has rendered them irrelevant. But when you are standing upright, a pool of blood courses slowly through the veins in the lower half of the body, waiting to be pumped back toward the heart by muscular activity. And if you are chronically inactive, fluids move so sluggishly out of this region that the processes of cellular nourishment and elimination are compromised. This gravity-induced congestion can affect any organ or tissue below the heart. Lying down for a night's sleep helps correct the situation, but we still often see the effects of gravity in chronically swollen ankles, varicose veins, and hemorrhoids. One remedy is vigorous movement in which muscles alternately contract and relax in order to propel blood through the venous valves. And this is one reason nurses try to get people up and about as soon as possible after surgery and why health practitioners constantly preach the benefits of exercise. Yoga teachers do not disagree, but suggest another alternative—inverted postures.

What happens specifically when you turn upside down? In the first moments of the headstand blood pools in the capillary beds and veins in the region of the body superior to the heart—in the head, neck, and shoulders—where it is kept until arterial pressure forces it back around to the heart. And because there are no valves in the veins of this region, skeletal muscle contraction cannot assist its return. This is not very important if you stay in the headstand only 2–5 minutes, but if you want to extend your time in the posture it can become a problem. We'll come back to this issue toward the end of the chapter.

[Technical note: There are many ways to affect venous return. Without naming the phenomenon, we looked in chapter 2 at the effects of a *Valsalva maneuver*—holding the breath and straining after an inhalation—in conjunction with hyperventilation. It also sometimes happens that X-ray technicians ask patients to hold their breath after an inhalation in order to get a more elongated and accurate profile of the heart, and if the patient gets overenthusiastic about this after locking the glottis, or if the technician dawdles, the unintentioned Valsalva maneuver impedes the venous return. If you hold the breath in this manner after a deep inhalation, the profile of the heart in a roentgenogram shrinks dramatically, and after 10-15 seconds, depending on how purposely you strain, the venous return is inhibited enough to cause you to pass out. One would not ordinarily think of trying the Valsalva maneuver in the headstand, and it's plainly inadvisable. It won't result in fainting because the inverted posture sends blood preferentially to the head, but it certainly causes a sharp and immediate rise in blood pressure. This is felt mostly prominently in the face, where it is disagreeable although probably not harmful—but for the brain and for the retina of the eye, look out: it is assuredly dangerous.]

THE PULMONARY CIRCULATION

The oxygenated blood that is pumped to the body from the left ventricle will be cycled straight back to the right side of the heart, and from there the passage of blood into the pulmonary circulation is like a slow-moving flood—5 liters per minute to the lungs. This pulmonary flow is the constant companion to the systemic circulation—5 liters per minute to the body. Inverted postures affect the pulmonary circulation very differently from the systemic circulation, and to understand how and why, we'll need a few more details.

If blood pressure in the pulmonary circuit were 120/80 mm Hg, as it is in the systemic circulation, blood would perfuse through all parts of the lungs fairly equally, but pressure in the pulmonary circuit is much lower—only 22/8 mm Hg—and because of this the pull of gravity will markedly affect the pulmonary flow and distribution of blood. As we discussed in chapter 2, if you are sitting or standing quietly in an upright posture, the lower parts of the lung are perfused with blood efficiently and the upper parts of the lung are perfused sluggishly. Although studies of pulmonary arterial pressure relationships and blood flow in inverted yoga postures have not been published, it seems certain that the patterns of pressure and flow of blood in the lungs will be reversed, and that inversion will cause the upper rather than the lower parts of the lungs to be perfused with blood most efficiently. Deep breathing in the headstand (to be examined in detail later in this chapter) can remedy this because it ventilates the lungs generously from top to bottom and insures that minor variations in circulation are insignificant.

BARORECEPTORS

In chapter 2 we looked at oxygen-sensitive peripheral chemoreceptors in the large arteries that lead from the heart to the head. We also have *baroreceptors* at those same sites for detecting blood pressure. Increased blood pressure in any posture stimulates the baroreceptors, which in turn affects both limbs of the autonomic nervous system: it increases parasympathetic nervous system input to the heart, and it reduces sympathetic nervous system input to both the heart and the arterioles—all of which tend to lower blood pressure. This is called *reflex hypotension*, and some people are especially sensitive to its effects, possibly even experiencing enough of a drop in blood pressure to produce fainting from the pressure of a tight collar or from mild pressure of someone's hands against the neck.

These reactions are pertinent to this chapter for several reasons. First of all, in the headstand, the baroreceptors are below instead of above the heart and will be subject to, and stimulated by, increased blood pressure. If you are entirely comfortable in the headstand, the input of the baroreceptors to the central nervous system will generally produce a lower heart rate and blood pressure than what is assumed simply on the basis of fluid dynamics. Second, if you have a general sensitivity to reflex hypotension, you might have an exaggerated response in the headstand and should approach it with caution and only after a lot of experimentation and preparation. Third, someone who has slightly elevated blood pressure in an upright posture might see that drop when they come into a comfortable headstand. Under such conditions, we would expect blood pressure to become elevated again upon standing. Last, if you are anxious and uncomfortable in the headstand, the accompanying increase in activity of the sympathetic nervous system could stimulate the heart, increase peripheral resistance, and elevate blood pressure excessively. These possibilities can only be checked out with a blood pressure cuff. In any case, unless you have a medical practitioner who is willing to take responsibility for advising you, the headstand is still contraindicated if you have elevated blood pressure in upright postures.

THE RECOVERY

If you stand up quickly after coming out of the headstand, you will feel a surge of blood falling from the veins in the upper half of the body. This will not hurt someone in good health, but conservative medical advice is to reverse these pressure and flow dynamics more slowly. Some instructors even recommend relaxing for a short time in the corpse posture (figs. 1.14 and 10.2) before standing. Whether you stand up immediately or cautiously, however, many authorities recommend that you remain upright, whether standing still, doing standing postures, or walking, for as long as you held the headstand.

CARDIOVASCULAR BENEFITS OF THE HEADSTAND

The literature on hatha yoga waxes eloquent on the wonders of the headstand. Kuvalayananda maintains that the posture benefits the special senses, the endocrine glands, and the digestive system, to name only a few, and Sivananda, in his usual style, calls the headstand "a panacea, a cure-all for all diseases." Anyone who has had a lot of experience with the headstand will agree that it's a marvelous posture, but it is not clear why this is. For a possible answer, the most obvious place to look is the inverted circulatory system. First, it is plain that when we are inverted, the venous return from the lower extremities is determined by the amount of blood pumped through the capillary beds, because once it gets into the venules it is quickly recirculated by the force of gravity. If you can remain in an inverted posture for just 3–5 minutes, blood will not only drain quickly to the heart from the lower extremities and the abdominal and pelvic organs, but tissue fluids will flow more efficiently into the veins and lymph channels, and this will make for a healthier exchange of nutrients and wastes between cells and capillaries.

It is also obvious that inversion increases blood pressure in the head and neck, the regions of the body that are filled with the body's regulatory mechanisms: the brain's hypothalamus, which regulates the autonomic nervous system and pituitary gland; the pituitary gland, which regulates many other endocrine glands; and the brain itself, which carries out all aspects of mental functioning. This region also contains the special senses that are so important in our interactions with the world: sight, hearing, taste, smell, and the sense of equilibrium.

Considered in isolation, the significance of this increased blood pressure in the head is uncertain. If peripheral arteriolar resistance were to remain constant when you come into a headstand, increased pressure would push more blood per minute into the capillary beds, increasing local blood flow, but without data to prove the point we cannot assume that will happen, because the increase in blood pressure might well be accompanied by enough increased peripheral resistance to keep blood flow the same. We do know that mental exercises and aerobic activities such as running increase blood flow to the brain, and if future research shows that the headstand produces the same result, it might help explain the intense but subjective feelings of well-being that accompany this posture.

THE TWO HEADSTANDS

If you were to watch a hundred hatha yoga teachers all doing the headstand at the same time, you would notice that they were not all doing it the same way. Some would be perfectly vertical, their eyes directed along a path parallel to the floor, their backs straight, and taking care to be balanced on the top of their heads. The other group, probably a minority, would be more relaxed, their eyes directed to the floor, their lumbar regions arched, and their heads balanced on a point a little more to the front than those in the first group.

These differences are lost on beginning students. When they first try to come into the headstand they are likely to be struggling, and one of the symptoms of their exertion is that they tend to balance much of their weight on their forearms with excess muscular tension in their arms and shoulders. Under those circumstances, where they position their heads against the floor is not critical. But after gaining more experience and building up their time in the posture beyond 1–2 minutes, they tire of supporting themselves so much with their upper extremities, and the question of where to position the head against the floor becomes more important. Many teachers are picky about this matter, some saying that the weight must be placed directly on the top of the head, and others saying that the weight should be placed more forward. Neither school of thought is incorrect. There are two ways to do the headstand, and these yield such different results that the two postures warrant separate names: the *crown headstand* and the *bregma headstand*.

For the crown headstand we place the weight of the body directly on the *crown* of the head, which is the topmost point of the cranium. For the bregma headstand we adjust the hand position slightly forward so the weight of the body is on a spot called the *bregma*, which is an inch or so in front of the crown, at the meeting point of the sagittal and coronal sutures (fig. 8.4). The *sutures* are fibrous joints: the *sagittal suture* links the two *parietal bones* in the midsagittal plane of the body, and the *coronal suture* links the two parietal bones to the *frontal bone*. Interestingly, the site in the head of a newborn baby where these two sutures have yet to meet and grow together is the soft spot—the *anterior fontanelle*—and the future bregma.

EXPLORING THE TWO POSSIBILITIES

Come into a kneeling position on a well-padded carpet or folded blanket, place the top of the head down, and brace it in the rear with the interlaced fingers. The forearms should be at about an 80° angle from one another. Now gently roll your head around and explore the possible spots on which you could place your weight. Notice that as you roll the forehead down, bringing the nose toward the floor and extending the head and neck, you tend to follow the rolling of the head with your interlaced fingers, and that if you place weight on the crown you move your hands more to the rear. To further explore, lift up the hips and walk the feet forward, keeping the

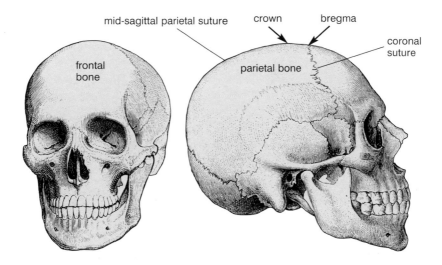

Figure 8.4. Human skull, viewed from the front (on the left) and from the side (on the right). For the crown headstand the weight of the inverted body is placed on the crown, the highest point of the cranium, and for the bregma headstand the weight is placed on the bregma, which is located about an inch in front of the crown at the intersection of the coronal and parietal sutures (Sappey).

knees bent as much as hamstring limitations require, and try different hand and head positions, supporting your weight only on the head and the feet. Now sit up and feel the top of the head. Sit perfectly straight and locate the highest point. This is the crown. Move the fingers forward about an inch. This is on or near the bregma. Behind the crown the cranium immediately rounds.

[Technical note: Notice that you probably have a clear preference for how to inter-lace the fingers, both in the headstand and in daily life. You will automatically place each of the fingers of the right hand on top of their counterparts on the left hand, or vice versa. In the headstand you should make sure that you alternate the interlacing, because always holding the fingers the same way will create subtle imbalances.]

The next thing to do, especially if you are a beginner, is to find a sandy beach or a spongy grass surface and practice turning somersaults. Do a few child's somersaults with the hands at the sides of the head, and then do them with the fingers laced behind the head and well to the rear. Notice that you do not need to make many adjustments in the head position as you roll over. Then try placing your weight on the crown. As you go into the somersault from this position you will find that you flex the head slightly before you flip over. Finally, place your weight on the bregma. Now you have to make a big adjustment in order to roll over. You have to either move your hands and flex the head forward, thus shifting the point of contact to the back of the head just before you roll into the somersault, or you will plop over backward instead of rolling. In the latter case you will have the option of falling flat on your back, which may not be comfortable even on a sandy beach, or of quickly extending the hips and flexing the knees as you start to fall, thus breaking the fall with your feet. The point of these experiments is to reduce your fear of the headstand by teaching you to roll down out of the posture gracefully if and when the need arises.

When you start to experiment with the headstand itself, it will become obvious that your main concern is that you might tip over backward. Even if practice somersaults have removed your fear of tipping over to the rear when you have to, you will still not do it by choice. So until you become confident, don't hesitate to come part way up over and over again, and each time drop back forward on your toes the same way you started up.

THE CROWN HEADSTAND

Although either posture can be learned first, the crown headstand is more elementary and simpler to explain than the bregma headstand. So even though there is a greater tendency to somersault in the crown headstand, try this one first. To do it, come into the preparatory kneeling position and adjust the forearms at an 80° angle from the clasped hands. This is important because if the forearms are placed at an obtuse angle (greater than 90°), your weight will be distributed over a base that is too short from front to back, and the posture will be unstable. And since the tendency to fall to the side in the headstand is not as great as the tendency to fall forward or backward, a long front-to-back base is more desirable than a wide side-to-side base. You can approximate the correct angle by placing the elbows alongside the knees, planting your head directly on the crown, and bracing the interlocked fingers just underneath the back of the head to maintain the position.

The headstand is best learned in four stages. To come into stage one, lift the buttocks and start walking the feet forward. Advanced students who are able to fold their chests down against their thighs with their knees straight in the posterior stretch may want to keep their knees extended for approaching this stage of the headstand (fig. 8.5), but most people will not have long enough hamstrings to allow this. So flexing the knees as necessary, walk toward your head, then tiptoe. Keep coming until less than 5% of your weight is still supported by the feet. Your back is probably rounded, you are close to the point of tipping over, almost no weight is on the elbows, and the tiniest nudge from the big toes could lift you off (fig. 8.6a). This is stage one, an inverted posture that is worth practicing in its own right—a forward bend with an inverted torso.

Going from stage one to stage two is difficult because now the hips have to be raised, the back has to become flattened, and the feet have to be lifted off the floor—all while the lower extremities are positioned well to the front of your future axis for balancing. It is a problem comparable to standing up and reaching forward at shoulder height to lift a weight straight up. In that case even a light weight would be difficult to lift because of your poor mechanical advantage. Here, when you are going from stage one to stage two, all of the burden is on your back and forearms. During the course of this transit the lumbar region will be rounded, much of the weight of the lower extremities will be to the front of the abdomen, and you will have to support the posture with the forearms as you are coming up. Fortunately,

Figure 8.5. Stage one of the headstand with the knees straight. This is a useful starting position for those who have excellent hip flexibility and lengthy enough hamstrings, but impossible for those who do not.

the situation is only temporary. When you finally settle into a relatively stable position in stage two, the lumbar region will be flatter, and depending on your hip flexibility, the thighs will end up at a 45–90° angle from the pelvis (that is, flexed 90–135°; fig. 8.6b).

Poor hip flexibility is the main obstacle when you are going from stage one to stage two. With the toes barely on the floor and the knees partially extended in stage one, short hamstring muscles keep tension on the pelvis and keep the back rounded posteriorly, and this prevents you from easily distributing the main bulk of your body weight above the head. The less flexible the hips, the more weight you will have to support on the forearms as you lift the feet. If you are quite strong this may not be a problem, but the average student will find it the most serious challenge to learning to do the headstand in stages.

In stage three you extend the hips, lifting the knees toward the ceiling while keeping them flexed. This is easy. As you extend the thighs the weight of the feet and legs shifts to the rear, and the lumbar region arches forward enough to maintain your balance (fig. 8.6c).

Figure 8.6a. Stage one of the headstand with the knees bent, a more realistic starting position for the average student. Keep walking your toes forward until you are almost ready to tip over. At this point only a slight nudge would cause you to somersault onto your back.

Figure 8.6b. Stage two of the headstand. This is a difficult stage to remain in for any length of time, because the weight of your lower extremities has to be supported by your deep back muscles.

The fourth and last step is to extend the knees. As you do this, the lumbar region will flatten as necessary to compensate for the fact that the feet and legs are now in line with the torso and thighs. You will gradually shift your weight off the forearms and balance on your head as you develop confidence in the final posture (fig. 8.6d).

Summarizing the four stages, first come into the preparatory position and walk forward until you are prepared to lift off without losing your balance; second, lift the feet and extend the back enough to bring the thighs to a 45–90° angle from the pelvis; third, extend the thighs while keeping the knees flexed and notice how this produces a pronounced lumbar arch; fourth, extend the legs while noticing that the lumbar arch decreases, and balance as much of your weight on the head as feels secure.

Figure 8.6c. Stage three of the head-stand with the hips extended is very stable, and you can stay in it as easily as in the headstand itself. Notice, however, that the flexed knee position drops the feet to the rear, and that keeping your balance will require a more prominent lumbar lordosis than stage four. Wearing ankle weights or heavy shoes will make this plain.

Figure 8.6d. Stage four of the crown headstand with the knees extended is a balancing posture with only a little extra weight on the elbows. Shifting more weight to the forearms can be accomplished conveniently only by thrusting the pelvis forward and the feet to the rear, thus increasing the lumbar curve.

Most hatha yoga teachers recommend coming up into the headstand in stages because they know that by doing this students will master each step in sequence and maintain control throughout the process. But if you have tried this for some time and are frustrated because you are not making progress, there is an alternative. Walk the feet forward as much as you can, and then simply lift one foot into the air at a time, coming into stage three with the hips extended and the knees flexed. From there it is easy to come into the final posture.

Even if you come up into the headstand one foot at a time you can still work on coming down in stages. Notice how your weight shifts when you flex the knees for stage three. Next, as you flex the hips for stage two, notice that your weight shifts forward as the knees come to the front and that you have to place extra weight on the forearms or lose your balance. At this point you may drop quickly to the floor. It takes a strong back to stop coming down and keep your balance when the thighs are flexed, but as soon as you are able to go back and forth between stages three and two, and even more obviously between stages two and one, you will have all the strength you need to come into the headstand using this four-part sequence. These efforts are particularly important for the many beginners who, without some preliminary coaching, will tend to come up tippy all the way, maintaining a shaky balance from beginning to end.

Finally, after you have had a year or two of experience with the headstand, try the four stages of the posture wearing heavy shoes or light ankle weights. This will show you clearly how shifting the pelvis and lower extremities in stages two, three, and four affects your balance.

THE BREGMA HEADSTAND

You come into the bregma headstand in the same four stages you used for the crown headstand, but this version is easier for many students because the head is more firmly supported. Plant your head on the bregma and cup your interlocked fingers underneath the head rather than posterior to it. From this position it is easy to thrust the upper back posteriorly as you walk the feet forward into stage one. There will be little tendency to somersault. And after you have learned the bregma headstand (fig. 8.7), you may find it to be more stable than the crown headstand.

CONTRAINDICATIONS

We have already discussed why the headstand is contraindicated for those who have elevated blood pressure. They usually become aware of discomfort as soon as they place the top of their head against the floor in preparation for stage one. And if the instructor has properly cautioned everyone against doing any posture which causes unusual discomfort, there will not be any question in the student's mind that they ought not complete even stage one, much less continue beyond that.

Most teachers also suggest that the headstand not be done during menstruation. Chronic or acute neck pain, excess weight, and osteoporosis are other obvious contraindications, as are glaucoma and other eye problems. We'll discuss the headstand in relation to right-to-left musculoskeletal imbalances later in this chapter.

THE STRUCTURE OF THE NECK

Many teachers favor the crown headstand, feeling that placing the weight further forward creates hyperextension and strain in the neck. Others claim that this is erroneous, that placing the weight forward need not have adverse effects on the neck at all. To explore the subtle differences between the two postures, and to decide which is most suitable for your own personal practice, we must look more closely at the structure of the neck.

The neck is different from the rest of the vertebral column in several ways: it permits extensive twisting as well as forward, backward, and side bending; it contains only synovial joints between the skull and C2 (figs. 4.8, 4.10, 4.13a, and 7.1–2); it ordinarily supports only the head; and its vertebral bodies and intervertebral disks are relatively small. These features might

Figure 8.7. The bregma headstand is characterized by having the hands a little more underneath the head than for the crown headstand, and this rotates the atlas (and with the atlas the rest of the body) on the cranium just enough to place the weight of the body about an inch in front of the crown at the bregma, the intersection point of the parietal and coronal sutures. This posture also pitches the pelvis forward, and one will have to arch the lumbar region more than in the case of the crown headstand to remain balanced.

lead an unbiased observer to caution against doing the headstand at all. And yet, we know that the posture is perfectly safe for those who are adequately prepared for it.

What is it, then, that allows the cervical spine of the well-prepared and average healthy person to bear the weight of the inverted body? Our first guess might be that the vertebral bodies and intervertebral disks support our weight by acting as a stack of building blocks. That point of view, from classical anatomy as it was understood until the mid-twentieth century, is now recognized as incomplete. The modern view is that in a healthy person without disk degeneration, it is the entire complex of vertebral bodies, intervertebral disks, vertebral arches, joints, muscles, and connective tissue restraints that is responsible for bearing the weight of the body. In that light, the relatively small size of the cervical vertebral bodies and intervertebral disks does not seem so critical, especially since the vertebral column as a whole in this region has considerable breadth (fig. 4.8).

To become aware of the total width of the cervical vertebrae, locate the lower rear corners of the mandible (the lower jaw) with your index fingers, find the mastoid processes just underneath the ears, and press deeply enough to find a bony point on each side between the mastoid process and the corner of the mandible. This is probably painful. The bony protuberance is the tip of the transverse process of C1 (chapters 4 and 7), and that's how wide the vertebral column is at that site. Watch yourself in a mirror as you locate both of these points, and you'll see that C1 is almost as wide as the neck (also refer to the drawing from the roentgenograms in fig. 4.8). Moving inferiorly, the transverse processes of C2 and below do not extend as far laterally as those for C1, but it's still impressive to feel how wide the vertebral column is in this region. The architecture of the cervical spine is more than capable of safely supporting the headstand.

As far as hyperextending the neck is concerned, remember from chapter 7 that we can extend the head almost 20° by rotating the skull on the atlas alone, and that we can do this without extending the spine between C1 and T1. In the bregma headstand, that 20° is more than enough to allow us to balance without additional bending in the cervical region (fig. 8.7). Unless someone is ignoring signals such as chronic pain and discomfort, worries about the neck's ability to withstand the stress of the headstand are usually misplaced.

THE CROWN VS. THE BREGMA HEADSTAND

The crown headstand is associated with a certain poise; the bregma headstand is associated with a certain zip. And the whole body seems to respond differently to the two postures. The natural response to the crown headstand is to hold the body straight, to keep the lower back flat, and to look straight ahead (fig. 8.6d). The sacroiliac joints will be in a neutral position between nutation and counternutation, or will be favoring counternutation. In the bregma headstand it is more natural to permit the lower back to relax and arch forward, allowing gravity to increase the lumbar lordosis (fig. 8.7) and to slip the sacroiliac joints into maximum nutation. The head and neck are slightly extended, and if you look straight ahead your eyes are directed toward the floor at a point several inches away from the head. You can also feel the extension of the head on the atlas that is so easily misinterpreted as extension of the neck between C2 and T1.

The bregma headstand has a more dynamic effect on your consciousness than the crown headstand. Its expression of energy seems to be related to the relaxed and arched lumbar region, which gives the bregma headstand the character of an inverted backbending posture. If you compare it to the crown headstand, which you can explore by moving your weight to the top of the head and flattening the back, the difference is obvious—the crown headstand is calm and poised; the bregma headstand is more dynamic.

Once you have learned both postures the bregma headstand is less tippy than the crown headstand. This is mainly because the fingers are braced under the back of the head and also because you do not have to maintain as much balance and tone in the postural muscles in the trunk to stay in the posture. You simply relax and allow moderate backbending to take place rather than keeping the lower back flattened with tension in the abdominal muscles. But be watchful. If you already have excellent flexibility for backbending, remaining passively in this posture for more than a minute or two can create lower back discomfort. The test: If you come out of the bregma headstand and have an urgent appetite for forward bends, it will probably be better for you to work mostly with the crown headstand.

THE UPPER EXTREMITIES

We know that the lower extremities form the foundation for standing postures: numerous muscles and ligaments attach the pelvic bones reliably to the sacrum and lumbar spine (figs. 3.4 and 3.7); the muscles of the hips, thighs, and legs flex and extend the hip joints, knees, and ankles; and the feet contact the earth. By contrast, the upper extremities are designed for touching, embracing, and handling tools. They are not foundations for any part of the body. Instead, it's the other way around: the torso is the foundation for the upper extremities, starting with the shoulder girdle.

THE SHOULDER GIRDLE

The *shoulder girdle* is formed from front to back by the sternum, the clavicles, and the scapulae (figs. 4.3–4). But unlike the *pelvic girdle*, whose pubic bones unite with one another at the pubic symphysis anteriorly and whose ilia unite with the sacrum posteriorly, the shoulder girdle is incomplete. The scapulae do not mate with or even come very close to one another posteriorly, and they have only tenuous and indirect connections to the sternum in the form of the small *acromio-clavicular joints* between each scapula and clavicle, along with the small *sterno-clavicular joints* between each clavicle and the sternum (chapter 4). Unlike the solid pelvic bowl and its appendicular-axial articulations that hold a tight rein on, and yet permit nutation and counternutation, the shoulder girdle is merely a framework. Even so, it still acts as a foundation for the arms, forearms, and hands; and for coming into the headstand that foundation must support the weight of the body. How can it do this? The scapula is the key.

THE SCAPULAE

The connection between the scapula and the torso is almost entirely muscular. This means that when we turn upside down for the headstand and expect the upper extremities to support that posture, we have to depend on muscular strength and flexibility rather than on robust bones and joints designed for bearing the weight of the body. It is thus not surprising to find that inflexibility, discomfort, and weakness in the shoulders, arms, and forearms prevent many people from getting very far in the headstand.

Envision two flat, triangular scapulae (shoulderblades) floating on the upper back. Each one provides a stable socket (figs. 1.13, 4.3–4, and 8.14) for the head of the humerus, a socket that is stabilized almost entirely through the agency of five muscles on each side of the body: the *trapezius*, the *rhomboid muscles*, the *levator scapulae*, the *pectoralis minor*, and the *serratus anterior*. Then envision flexor, extensor, abductor, adductor, and rotator muscles taking origin from the scapula and inserting on the humerus, much as comparable muscles take origin from the pelvis and insert on the femur. There's one big difference, however: while the pelvis is bound to the spine at the sacroiliac joints and forms a relatively immovable source from which muscles can move the thighs, the scapulae themselves participate in movements of the arms. Accordingly, its movements are critical for understanding all inverted and semi-inverted postures that are partially supported by the upper extremities.

To see how the scapula is held in place on the back of the chest wall, we'll work from the inside out, starting with the deepest muscles, two on the front side of the chest and three on the back. The serratus anterior muscle takes origin from a broad area on the front of the chest (fig. 8.11), runs laterally around the rib cage (fig. 8.9), passes underneath the scapula to insert on its medial border (fig. 8.12), and acts to abduct it, that is, to pull it laterally. This action of the serratus anterior is crucial for many postures, as in holding the scapula in place for completing the peacock (fig 3.23d), where its serrated edge often becomes sharply outlined, especially in bodybuilders. One more scapular supporting muscle, the pectoralis minor, also takes origin from the front side of the chest, but this one inserts on the acromion of the scapula (fig. 8.11), and from that position pulls it forward. The rhomboids and the levator scapulae stabilize the scapulae posteriorly. The rhomboid muscles adduct the scapula, pulling its medial border toward the midline from its origin on the thoracic spine (fig. 8.12), and the levator scapulae elevates the scapula, as its name implies, pulling on its upper border (fig. 8.12) from an origin on the transverse processes of C1–4.

The most superficial muscle that supports the scapula is the trapezius, so-named because the two trapezius muscles viewed together from the rear form a trapezoid (fig. 8.10). Each muscle is flat and triangular-shaped, takes origin medially from a line that runs from the skull to T12, and inserts on the clavicle and the spine of the scapula, the hard bony ridge you can feel on your upper back (fig. 8.10). Depending on which fibers of the muscle are active, the trapezius exerts traction to pull the scapula up, down, medially, or all three at the same time.

These five muscles not only stabilize the scapula, making a dependable foundation for movements of the arms, they move the scapula around on the surface of the back. The scapula can be moved laterally (the serratus anterior) and medially (the rhomboids); it can be elevated (the levator scapulae) or depressed (the lower fibers of the trapezius); and its pointed lower angle can be rotated out and upward (the trapezius and serratus anterior) or in and downward (the rhomboids, pectoralis minor, and levator scapulae). All of these movements are crucial to inverted postures and to the exercises that prepare us for them.

To check for yourself how the scapula works, ask someone who is slender, lightly muscled, and flexible to stand with their arms hanging alongside their thighs. Then trace the borders of both scapulae visually and by feel. The medial borders for each of these triangular bones are parallel to one another, an inch or so to either side of the midline, and the lateral borders angle up and laterally. The prominent bony landmark on top is the *scapular spine*. Next, while feeling the inferior angle of each scapula (its lowermost tip), ask your subject to slowly lift both hands overhead. Notice that as the arms are lifted each scapula rotates on an axis that runs roughly through the middle of the scapular spine, and from that axis you can feel that the inferior angle is carried in an arc out and up. This is called *upward rotation*. If this movement is constrained it will be harder to do the headstand, not to mention any other posture requiring an overhead stretch. Now ask your subject to lower both arms, spread the tops of the shoulders, and at the same time pull the inferior angles of the scapulae toward the midline of the body. This is the opposite movement—rotation of the inferior angle of the scapula down and medially, or *downward rotation*.

Other movements are self-explanatory. Still feeling the scapulae, ask your subject to do shoulder rotations. Watching carefully, notice that lifting the shoulders elevates the scapulae, pulling the shoulders downward depresses them, pulling the shoulders to the rear adducts them, and pulling the shoulders forward abducts them.

THE PECTORALIS MAJOR AND LATISSIMUS DORSI

Although most of the muscles that act on the arm take origin from the scapula and insert on the humerus, there are two major exceptions—the *pectoralis major* and the *latissimus dorsi*—both of which bypass the scapula on their way to insert on the humerus. The pectoralis major is the largest muscle that takes origin from the front of the chest (figs. 8.8–9). If you press your left hand against the side of your head with the arm angled out to the side, brace your left elbow with your right hand, and try to pull your left elbow forward and to the right, the pectoralis major tendon will tighten just above the axilla as it passes laterally to insert on the humerus. Its most powerful action is to pull the arm forward from behind, as when you try to do a push-up with the hands spread out laterally from the chest.

The latissimus dorsi takes a broad origin from the lower thoracic and lumbar spine, the sacrum, and the crest of the ilium (fig. 8.10), and from there it courses around the chest wall just lateral to the scapula (fig. 8.9), runs through the axilla, and inserts on the front of the humerus. It's unique—the only muscle in the body that connects the lower and upper extremities; its most powerful action is extension—pulling the arm down from above as in swimming, or lifting the body in a chin-up (chapter 1).

SEVEN MUSCLES THAT ACT ON THE ARM

Seven muscles take origin from the scapula and insert on the humerus. The middle segment of the *deltoid* (figs. 8.8–10) and the *supraspinatus* (figs. 1.1 and 8.12) abduct the arm, lifting it to the side. The *teres major*, acting synergistically with the latissimus dorsi, extends the arm, pulling it down and back (figs. 1.1, 8.12 and 8.14). The *coracobrachialis*, acting synergistically with the pectoralis major, is a flexor and acts to pull the arm forward (figs. 8.11 and 8.13). The *infraspinatus* and *teres minor* (figs. 1.1 and 8.12) pass to the rear of the head of the humerus and serve lateral rotation, and the *subscapularis* (figs. 1.13, 2.8, 8.11, and 8.13) passes from the underneath side of the scapula to the front of the head of the humerus. There the subscapularis serves as an agonist for accomplishing medial rotation of the arm (which also happens to be another powerful action of the latissimus dorsi that can be sensed—along with extension—in a swimming stroke).

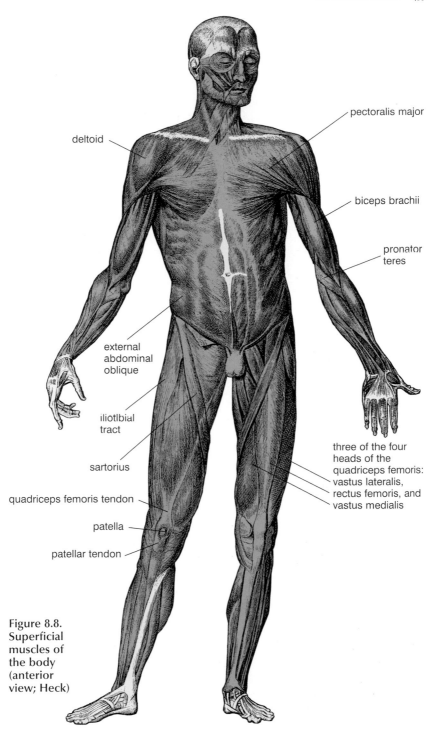

pectoralis major

deltoid

biceps brachii

pronator
teres

external
abdominal
oblique

iliotibial
tract

sartorius

quadriceps femoris tendon

patella

patellar tendon

three of the four
heads of the
quadriceps femoris:
vastus lateralis,
rectus femoris, and
vastus medialis

Figure 8.8.
Superficial
muscles of
the body
(anterior
view; Heck)

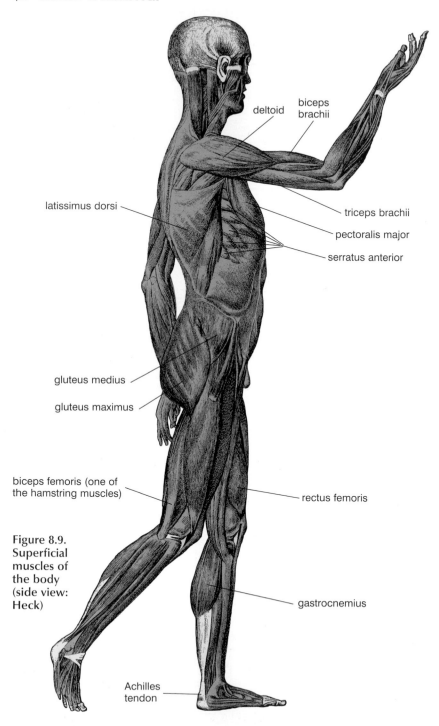

deltoid

biceps
brachii

latissimus dorsi

triceps brachii

pectoralis major

serratus anterior

gluteus medius

gluteus maximus

biceps femoris (one of
the hamstring muscles)

rectus femoris

Figure 8.9.
Superficial
muscles of
the body
(side view:
Heck)

gastrocnemius

Achilles
tendon

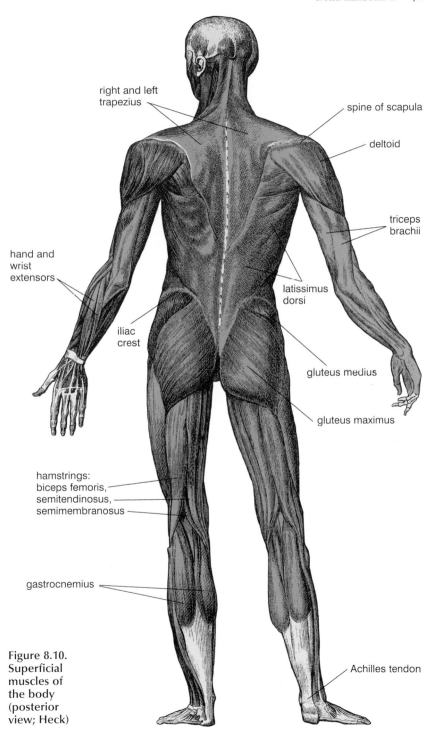

right and left trapezius

spine of scapula

deltoid

triceps brachii

hand and wrist extensors

latissimus dorsi

iliac crest

gluteus medius

gluteus maximus

hamstrings:
biceps femoris,
semitendinosus,
semimembranosus

gastrocnemius

Figure 8.10.
Superficial
muscles of
the body
(posterior
view; Heck)

Achilles tendon

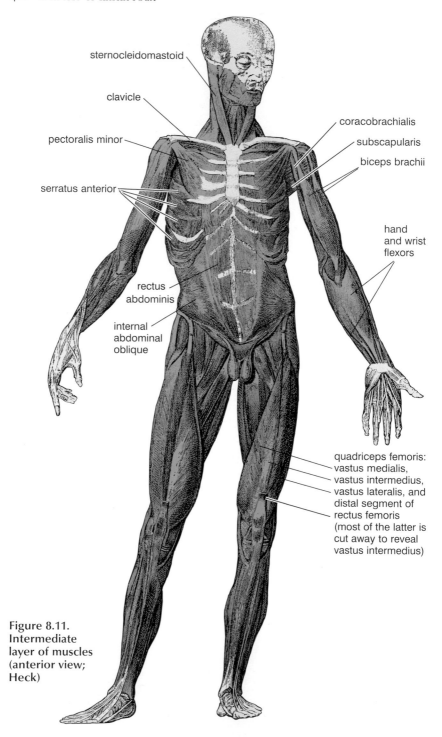

sternocleidomastoid

clavicle

pectoralis minor

serratus anterior

coracobrachialis

subscapularis

biceps brachii

hand
and wrist
flexors

rectus
abdominis

internal
abdominal
oblique

quadriceps femoris:
vastus medialis,
vastus intermedius,
vastus lateralis, and
distal segment of
rectus femoris
(most of the latter is
cut away to reveal
vastus intermedius)

Figure 8.11.
Intermediate
layer of muscles
(anterior view;
Heck)

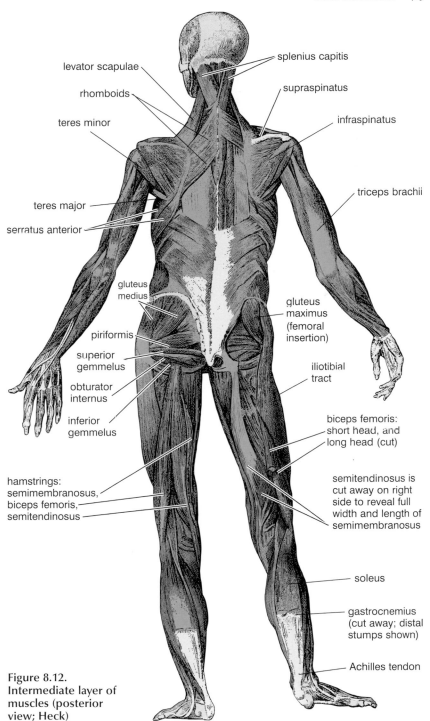

levator scapulae

rhomboids

teres minor

teres major

serratus anterior

gluteus
medius

piriformis

superior
gemmelus

obturator
internus

inferior
gemmelus

hamstrings:
semimembranosus,
biceps femoris,
semitendinosus

splenius capitis

supraspinatus

infraspinatus

triceps brachii

gluteus
maximus
(femoral
insertion)

iliotibial
tract

biceps femoris:
short head, and
long head (cut)

semitendinosus is
cut away on right
side to reveal full
width and length of
semimembranosus

soleus

gastrocnemius
(cut away; distal
stumps shown)

Achilles tendon

Figure 8.12.
Intermediate layer of
muscles (posterior
view; Heck)

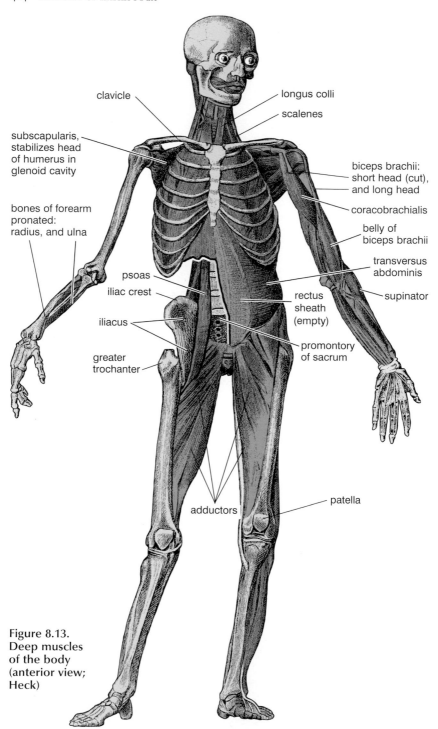

clavicle

longus colli

scalenes

subscapularis, stabilizes head of humerus in glenoid cavity

biceps brachii: short head (cut), and long head

coracobrachialis

belly of biceps brachii

bones of forearm pronated: radius, and ulna

transversus abdominis

supinator

psoas

iliac crest

rectus sheath (empty)

iliacus

promontory of sacrum

greater trochanter

adductors

patella

Figure 8.13. Deep muscles of the body (anterior view; Heck)

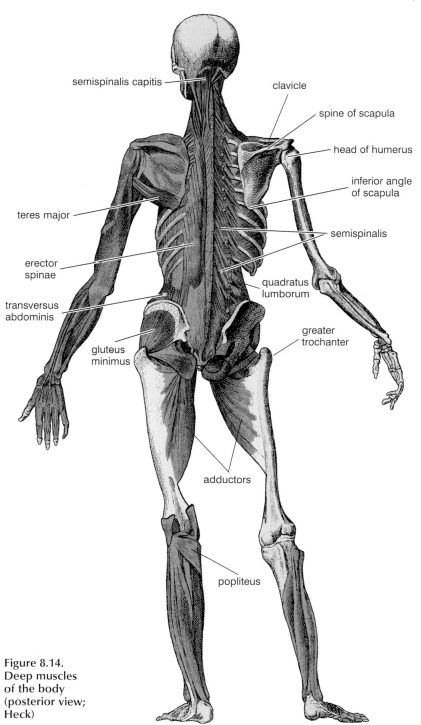

semispinalis capitis

clavicle

spine of scapula

head of humerus

inferior angle
of scapula

teres major

semispinalis

erector
spinae

quadratus
lumborum

transversus
abdominis

greater
trochanter

gluteus
minimus

adductors

popliteus

Figure 8.14.
Deep muscles
of the body
(posterior view;
Heck)

Four of these seven muscles—the teres minor, the supraspinatus, the infraspinatus (figs. 1.1, 1.13, and 8.12), and the subscapularis (figs. 1.13, 2.8, and 8.13)—form the well-known and important *rotator cuff*, which stabilizes the head of the humerus in the glenoid cavity (figs. 1.13 and 8.13). Without these, the action of other powerful muscles such as the pectoralis major and latissimus dorsi would quickly dislocate the shoulder. The rotator cuff muscles, in addition to acting from the scapula to the humerus, can also act in the opposite direction—from the arm to the scapula—assisting the trapezius, rhomboids, levator scapulae, pectoralis minor, and serratus anterior in stabilizing the scapula from one of the many fixed arm positions in the headstand and other inverted postures.

THE FOREARM, WRIST, AND HAND

The muscles that act throughout the rest of the upper extremity are easy to envision. The biceps brachii originates from the front of the arm and inserts on the forearm (figs. 1.1, 8.8–9, 8.11, and 8.13); its action is to flex the elbow concentrically, or resist its extension eccentrically. It is this muscle that powerfully resists extension of the forearm as you come forward into the peacock posture (fig. 3.23d). The triceps brachii originates from the back of the arm and inserts on the olecranon (figs. 1.1, 8.9–10, and 8.12); it acts to extend the elbow and resist its flexion. In the headstand it limits forearm flexion, and in the scorpion it contracts isometrically, keeping the elbows from collapsing (fig. 8.31). Other muscles in the forearm supinate (fig. 8.13) and pronate (fig. 8.8) the forearm. Movements of the wrists, hands, and fingers are accomplished by flexors on the anterior side of the forearm and hand (fig. 8.11), and by extensors on the posterior side of the forearm and hand (fig. 8.10). The flexors are activated any time you make a fist or plant your hands on the floor for postures such as the down-facing dog (figs. 6.17 and 8.26), the upward-facing dog (figs. 5.13–14), or the plank posture (fig. 6.16).

THE MOVEMENTS OF THE ARM

The movements of the arms at the shoulder joint are more complicated than the movements of the thigh at the hip joint because the range of possible movements is greater, and also because the separation of the shoulders by the width of the rib cage allows the arms to be pulled across the chest in a manner that has no counterpart in the lower extremities.

In their simplest form flexion and extension of the arms are movements in a sagittal (front-to-back) plane of the body, abduction and adduction are movements of the arms in a frontal (side-to-side) plane of the body, and medial and lateral rotation of the arms are movements of axial rotation.

These can all be superimposed onto one another: you can flex, adduct, and rotate the arm all at the same time. And because the scapulae are also involved, the movements can best be understood by checking out the accompanying shifts of the scapulae on a partner.

FLEXION AND EXTENSION

First considering *flexion*, if you start from the anatomical position (fig 4.2) with the hands alongside the thighs and then lift your arms up and forward until they are straight out in front of you, you will be flexing them 90° (fig. 8.16), and you can continue this movement (flexion) up through an arc of 180° overhead, stopping anywhere along the way. Next considering *extension*, if you pull the arms straight to the rear from the anatomical position, you will be extending them. This movement does not occur in isolation, however; it also requires adduction of the scapulae (pulling them toward one another medially). Most people can extend their arms in a sagittal plane about 45° to the rear from a neutral position alongside the chest (fig. 8.15; figure also shows adduction superimposed on extension).

ABDUCTION AND ADDUCTION

For *abduction*, first envision a frontal plane running through the ears, shoulders, chest, and lower extremities. Moving the arms from the anatomical position within such a plane, they will have been abducted 90° if you lift them straight out to the sides (fig. 8.17). Then if you continue to lift them until they are straight overhead, they will have been abducted 180°, in the same final position, incidentally, as when they are in 180° of flexion. The 180° of abduction and/or flexion, strictly speaking, always includes 60° of upward rotation of the scapula, which we considered earlier.

Depending on your starting position, *adduction* is more complicated than flexion or extension. In the simplest situation, if you start with the arms straight out to the sides and then drop them down to a neutral position alongside the chest, you will be adducting them 90°. And if you start from 180° overhead, as in the final position for the tree (chapter 4), adduction will first swing the arms away from overhead to the spread-eagled 90° position (straight out to the sides) before coming back to the fully adducted position alongside the chest.

Adduction can also be superimposed on other movements. You can start with the arms flexed forward 90° (straight out in front) and then adduct them across the chest past one another, bringing the elbows together (fig. 8.18). You can also start with the arms abducted 90° and then adduct them, not only straight back down into the anatomical position, just described, but toward one another behind your back, at least minimally (fig. 8.19).

Figure 8.15. Arms extended 45° and then adducted.

Figure 8.16. Arms flexed 90°.

Figure 8.17. Arms abducted 90°.

Adduction of the arms to the front from a flexed position also rounds the shoulders, which includes *abduction* of the scapulae (fig. 8.18); adduction of the arms to the rear from an abducted (fig. 8.19) or extended (fig. 8.15) position includes pulling the shoulders to the rear, which in turn includes *adduction* of the scapulae.

CIRCUMDUCTION

You can combine flexion and extension of the arm with abduction and adduction to yield the sequential movement called *circumduction* (see chapter 6 for circumduction of the thigh). For circumduction of the arm, flex it forward 90° while adducting it toward the midline, lift it overhead 180°, pull it around to the rear in an extended and adducted position, and then return it to a neutral position alongside the chest. Feel how circumduction of the arm affects the scapula: in the above sequence, circumduction of the arm first abducts the scapula as a result of flexing the arm forward and pulling it across the chest, then it elevates the scapula by lifting the arm overhead, adducts the scapula by pulling the arm to the rear, and depresses the scapula by bringing the arm back alongside the chest.

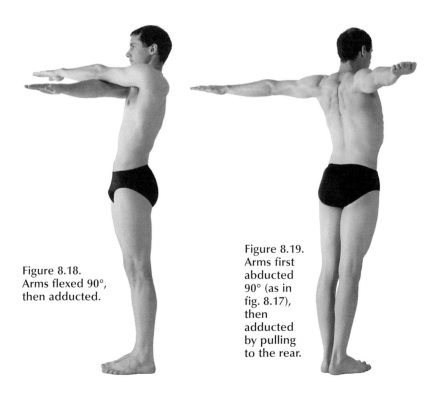

Figure 8.18.
Arms flexed 90°,
then adducted.

Figure 8.19.
Arms first
abducted
90° (as in
fig. 8.17),
then
adducted
by pulling
to the rear.

MEDIAL AND LATERAL ROTATION

Medial and lateral rotation of the arms in the shoulder joint is comparable to medial and lateral rotation of the thighs; they are movements of axial rotation around an imaginary line through the center of the humerus. If you stand in the anatomical position with the elbows extended and rotate the arms so that the palms face to the sides as much as possible, you will be rotating the arms laterally about 30°. Medial rotation of 60° is also possible but harder to isolate because it is easily confused with pronation of the forearms.

Medial and lateral rotation of the arms can be carried out in any position in combination with flexion, extension, abduction, or adduction. For example, let's say you abduct the arms 90° straight out to the side. Next, to avoid confusing arm rotation with supination and pronation of the forearms, flex the elbows 90°, pointing your hands straight to the front. Now swing the hands down through an arc of 30°: this motion has just medially rotated the arms that amount. Or, swing the hands up, with the fingers pointed toward the ceiling (the "get your hands up" gesture in a grade B Western). This motion has just laterally rotated the arms 90°. Notice that the scapulae participate extensively in all of these movements.

COMING INTO THE HEADSTAND

Now we can describe the movements of the upper extremities as we come into the headstand. The forearms remain flexed throughout, but the position of the arms varies: they are flexed about 90° in the starting position, about 165° in stage one (fig. 8.6a), and about 135–150° in stages two, three, and four (figs. 8.6b–d). These figures reflect the need to balance the weight of the lower extremities by adjusting the angle of the torso to the floor. At one extreme (stage one) the torso is pitched backward maximally, and at the other extreme (stage three) the position of the feet hanging to the rear requires the torso to be pitched slightly forward. When you are balanced in a straight line from head to toe in stage four and the arms are no longer supporting much weight, they are flexed about 150°.

As expected, flexion of the arms in the headstand is accompanied by shifts in the positions of the scapulae. If someone both watches and feels your back as you walk your feet forward toward stage one of the headstand (that is, as your arms go from 90° of flexion to 165° of flexion), they will notice that the scapulae become abducted, depressed, and rotated upward. Then, as you initiate the effort to move from stage one toward stage two, they will notice a gathering of strength in all the muscles that attach to the scapulae in the form of a rippling effect that accompanies the act of lifting the lower extremities off the floor toward stage two. Most of the weight of the body is now being supported by the scapulae and their surrounding muscles. The effort subsides slightly in stage two but does not diminish markedly until stage three. Finally, in stage four you are balancing with the least effort. If you are an experienced student, only slight adjustments and moderate isometric tension in the muscles are needed to maintain the final posture. If you start to tip backward, the tension eases, and if you start to tip forward, placing more weight on the forearms, more isometric tension develops.

Beginning students should be watched carefully as they are learning the headstand because they tend to allow the scapulae to become elevated and adducted rather than depressed and abducted. And they often have trouble achieving enough upward rotation of the scapulae, which is essential for the abducted arm position in the headstand. Any of these errors or deficiencies produce an unattractive, inelegant posture. Instructors make corrections by saying "support the posture with conscious tension in the shoulders," or "lift the body away from the floor with the shoulders." But once the scapulae are stabilized in their final position, the isometric tension can be eased.

STRUCTURAL IMBALANCES

When yoga instructors talk about structural misalignments of the body, they usually mean side-to-side imbalances—distortions of our bilateral symmetry. And for this reason they often suggest that students either practice in front of a mirror to search out right-left discrepancies, or feel experientially if they can bend or twist to one side more easily than the other. Only when teachers make such comments as, "Square your shoulders, stand up straight, pull your head more to the rear, tuck the pelvis, or don't tuck the pelvis," are they referring to front-to-back imbalances. You can't see those yourself except with a set of mirrors arranged to allow you to watch your posture from the side. Until now almost all of our focus has been on side-to-side imbalances, but we must be concerned with both possibilities when we consider the headstand.

SIDE-TO-SIDE IMBALANCES

The headstand is a balancing pose, and as such, it is not designed to correct side-to-side imbalances. That is best accomplished by postures in which you use a whole-body muscular effort. Watch yourself in a mirror while you are doing the headstand. If your head is at an angle, if one hand is covering your ear on one side, or if you see plainly that your body is tilting to one side, you have side-to-side imbalances and should forget about the headstand until these have been corrected by other postures. The headstand will only make them worse, and a close look at the muscles of the neck will make the reason for this plain.

THE SUBOCCIPITAL MUSCLES

Working from the inside out, the deepest muscles of the neck and shoulders are four pairs of *suboccipital muscles* (fig. 8.20): the *rectus capitis posterior major* runs from the spinous process of the axis (C2) somewhat laterally up to the skull; the *obliquus capitis superior* runs from the skull to the transverse process of the atlas (C1); the *obliquus capitis inferior*, which completes the *suboccipital triangle*, runs from the transverse process of C1 to the spinus process of C2; and the *rectus capitis posterior minor* runs straight up from the spinous process of the axis to the skull.

These muscles are responsible for the small movements at the top of the neck which we explored in chapter 7. For example, tipping the head to the right 5°, rotating to the right another 5°, and then extending the head back 5° are all accomplished between the cranium and C2 by concentric shortening of the suboccipital muscles on the right side and by eccentric

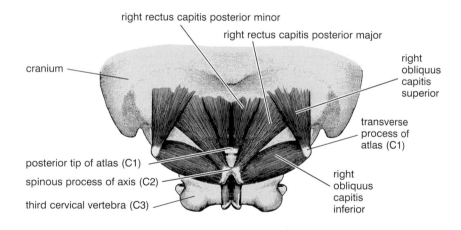

Figure 8.20. A deep dissection of the back of the upper cervical region showing the suboccipital muscles. The trapezius and large strap muscles of the neck have all been removed. Starting with an awareness of the movements possible at the joints between the cranium and the atlas, and between the atlas and the axis (chapter 7), the actions of the suboccipital muscles can be inferred by their anatomical arrangements. For example, the two rectus capitis posterior major muscles acting in unity with the two obliquus capitis superior muscles can only rotate the cranium backward on the atlas. Another example is that the right obliquus capitis inferior acting alone (but probably resisted by the left obliquus capitis inferior) can only rotate the atlas (and with the atlas the entire head) on the axis to the right, making use of the dens as a pivot joint. A last and more complex example is that the right rectus capitis posterior major acting across two joints between the cranium and the spinous process of the axis will combine two actions: it will rotate the head to the right (which is a result of the atlas and therefore the head as a whole rotating around the axis), and at the same time it will lift the chin and drop the back of the head to the rear (which is a result of the cranium as a whole rotating on the atlas). (from Sappey).

lengthening of those on the left. In the headstand these muscles all become isometrically active as extensile ligaments, and if you come into the headstand with the head always tipped off axis in the same direction, even a little, the suboccipital muscles will dutifully hold you in that position and accentuate the imbalance.

THE STRAP MUSCLES OF THE NECK

Superficial to the suboccipital muscles are the larger strap muscles of the neck (figs. 4.14 and 5.5). Three of the big ones are the *semispinalis capitis* (fig. 5.5), which runs from the transverse processes of C7–T6 straight up to the skull, the *splenius capitis* (fig. 5.5), which runs from the spinous processes of C7–T6 to the mastoid process (the bony protuberance just below the earlobe), and the *sternocleidomastoid muscle* (fig. 8.11), which runs from the sternum and clavicle up to the mastoid process. The first two of these are visible posteriorly, and the last is visible anteriorly.

The strap muscles of the neck also act as extensile ligaments in the headstand and react like the suboccipital muscles to harden right-left imbalances and make them even more deeply ingrained. One way to know if this is happening is if you feel soreness or tension more on one side than the other after you come out of the posture. Balanced, the paired muscles act equally for maintaining the pose, and if they get sore they get equally sore on both sides. The paired sternocleidomastoid muscles, for example, act together in an upright posture to pull the head forward. These muscles resist extension in the bregma headstand and get stronger if you practice that posture regularly. But if they get imbalanced, unilateral stress becomes apparent, and the headstand will only make the situation worse.

FRONT-TO-BACK IMBALANCES

Front-to-back imbalances are another matter, and making a conscious choice to correct them with one or other of the two headstands makes a lot of sense. If the back is relatively flat, with a lumbar lordosis that barely arches forward in the upright posture, you can gradually correct this by relaxing in the bregma headstand every day as a part of a balanced practice. Over time the lumbar region will gradually increase its arch. The opposite problem is swayback, in which the lower back is overly arched to begin with, and under those circumstances the bregma headstand is contraindicated and the crown headstand will be more useful.

BREATHING ISSUES

For those who can do it easily, the headstand is the best possible training posture for abdominal (abdomino-diaphragmatic) and diaphragmatic (thoraco-diaphragmatic) breathing (chapter 2). It invokes the most complete use of the diaphragm of any posture, and it does so automatically because the headstand both encourages abdominal breathing and restricts chest breathing.

THORACIC AND PARADOXICAL BREATHING

We saw in chapter 2 that the first requirement for learning abdominal and diaphragmatic breathing is to discard the habits of chronic chest and paradoxical breathing, and the headstand makes this possible. If you are familiar with the four methods of breathing in upright postures, and if you practice the headstand regularly, come up in the posture and try to breathe thoracically and then paradoxically. It's not so easy, because the rib cage is held in place by isometric contraction of muscles throughout the torso. Even the intercostal muscles participate in stabilizing the headstand isometrically, and it is difficult for them to do that and at the same time mobilize the rib cage for chest breathing. What is more, four of the five muscles that hold the scapula in place (excepting only the levator scapulae) have their origins on the chest and tend to hold it in a fixed position. And finally, the pectoralis major stabilizes the chest from the front, and the latissimus dorsi envelops it from behind. Taken together, these restrictions bind the rib cage so firmly that lifting it toward the floor for chest and paradoxical breathing would be unthinkable.

ABDOMINAL AND DIAPHRAGMATIC BREATHING

If the headstand prevents thoracic and paradoxical breathing, it necessitates abdominal and diaphragmatic breathing by default. Abdominal breathing should be the initial choice for novices. At the end of exhalation the abdominal organs are pushed superiorly (toward the floor) by the force of gravity, and this lengthens the muscle fibers of the diaphragm nearly to their working maximum. This means that the inhalations that follow will be deep and satisfying. The diaphragm will not only be drawing air into the lungs, it will be pressing the abdominal organs to a more inferior position in the trunk (toward the ceiling) from fixed origins on the base of the rib cage and the lumbar lordosis. And even though it is working against the force of gravity, this is the easiest way you can inhale. It's also extra exercise, it strengthens the diaphragm, and it creates the purest form of abdomino-diaphragmatic breathing.

What about diaphragmatic (thoraco-diaphragmatic) breathing in the headstand? This happens if you take deep inhalations: if you take as few as three breaths per minute, which is easy in the headstand as well as rewarding for anyone who has good respiratory health, you will feel the diaphragm flare the base of the chest during the last half or third of inhalation, which is the defining characteristic of diaphragmatic breathing (chapter 2). This is feasible because the muscles that suppress expansion of the rib cage from above in thoracic and paradoxical breathing do not provide nearly as many restrictions to its expansion from below.

When you exhale in inverted postures, the muscle fibers of the diaphragm don't just relax and allow its dome to move passively as usually happens when you exhale in upright postures (chapter 2). Upside down, the diaphragm stays in a state of eccentric contraction throughout exhalation to restrain the abdominal organs from a free fall toward the head.

You can see this for yourself if you come into the headstand (or shoulder-stand) and observe your cycle of breathing. Just focus on breathing evenly and naturally. Then at the end of a normal inhalation, relax suddenly. What you notice on your own will also be apparent if you ask a room full of students to try the same experiment. You, and most of them, will exhale with a sudden whooshing sound as the diaphragm relaxes, the organs drop toward the floor, and the lungs deflate. This does not happen in normal upside-down exhalations because the more fundamental impulse in the headstand, at least in a yoga practice, is to restrain exhalation.

THE FUNCTIONAL RESIDUAL CAPACITY

Because the muscle fibers of the diaphragm are lengthened to their working maximum during an inverted exhalation, the functional residual capacity of the lungs (chapter 2) will be substantially reduced. You can test this if you first sit upright and breathe in and out the tidal volume associated with normal relaxed breathing. Then, at the end of exhalation breathe out as much as you can (your expiratory reserve volume). In chapter 2 we estimated this to be around about 1,000 ml, or about two pints.

To continue the experiment, come into the headstand (or shoulderstand) and breathe normally for a minute or so to establish equilibrium. Then, at the end of a normal exhalation try breathing out as much as you can. You will see instantly that you cannot breathe out nearly as much as you could in the upright position. Your tidal inhalations and exhalations have shifted much closer to your residual volume, closer even than we saw in the corpse posture (chapter 2). Let's say for illustration that your tidal volume when you are upside down is a standard 500 ml, and that your expiratory reserve volume decreases to 200 ml (instead of 1,000 ml in an upright posture and 500 ml in the corpse posture). Your resulting functional residual capacity will be 1,400 ml instead of the 2,000 ml in an upright posture and the 1,700 ml in the corpse posture (fig. 8.21). What all this means is that a constant alveolar ventilation of 4,200 ml/minute will be more efficient in transferring oxygen and carbon dioxide to and from the blood in the headstand than in an upright posture or in the corpse posture. Inverted, you will either transfer gases more efficiently, thus increasing blood oxygen and decreasing blood carbon dioxide, or you will slow down the rate and or the depth of your breathing to keep your blood gases within a normal range (fig. 8.21).

The richness associated with how one tends to breathe in the headstand probably accounts at least partially for why the posture is praised so fulsomely by experienced teachers in the literature of hatha yoga. In the headstand the diaphragm is in a state of contraction during both inhalation and exhalation; it acts from a mildly stretched position at the end of deep exhalations; it operates as a piston smoothly and independently within the chest wall; and it is exercised more than usual because it has to push the abdominal organs toward the ceiling in addition to drawing air into the lungs. Even students with the worst breathing habits in upright postures will have to use their diaphragm for respiration in the headstand.

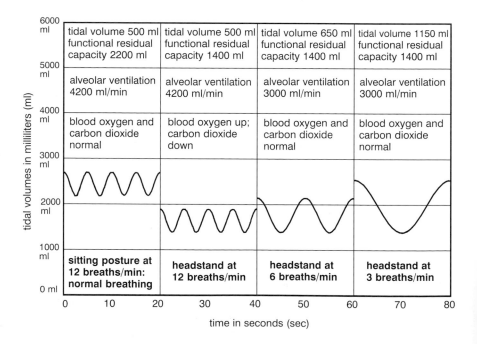

Figure 8.21. Simulated ventilation, sitting upright (far left in above figure, and repeated from first panel in fig. 2.14), and three possible modes of breathing in the headstand.

DEVELOPING STRENGTH AND FLEXIBILITY

Many people practice and even teach hatha yoga for years without being able to do the headstand themselves. Women generally face two challenges: less upper body strength than men and a greater proportion of their weight in their hips and thighs. To manage and balance this weight they will have to develop more strength in their upper extremities, back, and abdomen. For men the commonest challenge is poor hip flexibility, which makes it difficult to lift up into the posture in stages. So men can try to develop more hip flexibility, and both men and women can offset their respective limitations with more strength in the shoulders and torso.

HIP FLEXIBILITY

We can analyze the problem of hip flexibility by looking at two extremes, first at how difficult it would be to come up in the headstand if you had no hip flexibility at all. If the thighs and pelvis were in a cast that held them in the same plane so that you could flex only the spine, knees, and ankles, the only way you could get up in a headstand would be to place your head on the floor, bend as much as possible in the vertebral column, and, with a stupendous effort from your ankle and knee extensors, throw yourself up into the air. With enough practice—probably after thumping over onto your back several hundred times—you might be able to do it.

To envision the other extreme, think how easy it would be to come into the headstand if you had 180° of hip flexibility with the knees extended, and if the length of your combined torso, head, and neck were exactly equal to the lengths of your thighs, legs, and feet. You could plant your head on the floor and walk yourself into a folded head-foot stand with your toes on the floor near the forehead. Then you would only have to tiptoe enough further forward to balance on your head and come up into stages two, three, and four of the headstand—all with minimal abdominal and back strength, and with minimal help from the upper extremities.

Between the two extremes your work is cut out for you. Any posture that develops strength in the abdomen, back, and upper extremities, and any posture that improves hip flexibility will bring you closer to a successful headstand. Dozens of postures are helpful, some for strength, some for flexibility. Backbending and prone boats, forward bending and sitting boats, standing twists and bends, standing triangles and lunges, leglifting, hip-opening exercises, and sitting spinal twists are all helpful. The peacock and wheel will do wonders. Certain shoulderstand variations (chapter 9) will also be helpful, such as coming slowly into and out of the bridge from the shoulderstand, and coming slowly into and out of the plow.

SELECTED STANDING POSTURES

If you have come to an impasse as far as progress toward the headstand is concerned, a good place to begin is with two of the arm positions in the series of standing twists and bends outlined in chapter 7 (figs. 7.17–18). Both the position with the forearms interlocked behind the head (fig. 7.18b) and the cow-face hand position (fig. 7.18e) are excellent exercises for developing mobility of the shoulders and arms and for preparing you to place them confidently when you negotiate the four stages of the headstand. These postures, as well as the rest of that series, also develop much needed hip flexibility, as well as back and abdominal strength.

For right-left balance, repeat both arm positions alternating the way the right and left upper extremities are placed. This is obviously crucial for the cow-face position. If you can't reach far enough to interlock the fingers in that posture, hang a washcloth or hand towel from the hand reaching down from above, and grab it from below with the other hand. Last, if the exercise is easier on one side than the other, do it repeatedly as a three-part series— difficult side, easy side, and difficult side again. Don't be obsessed or impatient, however, because if you traumatize the elbow joint on the tight side with repetitive stress, it could set you back for a year or more.

THE CROW

Now we turn to a series of postures that build strength in the upper exremities, starting with the crow, a moderately inverted balancing posture. Start with the hands on the floor about eight inches apart, the wrists extended 90°, and the elbows locked. Most of the weight of the body is on the balls of

a.   **b.**

Figure 8.22. Crow posture (b) and starting position for the crow (a). The crow is a moderately inverted balancing posture that will develop strength in the upper extremities, get you accustomed to inversion of the torso, and give you confidence to begin work with the headstand.

the feet; the ankles, knees, and hips are all flexed; and the thighs are moderately abducted with the lateral sides of the elbows against the medial sides of the knees (fig. 8.22a). To come into the posture, lift the hips, bend the elbows, and take your weight forward. As you do that the knees will remain flexed, the feet will be lifted off the floor, and you will end up balancing in a plane that passes through the hands, the mid-section of the arms, and the upper parts of the legs just below the knees (fig. 8.22b).

Like the headstand, the crow requires courage, flexibility, and strength—courage to risk falling on your nose, enough hip flexibility to bring the thighs alongside the chest, and enough upper body strength to support yourself entirely with the upper extremities. To do it you have to have good strength in the triceps brachii, the five muscles that stabilize the scapula, the seven muscles that stabilize the posture between the arms and the scapula, and the pectoralis major and latissimus dorsi. As with the headstand, the chest is immobilized so much that you can only breathe abdominally.

THE STICK POSTURES AND THE TWO-HANDED COBRA

It is plain that good hip flexibility and upper body strength are needed for the headstand, but this posture also requires all-around strength in the torso. More specifically, going from stage one to stage two, as well as remaining in stage two for more than a moment, requires superb back strength. But back strength in isolation is not enough. To keep excess tension off the intervertebral disks we have to maintain intra-abdominal pressure, and this means that back strength must always be matched by strength in the abdominal muscles and in the respiratory and pelvic diaphragms—if it ever happens that you have a sore abdominal wall, you will find that you have little zest for the headstand. Numerous postures and exercises for developing abdominopelvic strength were outlined in the first half of chapter 3, but here are three more: the stick pose, the two-handed cobra, and the celibate's pose.

Figure 8.23. The stick pose is an isometric whole-body exercise without movement, with special emphasis on the upper and lower extremities, sacroiliac nutation, 90° of hip flexibility, and scapulae that are adducted and depressed. If your arms are not long enough for the heels of your hands to reach the floor even with the scapulae depressed, you can use a thin block under your hands for a prop.

To do the stick pose (fig. 8.23), sit on the floor with the thighs flexed 90°, the knees, ankles, and toes extended, and the feet together. Place the hands alongside the hips with the wrists extended 90° and the elbows locked. Pull the shoulders to the rear by adducting and depressing the scapulae, thus pressing the heels of the hands against the floor. Arch the lumbar region forward, establish your limits of nutation for the sacroiliac joints, and lift and thrust the chest forward as much as possible, immobilizing it in that position with all the muscles of the upper extremity. This is another pose in which you can only breathe abdominally, but even that is a challenge because now the abdominal wall is taut. The stick pose seems simple, but settling into it properly requires isometric contraction of most of the muscles of the upper extremity, a full 90° of hip flexibility, and intense concentration— all skills that are helpful for the headstand.

The two-handed cobra is a natural extension of the stick posture in that it requires even more strength in the abdomen as well as excellent hip flexibility and long hamstrings. The simplest starting position is to squat with the feet about eighteen inches apart, placing the palms on the floor between the thighs and as far to the rear as possible (fig. 8.24a). Keeping the elbows extended, lean the upper body forward by bending at the hips enough to support your weight on your arms while you lift the feet (fig. 8.24b). This will require abducting the thighs beyond their starting position of moderate abduction, and it will require extending the knees. This is a difficult combination because as you start to extend the knees you will

a.   **b.**

Figure 8.24. The two-handed cobra (b) and its starting position (a) are excellent preparations for the headstand because they require a combination of balance, courage, hip flexibility, and upper body strength. The key requirement for completing the posture is leaning forward as you extend the knees, and the key to leaning forward under these circumstances is good hip flexibility. Although this posture looks simple, it's a big surprise to many people who think of themselves as accomplished all-around athletes. In general, the moment they try to lift their feet they fall onto their backsides.

have to lean further forward to keep from falling to the rear, which in turn brings the hamstrings under even more tension and makes hip flexion even more problematical.

The two-handed cobra, like the headstand, requires a practical working combination of hip flexibility and upper body strength. If your hip flexibility is minimal you will not be able to lean forward enough to keep your balance, and the strongest person in the world cannot compensate for that deficit. If on the other hand you have the ability to flex your hips 120°, the posture is not much more difficult than the crow. If you are somewhere in between, your upper body strength makes all the difference in your ability to complete the posture.

The lifted stick, or the celibate's pose, is the most challenging of these three postures. Starting in the stick position, depress the scapulae enough to lift all of your weight off the floor. What first happens is that you can lift the buttocks easily but you can't even begin to lift the heels. You will have to do several things at the same time: lean slightly forward while keeping your back ramrod straight and bending perhaps 110° at the hips; keep the knees fully extended while lifting the thighs, legs, and feet with the iliacus, psoas, and quadriceps femoris muscles; and place your hands as far forward as necessary for supporting all of your body weight (fig. 8.25).

The celibate's pose is difficult for at least five reasons: you have to have excellent hip flexibility because the extended knees keep the hamstrings stretched to their limits; you have to have exceptionally strong hip flexors to lift the extended legs and feet from a pelvis that is floating in mid-air; you have to have a strong back to keep yourself sitting upright, again in mid-air; you have to have strong respiratory and pelvic diaphragms as well as strong abdominal muscles to support the effort with the back muscles; and you have to have excellent strength in the upper extremities to hold yourself in the posture. If you can do all of that, you will certainly be able to do the headstand.

Figure 8.25. The lifted stick, or celibate's pose, challenges your abdominal muscles, as well as your pelvic and respiratory diaphragms, like no other posture. Like many other poses, this one is impossible unless you are capable of at least 90° of hip flexibility.

THE DOWN-FACING DOG AND VARIATIONS

After the intense abdominal work with the two-handed cobra and the lifted stick postures, everyone will be ready to do something that provides a modicum of relief, and one of the best postures for this is the down-facing dog (see also chapter 6). Along with the crow, this pose is another excellent semi-inverted training posture for the headstand. The arms end up flexed 180° overhead, or even a little more, and this movement is accompanied by about 60° of upward rotation of the scapulae. And since the scapulae are supporting much of the weight of the upraised body, the five muscular attachments between it and torso must all be actively engaged. Allowing them to relax will cause adduction and elevation of the scapulae, as well as an unsightly jamming of the shoulders toward the floor. To counteract this tendency, instructors usually urge students to press the hands strongly against the floor, lift the buttocks, flatten the back, and press the shoulders toward the feet (fig. 8.26). They may not recognize it, but those adjustments also abduct and depress the scapulae as well as hold them isometrically in their upwardly rotated positions, and this happens to be exactly what is needed for coming up into the headstand. To compromise the posture, bend the knees and lift the heels rather than relax the shoulder muscles (fig. 8.27).

Textbooks usually discuss movements of the arm in reference to muscles that insert on the humerus, but in the down-facing dog the arm is relatively fixed and acts as an origin instead of an insertion. The teres major muscle (figs. 1.1, 8.12, and 8.14), for example, is ordinarily listed as an arm extensor, taking origin from the lateral border of the scapula and inserting on the humerus. But in the down-facing dog the teres major acts to abduct the scapula laterally and rotate it upward from a stabilized arm instead of acting as an arm extensor from a fixed scapula.

Figure 8.26. The down-facing dog posture, in its ideal form with 45° of ankle flexion and 110° of hip flexion (see fig. 6.17 for an intermediate-level pose). The isometric contraction of the muscles needed for stabilizing the scapulae in their correct positions comprise excellent training for the headstand.

The rotator cuff muscles are also important in the down-facing dog. Again, with their origins and insertions reversed, they abduct the scapulae from fixed arm positions while taking nothing away from their classic role in stabilizing the head of the humerus in the glenoid cavity. The subscapularis muscle (figs. 8.11 and 8.13) has a slightly different action from the other three rotator cuff muscles. Since it attaches to the front of the humerus rather than to its posterior side, it also acts to pull the scapula anteriorly as well as abducting it laterally, which assists in keeping it flat against the chest wall.

In the down-facing dog one other muscle, the serratus anterior, acts to pull the medial border of the scapula laterally—not from the arm, but from the front of the chest—and this action is especially helpful because it slides the scapula directly against the chest wall rather than pulling it to the side. And since the serratus anterior attaches near the inferior angle of the scapula (figs. 8.9 and 8.11), it is positioned to powerfully assist upward rotation by pulling the inferior angle of the scapula laterally.

Although the simple down-facing dog is by itself a good preparation for the headstand, a variation that will develop arm and shoulder strength through the ranges of movement needed for the headstand is to first come into the basic posture and then slowly lower the shoulders, slide your nose forward close to the floor, straighten the knees, and hold the posture isometrically in whatever position is especially difficult for you (fig. 8.28). Finally, let the elbows swing out, and (with considerable relief, at least for most of us) extend the elbows slowly into a simple upward-facing dog supported between the hands and the flexed toes (fig. 5.14).

Figure 8.27. This easy down-facing dog pose is within reach of almost everyone, and looks better as hip and ankle flexibility improve. Even from the beginning, however, it is important not to hang from the shoulders. To that end the serratus anterior muscles keep the scapulae rotated upward and stabilized.

THE DOLPHIN

If a hatha yoga instructor were to pick only one all-around training pose for students who are almost able to do the headstand, it would have to be the dolphin. This posture is related to the down-facing dog, but it is also a well-known posture in its own right. It's helpful for developing upper body strength, hip flexibility, and abdominal and back strength. To do it, begin in the child's pose (fig. 6.18) with the body folded onto itself on the floor. Then lift up enough to place the forearms on the floor in front of you with the hands interlocked. With the hips still resting near the heels, the forearms are positioned at a 90° angle from one another. Next, keeping

Figure 8.28. Bringing the nose down and forward (and more importantly, coming back up) from the down-facing dog is an excellent floor exercise for building enough strength in the upper extremities to begin practice of the headstand. You'll want to come forward only a little at first, so you can push yourself back up into the down-facing dog. This is no pushup—it's much more difficult.

Figure 8.29. The dolphin posture is the most famous preparatory posture for the headstand. It strengthens all the muscles that stabilize the scapulae and arms, and does so from the V-shaped position of the forearms that is similar to the customary starting position for the headstand.

the forearms on the floor, press up into a piked position ending with the hips flexed 90° (fig. 8.29).

Coming into the dolphin is accomplished by a combination of whole-body muscular efforts: lifting the head, straightening the knees with the quadriceps femoris muscles, and then pushing your weight back with the shoulders and triceps brachii muscles until the head is in the V between the forearms. If the hamstrings are so tight that you cannot push back with the knees straight, bend the knees as much as necessary to permit the movement, or adjust the feet slightly to the rear.

If you are able to push back into the dolphin, you will be stabilizing the scapulae in upwardly-rotated positions with the same muscles that assist the down-facing dog: the teres major and minor, the infraspinatus, the sub-scapularis, and the serratus anterior. Pushing the torso back and keeping the arms braced in the 180° flexed position overhead is strongly resisted by the pectoralis major and latissimus dorsi, so this posture gives those muscles an excellent workout in the stretched position. Finally, the triceps brachii is strongly engaged for extension of the forearm.

To build strength for the headstand, lift the head over the hands and stretch forward enough to touch the nose or chin to the floor in front of the hands (fig. 8.30). If the feet were well back in the first place, the body will now be almost straight. This makes the exercise too easy, so come back to the first position with the head in the V made by the forearms and walk the feet forward to reemphasize the piked position. Repeat the exercise, lifting the head over the hands and then pulling it back behind them, over and over again. The closer the knees are to the elbows in the preparatory position, the more strength and hamstring flexibility you will need to accomplish the movement gracefully. Finally, as your strength and flexibility continue to improve, you can take the head even further forward, barely touching the floor with the chin.

Figure 8.30. Bringing the nose forward from the dolphin posture and then back up creates a different exercise than in the case of coming down and forward from the down-facing dog. This one is relatively easy if your feet are far enough back, but quite difficult if you start the dolphin with an acutely angled pike position, especially if you do not have the ability to flex your hip joints 90°.

ALTERNATING THE UPWARD- AND DOWN-FACING DOG

Another good upper-body exercise is to alternate between the upward-facing dog (fig. 5.14) and the downward-facing dog (figs. 6.17 and 8.26). The easiest way to do this is to do it fast, by using the hip flexors to quickly swing the hips up into the down-facing dog from the upward-facing dog and let gravity drop them back down. But that's not so useful, and the better exercise is to do it slowly, maintaining abdominal tension at all times and never allowing the body to merely hang between the shoulders. And for another refinement that is custom designed to develop upper body strength, start with the upward-facing dog, slowly lower down into a straight push-up position with the body an inch or so from the floor, touching the floor only with the hands and the flexed toes, and then instead of using the powerful iliacus and psoas muscles to launch flexion of the hips into the down-facing dog, initiate the movement from the shoulders, pushing to the rear with the arms while sliding the nose along the floor until you are ready to complete the piked position. In this way you will be using the iliopsoas muscles as synergists for completing the posture instead of using them as prime movers to initiate it. Reverse everything to come back down, brushing the nose against the floor until you are again in a low push-up position (keep the body only an inch or so away from the floor) before lifting up into the upward-facing dog.

THE SCORPION

The scorpion posture looks like a scorpion, with a front pair of nipping claws and a long, slender, jointed tail ending in a curved poisonous stinger. The posture incorporates gravity-driven passive backbending with extreme hyperextension of the head and neck, and thus it requires more athletic ability than the headstand. Even a little practice of the scorpion will give the student enough confidence to try the headstand. And remaining in the posture for 30–60 seconds is a real wake-up—but it is not for the timid.

You can come into the scorpion in one of two ways: either by kicking up into the posture with the head lifted or coming into it from the headstand. Kicking up is more athletic. Start in the same position that you used for the dolphin, except that the forearms are at a 60–75° angle from one another and the palms are facing down with the thumbs touching (fig. 8.31a). You can also make the posture more difficult by keeping the forearms parallel. To come into the posture lift the pelvis up into the air, and kick up with both feet, one immediately after the other, adjusting the kick so that you get into the posture but do not overshoot and fall to the rear (fig. 8.31b). Be careful not to try this in a confined region where you might crash into something if you fall. The knees end up in a flexed position, which makes it easy for you to support your feet against a wall behind you until you gain confidence. In the final posture the weight is on the forearms, the head is lifted, the nose is fairly close to the hands, and the feet are as close to the head as the arch in your back permits.

The sacroiliac joints will be in full nutation for the scorpion, and the posture may not be comfortable for more than a few seconds for those who have a lot of sacroiliac mobility. In any case, anyone with good flexibility for backbending can easily touch their feet to their head. Come down by first straightening the body, then flexing the torso, and finally dropping forward onto the feet.

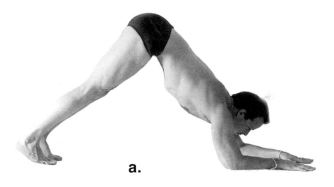

a.

Figure 8.31. Scorpion (b) and starting position (a). To come into the posture by kicking up, you toss your feet up from the starting position and balance your weight making use of a substantial back-bend. Until learning how much energy to put into the initial kick, most people use a wall as a prop so as not to fall over backward. With more experience you can forgo the wall. For the final posture you can keep the knees straight, or you can bend your knees and drop your feet toward your head. You can also come into the scorpion from the headstand, but if you do that, don't delay, because coming into the scorpion after being in the headstand for more than a few seconds creates excess pressure in the arterial circulation to the brain.

b.

When you come into the scorpion from the headstand, you arch the back, flatten the palms against the floor, transfer your weight to the forearms, lift the head until you are looking forward, and bend the knees. If you take this route to the scorpion, however, do it quickly before too much blood and tissue fluid has accumulated in the head. If you stay in the headstand too long before converting that posture into the scorpion, the feeling of pressure in the head is greatly intensified: it's much more pronounced than what you experience by simply kicking up, and it's also unnerving.

BENDING AND TWISTING IN THE HEADSTAND

When you are in the headstand and the hip joints are bearing only the weight of the lower extremities, you can do much of what you can do standing, except more creatively. What is more, certain poses that involve complex combinations of hip flexion or extension with knee flexion and rotation can be done only in the headstand. In this posture you can selectively stretch the adductors and hamstrings; you can work with hip opening exercises when the adductors and hamstrings are not under tension; you can twist, flex, and extend the torso alone or in combination with many creative stretches for the lower extremities; or you can fold the lower extremities into the lotus posture and flex and extend the thighs from that position.

WORKING WITH THE ADDUCTORS

The various adductor muscles take origin all along the inferior pubic rami from the pubic symphysis to the ischial tuberosities (figs. 1.12, 2.8, and 8.13–14). We have generally been concerned with the adductors that take origin posteriorly, and have noted that these muscles have a hamstring character that limits forward bending (chapter 6). It is less common to find postures that are effective in stretching the adductors that take origin anteriorly. The only pose so far mentioned that does this involves a standing backbend (chapter 6) with the feet wide apart. To be successful, any such stretch must also require that the spiraled ischiofemoral, iliofemoral, and pubofemoral ligaments be slack enough to limit extension (fig. 3.6) only after the anterior-most adductors have come under tension. Although any such standing posture should be approached with care, in the headstand it is easy to bring these specific muscles under an intense but controllable stretch simply by extending the abducted thighs with the knees bent. The next three sequences all make use of a relaxed inverted backbending pose (fig. 8.23b) that accomplishes this aim, in addition to rotating the sacroiliac joints into full nutation. This home-base posture alternates with three positions that build strength in the deep back muscles and that shift the sacroiliac joints either into counternutation or less extreme nutation.

Because the next three sequences all involve backbending, they go best with the bregma headstand. To begin, come into the third stage, the one with the thighs extended and the knees flexed (fig. 8.6c). Start with a relaxed and neutral position with the legs more or less parallel to one another and with the feet and knees slightly apart. Without shifting the positions of the lower extremities too much, adjust your posture, including head position, so that you can produce the maximum lumbar lordosis. After appraising exactly how much of a lumbar curve this posture permits, abduct the thighs maximally while keeping the feet fairly close together, and then, keeping the knees flexed and the thighs both extended and abducted, let the feet come apart, sensing the position that permits the lumbar arch to become the most pronounced. The sacroiliac joints will be fully nutated in this relaxed position (fig. 8.32b). This is the home-base posture. As a passive lumbar backbend, this posture complements standing backbends in two ways: the lower extremities are not confined by static foot positions as they are in standing postures, and the knees are flexed maximally, which is obviously not possible when you are standing.

For the first sequence, from the home position in the modified bregma headstand (fig. 8.23b), adduct the thighs, bringing the knees and feet tightly together, and notice that this flattens the lumbar region and draws the knees forward (fig. 8.32a). You can go back and forth, abducting the extended thighs to deepen the lumbar lordosis and establish maximum nutation, and then adducting the extended thighs tightly to flatten the back and ease the sacroiliac joints back into counternutation. The adducted position is peculiar. It creates intense tension in the rectus femoris muscle as well as in the lateral portions of the quadriceps femoris muscles, and this is what, in a round-about way, flattens the lumbar region. The abducted home position, on the other hand, places intense stretch on the adductors whose origins are located anteriorly along the inferior pubic rami.

The second alternative is to start with the same relaxed home position that permits the maximum lumbar arch (fig. 8.32b) and alternately flex and again hyperextend the thighs while keeping the knees flexed and the thighs abducted. This is similar to moving back and forth between stages two and three of the headstand except that now the thighs are kept fully abducted. To keep your bearings, you may wish to touch the big toes together for this particular back and forth sequence, especially as you flex the thighs forward. Extension of the abducted thighs (fig. 8.32b) makes this posture an easy one in which to rest. Flexion of the abducted thighs is more challenging and will probably be limited by your upper body strength because you have to support more of your weight with the forearms as you lower the knees forward (fig. 8.32c). This exercise is easier than moving back and forth between stage two and three of the headstand with the thighs in a more neutral adducted position, however, because some of the weight of your lower extremities is pitched out to the side rather than being held straight in front of you. Even so, flexing the abducted thighs while keeping the knees bent is one of the most rewarding exercises for developing strength in the deep back muscles that you can do in the headstand (fig 8.32c).

Last, come all the way up into stage four of the headstand, that is with the thighs and legs extended. Then abduct the thighs to the side, and hold the posture (fig. 8.33). Because the hips are not hyperextended, adductors that take origin posteriorly along the inferior pubic rami, as well as internal

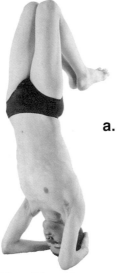
a.

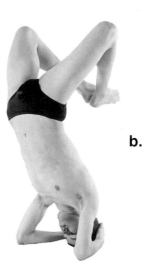
b.

c.

Figure 8.32a–c. These three postures illustrate musculoskeletal dynamics and train you to be inverted and balanced under varying circumstances. In 8.32a, the combination of adducted thighs and flexed knees severely limits the lumbar lordosis. In 8.32b, allowing the knees to come apart (thighs abducted) permits extreme hyperextension of the thighs, the deepest possible lumbar curve, and maximum sacroiliac nutation. Go back and forth between 8.32a and 8.32b several times to feel and understand what happens and why. Shifting from 8.32b to 8.32c requires a major shift in awareness from an acute, maximum backbend (but one in which it is easy to balance) to the necessity of supporting much of your weight on the forearms, which is similar to stage two of the headstand except that this pose with the thighs abducted is easier. Again, go back and forth between b and c to both feel and understand what happens and why.

structures of the hip joint, limit this particular stretch. You can again alternate this pose with the home position in which the knees are flexed, and the thighs are abducted and extended (fig 8.32b). This latter position takes tension off the adductors that take their origin posteriorly, allows you to abduct the thighs more fully, and by default brings the stretch to the adductors that have their origin more anteriorly on the inferior pubic rami. Go back and forth repetitively for clarification of these principles.

STRETCHING THE HAMSTRINGS

In the headstand you can stretch the hamstrings by extending one thigh posteriorly to approach the limits of hip extension and at the same time bring the other thigh forward to stretch the hamstrings. To come into this position you can either abduct the thighs and then swivel them around, which leaves you with one thigh extended and the other one flexed, or you can start from stage four of the headstand and hyperextend one thigh to the rear and flex the other one forward. If you also flex the back knee, this will drop more weight to the rear and you will not have to readjust your balance so much when you flex the other thigh (fig. 8.34). Then, as soon as you are balanced you can pull isometrically in opposite directions, allowing the forward knee to bend according to your capacity and inclination for stretching the hamstrings. Repeat on the other side.

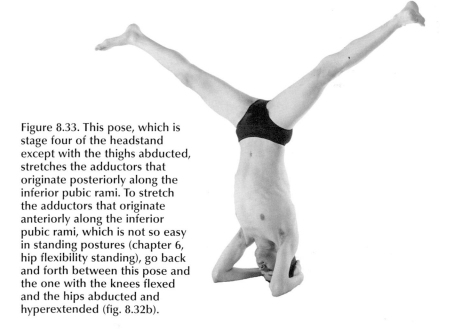

Figure 8.33. This pose, which is stage four of the headstand except with the thighs abducted, stretches the adductors that originate posteriorly along the inferior pubic rami. To stretch the adductors that originate anteriorly along the inferior pubic rami, which is not so easy in standing postures (chapter 6, hip flexibility standing), go back and forth between this pose and the one with the knees flexed and the hips abducted and hyperextended (fig. 8.32b).

The headstand is one of the best postures in which to work with hamstring stretches because the tension on the base of the pelvis that results from hip flexion and from stretch of the hamstrings on one side is countered by hyperextension of the opposite hip. The resulting asymmetrical stretch keeps the pelvis more in line with the spine than the same stretches in standing forward bends, which often place unwelcome additional tension on a region that is already being stressed to its limits.

INVERTED TORSO TWISTS

When you are in the headstand you can do inverted torso twists that are limited only by your imagination, strength, and balance. You can start with a twist in a simple headstand and go from there to a twist with one thigh back and the other forward. Simple twists such as these can be done in the open, but placing yourself near a wall adds to the possibilities. One is to position the back of the head about two feet from the wall, come up into the headstand, and twist your lower body to the right so that the lateral edge of the right foot ends up against the wall. From that position you can pry yourself around even more. This brings the left hip closer to the wall and the right hip further away. The right thigh is hyperextended, the left thigh is flexed about 110°, and both knees are flexed comfortably. If you are fairly flexible you will be stretching the abductors on the lateral aspects of the thigh and working directly within the hip joint. Repeat the exercise on the other side. This is an excellent whole-body twist, and you can feel the results from the knees to the neck, although the most twist will be created in the chest, as expected from the discussion in chapter 7.

Figure 8.34. You can come into this pose easily from the one shown in fig. 8.33 by swinging one foot forward and the other one back. The asymmetrical stretch of the hamstrings in the flexed thigh is balanced by hyperextension of the opposite thigh and tension in its hip flexors, all of which keep the pelvis stable and the lower back protected.

TORSO EXTENSION AND HIP FLEXION

A wall is also a good prop for working with passive extension in the lumbar region. From a simple headstand again facing away from a wall you can place both feet against the wall and walk them slowly down, or you can simply hold them within your reasonable limits, making sure you don't go so far down that you cannot comfortably walk them back up. Alternatively, you can stabilize one foot against the wall and bring the other one forward (away from the wall). If you pull down vigorously on the forward foot using the rectus femoris muscles and the hip flexors on that side while keeping the knee fairly straight with the quadriceps femoris muscle as a whole, you can stretch the forward hamstrings at your leisure; and unlike most standing and sitting forward bends, you can work with the stretch safely but insistently and without stressing the back in the slightest. Repeat the exercise on the other side.

THE LOTUS POSTURE IN THE HEADSTAND

If you are flexible enough to do the lotus posture in the headstand, you have many options for developing flexibility and a strong back. Just being in this posture stretches the adductors and makes the hip joint itself more flexible. An excellent exercise is to lift the knees as far as possible toward the ceiling to hyperextend the back (fig. 8.35a) and then slowly lower them as far as possible toward the floor (fig. 8.35b)—without falling, of course. This is similar to going back and forth between stages three and two of the headstand, except that it is easier because the legs and feet are folded in and because a smaller proportion of your lower body weight is carried forward. It is also rewarding to twist and bend from side to side in the lotus posture. With the knees up, whatever stretches you do along those lines will be combined with backbending, and with the knees down in a more neutral upside-down sitting lotus, whatever stretches you do will build strength in the back. Be sure to repeat all exercises you do in the lotus—no matter what the posture—by alternating the foot you fold in first.

EXTENDING YOUR TIME

When you are completely at home in the headstand, you may eventually want to increase your time in the posture. If done carefully and systematically this is safe, but because of the headstand's special effects on the brain and circulation there are certain guidelines that should be followed, not the least of which is consultation with someone who is experienced in the practice.

THE SURFACE

The surface on which you do the headstand is not very important if you stay in the posture for less than five minutes, but if you are going to hold it for a longer time, the softer the surface the better so long as your forearms do not spring this way and that when you are trying to adjust. A 2-inch thickness of high-density foam rubber is so springy that it is hard to keep your balance, but a heavy woolen blanket folded three times (eight thicknesses) or a 1-inch thickness of high-density foam is fine. Mattresses are nearly always too springy. If you use a futon make sure that you place your head in a region that is flat. You will certainly create problems for yourself if you place your head on an irregular surface that always favors one side.

THE CRANIAL VAULT AND SUTURES

Anyone who is serious about doing the headstand should be aware of danger signals from their cranial bones and sutures (fig. 8.4). You should feel the top of the head for any localized soreness as soon as you come down from the posture. This may not be directly on the region where you rested your weight; it could be on one side or the other or it could even manifest as vague internal discomfort such as headache.

a.

b.

Figure 8.35. For anyone who can do the lotus posture comfortably, working with this pose in the headstand offers many possibilities for improving back strength and hip flexibility. Lifting the knees toward the ceiling (a) extends the back and stretches the quadriceps femoris muscles, and bringing your knees forward and down (b) strengthens the back muscles as in stage two of the headstand, except doing this in the lotus is easier because the feet are tucked in and are not necessitating that you support so much of your weight on the forearms. Like many other stretches, these can be done only in the headstand.

If this is the case, you should either wait for the discomfort to pass before doing the headstand again or seek professional advice from any number of specialized therapists who are interested in such matters.

Another way to explore for excess sensitivity is to come into a hands-and-knees position on a soft surface and roll the head around from front to back and from side to side. Roll all the way forward, bringing the chin to the sternum, and then roll all the way back so the nose touches the floor. Roll from side to side, from ear to ear, diagonally, and around in a circle. If you do this routinely you will become sensitive to whether or not the headstand is creating difficulties. For example, you might have done the headstand fifty times in a row for five minutes each morning with no problems. Then one night you miss sleep, do the headstand the next morning in a cranky state of mind, and suddenly have a vague feeling that the posture doesn't feel right. Under these circumstances, if you have been exploring for excess sensitivity routinely, you are likely to find some localized tenderness on your cranium. And if that is the case you can give yourself a day or two of rest for repair and recuperation.

PAIN AND DISCOMFORT

Any time you are not comfortable in the headstand you should come down. In twisting, forward bending, or backbending postures you can explore the edges of minor aches and pains without too much worry, but in the headstand this is inadvisable because your frame of reference to what is normal is too fundamentally askew.

Pain in your shoulders usually means that you are making too much effort in the posture. If you slowly learn to balance and build strength and flexibility, shoulder pain should diminish. Pain in the neck region is always disquieting. It may be caused by imbalances in muscular tension or inflammation of vertebral joints. If you feel deep pain, stop doing the headstand until it has gone away. Muscles usually respond to joint problems by tensing at the fulcrum of the joint—if you turn your head to a certain point and find that it is painful to move it further, it indicates that your nervous system is objecting. So be conservative, listen to the body, and stop doing the headstand until you can turn your head freely through a normal range of motion. Get help from knowledgeable therapists if the problem does not go away of its own accord.

A recurring theme in the oral (as well as written) tradition of hatha yoga is that any extensive practice of the headstand should always be followed by the shoulderstand and related postures. Commentators tell us that practicing the headstand alone results in an imbalance that manifests as edginess and irritability. They also say that if you practice the headstand for more than twenty minutes, it is good to have some food afterward, or at least some hot, boiled milk. If you don't, they tell us, you are likely to experience a raw, uncomfortable feeling in the abdomen later in the day.

Another caution: the headstand does not work well after aerobic exercise. Hatha yoga postures are fine, but not dozens of sun salutations, walking briskly, or running. If you do the headstand regularly for more than five minutes during the course of a regular hatha practice, and then try it sometime after aerobic exercise, you will quickly sense an impulse to come down, an impulse it will be wise to follow. An old hatha yogi—a centenarian—from India once told me in utter seriousness that doing the headstand regularly after aerobic exercise would cause the skull to soften. Yes? And? The biological basis for such observations (assuming they might be valid) is uncertain. What is certain is that you should use common sense with this posture and honor all input from the senses that tells you to be moderate or even not do it at all.

EXTRACELLULAR FLUID AND MUCUS

When you are in the headstand, extracellular fluid accumulates in the tissues of your head and neck, and as you begin to hold the pose for longer periods, these tissues start to swell. You'll turn red in the face, which is obvious, but the swelling also starts to impede the flow of air through the nose and pharynx, making your breathing more labored and either forcing you to breathe through the mouth or come out of the posture. This is usually temporary. If you continue to practice the headstand every day, the swelling becomes less of a problem and you will be able to breathe easily in the posture for longer times. And once you have acclimated to staying in the posture for ten minutes or so you may find that you can continue to increase your time. Mucus is a separate but related problem. If you have a tendency toward colds or to chronic respiratory problems, doing the headstand for even a minute may be uncomfortable. Don't press the issue. If you have too much mucus, solve that problem first, and then come back to the headstand.

HOW LONG TO HOLD THE HEADSTAND

I used to have a habit of asking instructors publicly how long the headstand could be held, and their answers, appropriately enough, reflected the level and experience of their audience. For a general class of young but inexperienced pupils, experts will ordinarily advise a 1- or 2-minute maximum, or they will avoid the issue by saying that you get most of the important benefits in 3–5 minutes. One yogi traditionalist suggested two minutes maximum for householders (an Indian euphemism for those who are sexually active), and any length of time for those who are perfectly celibate. Many classical texts affirm this, hinting not only that orgasm and frequent headstands do not mix well, but that doing the headstand for long periods of time helps to maintain celibacy for those who wish to practice that discipline. The *Hatha Yoga Pradipika* extols the ability to hold the headstand for three hours. And finally, yet another elderly hatha yogi from India had a brilliant answer to my question. He said that you can do the headstand for any length of time—three hours, six hours, and that you can even sleep in the posture—but he added cagily that there should never be the slightest discomfort.

BENEFITS

The headstand lifts your spirits wonderfully. If something is drawing you down, turn upside down, and *voila*—the downward flow is upended into your head. The headstand is also a great morning wake-up. It increases digestive fire, counters depression, and fills you with enthusiasm for meeting your day. But doing this posture to excess is like increasing the voltage in an electrical circuit. Be careful.

"If you watch the breath, you will notice that it becomes finer and finer. In the beginning of practice, there will be slight difficulty in breathing. As you advance in practice, this vanishes entirely. You will find real pleasure, exhilaration of spirit, in this Asana."

— Swami Sivananda, in *Yoga Asanas*, p. 16.

CHAPTER NINE
THE SHOULDERSTAND

"The pose is called Sarvangasana because it influences the thyroid and through it the whole body and its functions. In Sanskrita, Sarva means the whole and Anga means the body."
— Swami Kuvalayananda, in *Popular Yoga Asanas*, p. 65.

The shoulderstand is the queen of postures and the headstand is the king, the yogis say—the former nurtures the body and the latter celebrates power and consciousness. These concepts will resonate with anyone who has had a lot of experience with both postures. Together they make a team. The headstand needs balance, and the shoulderstand, with its variations and sequelae, makes the best complete practice for providing that balance.

The Sanskrit name for the shoulderstand is *sarvangasana*, which means the "all-member's pose." Not only do all four extremities participate in creating it, the posture, at least in its fullest expression, also requires muscular effort throughout the body. This gives it an entirely different character from the headstand, which is a balancing pose. Placing your weight on a combination of the shoulders, neck, and head, as you must in the shoulderstand, requires that the full posture be supported either with your upper extremities or with a powerful internal effort.

We can learn a lot about the shoulderstand by looking at how it differs from the headstand. The most obvious point of contrast is that in the headstand the weight of the body is on the top of the head and has its primary skeletal effect on the neck. The headstand compresses its vertebrae axially; the shoulderstand stretches the neck. Put another way, the neck acts to support the headstand, and is acted upon by the shoulderstand.

Another difference is that in the headstand the entire spine from C1 to the sacrum is inverted but straight, and the posture is balanced simply by standing up. By contrast, all the variations of the shoulderstand and its associated postures include forward bending somewhere in the body: the cervical region is flexed in the shoulderstand, and the cervical region, lumbar region, and hips are flexed in the plow. This continuing theme of forward bending explains why these postures are often followed with back-bending in the bridge, the fish, and the wheel.

One last way in which the shoulderstand differs from the headstand is that significant time and commitment is required to learn about the nature of the posture and do it justice. We can get most of the common physical benefits from the headstand by practicing that posture 3–5 minutes a day, but any serious student who wishes to get acquainted with the postures in the shoulderstand series is well advised to practice them for 20–30 minutes a day for at least three months. After that a more abbreviated practice will suffice.

In this chapter we'll first summarize the anatomy that is pertinent to the most advanced expression of the shoulderstand. Next we'll discuss the entire shoulderstand series, starting with the easiest postures, and then we'll examine the plow series. We'll then cover the aspects of circulation and respiration relevant to these poses. Finally, we'll look at exercises and postures that usually follow and balance the shoulderstand and plow, and end with a brief discussion of benefits.

ANATOMY OF THE SHOULDERSTAND

To understand the complex anatomy of the shoulderstand, we'll begin with a brief description of the posture in its most extreme expression—the candle posture. This is an advanced pose, however, and should not be attempted until you are warmed up and have mastered the preliminary postures which follow. In this pose the feet are swung overhead from a supine position, and the arms, forearms, and hands are placed in a neutral position alongside the thighs (fig. 9.1). The body is balanced on a triangular-shaped region comprised of the back of the head, the neck, and the shoulders. The chin is pressed into the sternum, and the pose is held internally, mostly with the muscles of the torso and lower extremities. This is not easy. Few students will have enough back strength to keep the sternum tightly pressed against the chin, but unless they can do that they will not attain the full benefit of the posture.

THE NECK

Even though the anatomy of the advanced posture is complex, it is straightforward and easy to analyze. In the neck, most of the structures in the cervical spine are stretched, including the posterior longitudinal ligament on the back of the anterior functional unit, the interspinous ligaments between the spinous processes, the ligamenta flava between the vertebral arches, and the ligamentum nuchae, a fibro-elastic ligament which runs from the back of the head to all the spinous processes between C1 and C7 (fig. 4.13a). The synovial articulations between the adjacent superior and inferior articulating processes will be stretched to their limits as well. Also stretched by this posture are the muscles that attach to the upper thoracic and cervical spine: the trapezius muscles; the levator scapulae; and the strap muscles between the head and upper back, especially the semispinalis capitis between the cranium and C7–T6, and the splenius capitis between the mastoid process and C7–T6 (figs. 4.14, 5.5, 8.12, and 8.14).

The spinous processes (figs. 4.10a, 4.10c, and 4.13a) are the first bony points of contact with the floor in the shoulderstand. They are easily located, and are the hard structures that can be palpated directly in the midline of the body at the nape of the neck. Two of them generally stand out from all the others. The higher of the two bumps belongs to C7, and the lower one belongs to T1, which is also called the *vertebra prominens* because it usually protrudes more than any other (fig. 4.13a). You can confirm their identities for yourself if you find someone on whom these two bumps are pronounced, ask them to flex their neck, and then in the flexed position to twist their head right and left several times. You can distinguish C7 from T1 because C7 moves from side to side as your subject's head twists back and forth, while T1 is relatively stable. Once you see and feel the relative mobility of the tip of C7 on someone else, you can easily locate it and T1 on yourself. And once C7 and T1 are located, the tips of the spinous processes in the rest of the spine can easily be felt, especially in those who are slender and not very muscular.

Figure 9.1. The candle posture is the most advanced expression of the shoulderstand. It is balanced on a combination of the back of the head, the neck, and the shoulders, and is maintained by muscular effort throughout the body. The body from the shoulders to the toes is stiff as a board, and acts as a pry bar to stretch the cervical vertebral column from a fulcrum at the junction of the chin and the sternum. Never attempt this advanced pose until you have mastered all the less extreme variations of the shoulderstand series and are accustomed to at least a 20 minute practice of the series.

How far the spinous processes protrude can be a practical problem if you are lying on a hard floor, and anyone who has little subcutaneous fat and who is lightly muscled should do the shoulderstand on a pad, especially if they wish to roll down from the posture one vertebra at a time.

THE TORSO

A combination of the spine, deep back muscles, proximal muscles of the extremities, abdominal muscles, and the respiratory and pelvic diaphragms support this version of the shoulderstand. More than any other, this middle segment of the body maintains the pose, and the brunt of the effort is carried by the erector spinae and other deep back muscles (figs. 4.14, 5.5, and 8.14), which are situated posterior to the ribs and transverse processes of the vertebrae. When these muscles are maintained in a strong state of isometric contraction, they hold the spine straight.

The most obvious role of the deep back muscles is to counter the tendency for forward bending in the lumbar region. The temptation is to swing the lower extremities enough overhead to balance the body without much muscular effort, but this obviously can't be done without forward bending in the lumbar spine. And even though that makes the pose easier, it deprives us of its main benefit. If you allow the spine to flex, you lose the essence of the posture, and it would be better to concentrate on the more elementary inverted postures.

THE EXTREMITIES

To do this advanced version of the shoulderstand successfully, the tendency for forward bending also has to be supressed at the hips, which means keeping the hips extended. The main muscle responsible for this is the gluteus maximus. As seen earlier (figs. 3.8, 3.10, 8.9–10, and 8.12), this muscle takes origin from the back of the ilium and sacrum and has two insertions, one into the iliotibial tract (which as suggested by its name bypasses the knee and attaches to the tibia; figs. 3.8 and 8.12), and the other directly onto the femur (fig. 3.8b, 3.10, and 8.12). The gluteus maximus is the heaviest muscle in the body, and you can immediately feel it tighten up on both sides as you try to hold the thighs extended in the advanced shoulderstand. The effort that tightens the gluteus maximus also squeezes the hips together, with the result that this posture holds the sacroiliac joints in a position of counternutation—that is, with the ischial tuberosities pulled toward one another, the ilia spread apart, and the promontory of the sacrum rotated between the ilia to the rear.

As you try to bring the body straight in the candle pose, you will not at first feel much tension on the front of the thighs, but as you increase your efforts to extend the thighs with the gluteus maximus, the quadriceps femoris muscles (figs. 1.2, 3.9, 8.8–9, and 8.11) finally counter that effort antagonistically, which you can easily confirm with your hands because they are nearby. From the knees down, you have options: if the feet are extended, you will be mildly stretching the muscles on the front of the leg, and if they are flexed, you will be stretching the soleus and gastrocnemius muscles in the calves (figs. 3.10a–b, 7.6, 8.9–10, and 8.12).

Since the arms and forearms are positioned along the chest and thighs, you wouldn't think they were contributing to the posture. But the upper extremities also include the scapulae, and when you come into the candle pose you are adducting and depressing these two triangular bones. This ultimately results in lifting your weight off the nape of the neck and taking some of the pressure off the spinous processes of C7 and T1.

This version of the shoulderstand is the definitive all-member's pose. From head to toe, muscles are either activated isometrically or stretched. Extensors of the hips and spine straighten the body, acting synergistically with muscles of the upper back that depress and adduct the scapulae. In combination, the back and hip extensors also resist flexion of the spine and hips. The body becomes like a pry bar pushing the sternum against the chin, and the resulting tension creates significant traction in the neck. How different from the headstand, in which you hold only enough muscular activity to balance on the top of the head.

INVERTED ACTION POSTURES

The candle posture described above is demanding, and should not be approached without a lot of preparation: gradually getting accustomed to being in postures in which the hips are higher than the shoulders; gradually getting accustomed to more and more flexion of the neck; slowly becoming confident in balancing the body as a whole in a posture that is more and more perpendicular to the floor; and becoming familiar with the different methods of actuating and supporting the dozen or so postures that make up the shoulderstand series and its sequelae. We'll begin with the inverted action postures.

Technically, inverted action means upside down, but in most yoga traditions, "the inverted action posture" refers to *viparitakarani mudra*, in which the lower extremities are perpendicular to the floor, the torso is at a 45–60° angle from the floor, and the pelvis is supported by the elbows, forearms, and wrists. We'll first examine some easier variations that can lead systematically to the shoulderstand.

Even though most of the inverted action postures are not as difficult as the shoulderstand, they confer some of the same benefits and are particularly useful for older people. The first two variations that follow are of special value to anyone who is fearful of being upended. And like the headstand, the shoulderstand and the inverted action postures are contraindicated for anyone with high blood pressure, for women who are pregnant or in their menstrual period, or for anyone with osteoporosis. Being substantially overweight is another obvious contraindication. Those who are uncertain as to whether or not they should proceed will find the following two variations safe for beginning experimentation.

PASSIVE INVERTED ACTION POSTURES

Safe means simple and safe means conservative, and a good place to begin to learn the shoulderstand is to squirm your pelvis onto the top of a bolster, draw the knees toward the chest, and simply lift the feet into the air, straightening the knees so that the thighs and legs end up perpendicular to the floor. Once you get your pelvis in position, you do not even need to use your arms to help you get your feet up. Those who are more adventuresome can try supporting the pelvis on the edge of a couch, positioning the torso at a 30–45° angle from the floor depending on the height of the support and length of the torso. These postures provide excellent training for the full inverted action pose and for the postures in the shoulderstand series because the hips are higher than the shoulders, the neck is slightly flexed, the lower extremities are perpendicular to the floor, and the posture is supported passively by a prop.

If your balance is good, you can also support yourself with an 8 1/2 inch playground ball (chapter 5), or better yet, a bigger one 10–13 inches in diameter. Placing a supporting bolster or ball at different sites creates different effects. If the support is placed under the lower part of the sacrum and coccyx, the back will be rounded to the rear and mostly against the floor; if it is placed under the upper part of the sacrum, the pelvis will be raised higher and the back will be straighter; if it is placed under L4–L5, the back will be straight; and if it is under the junction of the lumbar and thoracic regions, the lumbar region will be arched forward in the other direction and the pelvis will drop, creating a passive backbend (chapter 6). This last position places an unusual stress in the lumbar region and is contraindicated for anyone with a tender back. With this exception, students can be fairly relaxed in all of these variations except for the effort needed to keep the knees extended.

An even more passive inverted action posture involves flattening the thighs and legs against a wall with the pelvis again supported on a bolster or ball. You do not have to use much effort to keep the knees straight, and you can combine the posture with a passive adductor stretch by letting the thighs rest in an abducted position.

THE RELAXED EASY INVERTED ACTION POSTURE

This next posture prepares you for both the shoulderstand and the plow. It is relaxing once you get into it but it requires more strength, flexibility, and athletic prowess than the propped postures just described. It is the logical next step for those who are trying to build confidence for doing more advanced inverted postures. Except for the fact that the legs are sticking out, it resembles a ball whose circumference is formed by the head, back, pelvis, thighs, elbows, and arms (fig. 9.2). To begin, lie supine on a padded surface with the top of the head about two feet from a wall, or a little less depending on your stature. Pull the knees toward the chest, place the hands against the floor below the hips, palms down, and in a single movement tighten the abdomen, push strongly against the floor with the hands and elbows, and lift the hips up and the feet overhead, straightening the knees slightly at the same time.

The feet should touch the wall lightly in the final position, and you may now have to adjust your distance from the wall to make that comfortable. The knees, hips, and back are all comfortably flexed. Interlock the fingers lightly at the top of the head, and brace the thighs with the elbows just above the knees, or place your hands against the lower back and pelvis (fig. 9.2). Adjust the posture for maximum comfort and relaxation.

This is a relaxed posture once you get into it, but getting there may be a challenge for those whose spinal and hip flexibility is poor. And another consideration for novices is that even though there is no pressure on the neck, and even though your body weight is so close to the floor that you do not have to worry about falling over, the weight of the lower extremities can compel so much flexion of the hips and spine that it shocks the uninitiated. Once in a while the extra weight on the chest prevents someone from inhaling in this posture, especially if an insensitive coach has lifted them into it. I witnessed that error once in a class of partnered hatha yoga for older but athletic beginners. Fortunately, other classmates were observing and quickly intervened, crying "Stop, stop, she can't breathe!" The opposite problem is a lack of weight from the waist down, as in barrel-chested men with skinny legs. If that is the case, you may want to try coming into the pose wearing heavy shoes or ankle weights in order to pull enough weight

Figure 9.2. This relaxed inverted action posture with the feet against the wall is easy for most people, and the head is freely movable for twisting from side to side, but the pose should still be monitored watchfully in the case of those who are trying it for the first time.

overhead to stabilize the posture. In any event, if you come into the pose but are not confident that you can balance gracefully, just roll down keeping the knees as close to the chest as possible.

THE INVERTED ACTION POSTURE

In the full inverted action posture (viparitakarani mudra), the weight of the lower part of the body is supported by the elbows, forearms, and wrists. Come into the posture from a supine position flat on the floor with the arms alongside the body, palms down. Lift the lower extremities by pressing the hands and forearms against the floor, tightening the abdomen, and pulling the feet overhead, all the while keeping the feet together and the knees straight. As you pull your weight to the rear, keep your arms against the floor and place your hands under the pelvis where they can steady you. Then, supporting yourself with your hands, complete the posture by bringing the thighs and legs perpendicular to the floor (fig. 9.3a).

Because you will have to support much of the weight of the torso as well as all of the weight of the lower extremities with the hands and forearms, the inverted action posture is difficult for many students. Depending on the length of the forearms and the exact placement of the hands, the torso

a.   **b.**

Figure 9.3. Viparitakarani mudra, or the inverted action posture (a), is a famous pose, but it places so much weight on the forearms that many people find it troublesome. The pose on the right with the feet slightly overhead (b) compromises the posture but is useful for beginning the process of getting acclimated to the formal pose.

will be at about a 45–60° angle from the floor. That is easy enough by itself, but instructors who want the posture done rigorously also insist that the thighs and legs be exactly perpendicular to the floor, and the combined weight of the extremities and lower torso may be hard on the elbows unless they are well cushioned. If you flex the thighs a little more, swinging the lower extremities overhead about 30° off axis from perpendicular (fig. 9.3b), the posture becomes easier but it begins to lose its original character.

THE SHOULDERSTAND

Now we are ready to look at the shoulderstand proper, in which the body (exclusive of the head and neck) is positioned more or less perpendicular to the floor. Since this requires that the cervical region be strong and flexible, we'll work up to it gradually. Before starting, however, you may want to explore and become familiar with the resistance neck exercises discussed in the last section of this chapter ("Sequelae"). Once you have done that, you are ready for the quarter plow.

THE QUARTER PLOW

To make a safe and easy transition to the shoulderstand, especially for beginners, the quarter plow (not illustrated) should come next. You come up into this pose exactly as you came into the inverted action posture, by pushing with the hands from a supine position, tightening the abdominal region, and swinging the feet overhead, all in a single coordinated movement. Then you simply let your feet hang far enough overhead to balance your weight while bracing the pelvis with the hands. The lower extremities will now be at an angle of 45° off axis from perpendicular instead of the 30° illustrated in the last posture.

The main point of this posture and what makes it good for beginners is that you will not have to support as much of your body weight with the arms as you do in the inverted action pose. You can steady the hips with the hands or, for a sharper-looking posture, you can brace the hands on the thighs just proximal to the knees. It is especially easy to support the legs in this position.

The quarter plow is another posture you can do with the feet lightly touching a wall, but whether you use a wall or not you can now begin to get the feeling of the shoulderstand. The torso is practically vertical and the sternum is pressing lightly against the chin. This is starting to become a balancing posture, but at the same time it is a pose that requires some musculoskeletal activity for resisting forward bending in the spine and hips. It's worth serious study.

THE BEGINNING SHOULDERSTAND

Now you can begin working with the shoulderstand itself. To come into the pose follow the same sequence to which you have become accustomed. From a supine position bend the knees comfortably and swing the hips overhead, using the hands to lift and steady the pelvis as it comes up. Then balance on the upper back and shoulders. You can best support the pose by wrapping the fingers around to the rear against the sacroiliac region and pushing the hands against the back, keeping the thumbs to the front just superior to the crests of the ilia. This is a comfortable position for most people because the neck is not under tension and because you are still not stretched straight up into the air. You'll end up with an obtuse angle of about 140° between the thighs and the torso (fig. 9.4a).

From this position you can start to explore. Gradually straighten the body, including the knees, hips, and spine, shifting your weight each day and placing the hands higher and higher on the back. Allow yourself several weeks of daily practice, making sure you are secure with each shift to a straighter position before going on. And be especially careful to monitor the feelings in the neck. In the initial position with your weight balanced mostly on the upper back, no stress is placed on the cervical region, but as you straighten the body the head and neck will have to become more sharply flexed. Soon you will be pressing the sternum against the chin, and as that happens your position becomes more tenuous. It is therefore important to adapt to the posture without haste.

THE CLASSIC SHOULDERSTAND

To transform the beginning shoulderstand into the classic shoulderstand you will have to make the pose more dynamic, and for this there are four requirements. First, instead of swinging up with bent knees, press the lower back against the floor, do a double leg left with straight knees (fig. 3.17), and lift your feet toward the ceiling (and only slightly overhead) using the abdominal muscles. While coming up, don't press your hands against the floor any more than you have to. Second, once you're up, straighten the body by pressing more insistently and with the hands higher on the back than in the beginning pose. Next, tighten the erector spinae and hip extensors, creating a forward thrust in the pelvis that complements the efforts from the hands. Finally, with the body supported in a straight line by the hands and the muscles of the trunk and hips, lean the sternum against the chin (fig. 9.4b). If this is uncomfortable, adopt a more moderate hand position for the time being. Alternatively, defer further work on the classic posture for several weeks and temporarily limit your efforts to the lifted shoulderstand, which will be described later in this chapter. The resistance neck exercises listed under "Sequelae" will also be helpful.

When you have completed the posture, pressure from the entire body is pushing against the chin. The sternum presses the lower jaw against the upper jaw, and the neck and skull as a whole comprise a unit that cannot twist or budge in any direction. And since the whole body is stiff, it acts as a lever that exerts traction on the cervical vertebrae. You do get many important benefits by simply lifting up with the body slightly curved, but this does not elicit the intense energy associated with the classic shoulderstand.

THE INTERNALLY SUPPORTED SHOULDERSTAND

If your strength and balance permit, assume the classic shoulderstand and then remove your hands from their supporting position on the upper back. You can place them in one of three positions: behind the back against the floor, with the arms and forearms extended as much as you can manage; overhead against the floor, with the arms flexed 90°; or alongside the thighs, with the arms adducted to a neutral position.

Figure 9.4. The beginning shoulderstand (a) is a straighter version of the inverted action pose, with the head still freely movable, the feet slightly overhead, little or no traction in the neck, and the hands situated comfortably to support the posture. The classic shoulderstand (b) is a more advanced posture. It is strongly supported by bracing the hands higher up on the back, and for the first time we see the sternum pressing firmly against the chin, thus creating traction in the neck. The hips should be tightly contracted, and the lower extremities held straight, so the body as a whole is not passive but aids the upper extremities in maintaining the posture.

For the first variation interlock the fingers behind the back, pressing the palms together; then straighten the elbows and press the arms and forearms against the floor (fig. 9.5). This places uncommon demands on the upper extremities from the scapulae to the hands, and if the position is too difficult you can just interlace the fingers leaving the palms apart. Those who have good strength and flexibility for extending the arms will find that this posture braces the back almost as effectively as bracing the posture in the conventional manner with the hands pushing against the upper back. And once you are in the posture with the arms only moderately extended, you can easily feel how extending them another 20–30° straightens the body into a vertical posture. Only those who are strong and flexible enough to press their arms and forearms forcefully against the floor will find this variation comfortable and rewarding.

After experiencing and analyzing your limitations for straightening the body with arm and forearm extension, you can start supplementing those efforts by tightening the deep back muscles (figs. 4.14, 5.5, and 8.14) and the gluteus maximus, the main hip extensor (figs. 3.8, 3.10, 8.9–10, and 8.12). The main difficulty with this is inadequate strength, and the only way you can work with the posture, apart from extending the arms more fully, is to try even harder to contract the hip extensors and deep back muscles. As soon as you reach your limit this becomes an isometric effort.

Figure 9.5. This pose is similar to the classic shoulderstand except that the extended arms and forearms are supporting the posture by pressing firmly against the floor. You can easily sense how important this support is by lifting the hands and noticing that the pose deteriorates immediately. By the same token, those who are unable to extend their arms a full 90° will find it difficult to keep their bodies straight and will almost certainly have to permit some flexion of their hips and backs.

One characteristic of this version of the completed shoulderstand is that the tips of C7 and T1 are now lifted away from the floor. The scapulae are adducted, and you are supporting the posture more on the back sides of the arms than on the shoulders and the nape of the neck. If you find this variation difficult, try to work up a little extra enthusiasm for it. Try it once, then rest in the relaxed easy inverted action posture, and try it again. Remember, this is the all-member's pose: it will augment your efforts for doing many other postures.

For the second variation bring the arms and forearms overhead in the opposite direction, along the floor behind the top of the head instead of behind the back, that is, with the arms flexed 90° instead of extended. Either interlock the hands or simply hold the arms and forearms against the floor (fig. 9.6a). This position is not as demanding of the upper extremities as the previous variation, but neither does it brace the shoulderstand, for the simple reason that the flexed arms lie passively against the floor. Trying to flex them further overhead will push them harder against the floor, and this can only push you out of the posture, but flexing them less can only mean lifting them away from the floor, which leaves you supporting the posture purely with the hip extensors and deep back muscles. So if it was difficult for you to hold your body straight in the last variation, it will be

Figure 9.6. With the arms flexed overhead, the pose on the left (a) is just as difficult as the candle pose (b and fig. 9.1). In both cases the postures must be maintained internally, and flexion must be resisted by the deep back muscles, gluteals, and hamstrings.

even more so in this one. If you cannot remain vertical, simply hold the back and gluteal muscles isometrically for a few seconds, and then either support the back again with the hands or rest in the relaxed easy inverted action pose.

For the most advanced shoulderstand—the one we used earlier to illustrate the anatomy of the posture—bring the hands up alongside the thighs after you have come into the classic pose. Be content at first with keeping the hips and back slightly flexed, with an obtuse angle of about 160° between the thighs and the chest. Balance in this position every day for a week without trying to complete the posture. Notice that you are not distracted by the upper extremities and that this pose follows naturally from placing the arms against the floor overhead. The final stage—straightening the body and pressing the sternum against the chin—is no different from what you have been doing all along except that now you are doing it entirely with the gluteal and deep back muscles, which acting together thrust the pelvis forward. Then pull the shoulders to the rear one side at a time by adducting the scapulae. With consistent effort over a period of time you can straighten the body like a stick and master this most advanced and purest variation of the internally supported shoulderstand (figs. 9.1 and 9.6b).

MUSCULAR ACTIVITY IN THE SHOULDERSTAND

The rationale for calling the shoulderstand the all-member's pose should now be clear. It's not a balancing posture: the body position is maintained by muscular effort. The internally supported shoulderstands in particular require a constant influx of nerve impulses to muscles throughout the body. You straighten the back by contracting the erector spinae, you press the front of the pelvis toward the wall behind the head by tightening the gluteal muscles, you extend the knees by tightening the quadriceps femori, and you press the heels toward the opposite wall by creating even more extension in the hip extensors.

If you analyze this effort from head to toe, you will find that extensor muscles throughout the body resist flexion and pull you straighter into the posture: the erector spinae extend the spine; the gluteus maximus muscles extend the thighs at the hip joints, aided in that effort by the hamstrings acting as synergists; the quadriceps femoris muscles act as agonists to keep the legs extended at the knee joints and also act as antagonists for countering the tendency of the hamstrings to flex the knees; the triceps brachii muscles extend the forearms; and extensors of the hands and wrists point the fingers toward the ceiling. The only option you have is deciding what to do with the feet. You can extend the ankle and toes toward the ceiling, flex them toward the head, or leave them relaxed. The only place where extensor muscles are both relaxed and stretched is in the neck.

STRENGTH AND FLEXIBILITY IN THE SHOULDERSTAND

If you work only with the basic shoulderstand postures you may find that you are improving only so much and that insufficient strength and flexibility continue to stop you in the same place. A serious commitment to a program of active backbending postures can help correct this situation, but the best remedy is to do additional twisting and bending exercises from within the shoulderstand itself. Some of these have already been discussed in chapters 5–7; working with them and with the exercises to follow will enable you to find new limits.

The simplest exercise is to twist in the classic (supported) shoulderstand. After coming into the posture twist to the right. (Here "twisting to the right" means from the perspective of the practitioner looking toward the ceiling.) Such a twist pulls the right side of the pelvis posteriorly and the left side of the pelvis anteriorly. Intensify the twist by pressing higher on the back of the chest and more forcefully with the left hand (fig. 9.7). This not only helps the twist, it also aids extension of the spine. Repeat on the other side, and then rest in the relaxed easy inverted action posture. If you come back up in the shoulderstand a second time, you may find yourself straighter.

Figure 9.7. In this simple twisted shoulderstand, the left hand is pushing the left side of the pelvis anteriorly and is assisting the effort to keep the body straight. The direction of the twist is referenced from the point of view of the practitioner looking toward the ceiling, since those are the terms in which someone in a class would follow directions. Here the model is twisting to his right.

You can also combine twisting and forward bending from the classic posture. The easiest and most natural exercise is to assume the position with the hands high on your back, and then twist to the left (pushing the right side of the pelvis anteriorly) and lower the right foot slowly overhead and across the body toward the floor while keeping the left thigh and leg extended. Ideally, the right knee will be extended and the right thigh will be flexed to its limit. Unless you are unusually flexible you will not be able to flex the right hip joint more than 90°, which would bring the thigh parallel to the floor. Slowly come back up and repeat on the other side, twisting to the right and lowering the left foot overhead.

This forward bend and twist stretches the hamstring muscles of the thigh you are lowering overhead, and because the upright thigh keeps the pelvis and lower back stable, the hamstring stretch does not compromise the back or sacroiliac joints. You can come into the posture slowly and stay there for a while, or you can go back and forth one foot at a time at a faster pace. As long as you use the best form you can manage most of the time, it is all right for those who are less flexible to bend the overhead knee and lumbar region enough to bring the big toe to the floor.

Now try a series of three (soon to be twelve) exercises that can form the basis of a comprehensive practice of the shoulderstand. The first one might be called a twisted half lotus one-legged plow. From the shoulderstand, twist right, pushing the left side of the pelvis anteriorly. Then flex the right knee and right hip, rotate the right thigh laterally, and place the right foot and ankle in the half lotus position against the left thigh as close as possible to the groin. To complete the posture lower the left foot overhead and across the body to the floor, or at least as close to the floor as possible, keeping the left knee straight (fig. 9.8). Repeat on the other side. This posture is rewarding if you want to rest between other poses in the shoulderstand series, and is also an excellent preparation for the full spinal twists that are accomplished from the half lotus position (fig. 7.33a) or from the full lotus (fig. 7.33b).

Next, and again from the classic shoulderstand, twist left, pushing the right side of the pelvis anteriorly, and flex the right knee and thigh while keeping the left knee and thigh fully extended. Then slowly bring the right knee diagonally across the body to the floor beside the left ear. Try not to twist your head to the right in order to reach the floor with the knee. This exercise requires a lot of concentration, a substantial torso twist, and excellent hip flexibility, but since the right knee is bent, the posture doesn't stretch the hamstrings. To bring the right knee all the way to the floor, most people will have to flex the torso to some extent, and perhaps the left thigh. As far as possible, keep the left thigh extended and the back straight (fig. 9.9). Come back up in reverse order, first extending the right thigh while keeping the knee flexed, then extending the right knee, and finally coming out of the twist. Repeat on the other side.

The last exercise in the series is to twist left, again pushing the right side of the pelvis anteriorly; then slowly flex both knees in isolation; and then flex the hips. Finally, bring both knees down together until the right knee ends up beside the left ear (fig. 9.10). This is easy in the supported posture because the upper extremities can easily accommodate the extra weight of the feet and legs when they are lowered. Come back up in reverse. Extend the thighs without extending the knees, then extend the knees, untwist, and end up again in the shoulderstand. Repeat on the other side.

Figure 9.8. To come into this twisted half lotus one-legged plow, twist right, pushing the left side of the pelvis anteriorly. Then flex the right knee and hip and place the right foot in the half lotus position. Last, drop the left foot overhead and across the body to the floor.

Figure 9.9. To come into this twisted knee-to-the-opposite-ear posture, twist left, pushing the right side of the pelvis anteriorly, flex the right knee and thigh, and bring the right knee across the body to the floor beside the left ear, all while bracing the right foot against the left knee, allowing gravity to flex the left thigh (which pushes the right knee toward the floor), and trying not to twist your head to the right.

Figure 9.10. To come into this twisted knees-to-the-floor pose, twist left, pushing the right side of the pelvis anteriorly. Then flex both knees, then both thighs, and finally bring both knees down so the right knee is beside the left ear. Come up in reverse order, first extending the thighs, then the knees, then untwist to the shoulderstand.

Now repeat the last three exercises, each one of them with the three different arm positions listed in the section on the internally supported shoulderstand: first with the fingers interlocked, palms pressed tightly together and forearms flat on the floor behind the back; second with the arms overhead (flexed 90°) and with the forearms extended; and third with the hands alongside the thighs. The main difficulty these exercises all share is that you are no longer supporting the back with the hands. Keep the sternum pressed against the chin, and do the nine additional exercises slowly, without compromising any more than you have to. They are challenging because it takes a great deal of strength in the extensors of the back and hips to keep the body straight. The first set of exercises with the arms and forearms extended is easiest because the arms support the posture. The second set is more challenging because the relaxed and passive arms are unable to prevent flexion of the torso. The third set is difficult for the same reason the internally supported shoulderstand with the arms flexed overhead or alongside the thighs is difficult: you have to maintain the posture entirely with the back muscles, keeping the body straight without the aid of either the hands pressing against the pelvis or the arms extended and pressing against the floor.

THE PLOW

The plow posture is named for the way the head and shoulders together resemble a plowshare cutting through soil. We might also characterize the posture as a shoulderstand forward bend because the hips and spine are flexed and the feet have been pulled overhead. And yet, as in the shoulderstand, you are balancing on some combination of the shoulders, upper back, neck, and head. In the classic posture the feet touch the floor and the knees are extended. This is an intermediate level pose that requires good flexibility of the spine and hips, but for beginners there are several variations that will lead up to it comfortably. We've already done two—the relaxed easy inverted action posture and the quarter plow—in order to introduce the shoulderstand. We'll continue with the half plow.

THE HALF PLOW

To come into the half plow, begin in the supine position and lift up as though you are going to come into the shoulderstand, but pull the feet further overhead. Try to keep the knees extended, at least on the start. The idea is to bring the lower extremities parallel to the floor, approximating a 75° angle from the chest (fig. 9.11). If that is beyond your capacity you can bend the knees slightly to take tension off the hamstring muscles. You can also flex the hips and spine a little less, so that the thighs end up at an angle somewhere between the horizontal position in the half plow and the

45° angle for the quarter plow. You can also place the feet against a wall and search out any comfortable position.

There are endless variations on this posture. If you lie supine on the floor and roll over backward with the toes making contact with a mattress or a thick pad, the half plow becomes a propped classic plow. Notice that the further to the rear you bring the feet the more you will be supporting your body weight on the shoulders and the tighter will be the bend in the neck. And if you permit the thighs to lift up from their position parallel to the floor, your weight will end up more on the upper back and there will be less flexion in the neck. Finally, if you bend the knees and take them closer to the floor, you will be approaching the knee-to-ear posture and *yoga nidrasana*, which will be described later. All in all it is easy and rewarding to spend twenty minutes or so just on the half plow and its many variations.

THE CLASSIC PLOW

The plow posture, like the shoulderstand, can be expressed in many different ways, but for most beginning and many intermediate students it can be done only one way. You come into the posture by lifting the feet overhead as though you were planning to come into the half plow, and then you keep going until the toes touch the floor. There is no middle ground for the knees. To qualify for the plow, they have to be completely extended. As with the shoulderstand, you can either extend the arms behind your back with the hands clasped or flex them overhead toward the toes. The latter is easier. You can either flex or extend the feet, but the posture is easiest with the toes and ankles flexed, for the simple reason that the heels are slightly higher and you do not have to flex the spine and hips quite as much to come into the posture. Try to keep the feet and knees together.

Doing the plow posture is comparable to touching your toes in a sitting forward bend, which you accomplish with some combination of hip and spinal flexion, with the addition here, of course, of forced flexion of the neck. If you have limited flexibility for forward bending in the hips, you won't be able to do the plow except by pushing the feet all the way to the

Figure 9.11. For the standard half plow pose, the thighs and legs should be parallel to the floor, but the posture can and should be modified to meet individual needs, such as moderate flexion of the knees, less flexion of the hips, or more flexion of the back.

rear and raising the chest so that the sternum is locked against the chin and the head is at a 90° angle from the sternum. If the chest is perpendicular to the floor, and if the thighs and legs run in a straight line from the hips to the floor, you will have to flex forward a total of about 110°, possibly combining 80° of hip flexion with 30° of spinal flexion, which is usually within the range of intermediate students. If you have to strain to reach the floor with your feet, prop the posture with a thick pad under your toes.

THREE VARIATIONS OF THE CLASSIC PLOW

The most flexible students can work with three variations of the classic plow. And because it's natural to follow a sequence from the first to the last, we can consider them stages of the plow as well as complete postures in their own right. In all three postures you can either extend the arms behind you, flex them overhead, or use them to support the back.

For the first stage lie supine on the floor and bring the feet overhead with the heels and toes together and the knees extended. Keep as much weight on the upper back and as little on the shoulders as possible, which means trying to keep the spine (including the neck) straight as you lower the toes to the floor. If you choose to keep the feet and toes flexed, you can stretch the hands overhead toward the feet and grasp the toes. Alternatively, you can extend the toes while extending the arms behind you and interlocking the fingers (fig. 9.12a). In any case, the more hip flexibility you have, the closer the thighs will be to the chest and face and the more limited will be flexion in the neck. Instead of 110° of flexion between the chest and thighs, as we saw in the half plow, this posture shows more like 160°, perhaps 120° in the hips and 40° in the lumbar region. The lower part of the chest will be lifted off the floor, creating an angle between the neck and the sternum of about 30° (fig. 9.12a). This is a useful posture in its own right, especially for stretching the hamstrings.

You can come into the second variation of the plow from the first. If they are not given specific directions, this is the one most students will do for the simple reason that it is the only one they can do. The posture will look different depending on hip and spinal flexibility. Those with excellent hip flexibility will ordinarily hold the back straight, flex the hips 120°, and lower the feet overhead (fig. 9.12b). The angle between the neck and the sternum will be about 80°. They can either grasp their toes with their fingers or extend their arms behind their back, fingers interlocked (fig. 9.12b). In any case, this posture allows those with excellent hip flexibility either to make full use of it or to flex their hips 70–110° and make up the difference with spinal flexion. In the latter case, the posture will obviously leave the back more rounded posteriorly and will push the feet further to the rear. In that case, students may not be able to reach their toes with their fingertips.

Compared with the first variation, the average intermediate student will reveal about 90° of flexion in their hips and about 70° of flexion in their lumbar spines, again for a total of 160°. More weight is on the shoulders, the chest is lifted to a more perpendicular position, and the angle between the neck and the sternum will be 60–90° depending on individual constraints.

The third variation takes you into a completely different posture. For this one you push your feet even more to the rear and press the sternum against the chin for the first time, creating a 90° bend between the head

Figure 9.12a. For the first variation (or stage) of the plow, the feet are pushed minimally overhead, the hips are flexed maximally, and the mid-back is kept as close to the floor as possible.

Figure 9.12b. For the second stage of the plow, the feet are pushed further overhead, hips are flexed moderately, and the back is now perpendicular to the floor.

Figure 9.12c. For stage three (third variation) of the plow, the feet are pushed maximally overhead with the toes extended, the hips are flexed minimally, and the back is flexed enough to permit the feet to reach the floor.

and the chest. It is now more convenient to extend the ankles (plantar flexion of the feet) and rest on the upper surfaces of the toes. In this stage both the hips and the lumbar region will be less flexed than in the second variation, perhaps 60° each in the hips and lumbar spine, for a total of 120° instead of the 160° that is characteristic of the first two variations (fig. 9.12c). The chest will now be fully perpendicular, at a 90° angle from the floor and from the neck, and if you want to do so you can take the option of pressing the hands against the back exactly as you did in the classic shoulderstand and for the same reason—to keep the sternum locked tightly against the chin. Alternatively, to sharpen the pose, instead of bracing the chest by pressing the hands against the upper back, flatten the arms and forearms against the floor behind the back with the fingers interlocked and the palms pressed together. That arm position lifts your weight even higher on the shoulders and pushes the extended toes even further to the rear (fig. 9.12c).

Notice that the first variation is more like the inverted action posture with respect to neck flexion, and that only in the last variation of the plow series is the head and neck in the full 90° shoulderstand position. Notice also that all three variations are relatively passive, even though you are stretching muscles on the back side of the body from head to toe, and that this makes the poses especially useful for those who have difficulty with forward bending. In contrast to the sitting forward bend, gravity aids the plow series whether you are flexible or inflexible, so be watchful that your body weight does not pull you further into the posture than is prudent.

STRENGTH AND FLEXIBILITY IN THE PLOW

No matter what version of the plow you are working on, if you wish to lengthen the hamstrings, keep the knees extended as a first priority; if you wish to work on adductors, abduct the thighs; and if you wish to improve hip flexibility without being impeded by either the hamstrings or the adductors, bend the knees, bring them together, and pull the thighs closer to the chest. Notice that if you support the thighs with the elbows in this last variation, you will be moving into what may now be an old favorite—the relaxed easy inverted action posture.

With the thighs abducted in the plow you can work with certain stretches that are not accessible in any other posture. First, to work specifically on hip flexibility, come into the pose with maximum abduction and flexion of the thighs: you will be stretching both the hamstrings and the adductors. Notice how far you can abduct the thighs. Then with the thighs abducted, push more to the rear into a posture that is comparable to variation two of the plow, and notice that when you do that the thighs can be abducted even further. The reason for this is simple. The shift from flexion in the hips to flexion in the lumbar region takes tension off the hamstrings and those adductors which have a posterior origin along the inferior pubic rami, and this in turn permits more abduction. This situation becomes increasingly pronounced the further to the rear you plant your feet. These are great stretches. After the struggles many students have with sitting forward bends in which the thighs are abducted and gravity is thwarting rather than assisting the bend, working with the abducted thighs in the plow or propped plow is a pleasure.

Next, keeping the feet together, bring them to one side and then the other as far as possible. This combination of a twist with a deep forward bend can be done from any stage of the plow, but try it first with the hands overhead in variation two, which is a comfortable intermediate balancing position. Slowly lift your toes just off the floor and slide them to one side as far as is comfortable. Then slide both feet slowly to the opposite side, again barely touching the floor. Keep the knees as straight as possible, don't hold your breath, and try not to bounce from one side to the other.

The only way you can develop the tension needed to raise the toes just off the floor in this posture is to tighten the back muscles and hamstrings. But as you raise the feet the hamstrings come under tension and tend to flex the knees, and that has to be resisted with the quadriceps femoris muscles. This is an exercise that can only be done inverted.

The last exercise in this series is to combine the plow with sitting forward bends and leglifts in a dynamic sequence. If you are slender and lightly muscled it should be done on a mat or soft surface. Start in the usual supine position with the hands beside the thighs, palms down. Then do a slow double leglift and come into the plow without using your upper extremities any more than you have to. This is an abdominal exercise. As soon as your feet touch the floor overhead come out of the posture by rolling slowly down, one vertebra at a time. Try to keep your head on the floor instead of raising it up as the middle segment of your back rolls down (which is easier said than done), and hold enough tension in the abdomen and arms to keep your pelvis from plopping down. Then, as soon as the pelvis reaches the floor, slowly lower the feet while keeping the knees extended. Keep going. When your heels touch the floor, roll slowly up into a sitting forward bend and reach forward with the hands. Don't try to bend from the hips or to establish sacroiliac nutation as a priority. It is more natural just to roll forward and down. Keep moving. As soon you are in an easy full forward bend, roll back down to a supine position one vertebra at a time, do another slow double leglift, and again come into the plow. Repeat the sequence as many times as is comfortable.

THE LIFTED SHOULDERSTAND AND PLOW

In the lifted shoulderstand and plow, you raise the shoulders higher than the head with a blanket or extra-firm mat to make the poses easier and to remove stress from the neck. This makes the lifted postures useful alternatives for students who are not prepared for the intensity of the full poses. For an initial trial, the mat should be an inch thick, long enough to accommodate the entire torso, and about three feet wide.

THE LIFTED SHOULDERSTAND

Try the lifted shoulderstand first. Lie supine with the edge of the mat at the mid-cervical region, the head off the mat and against the floor. Ever so slowly press the hands against the mat and lift the torso up and the feet overhead. Support the back with the hands as usual, but be watchful of your balance. The posture is unstable because the sternum is not compressed against the chin, and if you're not attentive you'll fall to one side or flip over into a backward somersault, which is difficult in the regular shoulderstand unless you are trying to do it.

After you have worked with a one-inch-thick mat you can try the lifted shoulderstand with one or more additional thicknesses. But the higher the support, the more unstable the posture, and the more care you must take not to fall. Many yoga teachers recommend and even insist on the use of a significant support—perhaps a 5-inch thickness of firm matting, or sometimes even more. But even though the lifted shoulderstand will protect the neck, the remedy can be worse than the cure if you lose your balance and fall. This is why, if you are having students use extra thick mats, you should lead them into the lifted shoulderstand guardedly, making sure that their shoulders are solidly placed and that they have done some preliminary experimentation with a thin mat. In the lifted shoulderstand illustrated here (fig. 9.13), a 2.5 inch thick wrestling mat provides substantial support.

We call this pose a lifted shoulderstand, and that is fair enough, but it is no longer sarvangasana, the all-member's pose. It is a shoulderstand because your weight is still placed mainly on the shoulders, but unlike the classic shoulderstand it is more of a balancing posture because the erector spinae and muscles of the lower extremities do not have to maintain nearly as much activity to stabilize it. In comparison with the classic shoulderstand, it takes almost no effort to straighten the body, especially if your mat is 2–4 inches in thickness. The three internally supported versions of the classic posture are also easier with this prop: since the chin is not compressed against the neck, the cervical vertebrae are not placed under traction and the sensations of intensity and energy associated with the classic shoulderstand are either absent or markedly reduced. It is not an exaggeration to say that in this posture we lose almost all of the defining characteristics of the classic shoulderstand.

In the lifted shoulderstand the torso is vertical but the neck angles off downhill, about 10° more than a right angle if you are minimally supported, and more like 45° beyond a right angle if your shoulders are lifted up by 3–4 inches of firm matting. And the more the head is directed downward at an angle, the more the neck will be supporting the posture (as in the headstand) rather than being acted upon by the posture (as in the classic shoulderstand).

The empty space behind the neck in the lifted shoulderstand presents one potential problem: it allows the cervical region to round to the rear, creating a reverse cervical curvature, in other words a curvature that is convex posteriorly instead of anteriorly. In general, the posture is not harmful for a limited time, but if you have this condition in your neck before you start working with this series, the lifted shoulderstand should not be done at all. It can only make a reverse curvature worse. The classic shoulderstand, on the other hand, can be therapeutic because that posture flattens and stretches the cervical region.

THE LIFTED PLOW

If you are not quite ready for the intensity of the classic plow, the lifted plow is a useful alternative. Find just the amount of lift needed to allow the feet to reach the floor. The lifted plow not only takes stress off the neck, it

Figure 9.13. To work with the lifted shoulderstand, start with a one-inch thick mat and then increase the thickness to 2–5 inches or even more. Here the shoulders are supported by a 2.5 inch wrestling mat. Be watchful. The lifted pose is much more of a balancing posture than the standard shoulderstand, and students who have never tried this before, even those who are experienced with the regular shoulderstand, may unexpectedly (and quite suddenly) tip over backward, just as sometimes happens if you try to come into the shoulderstand on a slope with the head down.

also makes the posture easier than the classic plow because you do not have to have as much hip and back flexibility to lower the feet to the floor while keeping the knees straight. The same cautions apply here as for the lifted shoulderstand. The higher the support, in this case two, 2.5 inch mats combined (fig. 9.14), the more unstable the posture. Proceed with caution.

CIRCULATION

Circulation and respiration go hand in hand. When athletes speak of cardio-respiratory fitness they are talking about both functions: getting air into the lungs and transferring oxygen from the lungs to the tissues of the body (chapter 2). Inverting the body affects these processes profoundly and in different ways depending on the specific posture. We'll look at six postures that illustrate some of the differences: the headstand, the shoulderstand, the inverted action posture, the lifted shoulderstand, stage one of the plow lying with the chest almost flat, and stage three of the plow with the feet pulled fully overhead.

Like the headstand, the shoulderstand and related postures drain blood and excess fluids from the lower extremities and abdominopelvic organs, and for this reason they are excellent practices for anyone with varicose veins or sluggish circulation in the lower half of the body. The effects on circulation in the head and neck, however, are different and more complex in the shoulderstand than they are in the headstand (fig. 9.15). The most obvious point of contrast is that blood pressure in the head is lower in the shoulderstand because the vertical distance between the heart and the brain is only a few inches, while it is roughly 12–16 inches in the headstand, depending on your body type. If we calculate that average blood pressure in the brain during the headstand is around 130 mm Hg (figs. 8.2 and 9.15)

Figure 9.14. The lifted plow is a special pleasure for those whose flexibility cannot quite accomodate to the classic plow. Here the shoulders are supported by two, 2.5 inch mats combined. This is not as much of a balancing posture as the lifted shoulderstand because the feet reach the floor, but there is still a tendency for the uninitiated student to tip over in a backward somersault.

instead of the 100 mm Hg (average) at chest level, we can estimate that blood pressure in the brain during the shoulderstand will be more like an average of 110 mm Hg (fig. 9.15).

It is obvious that blood pressure in the head decreases in the shoulderstand in comparison with the headstand, but the situation in the neck is a separate question. One might expect a decrease in blood pressure here as well, and for the same reason—because in the shoulderstand, the neck is not as far below the heart as it is in the headstand. This is not, however, borne out experientially. Unlike the headstand, when you are in the classic shoulderstand there is a localized sensation of extra, rather than reduced, pressure and tension in the neck. Exactly what these sensations mean has not been tested in the clinic or laboratory, but the lore in hatha yoga is that some of the major arteries supplying the brain are slightly occluded in the shoulderstand because of the severe flexion in the neck. And if that is what happens, constrictions in those major arteries could cause increased blood pressure in any nearby region that is supplied by arteries that branch off just before the hypothesized constriction.

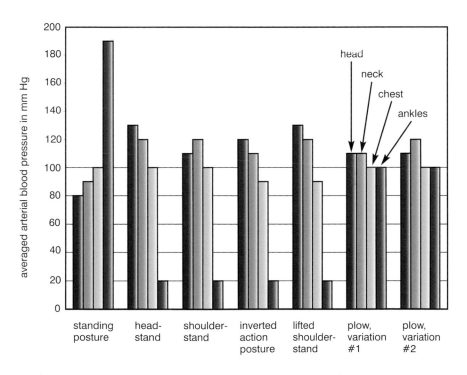

Figure 9.15. Comparisons of postulated regional blood pressures in standing and various inverted postures. For each posture (from the left), the average arterial blood pressure (systolic/diastolic, over time) is estimated locally for the head, neck, chest, and ankles.

We can gain insight into this puzzle and find a possible solution to it by comparing the inverted action pose with the shoulderstand. The distance between the heart and the neck, and between the heart and the head, are almost the same in both postures, so any differences in blood pressure due to the pull of gravity should be minimal. But anyone who has compared the postures experientially notices two things about the inverted action pose: in the neck there is dramatically less local tension and pressure than in the shoulderstand, and in the eyes, ears, and face, there is an increased sensation of pressure. There is only one way to explain these findings easily—by postulating that, compared with the classic shoulderstand, the inverted action posture releases constrictions in the great vessels of the neck, which in turn allows blood to course more easily into the head.

The lifted shoulderstand has its own special effects on circulation (fig. 9.15). With the shoulders elevated the heart is lifted even higher than in the shoulderstand, and this will increase the blood pressure in the brain in accordance with the height of the lift. You do not notice this so much if you are using a one-inch mat, but it becomes pronounced as you raise yourself higher. Lifted up five inches you feel a rush of pressure in the head which is almost identical to that felt in the headstand.

The plow postures, with the feet touching the floor overhead, have still a different effect on circulation. Here there is no pronounced drainage of blood from the lower extremities, but once blood is in the abdominal region it will be recirculated quickly back to the heart. As far as effects in the head and neck are concerned, if your hamstrings and hip flexibility allow you to draw the feet overhead without lifting the chest very far off the floor (as in variation one of the plow), the heart will be just a little further off the floor than it is in the corpse posture. The feet and lower extremities will not be very far up in the air, and the posture will affect blood pressure in the head only mildly (fig. 9.15). But if you are flexible enough to take the plow to variation three by pushing the feet to the rear and flexing the neck maximally, the expected effect on blood pressure in the head and neck will be similar to what we see in the shoulderstand (fig. 9.15).

RESPIRATION

In the shoulderstand and its related postures, some of the effects of breathing are similar to what we found in the headstand, but we also see several important differences. For one thing you are generally more at your leisure in the shoulderstand and plow series; for another you can watch your abdomen as you breathe, evaluate the character of exhalation and inhalation, and time your rate of breathing with the second hand of a watch.

Starting with the classic shoulderstand, repeat the experiment we did earlier with the headstand: breathe normally for several cycles and then relax the respiration suddenly at the end of a normal inhalation. As in the headstand, you will notice that air is expelled with a whoosh and that the abdomen caves in suddenly as the diaphragm relaxes and the abdominal organs drop without restraint toward the head and neck. Anyone who understands how the diaphragm works can then return to normal breathing and sense how it ordinarily restrains exhalation in the shoulderstand by lengthening eccentrically and restraining the fall of the abdominal organs toward the floor.

The most important difference between breathing in the headstand and in the classic shoulderstand is that the headstand allows slower, deeper breathing. As we saw in chapter 8, it is easy to acclimate to as few as 3–4 breaths per minute in the headstand; in the shoulderstand it is inconvenient to breathe at rates of less than 6–8 breaths per minute, and 20 breaths per minute feels more comfortable. Why this happens is a mystery. As in the headstand, you can't breathe thoracically or paradoxically, but abdominal breathing feels free and easy. For whatever reason, in the end, the tidal volume seems to be reduced in the shoulderstand, and the more you reduce the tidal volume, the faster you have to breathe to get enough air. The question is, why is the tidal volume reduced? One possible answer is that expansion of the chest is even more restricted in the shoulderstand than it is in the headstand. We know that diaphragmatic breathing cannot take place if the base of the rib cage is constricted (chapter 2), and we know that we depend on diaphragmatic breathing to reduce our respiration to 3–4 breaths per minute in the headstand. The confounding element in the classic shoulderstand is possibly that your hands are pushing so insistently on the lower border of the rib cage that the diaphragm cannot easily enlarge it from its base.

The lifted shoulderstand is still different. This is a more relaxed posture than the ordinary shoulderstand, but for reasons that are not entirely clear the rate of breathing increases, especially if the posture is lifted 3–4 inches. Come into this pose after timing your rate of breathing in the shoulderstand, and you will suddenly feel a sense of urgency to breathe faster. If your normal rate of breathing in the classic shoulderstand is 20 breaths per minute, it may go up to about 30 breaths per minute in the lifted shoulderstand. The source and neurologic mechanisms for the increased rate of breathing are not clear, but it feels like a reflex, and it may have something to do with the fact that the neck is no longer flexed 90°.

Respiration in the plow is similar to that in the shoulderstand but somewhat slower, especially if you are able to make yourself comfortable. The whole-body forward bend creates a situation similar to that seen in the sitting forward bend in which each inhalation lifts the body and each exhalation lets you drop further forward. The same thing happens in the plow except now it is the body that is fixed in position and the lower extremities that can be lifted. And that's what we find. Come into the second stage (variation) of the plow posture, and you'll find that each inhalation tends to lift the toes off the floor. If you are not convinced, inhale as deeply as possible in such a posture and the picture will become clear.

Although there are differences in breathing among the various inverted action postures, most of them increase the rate of breathing to around 30 breaths per minute. This is particularly noticeable in the passive inverted action postures supported by a ball or by the edge of a couch. In most of these postures you are so relaxed that you breathe out most of your expiratory reserve in a quick burst as the abdominal viscera press against the diaphragm from above. It's like a mild, automatic kapalabhati (chapter 2) in which the short bursts of exhalation are followed by longer inhalations. The difference here is that the exhalations are passive instead of active. You can breathe evenly if you want in passive inverted action postures by consciously restraining exhalations, but doing so requires constant attention.

SEQUELAE

Now we turn to a variety of exercises and postures that in one way or another closely relate to the shoulderstand and plow series. Some of them strengthen the neck and others pamper it; some prepare you for the formal postures and others counteract their stresses; and often the same exercise confers more than one of these benefits. The threading-the-needle and knee-to-ear poses are good training tools for both the plow and the shoulderstand, limbering the upper body to prepare you for the plow, and accustoming you to stress in the cervical region. The arch and bridge are also excellent training postures which can either be used in preparation for the shoulderstand and plow or as follow-up postures, along with the fish, to provide counterstretches for the back. We'll end with yoga nidrasana, the last posture before embarking on relaxation and meditation poses in chapter 10, and a supremely comfortable forward bend for those who can do it without stress and strain.

RESISTANCE NECK EXERCISES

In chapter 7 we looked at neck exercises in which you simply take the head through its full range of motion, differentiating among the movements that are possible between the cranium and C1, between C1 and C2, and between C2 and T1. In all such exercises minimal muscular activity is needed until you come to the end of the excursion, at which point joint and ligament restrictions permit no more than an isometric effort. Such work is useful, but it is even more effective to create some form of resistance to neck movements from beginning to end, and this is the definition of resistance neck exercises.

These exercises can be done at any time during the course of inverted postures. Doing them before the shoulderstand will prepare the neck muscles, joints, and ligaments for the unusual stresses and stretches to follow. They are also helpful after the headstand because that posture places a constant, isometric stress on the muscles of the neck, and the best way we can counter-act that static condition is to challenge the muscles throughout their full range of motion. Use moderation, however, especially in the beginning. Although these exercises are safer than ordinary neck movements, it can't hurt to be cautious.

Ten resistance neck exercises will get you started. First press the right hand against the right side of the head and at the same time bring the right ear toward the right shoulder by tightening the muscles on the right side of the neck. Resist this movement with the right arm. Then slowly raise the head, still pushing with the hand and resisting with the neck. As soon as you reach the upright position keep going to the other side by pressing the left ear toward the left shoulder with the right hand, still resisting all the way with the muscles on the right side of the neck (fig. 9.16). Go from side to side as far as possible two or three times. Second, repeat the exercise from the other side with the left hand pushing against the left side of the head as the muscles on the left side of the neck resist. Third, bring both

Figure 9.16. This is the first of ten or more resistance neck exercises. Here the model is resisting lateral flexion of the neck and head to his left with his right hand. That is, he is pushing with his right hand and resisting with the muscles on the right side of his neck. This is followed by slowly pulling his head to his right as far as possible, still resisting all the way with his right hand. The movements should be done slowly enough that they have an isometric character. Number two in the series is done with the left hand, and the rest follow logically.

hands to the forehead and slowly pull the chin toward the sternum and then to the rear as far as possible, creating resistance with the hands in both directions all the way. Fourth, repeat with both hands behind the head. In all of these exercises move slowly enough so that the tension has an isometric character at any given moment.

For exercise number five, twist the head 45° to the right and then move it from right front to left rear, resisting the movement with the right hand placed on the forehead; for number six twist the head 45° to the left and resist the movement from left front to right rear with the left hand. For number seven and eight repeat with the fingers interlocked behind the head, first with the head twisted 45° to the right and then 45° to the left.

For number nine twist the head and neck axially as far as you can in both directions, resisting the movement with the right hand on the left temple and with the left hand behind the head almost to the right ear. Your hands are trying to twist the head to the right and the neck muscles are resisting; you can go all the way to the right, but you are limited when you go to the left because the left forearm stops the twist before you reach your limit. For number ten, switch hands and again twist while resisting. Now you can go all the way to the left but are limited when you go to the right.

From here on you are limited only by your imagination. You can twist first, and then go forward and backward, or you can take your head forward, backward, or to the side, and then resist a twisting movement. Or you can resist large arcs of movement, looking under the axilla on one side and creating an arc up and back to the other side.

Muscular resistance is a time-tested agent for protecting the body as well as strengthening it, and combining strength-building exercises with those that improve flexibility can help a lot in building confidence for doing the headstand and shoulderstand. The exercises are safe provided you build your capacity for them slowly. Interestingly, after you have done them for a few months the neck will be stronger than the upper extremities, and no matter how hard you push with your hands you will be able to stop the movement with the neck muscles.

SELF-MASSAGE OF THE NECK

Most of the inverted postures place unusual stresses on the neck, and anyone who spends a lot of time with them should coddle the associated muscles, ligaments, and bony structures. One way is to administer some self-massage by pressing the neck down against the same 8 1/2 inch playground ball discussed in connection with passive backbending in chapter 5.

Start by placing the ball under your upper shoulders just below the nape of the neck at T1. From that position you can roll the body down so that the ball is in contact with successively higher regions of the neck and head, ending with the back of the skull where the strap muscles insert on the occipital region of the cranium. Or you can start in the middle and work both down and up. Starting in the mid-cervical region, you can either press your weight straight backward into the ball or twist the neck one way or the other while pressing the head to the rear. The former will affect the insertions and origins of muscles that attach to the back of the head and to the spinous processes of the vertebrae; the latter will massage muscles whose tendons attach both to spinous processes and to transverse processes. In either case you will be manually stimulating the Golgi tendon organs and relaxing their associated muscles.

In company with the resistance neck exercises, this self-massage not only strengthens and relaxes the neck, it also gives you effective feedback as to right-left imbalances. You will sense differences in tenderness or mobility on the two sides, and this will enable you to plan a more effective practice. After you have toughened up the joints, muscles, and ligaments, and after you have adapted to an 8 1/2 inch playground ball, you can graduate to a harder volleyball, soccer ball, or basketball, which is where wrestlers and bodybuilders can begin.

No matter what your stage of practice, as you become familiar with the sensations created by the resistance neck exercises and self-massage with a playground ball, you will gain more sensitivity to the inverted postures that you need to approach with caution as well as to those you can approach with more enthusiasm. These are great exercises for engendering self-awareness.

THREADING-THE-NEEDLE

The threading-the-needle posture and its variants can be done at any time in a sequence of inverted postures. They prepare you for what is yet to come and relax you from what you just did. The threading-the-needle posture could also be called the moldboard plow posture, named for a modern plow that throws the dirt in only one direction. Another name for it is the trapezius stretch, named from how it stretches and stimulates that muscle (fig. 8.10). It is also one of the best postures for stretching the muscles on the side of the neck, for flexing the spine in a moderately twisted position, for working with hip flexibility unencumbered by hamstrings and adductors, and for preparing you for advanced spinal twists. And if you are stressed out by your job and find yourself wanting a massage for the muscles in your neck and upper back, you will find this posture a special friend.

Start out on a mat or other padded surface in a hands-and-knees position. Then bring the point of the right shoulder against the pad, and stretch the right arm and forearm under the body to place the right elbow against the lateral edge of the left knee. Place the right temple in a comfortable position against the pad and roll the head against the floor, twisting it to the left and looking up toward the ceiling. Roll the head back to a neutral position with the weight against the temple on the right side. Twist it back and forth as far as is comfortable, rocking and soothing the cranium. Repeat on the other side with the left elbow beside the right knee. These exercises are preliminary to the next stretch.

Begin as before. Reach down with the right hand again, but this time lift the left knee and foot over the right forearm and then push the body forward with the left foot so that your weight settles on the back side of the right shoulder. If you can keep both knees down, fine, but if you are not very flexible you may not be able to keep your balance while you roll onto the back of the shoulder. If this is the case, you may have to stretch the left foot out further to the side and then forward to keep yourself from toppling over onto your right side. You also have to keep the right knee bent (the right knee represents the rolling cutter of the moldboard plow). Now stretch the left foot out to the side so that the right arm and forearm (the thread) end up lying on the floor between the two legs (the eye of the needle). Alternatively, you can ease yourself slightly out of the posture and place the right elbow to the right side of the right knee (fig. 9.17a). From that position swing your right hand even further around behind you, and that will both keep you from toppling over and help you roll over onto both shoulders combined. Explore the posture to find the position in which you are most comfortable.

Flattening both shoulders against the floor is the most rewarding position for stretching the trapezius and strap muscles of the neck. If you topple over, the main reason is that your outside knee (or foot) is not pulled far enough out and forward. And if, in spite of all your efforts you still keep falling over, try doing the posture on a soft mattress. The shoulder you first apply to the mattress will dig in, and it will be easier to press the other one down as well. This way you will be able to feel the essence of the posture, and as you become more flexible you can use a harder surface. Repeat on the other side.

If you find this posture difficult, don't rush. Approach it in a spirit of play. Once it starts feeling comfortable you will value it greatly for relieving stress in the shoulders and neck and for serving as a bridge between more demanding postures. It is a special help for those who are not quite able to do the classic plow. It is also valuable as a counterbalance to the headstand, especially if that posture is held for more than 3–5 minutes. And after you have acclimated to the simplest threading-the-needle version you can search out small variations. You can work on lowering the outside knee closer to the floor, and after you are confident you can swing around from one side to the other. First bring the legs in line with the body, remain for a moment in any relaxed easy inverted action posture with the knees bent (fig. 9.2), and then keep on going by swinging the knees to the other side.

This next variation of threading the needle could be called the half lotus threading-the-needle (fig. 9.17b) because one foot is pulled into the half lotus position in the final pose. It is especially valuable as preparation for the full spinal twist (fig. 7.33b). From the beginning hands-and-knees position, again bring the right elbow to the lateral side of the left knee and place the right temple against the floor. Lift the left knee over the right forearm and thread the needle by pushing over onto the right shoulder with the aid of the left foot. So far everything is the same as in the basic

Figure 9.17a. This is one of several possible variations of the trapezius stretch. In this case the right elbow is located to the right side of the right knee (from the practitioner's perspective). For the mulboard plow, or threading the needle, the right elbow and forearm would be between the two knees. Everyone should experiment freely to make this posture a special, personalized joy. In time, many variations will suggest themselves to you.

Figure 9.17b. This pose might well be called the half lotus trapezius stretch. It is exactly like the one shown in fig. 9.17a except that the right foot and ankle are placed in a half lotus position near the top of the left thigh. Because the body is twisted markedly, this posture is especially valuable as a training tool for the full spinal twist.

posture. Now grab the right foot with the right hand and pull it forward, and at the same time straighten the left knee and pull the right foot to the front side of the left thigh in a half lotus position. Pull the right foot as close to the top of the thigh as possible and slowly flex the left thigh, lowering the left knee to the floor or as far down as it will go (fig. 9.17b). If you can't get the right foot all the way up to the crease between the thigh and the torso, you will obviously not be able to flex the left thigh. Don't rush it.

THE KNEE-TO-EAR POSE

It is natural to follow the threading-the-needle posture with the knee-to-ear posture, although if your flexibility is limited, you may wish to delay this until after you have gotten comfortable with the plow, the bridge, and the fish. To do the posture you simply lower down from the shoulderstand or plow, or swing over into the midline from the threading-the-needle posture and place the knees on the floor beside the ears. You can simply reside there or you can wrap the hands around the thighs and pull them further into the posture (fig. 9.18). The hamstring muscles do not limit the knee-to-ear pose because the knees are bent. And because the adductors are also not stretched, the posture is limited solely by the hip joint and the soft tissues of the groin. If you are close to completing the pose you can temporarily interrupt your efforts, do some forward bending and hip-opening postures, and the knee-to-ear pose will then come more easily.

The threading-the-needle and knee-to-ear postures complement the shoulderstand and plow in that they are mild but effective neck stretches and do not require too much of the lower extremities. They are especially helpful as preparatory postures for the plow, limbering the upper body so that it is flexible enough for you to be able to attend more single-mindedly to stretching the hamstring muscles.

Figure 9.18. The knee-to-ear posture is not limited either by the hamstrings (because the knees are flexed), or by the adductors (because the thighs are mostly adducted), and this means the pose is especially valuable for pushing the hip joints to their maximally flexed positions. You can come into this posture from the shoulderstand, from the plow, or from threading the needle.

THE FISH

We discussed the fish pose with backward bending (fig. 5.28) because it is a backbending posture, and we described the superfish leglift with abdominopelvic exercises (fig. 3.19b) in relation to abdominal strength. Traditionally, however, the fish is practiced after the shoulderstand and plow because it gives the neck an effective counterstretch and because it opens and releases the chest after the stress of those two postures. Several variations are common. The lower extremities can be crossed in either the easy posture (fig. 9.19) or the lotus posture (fig. 5.28), but the fish posture that is usually taught to beginners is sharper. It simply involves keeping the feet outstretched, with the heels and toes together, lifting up on the forearms, arching the back and neck, and placing a little of your upper body weight on the back of the head (fig. 3.19a). As soon as you are confident that the neck is strong you can reside in the posture with the hands in the prayer position. And as soon as the neck gets really strong you can do a wrestler's bridge, supporting the entire body between the feet and head, arching up as high as possible, and extending the head and neck (fig. 9.20).

The superfish leglift (fig. 3.19b) is excellent both for building strength and for complementing the shoulderstand and plow. It counters your inclination to flex the back in the shoulderstand and to flex both the back and the hips in the plow, and it balances the emphasis on counternutation of the sacroiliac joints in both postures. From a supine position lift the torso

Figure 9.19. This fish pose using the easy posture is supported mostly by the forearms; little weight is borne by the head and neck (also see figs. 3.19a and 5.28).

Figure 9.20. The wrestler's bridge is excellent for improving flexibility for backbending, for strengthening the neck, and for stretching the hip flexors and quadriceps femoris muscles.

partially up, supporting yourself with the forearms, and place the palms under the hips. Then lift the lower back and chest maximally, draw the toes toward the head, extend the knees fully, and raise the heels an inch or two. As in the fish, your head should barely touch the floor.

Raising the thighs is accomplished by the psoas and iliacus muscles, aided by the rectus femoris muscles acting as synergists, and since the psoas lifts directly from the lumbar lordosis with little help from the abdominal muscles, the spinal origin of the psoas should be stabilized in the forward position before you attempt to lift the lower extremities. To get the most benefit from this exercise, be sure to start the leglift with the sacroiliac joint in full nutation. Keep that attitude along with the deepest possible lumbar arch as you start to flex the thighs. You probably can't go very far: unless you have excellent hip flexibility, you will not be able to lift up more than an inch or two before you feel the back begin to flatten down against the floor.

THE ARCH

The arch and the bridge postures extend the back from the chest down, and except in the neck they counter the forward bending tendencies of the shoulderstand and plow. You will find that after doing either the bridge or the fish for a minute or so you can come back up into the shoulderstand and use your back and gluteal muscles for that posture with renewed energy. Although the arch and the bridge look as if they are related, their musculo-skeletal dynamics are quite different.

The arch is the simplest of the two postures and the best one for beginners. Begin in a supine position and grasp the feet or the heels with the hands and lift the pelvis as high as possible with the deep back muscles, the gluteus maximus muscles, and the hamstrings (fig. 9.21). The deep back muscles are contracting concentrically between the chest and the pelvis, the gluteus maximus muscles act between the pelvis and the thighs, and

Figure 9.21. The arch posture is a good preparation for the more demanding bridge pose. Lift the pelvis as high as possible to stretch the hip flexors and the quadriceps femoris muscles.

the hamstrings act between the pelvis and the legs—and all are working in concert to lift the pelvis toward the ceiling. In those who are least flexible, the arch is resisted by the rectus femoris muscles, which act between the pelvis and the knees to restrict the pelvis from being lifted very far. In those who are stronger and more flexible, the arch is resisted by the hip flexors and abdominal muscles. You produce a natural ashwini mudra in this posture, pulling inward with both the gluteal muscles and the pelvic diaphragm.

We have seen that anyone who has knee problems may find that placing tension on the knees when they are flexed causes pain, so the arch posture is contraindicated in that case. Except for that, it is safe for beginners because it is supported so completely with muscular activity. To come out of it you just drop back down to the floor.

THE BRIDGE

Like the arch, the bridge posture counteracts some of the effects of the shoulderstand and plow by arching the back. But unlike the arch, the bridge supports the lower back with the forearms and hands. This has two effects: it allows the deep back muscles to relax, and it requires additional extension. The posture is also a good preparation for the shoulderstand and plow, and it can even be used as a substitute by those who are wary of being fully inverted. The main difference is that the bridge does not require you to pull the lower half of the body straight up in the air (as in the shoulderstand) or overhead (as in the plow). Even so, students will experience many of the same sensations in the neck, shoulders, and upper back that will be found in more intense form in the other postures.

Figure 9.22. The bridge posture in its final form demands a healthy back. Placing the hands with the thumbs in and the fingers out is easier than the position shown here with the fingers directed toward the center of the back and thumbs coming up the sides. At first keep the knees bent to make the posture easier, and come into the pose from the shoulderstand. Come back up into the shoulderstand one foot at a time before trying to come in and out symmetrically (both feet down and back up at the same time).

The easiest, although the least elegant, way to come into the bridge is to start by lying flat on the floor, lifting the pelvis, and working your hands into position under the lower back. At this point you have two options for the hand position and two for the feet. Those who have wrists they can trust for maximal extension can point their fingers medially; if that is too difficult, they can support the posture with the heels of the hands and let the fingers extend laterally around the lower back. For foot position, those who have excellent strength and flexibility can stretch their feet straight out in front of them, keeping the knees extended (fig. 9.22), but a more moderate position is to bend the knees 90° and plant the feet flat on the floor.

If you want to come into the bridge from the shoulderstand, which is the usual choice, you will probably need to learn it in stages. And the way to do this is to assume the shoulderstand and then simply bring one foot down as far as you can, going slowly and remaining certain that you will be able to come back up under control. Do the same with the other foot, and repeat the exercise over and over until you are confident that you can reach the floor with one foot and still return to the shoulderstand.

The next stage is to lower the second foot to the floor and then bring both feet (but still one at a time) back up into the shoulderstand. There is no substitute for trial and error in doing this exercise. Once the lead foot is most of the way down, you are committed to bringing it all the way down. And once you get one foot down, you will want to get the other foot down gracefully and then get both feet back up without having to toss them into the air. If it is beyond your capacity to come back up into the shoulderstand the same way you came down (meaning slowly and under control), you can always release the hands from the back and support the pelvis instead, or as a last resort you can lower down into the arch and from that safe position ease yourself to the floor.

The most advanced bridge is to lower both feet at the same time toward the floor in front of you. Then, after remaining in the pose for a minute or so, gather your energy and slowly raise both feet straight up in the air, again at the same time, in order to come back into the shoulderstand. To make this easier, you can keep the knees bent and drop down only on the balls of the feet, which requires keeping the ankles extended and the toes flexed. Both this and the hand position with the fingers pointing laterally will lessen the extension required of the back.

YOGA NIDRASANA

In *yoga nidrasana* you are not resting on the triangular region formed by the back of the head, the neck, and the shoulders (as in the shoulderstand), but on the upper back (as in variation one of the plow). In fact, if you envision that posture with the sacroiliac joints in full counternutation, the knees slightly bent, the thighs placed against the sides of the chest, the legs placed behind the head and neck, and the ankles interlocked, you will be envisioning yoga nidrasana.

To come into the pose, you can either begin with variation one of the plow, roll down a little more onto the back, and pull the legs behind the head from that position, or you can come into a supine position, cradle the legs one at a time, and again pull them into position. In either case, rest the head on the crossed ankles and complete the posture by bringing the arms outside the thighs and interlocking the hands behind the back (fig. 9.23).

Yoga nidrasana is challenging. To do it you must have enough hip and sacroiliac flexibility to press the knees to the floor alongside the chest in the supine position, and your hamstrings must be long enough to permit the ankles to be interlocked behind the head. This posture is named after the meditative practice of *yoga nidra* (yogic sleep) although it will probably be rare than anyone will be able to do that practice in this demanding pose. Most pupils will attempt exploration of yoga nidra in the corpse posture. Even so, the association of yoga nidrasana with the practice of yoga nidra makes the pose a fitting conclusion to what we have learned so far and points the way to the next chapter on relaxation and meditation.

Figure 9.23. Yoga nidrasana is especially demanding of hip flexibility, and is a fitting end to a practice session of hatha yoga and to beginning a practice of relaxation and meditation.

BENEFITS

The shoulderstand creates intense sensations throughout the neck, but the medical correlates of this are not obvious. It is an article of faith and experience among hatha yogis that this posture has beneficial effects on a wealth of functions: regulation of metabolism and mineral balance by the thyroid and parathyroid glands, beneficent influences on the larynx and speech, and salutary effects on immune function in the thymus gland and tonsils. But aside from anecdotal reports that practicing the shoulderstand is a good remedy for sore throats and nervous coughs (especially in children who are unable to fall asleep), and that vigorous practice of the classic posture and its variations enhances thyroid function in those who have mild hypothyroid conditions that are not due to iodine deficiency, we do not have much to go on. And without data reported in the peer-reviewed literature, there is little we can say other than agree that such effects are possible but unsubstantiated.

"*Symptoms of old age due to the faulty functioning of the thyroid are counteracted by means of Sarvangasana. Seminal weakness arising from the degeneration of the testes in the case of males and sexual disorders arising from the degeneration of the ovaries in the females can be extensively controlled by the practice of Sarvangasana. Dyspepsia, constipation, hernia, visceroptosis can be treated by Sarvangasana as well as by Sirshasana.*"

— Swami Kuvalayananda, in *Popular Yoga Asanas*, p. 67.

CHAPTER TEN
RELAXATION AND
MEDITATION

"Contemplation, especially the contemplation of inspiring concepts or ideals—such as truth, peace, and love—can be very helpful, although it is distinct and different than the process of meditation. In contemplation, you engage your mind in inquiry into this concept, and ask it to consider the meaning and value of the concept. In the system of meditation, contemplation is considered a separate practice, although one that also may be very useful at some times. When you engage in meditation, you do not ask the mind to think about or contemplate on any concept, but rather, to go beyond this level of mental activity."

— Swami Rama, in *Meditation and Its Practice*, p. 11.

Relaxation in yoga means lying down, most commonly in the corpse posture, and it means settling yourself in a circle of quiet, emptying the mind of all outside concerns, and relaxing from head to toe. Meditation is a stage beyond. It means sitting straight, most commonly in one of several classic meditative postures, and it means schooling your psyche and quickening your existence in consciousness. Relaxation is a concentration exercise for the body-mind; meditation is an experience for the mind-body. Both complement the rejuvenating effects of hatha yoga—relaxation because it releases deep tensions, and meditation because it pulls the mind inward and introduces it to higher states of yoga. We'll look at relaxation first because meditation cannot begin without a relaxed body.

Relaxation takes us beyond simply letting go of tension in skeletal muscles. It is a multifaceted process involving conscious control of the somatic nervous system and its innervation of skeletal muscle (chapters 1 and 2), regulation of the autonomic nervous system and its control over smooth and cardiac muscle (chapters 2 and 10), and the reining in of emotion and mental activity. The plan of this chapter is to look at how skeletal muscles relax and to examine the postures that are best suited for this. Next we'll turn to the role of the autonomic nervous system in relaxation, and then we'll look at advanced relaxation practices whose physiological correlates are unknown. Finally, we'll consider the six meditative postures, leaving discussion of meditation itself to the experts in that field and to the many texts that delve into that topic.

THREE AXIOMS

After beginning students have experimented for a while with relaxation, meditation, and other hatha yoga practices, certain principles become axiomatic. The first is that relaxation and meditation usually work better after a session of hatha yoga postures than before. Done first or by themselves these practices can intensify tendencies to lethargy and sleepiness, and many people can avoid dozing off only by priming their nervous systems with postures. Experienced students in good physical and mental condition are exceptions: if they are clear, calm, and alert even after a sound sleep, their teachers often advise them do their meditation immediately upon waking.

Experience also makes it apparent that even though relaxation and meditation go beautifully after hatha yoga, they do not work as well after a lengthy practice of aerobic exercise or heavy musculoskeletal workouts. Hatha yoga postures pull the mind inward, and physical exercises tend to scatter it. So if you are an enthusiast for sports or conventional exercises, try to do them at a different time of day from the practices of yoga. If you can't do that, if your schedule requires you to do everything in a sequence, do the conventional exercises first, follow these with hatha yoga, and end with relaxation and meditation.

The third axiom is more subtle. Unless you are compulsive, you can't overdo conventional exercises, hatha postures, or meditation. They are all self-limiting for one reason or another—aerobic exercise and musculoskeletal workouts because you know you'll get sore, hatha postures because they self-regulate from beginning to end, and meditation because you'll be uncomfortable if you sit too long. None of this is true of yogic relaxation. If you do too much of that practice the motor neurons seem to lose their edge: they get lazy. And as that condition develops you can lose your capacity for controlling them. Even worse, you do not get immediate feedback that this is happening. For this reason most of us would do better to practice relaxation only once or twice a day for no more than twenty minutes per session.

MUSCULAR RELAXATION

We'll begin our discussion of relaxation with skeletal muscles for two reasons: they are linked directly to the somatic nervous system, whose neurological ·circuitry has been widely studied and is relatively simple (chapters 1 and 2), and because skeletal muscles can usually be brought under conscious control. In chapter 1 we saw that the dendrites and cell bodies of the somatic motor neurons are located in the central nervous system (the spinal cord and brain), that the axons of motor neurons exit the central nervous system and travel in peripheral nerves, and that the axon terminals of motor neurons stimulate contraction of the muscle cells. The more nerve impulses per second that travel down the axon, the more vigorous the contraction of the muscle; the fewer nerve impulses per second, the weaker the contraction. We also saw in chapter 1 that skeletal muscles execute our speech, respiration, and every willful and habitual movement of the body, and that the commands of motor neurons (figs. 1.3–9 and 2.12) are an absolute dictatorship: they control skeletal muscles just as a puppeteer controls puppets. When the axons of motor neurons transmit nerve impulses, the muscle fibers on which they impinge contract, and when the axons don't transmit nerve impulses, the muscle fibers relax. You might think that muscular relaxation involves some complex planned activity, but that is not the case. All you have to do is silence the motor neurons. What we want to know here is how to do this consciously.

THE MOTOR UNIT

The functional entity for both relaxation and muscular activity is the *motor unit*, which is defined as a single motor neuron plus all the individual muscle fibers it innervates (fig. 10.1). The fewer the muscle fibers in the motor unit, the more finely we can control the muscle. The fact that the smallest muscles of the fingers contain only 10–15 muscle fibers in each of their motor units is what allows us to perform precise and delicate work with our fingers. As motor units become larger, ranging up to 500–1000 muscle fibers in the individual motor units of the largest postural muscles, our capability for fine movement diminishes. This is the reason for establishing distal-to-proximal priorities in standing postures (chapter 4). If you first establish control distally, it is possible to keep awareness of the small muscles in the background while you attend more consciously to the bulkier and less easily controlled proximal muscles.

The motor unit, in short, is the sole link between the central nervous system and skeletal muscular activity. Every time a nerve impulse reaches an axon terminal, all the muscle fibers in the motor unit contract. And when many motor units in an individual muscle contract repetitively and in concert, the entire muscle becomes active. What is important to us here is that relaxation requires us to silence the individual motor units one or more at a time. And this is indeed possible. Studies with electromyography carried out with needle electrodes have demonstrated since the 1950s that we can train motor units in most parts of the body to become totally silent.

Although motor neurons exert absolute control over muscles, they themselves are only agents of the body and mind as a whole. They are prisoners of our habits, addictions, and the willful decisions of the mind; they are prisoners of our hearing, sight, taste, smell, and touch; and they are prisoners of internal signals from stretched muscles, pain, or an overloaded stomach. Data from all of these sources are integrated and funneled into the final common pathway of the motor neurons.

FACILITATION, INHIBITION, AND RELAXATION

The key issue for someone trying to relax their skeletal muscles is to know something—knowledge is indeed power—about the specific form of the marching orders that regulate the rate of firing of motor neurons. These orders take the form of signals from thousands of other neurons whose cell bodies are located throughout the nervous system (mostly in the brain and spinal cord) and whose axon terminals impinge on motor neuronal dendrites and cell bodies that are specialized for receiving this information. Some axon terminals signal the motor neuron to increase its number of nerve impulses per second (facilitation), and other axon terminals signal it to decrease them (inhibition). It is in this manner that orders to the motor neurons are translated into simple stop and go inputs. The motor neuron integrates the sum of these often conflicting signals to increase or decrease the firing rate of its axon, and in that way it dictates contraction or relaxation of the individual muscle fibers of the motor unit (fig. 10.1). To relax we can conceive of decreasing the firing rate of motor neurons in three ways: by decreasing the rate of firing of the facilitatory neurons whose axons impinge on the motor neurons, by increasing the rate of firing of the inhibitory neurons, or by both in combination. Speaking simplistically, that's what is happening every time you do not respond to some desire or sensory signal.

What we see recorded in our movements is the summed activity of motor units, but something more subtle happens in the central nervous system during deep relaxation: the motor neuron cell bodies in the brain and spinal cord become so inhibited that a large facilitatory input is required to fire them back into activity. This can have an unexpected result: if you are in deep relaxation you may not be able to move on command. The telephone may ring, and when you try to jump up and answer it—surprise. You can't do it. You may experience several seconds of temporary paralysis, which can be startling and even alarming. With practice, however, you can learn

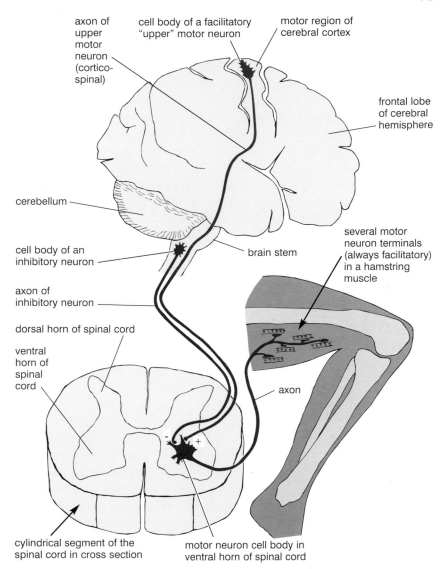

Figure 10.1. Two possible pathways for relaxation of skeletal muscles are shown here. In the cerebral cortex, if the activity of a facilitatory upper motor neuron is diminished, it will send fewer nerve impulses per second to spinal cord lower motor neurons (which in their turn drive the contraction of skeletal muscles). Conscious relaxation might also involve a pathway for stimulation of inhibitory pathways, as exemplified here by an inhibitory neuron in the brain stem. If such a neuron starts firing more nerve impulses per second than usual to the lower motor neuron, it could help silence the motor neuron independently of the reduced input from the facilitatory neuron in the cerebral cortex. In the spinal cord, the + indicates the corticospinal neuron's facilitatory synaptic effects, and the - indicates the brain stem inhibitory neuron's inhibitory synaptic effects (Dodd).

to speed the process of facilitation and gear the system up for activity in less than a second. You can feel it happening. First you can't move, a half-second later you feel the nervous system preparing itself, and a half-second after that you can spring up like a grasshopper. This should never be done as a classroom exercise, of course, because it is too jarring. It's a physiological experiment, not a yoga practice.

The temporary paralysis induced by deep muscular relaxation underscores and documents the need for honoring the third axiom—not to overdo relaxation. It is tempting to be excessive, relaxing before and after meditation, before and after meals, and before and after a night's sleep, but doing this without the balancing effect of other hatha practices diminishes your conscious control over the motor neurons. So get a good night's sleep, go to the bathroom, take a shower, and practice asanas enthusiastically for an hour. Then relax thoroughly while you are still full of energy.

Not everyone can will their muscles to relax. What goes wrong? We can only make inferences. If the circuits from the cerebral cortex are intact (fig. 10.1), the potential for willed relaxation as well as willed movement is available, but if those higher circuits are not used regularly, they gradually become dysfunctional and the unconscious input from other regions of the brain gets bossier. And because these non-cortical circuits are not under conscious control, their continuing activity prevents conscious relaxation. More encouraging, as long as the circuits from the cerebral cortex are intact, hatha yoga can help train them.

DISTAL AND PROXIMAL MUSCLES

As we have seen, the control of movement becomes more precise as the motor units become smaller in the more distal muscles in the extremities. Your ability to relax is affected in the same way and for the same reason. It is easy for almost everyone to relax the muscles of their hands, but shoulders and backs are another story. Relaxing larger and more proximal muscles of the body may require some additional help. A regular practice of hatha is the most certain tactic—stretching, contracting, and relaxing the deeper muscles of the body over and over again until the motor neurons get accustomed to obeying your conscious will and becoming silent on command. Another useful trick is the tension-relaxation sequence to be discussed in the next section. As a last resort, those who have chronic muscular stress in the deep postural muscles of the body can try biofeedback with electromyography. In such conditioning, electrodes are placed near hard-to-train muscles that operate just beyond the edge of your conscious awareness; you learn to relax and gain control of them by gaining control of their motor units.

TWO RELAXATION POSTURES

There is one classic relaxation posture in hatha yoga for the supine position, the corpse posture, which was discussed briefly in chapter 1 (fig. 1.14), and one for the prone position, the crocodile, two versions of which were discussed extensively in chapter 2 (figs. 2.23–24). We'll begin here with the corpse because it is the most common and because it permits the most complete relaxation.

THE CORPSE POSTURE

The corpse posture can both precede and follow a session of hatha yoga postures. At the beginning it calms the body and focuses the mind in preparation for the postures to follow; at the end it relaxes you from head to toe and integrates that experience into your conscious and unconscious awareness.

Lie supine on a padded but relatively firm surface, one that allows the spine to flatten slightly against the floor but still provides some support for the lumbar and cervical region (fig. 10.2). The surface should be flat. If you are on a surface that slants, lie with the head down. Otherwise, in the deep relaxation that minimizes or even eliminates most skeletal muscular activity, blood will pool in the lower half of the body, possibly diminishing venous return enough to jeopardize the blood supply to the brain and other vital organs.

A soft pillow to support the head and neck may be helpful. The arms and thighs should be comfortably abducted, leaving the feet spread apart and the hands swung 12–18 inches away from the thighs with the forearms supinated (palms facing up). The upper and lower extremities will keep the entire body in a state of mild tension if they are kept adducted, and pronating the forearms by facing the palms of the hands to the floor creates a closed-off feeling that is also counterproductive to relaxation.

Stretch the body out before settling down. Thrust the hips down from the shoulders, the hands down from the shoulders, the heels down from the

Figure 10.2. The corpse pose is the most famous relaxation posture in hatha yoga. Ideally, muscular activity is greatly diminished, if not eliminated, in skeletal muscles throughout the body, the most notable exception being the respiratory diaphragm, which is required for breathing.

hips, and lift the head away from the body, stretching the neck. Then adjust this way and that without losing too much of the stretched-out feeling. You should start in a slightly elongated position because when the muscles of the extremities and torso relax, they lengthen, and if you relax without first doing such stretching you may end up crowded in a foreshortened position and feel crimped at the end of the exercise.

Now let go. Just relax and breathe abdominally. Notice that the abdomen rises with each inhalation and falls with each exhalation. You can check this by temporarily placing one hand on the abdomen and the other hand on the chest, and then inhaling and exhaling in such as way that the chest does not move. As you become more relaxed, your rate of breathing will slow down.

After getting settled, don't move again until you are ready to break the relaxation. If you feel a need to move you'll know you didn't get positioned ideally in the first place. You should hold the posture no more than 3–5 minutes at first, and then gradually build up your time. The longer you hold the pose, at least within your personal limits, the quieter the motor neurons become. Finally, when you are ready to come out of the relaxation, bring the arms overhead and stretch from the tips of the fingers to the tips of the toes. Or you can first wiggle the fingers and toes, slowly recruit the proximal muscles of the extremities and the deep muscles of the torso, roll over to one side, and sit up slowly.

To avoid fainting or dizziness, those who have low blood pressure may need to turn onto their left sides before sitting up; if they rise suddenly, their venous return and cardiac output may not be sufficient to supply the brain. Turning onto the left side before sitting up will momentarily increase blood flow from the vena cava (situated on the right side of the heart) into the right atrium and tend to keep cardiac output high enough to prevent fainting.

What exactly do you do to quiet the activity of the motor neurons? Most important are two things you don't do. The first is don't move. Even the thought of movement or of responding to some sensory stimulus results in neuronal activity which cascades throughout the nervous system, stimulates the motor neurons enough to activate at least a few motor units, and sabotages relaxation.

The second thing you should not do is sleep. Learning not to sleep during relaxations is just as important as not moving, because if you lose consciousness and doze off, motor neurons throughout the nervous system become active. If falling asleep during relaxation exercises is a persistent problem for you even though you are rested and have done hatha postures before you try to relax, you may be able to prevent this by holding the root lock mildly or by simply moving move your feet closer together, which will accomplish the same end. Or you can bring the upper extremities closer to the body. It is true that these measures create a small amount of tension, but this is usually manageable and does not interfere with relaxation nearly as much as losing consciousness. Finally and simplest of all, you can adjust your breathing to keep you awake, either by breathing faster or by breathing diaphragmatically. It is difficult to fall asleep if you are breathing on a 1- or 2-second cycle of inhalation and exhalation. We'll examine the consequences of diaphragmatic breathing on relaxation in the corpse posture later.

Be prepared for one other potential problem. If you do not have a sense of calm and clarity when you lie down to relax, but instead feel as if you might jump out of your skin, you may be experiencing "relaxation-induced anxiety." If you are not accustomed to relaxing your skeletal muscles, any number of mental, psychological, and spiritual concerns may arise during the course of the practice, and if you are unable to come to terms with them, you may be uneasy in the exercise and find some excuse not to continue. A course of hatha yoga postures before you lie down is the most helpful remedy because it produces pleasant mental states and reintegrates body, mind, and spirit. If that doesn't help, don't lie there and suffer. Get up and get moving. This is not for you, at least not yet.

SUPINE TENSION-RELAXATION EXERCISES

If you consistently have trouble relaxing, one of the best remedies is an isometric tension-relaxation exercise. Lying in the supine corpse posture, try a sequence of tensing and then relaxing both upper extremities, both lower extremities, the right upper and right lower, the left upper and left lower, the right upper and left lower, the left upper and right lower. Finally, create whole-body tension; then relax.

These exercises not only train the motor units to respond to your will, they also place tension on the Golgi tendon organs in the tendons connected to the muscles being tensed. Stimulation of those receptors reflexly relaxes the muscles involved. Students can safely be encouraged to go to extremes in this exercise: the more vigorously they hold tension, the more the Golgi tendon receptors will stimulate relaxation in the affected muscles.

COMING OUT OF RELAXATION IN SLOW MOTION

When you have gotten to the point at which you can sustain relaxation for a while, you can try this experiment. At the end of a long relaxation, and before you have moved any part of the body, think of coming out of the posture in slow motion. Try to flex only one finger in the most minuscule possible movement. What you notice if you are completely relaxed is that each movement comes as a tiny jerk, like a cursor moving across a computer screen one pixel at a time rather than in a smooth sweeping motion. You do not ordinarily notice either the jerks in the body or on a computer screen because they are so tiny and numerous—you are aware only of the totality of the motion.

After you have made yourself aware of the jerks in one or two fingers, try the same thing with a shoulder. Concentrate carefully on making the smallest possible movement to lift the shoulder, and again you will see that the first movement comes as a jerk, larger in the shoulder than in the finger. These jerks may result from single nerve impulses impinging on the muscle fibers of the motor unit or they may result from overcoming resistance within the connective tissues—resistance that for some reason yields in spurts. I think it is more likely the former, but the question can't be settled without electromyography using thin-wire electrodes embedded directly into the appropriate muscles.

THE CROCODILE

The crocodile is the standard prone relaxation posture. It may not allow you to relax as deeply as the corpse, but the first variation that follows comes close. We discussed the crocodile in chapter 2 with respect to breathing, but here we'll look at it in its entirety, as a mild backbending relaxation posture that also stretches the arms overhead. First try the simplest crocodile with the arms stretched out in front of you, the hands catching the elbows, and the forehead on the uppermost wrist (fig. 2.23). If that position does not overstretch the muscles on the undersides of your arms, it will probably be comfortable. If it is not, try supporting the chest with a pillow, and notice how easing the stretch in the arms allows those muscles to relax. Use your personal preferences for foot position: heels in, heels out, or toes straight back. Those who have limited ability to extend their ankles may need to support them with a pillow. Then, lying still, try to notice if you are still holding tension, and if so, where.

Next try the second position for the crocodile with the arms at a 45–90° angle from the floor (fig. 2.24). This will be more of a challenge for many people. The abdominal muscles may resist the backbending that is defined by the posture, and the flexion in the neck may not be entirely comfortable, but if you practice this posture regularly, you will sooner or later be able to relax in it.

Last, try the full crocodile with one elbow on top of the other, each hand gripping the opposite shoulder, and the head tucked into the combined crooks of the two elbows (fig. 10.3). This is the only true crocodile; its "jaws" are formed by the upper extremities. Calling it a relaxation posture may be something of a joke at first. Many people will have to arch their upper back and neck substantially to the rear in order to make room for the head and neck, and it may not be possible to use this posture for relaxation until muscles and joints throughout the body get acclimated.

PAIN

Pain stops relaxation in a hurry. It intensifies undesirable mental states, disrupts autonomic functions, and prevents muscular relaxation. If you have low back pain, shoulder pain, neck pain, gas pain, arthritis, or any other physical discomfort, your mind will be diverted in that direction as soon as you lie down and make yourself still. Another problem is pain induced by a particular posture, but this can sometimes be circumvented with props. If you do not have enough subcutaneous fat to lie comfortably in the corpse posture on a hard surface, you can lie on a mat that is soft enough to keep your bony protuberances from pressing directly against the floor; if that is not enough, you can place pillows under the knees, the head, and even the arms and forearms.

Even if you do not use a prop elsewhere, it is important for everyone to experiment with a pillow under the head. If you have good mobility in the neck, a soft down pillow provides important support for the mid-cervical region (which will tend to prevent and be therapeutic for reverse cervical curvatures) and yet allows the back of the head to drop to the same level as the rest of the body. A more substantial pillow creates a very different experience. It establishes traction in the neck, pulls the head both anteriorly and superiorly, and induces a well-known relaxation response all by itself. And more problematic, older people often have semi-permanent skeletal forward bends in their necks and chests which may leave their heads lifted 2–8 inches away from the floor even before the attempt to relax gets under way. The only way to make such students comfortable is with one or more thick supporting pillows; such props are especially important for gentle yoga classes.

Figure 10.3. The advanced crocodile pose is a relaxed backbending pose, but many beginning and even intermediate students will find that the position of the head, neck, and upper extremities creates so much discomfort that the front of the body cannot relax. Such students may find the beginner's crocodile (fig. 2.24) more useful, at least for the time being.

BREATHING AND RELAXATION

Even breathing in both the corpse pose and the crocodile has a soothing effect and seems to diminish activity generally in motor neurons. In either posture we have a choice of breathing abdominally or diaphragmatically.

ABDOMINAL BREATHING IN THE CORPSE POSTURE

Abdominal breathing in the corpse pose is the most relaxing of any mode of breathing in any posture. In this position the abdominal wall yields so easily to the movement of the diaphragm that respiration produces little movement in the rest of the body, and the breath becomes so delicate that it almost seems to stop. The motor neurons whose axons travel in the phrenic nerves to the respiratory diaphragm supervise this activity from beginning to end, resulting in concentric shortening of the diaphragm during inhalation and eccentric lengthening during exhalation (chapter 2). The number of nerve impulses per second to the diaphragm will peak at the end of inhalation when the abdominal organs are pushed most inferiorly, diminish as exhalation proceeds, and drop almost to zero at the end of exhalation—the only time the system is at rest. You can feel it. You have to make an effort to stop breathing at any other time in the cycle, but at the end of your tidal exhalation you can stop breathing effortlessly. Please note: this is fine to try as an experiment, but it's a risky habit pattern. As mentioned in chapter 2, constantly pausing the breath at the end of exhalation is thought in yoga to have an adverse impact on your cardio-vascular health.

Even though abdominal breathing during deep relaxation is minimal, it is still desirable to cultivate even breathing because every inhalation, no matter how fine, still pulls on the lumbar region by way of the crus of the diaphragm and on the lower part of the rib cage by way of the costal attachment of the diaphragm. As a result, tension spreads throughout the body by way of those attachments, and if we can minimize such tensions we can minimize their impact on relaxation.

There are two ways that feedback from breathing will tell you if you are relaxed. First, your breathing can become so quiet during the course of 10–15 minutes of stillness in the corpse posture that there is a diminished need for the intercostal muscles to maintain the muscular tone that ordinarily keeps the chest from collapsing inward during inhalation (chapter 2). The chest can actually become frozen, not from muscular activity but from inertia as the motor neurons supplying the intercostal muscles become silent. You may not notice this until you have had some experience with the posture and have remained quiet for twenty minutes or so, and even then you may not be aware that the chest has become frozen until after you have felt the sudden gathering of activity in the intercostal muscles as you start to come out of the relaxation. This situation is not found in any other posture or with any other method of breathing.

Second, when you are relaxed, the eyes, cheeks, and other soft tissues of the face are pulled inward during each inhalation and bulge out during each exhalation. This is because fasciae are interconnected throughout the body, and the slight vacuum in the chest produced during inhalation is translated through the fasciae to the face. You may not be able to feel this unless you are completely relaxed, but it's there.

DIAPHRAGMATIC BREATHING IN THE CORPSE POSTURE

Diaphragmatic (thoraco-diaphragmatic) breathing in the corpse posture is not as relaxing as abdominal breathing, but it is a powerful energizing technique that should be explored by intermediate and advanced students. It engages the chest and abdomen without significantly disturbing relaxation in the rest of the body, and it is more effective for keeping people alert than abdominal breathing. Anyone who has a tendency to chronic chest breathing should avoid this for the time being, however. As we saw in chapter 2, the moment chest breathers stray from the tried and true abdominal breathing that is so important at the early stages of their training, they will automatically revert to constricted thoracic breathing.

To explore diaphragmatic breathing in the corpse posture, first lie down and breathe abdominally for a few minutes. Make sure your shoulders are relaxed and that your forearms are relaxed but supinated, with the palms facing up and situated about 12–18 inches from the thighs. Then breathe diaphragmatically, holding a little tension in the abdomen during each inhalation. You can breathe as slowly as five breaths per minute or as fast as 12–15 breaths per minute. The concentration required for either tactic will help keep you awake if you tend to doze off.

RELAXED BREATHING IN THE CROCODILE

Although the crocodiles are rightfully included among relaxation postures, their architecture, both with respect to arm position and to breathing, does not lend them to the same kind of deep relaxation as the corpse posture. It would be rare for a class of students to hold the crocodile for more than a minute or so, but an extravagant teacher might hold a class of experienced students in the corpse posture for half an hour or more.

One reason the stretched-out crocodile is less effective for complete relaxation than the corpse posture is that it produces a more complex mode of breathing (fig. 2.23). To reiterate, in the prone position the abdomen is pressed against the floor, and the downward thrust of the dome of the diaphragm has to push the abdominal organs inferiorly and lift the entire body, rather than push the abdominal wall forward as happens in the corpse posture. This is a type of abdominal breathing, but the movement of the entire body in the crocodile, together with the extra effort needed to breathe, makes it more difficult to relax.

The beginner's crocodile, with the arms at a 45–90° angle from the floor, also prevents complete relaxation. In addition to the effort required to press the abdominal organs inferiorly (which also happens in the stretched-out crocodile), the beginner's crocodile requires diaphragmatic breathing (chapter 2), and this creates movement in the chest that in turn activates even more of the body (fig. 2.24). And that applies to someone who is relatively comfortable. The problem is compounded in those who are not.

The full crocodile (fig. 10.3) yields either abdominal or diaphragmatic breathing depending on the length of the arms. If they are long enough to lie flat while you tuck the head into the crook of the elbows, we see a situation similar to the easy stretched-out crocodile, which yields abdominal breathing. In this case the posture is very relaxing, at least for those who have thoroughly accustomed themselves to the pose. But if your arms are shorter, you will have to arch the upper back and neck upward in order to place the top of the head in the crook of the elbows, and in that case your chest will be lifted as in the beginner's crocodile, which requires you to breathe diaphragmatically and which again checks relaxation.

One indirect way in which the crocodile posture furthers relaxation is that the arm position tends to prevent thoracic breathing, which as we saw in chapter 2 stimulates the sympathetic nervous system and prepares the body for activity. Because of the anatomical restrictions to chest breathing, the easy stretched-out crocodile pose is an excellent posture for compelling chest breathers to come in touch with relaxation and abdominal breathing, and the beginner's crocodile is possibly the best all-around posture (as well as the only relaxation posture) for teaching chest breathers diaphragmatic breathing.

THE AUTONOMIC NERVOUS SYSTEM

Our discussion of the autonomic nervous system in chapter 2 focused on its associations with breathing, but its main role is helping to regulate functions of internal organs in general, including non-somatic structures such as blood vessels throughout the body and sweat glands in the skin (figs. 10.4a–b). Even though these regulatory functions are mostly outside our conscious awareness, some of their effects are important to relaxation. If you lie down to relax and notice that your hands and feet are cold and sweaty or that you are dizzy and headachy or that your heart is beating too fast, autonomic functions are not being optimally managed. In such cases training in relaxation may help you diminish the symptoms. How this happens is not clear because the autonomic nervous system is not directly under the control of our will. Its regulation has to take place indirectly, and the only way we can manage it is to capitalize on everything in our bag of tricks—mental attitude, exercise, postures, and relaxation—and then trust the wisdom of the body to keep everything operating smoothly. And that's what you want. You would no more want to micromanage internal functions of the body than you would want to enter into a debate with a pilot of a commercial airliner about how fast to fly. If authority for management of internal organs is delegated to the autonomic nervous system, internal functions will not annoy us when we try to relax, and we'll be able to work with those aspects of mind and body that are more obviously within our grasp.

THE SYMPATHETIC NERVOUS SYSTEM

The idea of the sympathetic nervous system goes back almost 2000 years, to Galen, who believed and taught that this complex of nerves permitted animal spirits to travel from one organ to another, creating "sympathy" among them. Here we are especially concerned with the sympathetic nervous system as a part of what we now recognize as the autonomic nervous system (chapter 2 and figs. 10.4a–b) because the sympathetic component is generally sedated by breathing and relaxation techniques. It activates the body globally when it is stimulated and it quiets the body down globally afterward. Stress is its stimulant, and preparation for "fight or flight" is its response. To that end it speeds the heart rate, opens the bronchial tree, stimulates the release of glucose from the liver, dilates the pupils, constricts arterioles in the digestive system and skin, dilates arterioles in the heart, and contracts sphincters in the gastrointestinal tract and in the urethra. All of these responses act together and in "sympathy" with one another—Galen had the right idea, after all—to gear up the body for emergencies. Only a few sympathetic reactions, such as orgasm and dilation of the pupils in dim light, occur in relative isolation. The parasympathetic nervous system, by contrast, affects the internal organs specifically and often in isolation.

We can see the operation of the sympathetic nervous system and how it orchestrates bodily processes during an ordinary sequence of postures followed by relaxation. The sympathetic nervous system, including one of its main endocrine components, the *adrenal medulla*, becomes activated by vigorous hatha yoga practices, especially standing stretches, sun salutations, and abdominopelvic exercises such as leglifting and agni sara. Relaxing in the corpse posture after an invigorating session of postures will then quiet this system and diminish the sympathetic effects on organs and tissues throughout the body.

RELAXATION AND BODY TEMPERATURE

The autonomic nervous system regulates body temperature, not only in the core of the body where it affects the chest, abdomen, and pelvis, but also in tissues of the skin and extremities. When the sympathetic system is activated, skin (especially in the hands and feet) becomes cold because its blood supply is diminished by vasoconstriction (chapter 2), and it becomes clammy because sweat glands flood the surface of the body with moisture. When you lie down to relax, the sympathetic nervous system calms down, decreasing muscular tone in the smooth muscle that encircles arteries and arterioles, which allows those vessels to dilate and causes blood supply to the skin to increase and skin temperature to rise.

You would think this would make you feel warm, but that is not the case because during relaxation, the skin temperature rises and radiates heat, and this causes the body to cool down. Ordinarily the internal temperature regulating systems would then prompt you to shiver, but you curb that impulse, knowing that any movement would break the relaxation. The result is that you are chilly when you come out of the exercise, and the next time you try to relax, your body does not remember the experience fondly. But knowing in advance that the body will lose radiant heat during a long relaxation, you can cover up—even a pillow placed on the chest and abdomen can make a difference. This is a good example of how we can anticipate and accommodate to the peculiarities of the autonomic nervous system—coddling it and letting it do its job rather than trying to bend it to our will.

BRONCHIOLAR DILATION AND CONSTRICTION

Breathing and relaxation go hand in hand. As we have seen, respiration slows down naturally during the course of relaxation, and under carefully monitored conditions 2:1 breathing (chapter 2) can slow the heart rate. Relaxation also has an impact on how easily air moves through the airways, and this is controlled by both the sympathetic and parasympathetic components of the autonomic nervous system.

Exercise stimulates the sympathetic nervous system and the release of epinephrine (adrenaline) from the adrenal medulla, and this dilates the bronchioles and makes breathing easier. As we saw in chapter 2, relaxation diminishes the activity of the sympathetic nervous system to the bronchioles and increases parasympathetic input. Both systems together act on the smooth muscle encircling the airways, causing them to constrict and thereby increasing resistance to air flow, and this neatly coincides with the fact that we make use of less alveolar ventilation when we are relaxed.

During an hour or so of hatha yoga, management of air flowing through the airways should take care of itself. Your session may start quietly with a short relaxation, build quickly into abdominal exercises and vigorous standing poses, proceed to postures done sitting and lying on the floor, continue with inverted postures, and end with a final relaxation. When you start, your airways are slightly constricted, corresponding to decreased activity in the sympathetic nervous system and increased activity in the parasympathetic input to the lungs. Then as you do more vigorous postures, the sympathetic nervous system prevails and the airways open. As you quiet down in your final relaxation, your bronchiolar tree constricts but remains open enough to accommodate air flow.

It should now be apparent that anyone who wants to master relaxation must have a full repertoire of breathing practices. If airflow is unimpeded, you can stick with abdominal breathing for maximum relaxation and stimulation of the parasympathetic effects on the bronchioles. If you sense some restrictions in the airways, it will be more appropriate to mildly stimulate the sympathetic nervous system with diaphragmatic breathing, which will not only open the airways but will also keep you awake. And for a more dramatic stimulation of the sympathetic nervous system, take 5–10 empowered thoracic breaths, or even complete breaths, which have a paradoxical component. This will open the airways significantly and prepare you for a peaceful relaxation.

THE PARASYMPATHETIC NERVOUS SYSTEM

It wasn't until the late 1800s that an anatomically and functionally separate division of the nervous system to the viscera was described. Since it was anatomically separate and yet supplied most of the same organs (fig. 10.4a), it was named the parasympathetic (para = alongside) nervous system. It soon became evident, however, that the mode of operation of this accessory system was entirely different from that of the sympathetic nervous system. In contrast to the operation of the sympathetic nervous system, "relaxation" of the parasympathetic nervous system has no meaning except in relation to each organ and each function. In contrast to the sympathetic nervous system, which causes global effects throughout the body, the parasympathetic nervous system is organ-specific. With only a few exceptions such as regulation of the heart and lungs, it does not "balance" the sympathetic nervous system. Nor does it become generally active or inactive as does the sympathetic nervous system. It presides over functions as disparate as stimulating salivary, gastric, and pancreatic secretions, stimulating peristalsis (the movement of food distally through the gastrointestinal tract), stimulating the synthesis of liver glycogen from glucose, accommodating for near vision (as for threading a needle), sexual arousal, slowing the heart rate, constricting the bronchial tree, contracting the wall of the bladder for urination, and relaxing the internal urethral and anal sphincters to facilitate elimination.

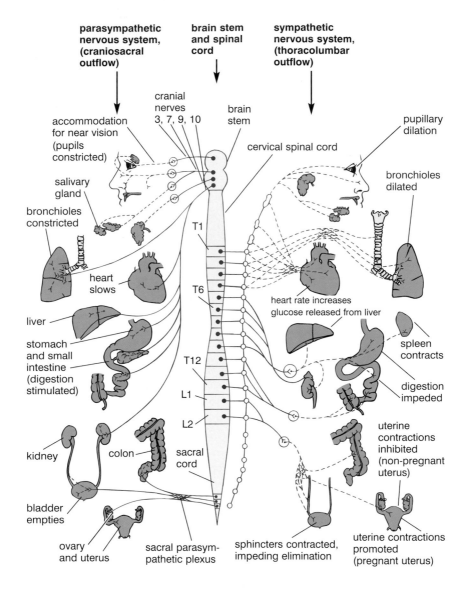

parasympathetic nervous system, (craniosacral outflow)

brain stem and spinal cord

sympathetic nervous system, (thoracolumbar outflow)

cranial nerves 3, 7, 9, 10

brain stem

accommodation for near vision (pupils constricted)

cervical spinal cord

pupillary dilation

salivary gland

bronchioles dilated

bronchioles constricted

T1

heart slows

liver

heart rate increases
glucose released from liver

spleen contracts

stomach and small intestine (digestion stimulated)

T6

T12

L1

L2

digestion impeded

sacral cord

uterine contractions inhibited (non-pregnant uterus)

kidney   colon

bladder empties

ovary and uterus

sacral parasympathetic plexus

sphincters contracted, impeding elimination

uterine contractions promoted (pregnant uterus)

Figure 10.4a. Parasympathetic nervous system (craniosacral outflow) outlined on the left, and sympathetic nervous system (thoracolumbar outflow) outlined on the right. (Both systems innervate both sides.) The cranial portion of the parasympathetic nervous system is included in cranial nerves 3, 7, 9, and 10, the latter being the famous vagus nerve that innervates most of the viscera, including the lungs, heart, liver, and the upper part of the digestive tract. The sacral parasympathetic plexus innervates the genitals, the bladder, and the lower part of the digestive tube. The sympathetic nervous system innervates the entire body from its source in the spinal cord between T1 and L2, which is why the sympathetic nervous system is called the thoracolumbar outflow (Dodd).

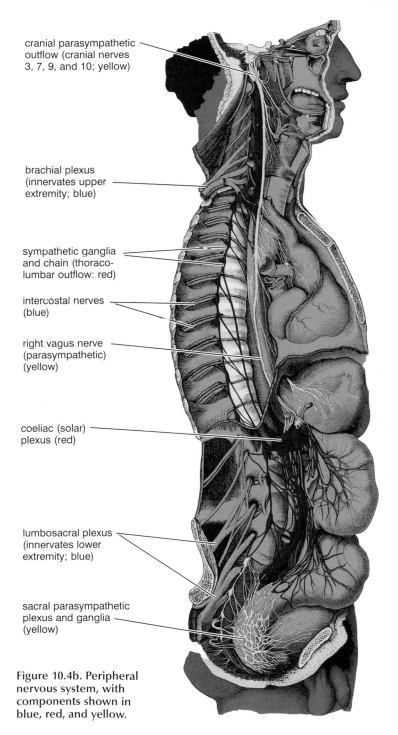

cranial parasympathetic
outflow (cranial nerves
3, 7, 9, and 10; yellow)

brachial plexus
(innervates upper
extremity; blue)

sympathetic ganglia
and chain (thoraco-
lumbar outflow: red)

intercostal nerves
(blue)

right vagus nerve
(parasympathetic)
(yellow)

coeliac (solar)
plexus (red)

lumbosacral plexus
(innervates lower
extremity; blue)

sacral parasympathetic
plexus and ganglia
(yellow)

Figure 10.4b. Peripheral
nervous system, with
components shown in
blue, red, and yellow.

Although it's not in the realm of ordinary experience, there is one way that you would be able to experience all parasympathetic reactions at the same time. Untutored and unwary mushroom aficionados have been known to ingest the poisonous fungus *Amanita muscaria*, which globally stimulates the muscarinic receptors of the parasympathetic nervous system (and which is where those receptors got their name). Don't be tempted. Depending on the dose, you'll have a most disagreeable experience. From the top down if you don't die your pupils will constrict to pinpoints, you'll slobber and froth saliva, your heart rate and blood pressure will plummet to near death, extreme bronchiolar constriction will make it almost impossible to breathe, your digestive tract will thump, thrash, and grind like a washing machine gone berserk, and you'll suffer from explosive diarrhea. *Amanita muscaria* is plainly not suitable as a mushroom sauce for your dinner guests, but it does provide us with the only naturally occurring illustration of a general parasympathetic reaction, and it also makes clear the altogether different ways in which the parasympathetic and sympathetic nervous systems act on internal organs.

Looking at a more typical scenario in ordinary life, if you eat a large meal at noon, the portions of the parasympathetic nervous system associated with the digestive system will preside over digestion during the course of the afternoon, and then its effects will recede into the background. Then if you do a session of hatha yoga followed by relaxation at 5 PM, the main autonomic effects will first be stimulation and then "relaxation" of the sympathetic nervous system. In another possible scenario, if you were to eat a meal at 5 PM, do a vigorous practice of hatha between 6 PM and 7 PM, and finally lie down to relax between 7 PM and 7:30 PM, the parasympathetic nervous system would initiate digestion after the meal, be partially stymied around 6:15 PM by the hatha session, and kick in again to re-stimulate digestion during the course of relaxation around 7:15 PM. And so it goes, day and night, year after year, without the necessity of our conscious supervision.

THE ENTERIC NERVOUS SYSTEM

It was not until 1921 that the physiologist Langley published his famous textbook in which he conceptualized and formalized the idea of an autonomic (autonomous) nervous system. He also suggested that the autonomic nervous system was made up of not two but three components—the sympathetic, parasympathetic, and *enteric* nervous systems. He defined the latter as being comprised of the vast system of nerve cells that resides in the wall of the gut (by definition, the entire digestive tube from the oral cavity to the anus). Although Langley publicized his ideas widely, only the idea of the sympathetic and parasympathetic components caught on, and the concept of an enteric nervous system was either forgotten or was simply included conceptually as a part of the parasympathetic nervous system. Finally, however, after the idea had languished for six decades, biologists started seeing in the 1980s that the enteric system was indeed an independent nervous system just as complex, with just as many neurotransmitters, and with even more neurons, than the spinal cord itself—a veritable "second brain" within the gut, one which is not only important to digestive system function but one that is probably important for the experience of many emotions. The story of this system has just begun, and textbooks that continue to ignore it will soon be obsolete.

Langley actually commented in his 1921 text that the enteric nervous system was probably more truly autonomous than either the sympathetic or parasympathetic systems. And neuroscientists ultimately found this to be correct. We now know that the enteric nervous system is capable of supervising the digestion of food, the propulsion of food through the bowel, and management of all other bowel functions, even when all nerve connections from the brain and spinal cord have been experimentally severed. This system is almost certainly important to relaxation, although attempting to say exactly how would be premature. Emotional connections may catch the attention of experimental psychologists in time, but it's anyone's guess.

So in summing up the widely disparate ways in which the nervous system affects relaxation, we see that several separate but related pictures emerge, one relating to muscular relaxation and the somatic nervous system (which we discussed earlier in this chapter as well as in chapter 1 and 2), another to the sympathetic nervous system, another to the parasympathetic nervous system, and possibly yet another to the enteric nervous system. First of all, relaxation requires skeletal muscular relaxation, and it requires that the sympathetic nervous system be quieted down. There is no way you can relax if you are fidgeting (somatic system) or if you are in the middle of a flight-or-fight response (sympathetic system). And second, before you do a relaxation, your activities and behaviors should be adjusted so that the parasympathetic and enteric nervous system are not overly burdened with housekeeping tasks that draw attention to themselves and prevent relaxation—tasks such as digesting a large meal or coping with constipation, diarrhea, a full bladder, asthma, menstrual cramping, or sexual excitation. And finally, if the purported emotional content of the enteric nervous system grips you with regrets for the past, anxiety for the future, fear of the unknown, or an overly joyful mood in the present, relaxation will be impossible. Let it go; it's useless for relaxation.

DEEPENING RELAXATION

Even if the somatic and autonomic systems are not creating obstacles, the mind may still be active, and this too can prevent deep relaxation. It has been documented that when the body is relaxed and the subject purposely entertains extraneous thoughts, a biofeedback monitor keeping track of skeletal muscular activity reacts immediately, showing clearly that quieting the mind is just as important to relaxation as quieting the body.

Hatha yoga includes many techniques for making the mind still, from concentration exercises on the surface of the body, to watching the breath, to more subtle abstract exercises such as "sweeping the breath up and down the body," or holding your attention on regions such as the navel center, the point between the eyebrows, or the pit of the throat, whose anatomical locations are not precisely defined. The results of these exercises are clear, the mechanisms obscure. We know only that the skeletal muscles relax, and that the autonomic nervous system fulfills its autonomous role in managing the internal organs and tissues of the body without the necessity of conscious input. That is quite enough.

One word of caution about abstract exercises: those who are flighty sometimes go off on odd tangents when they try them. Anyone who is psychologically fragile should stick with more grounding concentrations such as focusing their attention on the rise and fall of the abdomen during inhalation and exhalation, or thinking of relaxing large body parts such as the head, neck, shoulders, arms, and forearms.

THE RISE AND FALL OF THE ABDOMEN

This is the beginner's first exercise for relaxation. You simply lie in the corpse posture and concentrate on the rise of the abdomen with each inhalation and the fall of the abdomen with each exhalation. Make the breath as even as possible and watch its pace gradually diminish. Notice that inhalation merges smoothly into exhalation, but that exhalation does not merge so smoothly into inhalation (chapter 2). After you have practiced this exercise for some time, it will feel natural to allow your breath to stop at the end of exhalation. It seems as if it could stop forever, as if you could literally expire. As discussed earlier, the end of exhalation is the only time this should happen because that is the only time the diaphragm is completely relaxed.

Paradoxically, this simple concentration exercise in even breathing is also one of the most advanced, and few people will be able to do it for long without falling asleep or letting their minds drift. Resorting to diaphragmatic breathing, taking deeper breaths, complete breaths, or breathing faster all minimize those two problems, but they also compromise relaxation. But even if these techniques are counterproductive as far as relaxation is concerned, at least they prove that concentrating on the rise and fall of the abdominal wall is so relaxing that you have to be hyperalert not to drift. This exercise is not easy.

SWEEPING THE BREATH UP AND DOWN THE BODY

There are many different relaxation exercises in the corpse posture that involve sweeping the attention up and down the body in coordination with the breath, usually from the toes to the crown of the head during inhalation, and from the crown to the toes during exhalation. You should breathe normally at first and then gradually lengthen the breath as relaxation deepens. Your greatest challenge will be to keep attentive.

There are many variations of this exercise. You can reverse directions and inhale your attention to the toes and exhale your attention to the top of the head. Or you can use your "breath" as a vehicle for traveling your attention to different parts of the body. The point of all such abstract relaxation exercises is less complicated than may at first be apparent: they are all tricks to hold the mind steady while the body relaxes of its own accord.

THE SIXTY-ONE POINTS EXERCISE

One of the simplest but most interesting of the abstract exercises carries your attention to a succession of 61 points on the surface of the body (fig. 10.5). If you can hold your focus for a moment at each spot, by the time you get all the way around the body and mind will be deeply relaxed. The exercise can be done in many different ways, but the simplest is to pretend that you exhale to each point. This means that at the beginning of each exhalation you lock your attention onto a specific point on the body, hold it there throughout exhalation and the ensuing inhalation, and then move on to the next point at the start of the next exhalation. A variation is to imagine a blue light at each point. As you become proficient you can lengthen the time you hold your attention at each point in one of two ways—by extending your time for each inhalation and exhalation, or by holding your attention at the same spot for more than one breath cycle.

Assuming that you are physically comfortable, the main challenge in this exercise is holding your attention on each one of the sixty-one points successively. You might get down to the right hand, or partially back up to the tip of the right shoulder, and suddenly realize that you have lost awareness of where you were. Or if someone is leading you through the exercise you may suddenly find yourself being asked to concentrate at some specific point and have no idea of how you got there. In any event, this means that you either went to sleep or let your attention wander, and this broke the relaxation. The most certain remedy for such problems is daily practice and better preparation, which can include a good night's sleep, an enthusiastic session of hatha yoga, and a less soporific diet.

If you have trouble keeping your concentration even under ideal conditions, try just focusing on the first 31 points (the upper half of the body) and come back to the beginning from there. Or you can try breathing faster, and thereby move more quickly through the exercise. If you take 1-second inhalations and 1-second exhalations the entire exercise will take only two minutes, and once you have memorized the points you can usually hold your attention throughout the cycle, at least at that rate. Then you can slowly lengthen your breaths. Don't be surprised, however, to learn that lengthening the exercise to just four minutes (2-second inhalations and 2-second exhalations) may create a real challenge. Keep trying. Keep improving the conditions under which you do the exercise, and keep gathering more experience. If you are determined and patient the exercise will finally work.

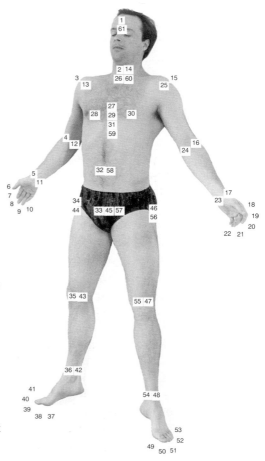

Figure 10.5. To practice the 61 point concentration and relaxation exercise, you travel your attention sequentially to 61 points on the surface of or within the body. #1 and 61 is the "point between the eyebrows." #2, 14, 26, and 60 is the pit of the throat. # 27, 29, 31, and 59 is the "heart center," between the nipples, which are in turn 28 and 30. #32 and 58 is the "navel center," and #33, 45, and 57 is the "pelvic center." The rest are obvious—shoulders, elbows, wrists, tips of the fingers, hips, knees, ankles, and the tips of the toes. The exercise is easier if you journey from beginning to end quickly (2–4 minutes total, or one breath and 2–4 seconds for each point); the exercise may challenge your alertness severely if you move through it more slowly. The main idea is to move quickly enough that you do not fall asleep or get confused about where you are in the sequence.

CONCENTRATING ON THE HEART CENTER

The more steadily you hold your focus on a fixed spot the more advanced the exercise, and one of the most interesting is to hold your attention at the heart center. Here we are not necessarily speaking of something structural. The "heart center" is an abstract concept. You simply hold your attention in the region of the mid-chest. You may notice that your concentration tends to wander once in a while, flitting away especially at the junction of inhalation and exhalation or at the junction of exhalation and inhalation. So be sure your breath is slow and even. Be especially aware of your eyes. Even though they are closed you must hold the gaze steady: tiny darting movements of the eyes beneath the closed lids correspond to lapses in concentration. Hold this exercise as long as you are comfortable, building up your time gradually.

If this practice is difficult for you, draw in your mind's eye a tiny ellipse in the heart center. Then inhale up the ellipse and exhale down the ellipse. Allowing your attention to move just an infinitesimal amount makes the exercise easier. And after making use of this trick for a few weeks you can go back and again try to hold your attention steadily on one point.

These exercises are usually done in the corpse posture, but they are concentrations that relate as much to meditation as they do to relaxation. They assume that you are relaxed; then they carry you beyond . . . .

MEDITATION POSTURES

According to many traditionalists, one of the main purposes of hatha yoga is to train the body and mind for meditation. The relaxation exercises just discussed are part of this training, and they can also be used in modified form when we sit in meditative postures. Other exercises that can prepare you for meditation include sitting perfectly still in the meditative posture of your choice; holding your attention within the body and maintaining awareness of muscles, joints, and the structure of the sitting posture; remaining calm, comfortable, alert, and still; and above all breathing evenly, silently, and diaphragmatically without jerks, pauses, or noise. Only when you have achieved some mastery of these seemingly simple exercises in the meditative sitting postures can actual training in meditation begin.

To get an idea of what's involved, sit cross-legged on the floor and focus for a few seconds on the body. Then prepare for meditation—going beyond the body to work with the breath and beyond the breath to examine the operations of the mind. To remain alert in meditation the body has to be erect, stable, comfortable, and still. So why not stand? Because you do not want to be concerned about tipping around in a gravitational field. Then why not lie in the corpse posture? Because you do not want to doze off instead of remaining alert. But what if you can stay awake in the corpse? That's harder to say, but something about meditation requires an erect spine—not merely erect, but erect under its own power. Lying down, it does not work for most people.

So we must sit and sit erect. There are six main postures that accomplish this—sitting in a chair, kneeling on the floor or a bench, sitting in the cross-legged easy posture, and sitting in what we'll call the classic meditative postures: the auspicious, the accomplished, and the lotus poses. Most students of yoga would like to use one of these last three, but nearly all Westerners will have difficulty doing that at first. In ancient India where the postures were refined, people sat on the ground or on their haunches with the feet flat, and they grew up with strong and flexible hip and ankle joints. They had no trouble sitting flat on the floor. But in the West we sit mostly in chairs, couches, and automobile seats, and because of this we have trouble acclimating to these three traditional meditative postures.

To resolve some of these difficulties, we'll first look at the architecture of a generalized classic yoga sitting posture and then we'll discuss the anatomical features required to hold it upright. Next we'll examine props to make the pose easier for Westerners. Then we'll look at the six standard sitting postures in detail, and we'll end with some general guidelines for mastering them.

MAINTAINING THE GEOMETRY

When you are sitting correctly in one of the three classic meditation postures, the pose is self-supporting. The spine is straight, the feet are locked into positions that keep the lower extremities in place, and the upper extremities stabilize the torso. When you have mastered them, the postures should be so comfortable that you are not aware of the body, and this permits you to focus your mind on meditation.

THE RIGHT TETRAHEDRON

It is the geometry of these three postures that makes them so stable. They are identical except for the position of the lower extremities, and in all of them the body assumes the shape of a tetrahedron—the simplest possible three dimensional structure. All tetrahedrons are made up of four triangular surfaces. In the classic meditation postures the base triangle lies against the floor and is formed by the thighs and an imaginary line connecting the two knees. The spine extends upward perpendicular from this triangle's posterior corner and forms the upright axis for both the second and third triangles, which are completed by the right and left upper extremities resting on the right and left lower extremities at the knees. The fourth triangular surface, which completes the tetrahedron, is formed by an imaginary line that connects the knees, and by two lines that follow the upper extremities from the knees to a point near the top of the head (fig. 10.6). The 90° angle between the axial skeleton and the lower extremities therefore gives the posture the shape of a right tetrahedron. The triangle formed by the lower extremities provides a firm base for the posture, the right angle between the base triangle and the upright axis of the spine is what makes the posture upright, while the upper extremities stabilize the upright axis and keep the torso from pitching forward.

THE AXIAL MUSCLES

The erector spinae, the muscle complex that attaches between various vertebrae on the posterior side of the body and to the pelvis (figs. 4.14, 5.5, and 8.14), holds the back axis of the tetrahedron in a straight line, flattening the thoracic kyphosis and providing lift to that part of the vertebral column. Since part of its inferior attachment is to the pelvis, the erector spinae also acts to increase the lumbar lordosis. In the classic cross-legged sitting posture, it is the foremost muscle group that counteracts slumping (that is, the flattening of the lumbar lordosis followed by its further rounding to the rear). The erector spinae also acts to create what we call an axial extension, which is an overall lifting of the axis of the body.

Two more muscles—the *quadratus lumborums*—lie anterior to (beneath, when approached in a dissection) the lowermost portion of the erector spinae. These muscles, one on each side, take origin from the posterior part of the iliac crest and the sacrum, and insert on the upper lumbar vertebrae and the 12th rib (figs. 2.7, 3.7, 5.5 and 8.14). They function as synergists to augment the function of the erector spinae, acting roughly like a string on a bow. When they contract, they bend the bow of the lumbar curve forward, and so they are most effective when the lumbar lordosis is already being maintained by the erector spinae and the iliopsoas muscles.

Figure 10.6. A generic meditative sitting posture (in this case the accomplished pose), with a right tetrahedron superimposed on the model. The base triangle rests on the ground, with its back corner underneath the coccyx. The middle upright line extends to the top of the head at a right angle (90°) from the plane (and back corner) of the base triangle.

The longus colli muscles on the anterior side of the cervical vertebrae act to increase lift in the region of the head and neck. And like the erector spinae in the thoracic region, these muscles are situated on the convex side of their vertebrae (fig. 8.13). If the longus colli muscles along with the scalenes acting as synergists are not fulfilling their role in supporting the posture, the head and neck droop to the rear.

THE ILIOPSOAS MUSCLES AND THE SACROILIAC JOINTS

The erector spinae, quadratus lumborums, longus colli, and scalenes operate to keep the general orientation of the spine in a straight line, and in principle this could mean a spine that is leaned forward, that is perpendicular to the floor, or that is tilted to the rear. It falls then to the psoas and iliacus muscles to maintain the rear axis of the tetrahedron at a 90° angle from the base triangle. The individual actions of the psoas and iliacus muscles differ, however, even though they share an attachment site on the upper part of the femur, and these separate roles are significant. Specifically, the iliacus muscles sit you up by pulling forward on the ilia and favoring sacroiliac counternutation, while the psoas muscles sit you up by pulling forward on the lumbar lordosis and favoring sacroiliac nutation (see chapters 3 and 6 for more background).

The iliacus muscles are the simplest to understand because they act solely across the hip joints, from the femur to the ilium, fanning up and out to the inner surface of the ilium on each side and acting as a pair to create an anterior pelvic tilt. You can feel these muscles in action if you sit upright in a straight chair and think of pulling the ilia forward to the exclusion of the sacrum. You want to leave the sacrum behind in a position of counternutation (fig. 6.2a), or at least in a neutral position between nutation and counternutation. The actions of the iliacus muscles are subtle. You'll get the idea if you feel a sense of lifting along with a sense of controlling the angle of the spine with respect to the floor. But try to avoid a direct pulling forward on the lumbar spine. And even more to the point, avoid creating a lateral spreading of the ischial tuberosities. Just to make sure of the latter, squeeze the hips together—and with the hips the ischia—to affirm counternutation.

The two psoas muscles have a more complex action than the iliacus muscles because they each act across three joints—the hip joint, the sacroiliac joint, and the lumbosacral joint—rather than just the hip joint. Since the psoas muscles attach to the lumbar spine, they not only act with the iliacus muscles to flex the two hip joints, they also place tension on the lumbosacral joint between L5 and the sacrum, and they support nutation at the sacroiliac joints (fig. 6.2a).

To feel the effects of the psoas muscles over and above those that are characteristic of the iliacus muscles, lift the posture as much as possible with the iliacus muscles alone (which also produces a slight anterior pelvic tilt, as mentioned above). Next, without releasing the iliacus tension, think along completely different lines and pull the lumbar region frankly forward. Look for a deep and peculiar feeling—an internal tension that is directed specifically to the lumbar spine from the femurs, a tension that ultimately pulls the promontory of the sacrum forward in relation to the pelvic bones. This is nutation. Also look for the other components of nutation—the squeezing together of the ilia and an even more obvious spreading apart of the ischial tuberosities. Full sacroiliac nutation is what to aim for and hold in a classic meditation posture because it permits the lumbar lordosis to be maintained and even accentuated without depending so much on the iliacus muscles and an anterior pelvic tilt. And it's also helpful that spreading the ischia apart from one another during nutation shifts the origins of the adductor muscles laterally. That is practical and significant for everyone who is struggling with tight adductors, which we'll soon see are the muscles that protest meditative sitting postures the most.

The lumbosacral joint and the sacroiliac joints are the weakest links in the classic sitting poses, so be watchful. In the absence of good hip flexibility, trying to perfect these postures forcefully by ratcheting the promontory of the sacrum forward with the psoas muscles may strain one or more of these weak links. And if this happens, pain emerges near the site of the strain— lumbosacral pain close to the midline in the lower back, and sacroiliac pain at the rear of the pelvic bowl (chapter 6) just an inch or so lateral to the midsagittal plane of the body. To avoid problems with all of these joints, you must work patiently with the exercises suggested for freeing up the sacroiliac joints in chapter 6. You are less likely to hurt the hip joint because it is designed for flexion.

THE GLUTEAL MUSCLES

The gluteus maximus muscles (figs. 3.8, 3.10, and 8.9–10) are hip extensors and antagonists to the hip flexors, and one would at first assume that they would inhibit sitting correctly in the cross-legged postures, but that is not the case. It's true that they act as antagonists to the hip flexors in standing postures, but paradoxically, in meditative sitting postures they can actually support the action of the iliopsoas muscles by acting as slings to lift the pelvis from underneath. They do this only temporarily, however, if and when you momentarily double your efforts to sit straighter. In contrast to the continuing and highly desirable isometric contraction of the psoas muscles, keeping the gluteus maximus muscles continuously under tension would be too distracting.

The gluteus medius and gluteus minimus muscles (fig. 3.8, 3.10, 8.9–10, 8.12, and 8.14) are abductors of the thighs, and will thus aid any sitting posture in which you are aiming to stretch the adductors and spread the thighs apart, which is necessary for all three of the classic yoga meditative postures. If you have a sudden urge to try harder to sit up straight in the classic sitting postures, and if you feel that effort in muscles deep to the more superficial gluteus maximus muscles, you are probably feeling the gluteus medius and minimus. As in the case of the gluteus maximus, such efforts can only be of a momentary nature—more for educating the adductors about new limits than for wrangling with them throughout a period of sitting.

THE ADDUCTORS

Along with flexors (the iliopsoas), extensors (the gluteus maximus), and abductors (the gluteus medius and minimus), we also have adductors for pulling the thigh medially, a movement which we considered in detail in chapter 6 in relation to forward bending. To understand why the adductors are important here, experiment with the movements required for sitting cross-legged. Sit on the floor with the knees up and the feet side by side about a foot from the buttocks. Now pull one foot in toward the perineum, flexing the leg maximally against the thigh, and lower the knee laterally toward the floor. Notice that the thigh has now been abducted, flexed, and laterally rotated, and recall from chapter 6 that each and every one of these movements is resisted by the adductor muscles. And not only that: since the adductors attach between the bottom of the pelvic bowl and the femur, they also act in the other direction, pulling forward on the base of the pelvis and causing the lumbar region to flatten or even round posteriorly. This creates resistance to the 90° of hip flexion necessary for maintaining the right tetrahedron.

In summary, the iliacus and psoas muscles are most important for creating 90° hip flexion in the three classic sitting postures, and the adductors are most important for preventing it. Accordingly, the iliopsoas must be strong enough and the adductors long enough to permit you to sit upright with a convincing lumbar lordosis, flex the thighs 90°, abduct them enough to bring the heels to the position required by the posture, and rotate them laterally enough to keep the knees pressed to the floor—and do all of this at the same time.

The most obvious (although long-term) solution is to strengthen the psoas and iliacus muscles with leglifts (chapter 3) and gradually lengthen the adductors with prolonged adductor stretches. The less obvious solution is to free up the sacroiliac joints so that full nutation permits the ischia— and with the ischia the origins of the adductor muscles on the inferior pubic rami—to be spread apart from one another. This has the same practical effect as lengthening the adductors.

THE HAMSTRINGS

What about the hamstrings? These muscles take origin from the ischial tuberosities, which are even further to the rear than the inferior pubic rami, and they certainly limit flexion from the hip in forward bending positions with the knees straight (chapter 6). In the classic sitting postures, however, in which the knees are bent, the hamstrings are not stretched at all because their insertions on the tibia and fibula are drawn toward the ischial tuberosities, leaving them slack. They are a problem in sitting postures only if they are injured.

THE PUBOFEMORAL LIGAMENT

Muscles establish the most important restrictions to hip flexibility, but there is another restriction within the hip joint itself—the pubofemoral ligament—which is one of the three extension-limiting ligaments that spiral down to the neck of the femur from each of the three parts of the pelvis (fig. 3.6). Tension comes off these ligaments during a forward bend, but when we abduct the flexed thighs in a sitting posture, the head of the femur is pulled away from the acetabulum, and this creates tension in the pubofemoral ligament, which runs straight laterally from the pubis to the femur in the flexed and abducted thigh. In an anatomical dissection, this ligament may even have to be cut to permit full abduction of the flexed thigh. The pubofemoral ligament is one of the few ligaments in the body that can and should be stretched over a period of time by those who wish to use the classic sitting postures. In the absence of that need, leave it alone.

THE KNEE JOINTS

The knee joint (figs. 5.24–25) also inhibits sitting properly in the classic sitting poses. As seen in chapter 5, in an extended position it can withstand extreme pressures because the joint capsule is taut and all of its components fit together perfectly. In flexing the joint for walking, however, the joint capsule and internal components have to become loose to permit motion. This is why almost all severe knee injuries occur when the joint is in the flexed position. A sitting posture is less hazardous than others, but still, folding the legs back towards the thighs, laterally rotating the legs at the knee joint (chapter 7) and pressing the knees to the floor, and then staying in that position for a long time, places an unusual strain on the joint. Not everyone can accommodate to this position overnight. Flexibility and strength has to be developed in connective tissues that may have been held in static positions, possibly for decades, and to accomplish this, the time-honored solution is to strengthen the muscles that insert into the joint capsule with standing postures (chapters 4 and 7).

RIGHT-LEFT BALANCE

The cross-legged sitting postures are inherently asymmetrical because one foot has to be tucked in place before the other one can be settled, and this will cause small but habitual imbalances in the muscles that act as extensile ligaments (chapter 1). If you always place the same foot down first, dozens of pairs of postural muscles will develop differing lengths on the right and left sides and ultimately lead to imbalances in the skeleton and other connective tissues. In time the pelvis will develop a permanent side-to-side tilt as well as a twist with respect to the rest of the torso. In the short run this is sometimes the reason for deep muscle pain in the hip on one side, but in time it will cause the right lower extremity to become shorter than the left. Needless to say, it is important to compensate for these imbalances by alternating the position of the feet. We'll look at this matter experientially when we examine the auspicious posture later in this chapter.

PROPS

The ideal posture for meditation is sitting flat on the floor with optimal curvatures in the vertebral column, but it may be some time (if ever) before you will be able to maintain such a position comfortably. One solution is to support the pose with one or more props. Even though many traditional teachers are ardently opposed to such devices, they can help get you started and are useful for pointing you in the right direction. Some props raise the hips off the floor and take tension off the adductors, some brace the lower back from behind and relieve tension on the erector spinae and hip flexors, and some hold the knees down and relieve tension on the hip extensors. The commonest and simplest device is one that lifts the hips.

RAISING THE HIPS

Most of the problems associated with the classic sitting postures can be alleviated by raising the hips 1–6 inches off the floor (fig. 10.7). A rolled-up blanket is ideal for this because it is firm and because you can vary the thickness to suit yourself. If the height is adjusted precisely, you feel as if you can sit up straight without straining and yet notice that some effort is required.

You can sit straighter with a prop under the hips for several reasons. The thighs are angled down, and as a consequence the reduced demands for flexion of the thighs means that adductor muscles that are otherwise too short have more slack. Second, since the hips are lifted, the legs end up further underneath the thigh muscles and will not force the thighs into as much lateral rotation. Third, with less stretch being placed on the adductors by flexion and lateral rotation, these muscles have more room to permit abduction. And last, the adductors are not pulling so insistently on the bottom of the pelvis, so the iliopsoas muscles will not have so much difficulty maintaining the back axis of the body perpendicular to the floor.

If your support is not high enough to suit your particular needs, and if you are determined to sit "straight" no matter what, a strained and inappropriate posture results. The vertebral column compensates for the stress by generating a lordosis that is higher in the back than normal. So instead of an anteriorly convex lumbar lordosis between the first lumbar vertebra and the sacrum (L1 to S1), the curvature flattens and deteriorates in this region and you develop what might be called a thoracolumbar lordosis, or one that is convex anteriorly between about the eighth thoracic and the second lumbar vertebrae (T8 to L2). The posture is awkward and unnatural; to make matters worse you can adjust to it easily and begin to sit that way habitually.

It is also important not to sit on a support that is too high. Your pose may be satisfactory as far as the basic curvatures are concerned, but after a few minutes it will be unstable and you will tend to pitch forward. This will soon cause you to weave around and lose your concentration. A support that frankly lifts the hips also causes you to lose the purity of the right tetrahedron. You may be upright, but the angle between the thighs and trunk becomes an obtuse rather than a right angle, and the proximal parts of the thighs are not solidly against the floor.

KEEPING THE KNEES DOWN

Unstable thighs can be a big problem: if you are pushing your limits trying to sit in one of the three traditional cross-legged postures with minimal or no elevation from a support, the knees will tend to float away from the floor, especially the knee on the same side as the foot that is on top. This situation is less than ideal in the three classic sitting postures because the base triangle of the tetrahedron depends on stable thighs. If the psoas and iliacus do not have a rock-solid foundation from the femur, the posture gets tippy.

Figure 10.7. The supported auspicious pose (see fig. 10.11 for details of the unsupported posture) is propped here with a sandbag that permits the model to maintain a convincing lumbar lordosis, which is required for a "straight" spine. Unless you can flex your hip joints 90° with your thighs abducted and feet folded in, you may not be able to sit comfortably and straight without a firm prop under your ischial tuberosities.

It is muscular tension from the iliopsoas muscles that either lifts the knees or pulls the torso forward at an unattractive acute angle, and one remedy for this is to place 10–40 pound weights on your knees to hold them down. This prop all by itself helps you maintain the correct lumbar lordosis and permits you to minimize or even eliminate the need for a cushion under the hips. A commercial apparatus has also been developed that will press the knees down painlessly and evenly on both sides, make a cushion unnecessary, and yet maintain the posture in an exact right tetrahedron.

Keeping the knees down allows you to place all of your attention on lifting the posture with the iliacus muscles and creating sacroiliac nutation with the psoas muscles, but it also takes the place of the work you would ordinarily do to maintain the pose. It improves the external appearance of the posture but changes its fundamental nature. The other disadvantage is that it is impractical to carry such props around with you, so they are generally fit only for home use.

BRACING THE BACK FROM BEHIND

An even more radical way to support the classic sitting postures is to use a special manufactured back support that pulls the lumbar region forward. These devices are quite comfortable, and they certainly place you in a state of ease. The main problem is that they also put you to sleep. They should probably be reserved for those who usually sit on the floor but who are temporarily decommissioned with low back problems. In classes taught by many traditional teachers, such supports are viewed with considerable distaste.

THE MIDDLE GROUND

In the last analysis the important question is not whether props are comfortable, but whether they are advisable. If you believe that much of the point of the meditation postures is to lift your energy and consciousness with your own internal efforts, props that make the posture too easy can only be counterproductive.

You have many options for training yourself to sit in the traditional postures. One is to forget all about them for a year or so and concentrate solely on improving your sacroiliac and hip flexibility (chapter 6), back strength (chapter 5), and abdominal strength (chapter 3). Then make a serious effort to construct (or re-construct) a respectable meditative posture. Or you can try to sit as straight as possible without any props at all while at the same time working to improve strength and flexibility with a rigorous and balanced program of hatha yoga postures. That may work if you are already fairly flexible, but if you are not it is a tough course indeed. Finally, you can take the middle ground with the judicious use of props:

you can use a cushion or folded blanket but adjust the height so that the knees remain down without external weight and with only moderate effort. The mild isometric effort needed to keep the lumbar region arched forward and the knees down keeps you aware of the dynamics of the posture and yet allows you to maintain an inward focus.

If you take this last option, you can gradually lower the height of the support as you lengthen your adductors and develop sacroiliac flexibility for nutation. Those who are almost able to sit flat on the floor comfortably may even be able to get by with only a thin cushion placed under the fatty portion of the hips—behind the ischial tuberosities rather than directly underneath them. This does not lift you significantly away from the floor but it does help you hold the lumbar region in its proper curvature. In any case there may come a time when you feel that you have reached your limit for stretching the adductors, opening the hip joints, and freeing up the sacroiliac joints (chapter 6), and in that event you should content yourself with the imperfections, use the least conspicuous prop possible, and attend to meditation.

THE SIX POSTURES

The three classic sitting postures can be challenging, but fortunately there are easier poses you can use while you work up to the traditional ones. We'll consider six meditation poses in an ascending curve of intensity and effort. The friendship posture (*maitryasana*) is excellent if you need (or prefer) to sit on a chair; the adamantine posture (*vajrasana*) while kneeling on a bench is probably the most comfortable; the easy posture (*sukasana*) is suitable for short periods of sitting before you are comfortable in the traditional poses; the auspicious posture (*swastikasana*) is the easiest of the three classical postures; the accomplished posture (*siddhasana*) is valued for stabilizing sexual energy; and the lotus posture (*padmasana*) brings a profound sense of repose.

THE FRIENDSHIP POSTURE (MAITRYASANA)

The friendship posture (*maitryasana*), in which you are sitting on a chair, is best if you are just beginning to practice hatha yoga, if you are among those who are not able to sit comfortably in any position on the floor, or if you usually sit on the floor but are in pain for one reason or another. And even if you do not intend to use the posture as your primary sitting pose in the long run, it is a useful learning tool. Start with a wooden chair cushioned just enough to enable you to sit still comfortably for 10–20 minutes. To do the posture simply sit on the edge of the chair with the head, neck, and spine in a straight line, the knees comfortably apart, the feet flat on the floor, and the hands resting on the thighs (fig. 10.8).

The greatest advantage to this pose is that the lumbar lordosis is easy to maintain. The thighs are at a 90° angle from the trunk, the feet are planted solidly on the floor, and the arms are resting on the knees where they can help stabilize the spine. The vertebral column can be held erect because there is no tension from the adductors and hamstring muscles on the underside of the pelvis, and also because the iliopsoas muscles exert only minimal effort to maintain a right angle between the thighs and the torso. This makes it easy to do breathing exercises, pranayama, and meditation.

The disadvantage to the friendship pose is that sitting on the edge of the chair requires you to be constantly alert to keep your balance. Unlike the classic postures, your base does not form a solid triangle against the floor, and without that stability a big part of your attention must remain on staying upright. If your awareness lapses, the posture will begin to sag and wobble.

Sitting toward the rear of the chair is a more secure option because in that position your back is steadied from behind. The pelvis and sacrum are braced, and the position of the ilium against the back of the chair stabilizes the origins of the erector spinae and quadratus lumborum muscles. So in this version of the friendship posture you can concentrate your attention on keeping a sense of lift to the spine and do not have to think about stability. There are two downsides, however: it is too easy to relax, get drowsy, and lose your concentration, and the pressure of the chair against your back impedes diaphragmatic (although not abdominal) breathing.

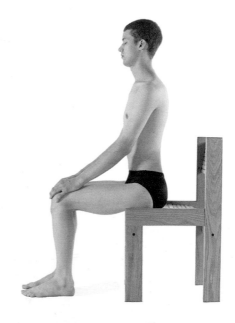

Figure 10.8. It is easy to sit straight in the friendship pose (maitryasana), because neither the adductors or the hamstring muscles create tension on the base of the pelvis that would cause a posterior pelvic tilt (a rotation of the top of the pelvis to the rear that would in turn degrade the lumbar lordosis). It is also easy to maximize sacroiliac nutation in this posture by selectively contracting the psoas muscles. The main disadvantage of the pose is that without a broad base it tips around (both front to back and from side to side) fairly easily, and you have to remain hyperalert to maintain stability.

THE ADAMANTINE POSTURE (VAJRASANA)

There are several variations of the adamantine, or warrior pose (*vajrasana*), but only two of them are suitable for sitting more than a few minutes or so at a time. For the basic posture start in a kneeling position with the thighs together and the head and torso vertical. The feet should be together with their upper surfaces facing the floor and the heels slightly apart. Now lower your body until you are sitting on the heels, which forces full extension of the ankles. If the posture is beyond your capacity you can use any combination of several props: supporting the ankles with a small pillow or folded towel, placing a soft pillow between the thighs and the legs, or placing a substantial pillow under the hips and between the feet.

To develop enough flexibility for the basic posture, you can spread your feet apart and sit between them in one of three possible positions: with the heels up, with the heels in and the toes out, or with the heels out and the toes in. In this last one you usually place some of your weight on the feet as well as on the floor. If your knee and ankle flexibility do not permit sitting squarely on the floor in these three variations, most instructors suggest that you place a supporting pillow between the feet just high enough to make the posture comfortable.

The basic adamantine pose can be used for brief periods of meditation and pranayama, but unless you have grown up with it in an Islamic culture, or have taken a decade or two of training in the formal Japanese tea ceremony, sitting in this posture for more than a few minutes should be approached cautiously. First of all, it may strain ligaments in the knee that are not accustomed to prolonged tension. Second, circulation may be cut off in the legs as a result of the extreme flexion of the knees, causing a pins-and-needles sensation. Finally, the pose places pressure directly on a superficial branch of the common peroneal nerve, which is subcutaneous just lateral to the head of the fibula and which supplies several muscles on the anterior side of the leg. If that nerve is traumatized by prolonged sitting in the adamantine pose, the muscles it supplies can be temporarily paralyzed. You will experience numbness and a clinical syndrome—someone cleverly but unjustifiably thought to call it "yoga foot drop"—in which you are unable to flex the ankle when you step forward. If the trauma is mild you will experience the symptoms for only a few minutes or at most a few days. But if you damage the nerve to the point at which its axons degenerate distally from the site of their injury, you will have to wait for the nerve fibers to regenerate from that site to the peripheral sensory receptors and muscles before you regain sensory and motor function. This regeneration happens slowly but surely, at the rate of about one mm per day.

All of these problems can be remedied by sitting 5–8 inches off the floor on a small bench with a tilted-forward seat (fig. 10.9). In this position the knees are incompletely flexed, and because of this, little pressure is placed on the common peroneal nerve, and blood circulation is less impeded. The biggest advantage, however, is that none of the muscular tension associated with cross-legged postures is present. It is easy for the iliopsoas muscles to tilt the back of the pelvis up and forward to create a strain-free lumbar lordosis, even for those with severe restrictions in hip flexibility. The pelvis is automatically placed in a forward tilt defined by the angle of the seat (the only sitting posture in which that happens) and this creates an automatic lumbar curve that keeps the abdomen open and yet taut, which in turn is helpful for experimenting with different methods of breathing. The other attractive feature of this posture is that it is possible to sit comfortably in it for much longer than is possible for any of the three classic sitting poses—all without pins-and-needles sensations, lower back discomfort, and cramped circulation.

The advantages to sitting in the adamantine posture on a bench are logical, obvious, and huge. So why is it not in widespread use, and why is it not included in the canon of traditional yogic meditation postures? The reasons fall into two categories. Structurally, the adamantine posture on a bench does not form a right tetrahedron, and it is therefore not as stable as the classic sitting poses. The thighs are not at a right angle from the spine, the knees are not very far apart, and the upper extremities do not brace the posture as efficiently as they do in the cross-legged postures. If you have gotten accustomed to one of the three classic sitting poses, and you try this one as an alternative, you may sense a lack of groundedness— a floating feeling and a sense that your energy is being dissipated. If you

Figure 10.9. The adamantine posture (vajrasana) on a bench is hard to match for long periods of sitting. The psoas and iliacus muscles work efficiently to keep you upright, the tilted-forward seat places the pelvis in a natural anterior pelvic tilt that encourages a comfortable lumbar lordosis, the feet and knees are not stressed, and problems with blood circulation and damage to the common peroneal nerve are minimized. The disadvantages, often mentioned by those who have long accustomed themselves to one of the three classic sitting postures on the floor, is that this posture leaves them with an ungrounded, floating sensation that distracts them from meditation.

cannot put this feeling aside or compensate for it in some manner, you will probably not be content with this pose in the long run. But still, if you keep having discomfort with the other postures, and especially if you wish to sit in perfect comfort for long periods of time, it's worth a try. Perhaps you can make it work.

THE EASY POSTURE (SUKASANA)

As the name implies, the easy posture (*sukasana*) is the first one beginners learn when they are ready to sit on the floor. To do it, simply fold the lower extremities so that each leg rests on the opposite foot and sit up as straight as possible. The lateral sides of the feet are against the floor and the legs and thighs may point up at an angle of 20–30° (fig. 10.10a).

The easy posture is appropriate for 2–5 minute periods of meditation or for breathing exercises at the beginning or end of a class, but it has several disadvantages as a meditative posture. To understand why, try the following experiment. Sit flat on the floor and assume the easy posture, making a moderate effort to hold the back straight. Remembering that the psoas and iliacus muscles insert on the femur, lift the posture by using those muscles. Notice that when you try to lift and straighten the vertebral column, the lumbar lordosis is pulled forward by the iliopsoas muscles, as expected, but the thigh is also pulled toward the torso, raising the knees (fig. 10.10b). This makes the posture unstable: the thighs float up and down and the lumbar region floats back and forth.

Another reason this posture is unstable is that it doesn't form a true right tetrahedron. Compared with the other cross-legged sitting poses, the base triangle is smaller, since it is supported only by the lateral sides of the feet rather than the full length of the thighs, and the elbows are bent 90°, making it difficult for the upper limbs to stabilize the posture.

A cushion that lifts you several inches off the floor modifies the posture to the point at which it becomes more stable: the thighs are closer to being horizontal and the hands can grasp the knees with the forearms extended. And because of this you can lift the vertebral column without springing the thighs into a flexed position. This posture, supported by a thick, firm, round pillow called a zafu cushion, is commonly used for marathon periods of sitting in zen meditative traditions, and is definitely worth exploring.

THE AUSPICIOUS POSTURE (SWASTIKASANA)

At some point serious students of yoga will want to try one of the classic sitting postures for meditation, and the one to start with is the easiest of the three—the auspicious pose (*swastikasana*). Place the left foot against the opposite inner thigh with the back of the heel to the right side of the genitals (fig. 10.11a). Notice a prominent bony knob on the medial side of the ankle. This is the *medial malleolus*, and it should be just to the right of the midline of the body. T*he lateral malleolus* (fig. 6.8), on the opposite side of the ankle, rests against the floor. Next, tuck the lateral side of the right foot between the left leg and thigh. The two heels are now separated by the width of about four fingers. The lateral malleolus of the right ankle is now to the left of the medial malleolus of the left ankle. In other words, they cross one another in the midline. If the heels are not far enough apart, these bones will be on top of one another and will cause discomfort. Now, reaching between the right leg and thigh, pull the left foot up so it is fixed between the calf and thigh muscles. Then sit straight and place the hands lightly on the knees with the palms down (fig. 10.11b). The body is now stabilized in the form of a right tetrahedron.

Figure 10.10a. The easy posture (sukasana) isn't easy for long because it doesn't have a stable base. It is fine for a few minutes of meditation or breathing exercises at the end of a class, during which time you can make a special effort to sit up straight. Otherwise, it can be used as a more relaxed posture for playing music or eating a meal. In any case, the foot that is pulled in first should be alternated regularly. The pose is shown here with the hands at the sides for comparison with the profiles of the thighs and back in figure 10.10b. The hands would ordinarily be placed on the knees.

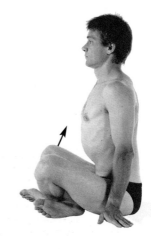

Figure 10.10b. When you try to sit up straighter in the easy pose, your psoas muscles act to pull the lumbar lordosis forward, and your iliacus muscles act in emphasizing an anterior pelvic tilt. That much is fine. The problem, however, is that the thighs are not stabilized, and the hip flexors create an unwanted side effect of increasing hip flexion (lifting the knees), which makes sitting in this posture for meditation a constant battle for stability.

As mentioned earlier, the right foot should be placed on the bottom every other time you sit in the auspicious pose, as well as in all other cross-legged sitting postures. After a few weeks, it becomes a simple matter of habit to alternate. As a reminder, you can place the left foot underneath on odd days of the month and the right foot underneath on even days of the month. Why do this? With nothing on from the waist down, watch yourself carefully in a mirror and notice that the foot on the bottom lifts the pelvis slightly on that same side. More specifically, if the left foot is on the bottom, the left leg is slightly further underneath the left thigh than the right leg is underneath the right thigh. The proximal portion of the left thigh will therefore be slightly higher than that of the right thigh, and as a result the crest of the left ilium will be slightly higher than the crest of the right ilium. And since the pelvic bowl is tilted to the right, the vertebral column has to tilt slightly to the right in the lumbar region, to the left in the thorax, and again to the right in the neck. If you switch the positions of the feet you will notice that the postural adjustments reverse themselves from head to toe, leaving you with an opposite set of right-left imbalances. It is impossible to eliminate them entirely, but it is a good idea to compensate for the imbalance on one day with an equal and opposite imbalance the next. It becomes a matter of routine and does not violate the classic injunction to stick with one sitting posture for meditation.

a.   **b.**

Figure 10.11. To prevent imbalances in all the cross-legged sitting postures, the foot that is placed first should be alternated daily, such as the left foot first on odd days of the month (a) and the right foot first on even days of the month. The key feature of the auspicious posture (swastikasana, shown here) is that the feet are placed against the opposite thighs so that the medial malleolus of the lowermost ankle and the lateral malleolus of the upper ankle are both situated beyond the midsagittal plane (b). In other words, the malleoli cross one another rather than being on top of one another. In the case pictured above, the left foot (big toe side) is pulled up between the right calf and thigh, and the right foot (little toe side) is inserted between the left calf and thigh (b).

The auspicious pose is the easiest of the three classic sitting postures because the feet cross one another in the midline of the body and end up in a natural and stable position planted against the opposite thighs. The adductors and pubofemoral ligaments are not stretched excessively, and the knees and hip joints are not placed under intense torque. Your main needs for this pose will be to develop hip flexibility and to acclimate to torques and pressures on the ankles. After that, almost everyone can be comfortable in this posture. If you are using a cushion, adjust it to the point at which it is high enough to support you but not so high that you do not need to make moderate efforts to keep the lumbar lordosis arched and the sacroiliac joints in full nutation. Decrease the height of the cushion over a long period of time until you are sitting closer to the floor. This pose is aptly named the auspicious posture. You can settle into it indefinitely and without regret.

THE ACCOMPLISHED POSTURE (SIDDHASANA)

The accomplished pose (*siddhasana*), also known as the perfect pose, is said to be the meditative posture of yoga adepts and renunciates. It is the most demanding—and some say the most rewarding—of all the sitting postures, but anyone who is reasonably comfortable with the auspicious pose can begin to learn it.

Several variations of the accomplished posture are given in the hatha yoga literature, but for even the commonest one which we will describe here, different authorities give slightly different directions for how to place the feet. Regardless of this, all agree that the backs of the heels are to be exactly in the midsagittal plane of the torso.

In men the base of the penis rests against the bottom heel, so when you first explore the posture it is best not to wear anything from the waist down. Lift the penis, scrotum, and testes up and out of the way, and then place the left heel underneath the inferior pubic rami (figs. 1.12 and 3.2) so that there is barely room for the penis to emerge above the heel. If you are sitting squarely on the floor with no support to lift you up, the back of the heel will be situated slightly in front of the center of the perineum (fig. 10.12). Your weight will be supported by the thighs, the ischial tuberosities, and by the left heel and inferior pubic rami.

The corpus spongiosum of the penis (chapter 3) will be slightly compressed from below by the medial aspect of the heel, but it will be protected by the depth of the upside-down V formed by the inferior pubic rami (fig. 10.13). The back of the heel should be pressing against the inner surfaces of the inferior pubic rami from which the corpora cavernosa arise. The skeletal foundation of the posture, which shows the contact points of the heel with the two ischiopubic rami, as well as space for the penis between the heel and the pubic symphysis, makes the nature of this pose apparent immediately (fig. 10.13). The corpus spongiosum should be exactly in the midline, locked between the heel and the top of the V. Be careful not to get any part of the spermatic cord (containing the ductus deferens, blood vessels, and nerves) caught between the pelvic bones and the heel. Make doubly sure of this after settling the first foot into position (in this case the left foot) by pulling the skin of the scrotum up and forward so that the skin and underlying tissues of the front of the perineum are pulled taut.

Notice that the medial malleolus of the left foot is to the left of the midline. Adjust the genitals to either side of the heel—penis and one testis to one side and the other testis to the other. Then place the right foot above the left so that the back of the right heel is exactly in line with the back of the left heel. You can lift the right heel above and to the front of the pubic bone, allowing the foot to angle downward, or if that is not possible you can fold a small, soft cloth and use it as a cushion between the two ankles without violating the essence of the posture. The malleoli do not cross beyond one another as they do in the auspicious posture. The lateral malleolus of the right foot is now to the right of the midline and also to the right of the medial malleolus of the left foot. Straighten the spine and rest the hands on the knees, allowing the thumbs and index fingers to touch one another. The body has now formed a right tetrahedron (fig. 10.14). As with other cross-legged sitting postures, the positions of the feet should be switched on alternate days or sittings.

Women, like men, should position the heel against the inner surface of the inferior pubic rami. But this means that the heel will have to be placed directly against the soft tissues of the genitals, well in front of the fourchette (the fold of skin which forms the union of the lower ends of the labia minora). The heel will be more intrusive in the female because the

Figure 10.12. The key feature of the unpropped accomplished posture (siddhasana) is that the back of the heel of the lowermost foot is placed at the perineum (between the anus and the genitals) exactly in the midline. The medial and lateral malleoli of the respective feet are both located short of the midsagittal plane of the body, so that if the top foot is laid down directly against the bottom foot, the two malleoli do not rest on top of one another. Most people will require a prop to make this posture comfortable.

upside-down V formed by the pubic rami is shallower than in the male. Some authorities describe a posture in which women sit with nothing on from the waist down and press the lowermost heel between the labia, calling the pose *yoni siddhasana*. Other authorities state heatedly that women should never sit in any form of the accomplished pose. Those who find it uncomfortable can use the auspicious pose.

It is said that the accomplished posture stabilizes and sublimates sexual energy because of the position of the feet with respect to the genitals. So to monitor the subtleties of the pose you have to monitor the position of the lower heel in relation to the structures lying within the confines of the urogenital triangle (figs. 3.4 and 3.27–29). If you are sitting straight in the

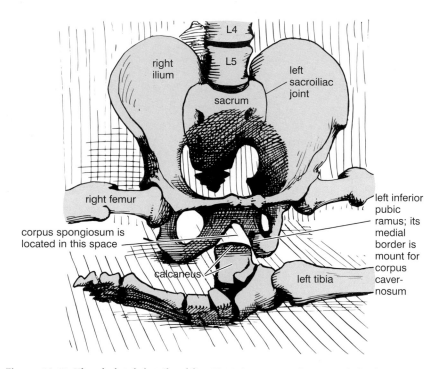

Figure 10.13. The skeletal details of fig. 10.12 (unpropped accomplished pose) reveal that the tarsal bone of the heel (the calcaneus) is locked squarely against the inferior pubic rami. The soft tissues at possible hazard in the male are the corpus spongiosum of the penis in the midline, and especially the two corpora cavernosa, which take origin along the medial borders of the inferior pubic rami. It is the possible damage to arteries within the corpus spongiosum and the corpora cavernosa that suggest contraindicating extensive practice of the accomplished posture in men who wish to maintain sexual activity. With three exceptions, a similar situation occurs for a woman: the exposed portion of the clitoris is well out of harm's way; the pubic arch is wider, thus easing contact of the heel with the corpora cavernosa; and the back of the heel will be placed squarely between the labia in the vaginal introitus (Dodd).

accomplished pose without a cushion, and if the sacroiliac joints are fused or locked, the pelvis will rotate forward in exact proportion to how far the lower back arches. And if or as the posture improves under these circumstances (which means establishing a fuller lumbar lordosis and a straighter posture), the inferior pubic rami are rotated to an even more acute angle with respect to the floor. In men this catches the base of the penis between two unyielding surfaces—the bones of the foot and the inferior pubic rami. In women, coming more fully into the posture presses the lower heel more deeply into the soft tissues of the external genitals.

It is at this point that sacroiliac joint mobility makes a big difference. If instead of being fused or locked, the sacroiliac joint is capable of 5–10° of movement between the extremes of nutation and counternutation, you will want to establish full nutation. This can help the posture circuitously in two ways: First, because nutation spreads the ischial tuberosities apart, it eases tension on the adductors, and this helps you sit straighter with an intact lumbar lordosis. This can help you in all of the classic meditative sitting postures, but nutation is especially important in the accomplished posture for yet another reason: with the sacroiliac joints in full nutation, the pelvic bowl as a whole does not have to be tipped quite so far forward to complete the posture, and this means that in relative terms nutation will have produced a slight posterior pelvic tilt and will have reduced the acuteness of the angle between the inferior pubic rami and the floor. In combination with spreading the ischia apart, this makes a little more room for the penis. Each of these effects is tiny, but the results add up. It is not an exaggeration to suggest that sacroiliac joint mobility for nutation is almost as important to this posture as adductor and hip flexibility.

Figure 10.14. The completed accomplished posture. As in the case of the unpropped auspicious pose, the foot that is placed first should be alternated daily for the accomplished pose. There are two options for the upper foot, one with a pad between the ankles (but with the malleoli short of the midsagittal plane), and the other with the upper leg rotated severely enough that the upper foot is placed entirely above the genitals. The former is easier; the latter, shown here, is more traditional. In any case, the toes are tucked in between the calves and thighs as in the auspicious pose.

As to props, a supporting cushion changes the position of the heel and alters the posture so completely that even calling it the accomplished pose becomes questionable. If you are four inches off the floor, the lower heel will probably not even be in contact with the body and even the upper heel will be below the base of the penis, or, in the female, midway in the labia. At three inches off the floor the lower heel can be positioned easily in the center of the perineum in either men or women, but it may not press firmly at that site; and the upper heel is still below the genitals in the male and at the level of the opening of the urethra in the female. One to two inches off the floor, the lower heel is situated in front of the center of the perineum, and the upper heel will now be slightly above the penis in men, and at the level of the clitoris in women.

The hatha yoga literature suggests that pressure of the lower heel against the penis is beneficial for men who are attempting to maintain celibacy. At the same time we see occasional warnings that the accomplished pose can cause impotence. There are two concerns. The first is that the posture can cause numbness in the penis as the result of direct pressure on the cutaneous nerves. If you wish to maintain sexual activity you should therefore sit without traumatizing those nerves. This is easy if you sit on a support which lifts the ischial tuberosities—and with the ischial tuberosities the inferior pubic rami—high enough so that the base of the penis is not compressed. Also take care to sit with the heel perfectly in the midline. Cutaneous nerves (nerves that distribute their fibers to the skin) never cross the midline, and if the posture is adjusted perfectly no serious problems are likely to be found.

A second concern is potentially more serious: impotence caused by traumatization of the central arteries of the corpora cavernosa, the erectile bodies of the penis (fig. 3.28a), from too much or too prolonged pressure from the lowermost heel. Urologists who specialize in sports medicine commonly see this problem in cyclists who fall against the top tube of a bicycle, a mishap which damages the arteries of the corpora cavernosa and interferes with their ability to dilate. Sitting on a support for the accomplished posture may prevent trauma to the central arteries of the corpora cavernosa as well as numbness because the rear of the heel is situated more posteriorly and will not lock the corpora cavernosa quite so firmly against the inferior pubic rami.

In summary, sitting up on a cushion that protects the vessels and nerves of the base of the penis changes the posture from one that significantly restrains sexual energy to one that moderates the sexual impulse more subtly and provides support and nurturance generally to the base of the body. A soft, supported posture seems more appropriate for men who are in an active sexual relationship or who wish to maintain that potential. For men at least, sitting for hours daily in the accomplished pose in its pure form—flat on the floor with the penis locked in place—is appropriate only for those who are in a state of celibacy and who wish to remain so for the rest of their lives.

If women sit directly on the floor, the back of the lower heel will be in the exact place where both the urogenital and pelvic diaphragms are interrupted by the vaginal introitus. We do not have enough data to say whether this heel position brings the same benefits (or problems) to women as it does to men. Some women report that it is uncomfortable but harmless, others say that it is beneficial for restraining sexual energy.

If you wish to master the accomplished pose, first sit in the auspicious pose regularly for a few months, gradually decreasing the height of your support. Then make it an inch higher and try the accomplished pose. Because the backs of the heels are aligned in the midline of the body for the accomplished pose, the adductors must be longer than for the auspicious posture (given a constant height of the support), and for this reason the accomplished pose will create more resistance to flexion at the hip joints. Adjust the height until you are stable and then gradually decrease it. Your final position will depend on exactly what you hope to learn and gain from this posture.

THE LOTUS POSTURE (PADMASANA)

The lotus posture (*padmasana*) is one of the most beautiful postures in yoga but it is not practical for most Westerners as a meditation pose. It places peculiar stress on the knee and hip joints, and unless you have done it in your formative years it is not likely to work satisfactorily. If mastered it is said to bring an incomparable feeling of repose and calmness to the mind. The lotus posture is also used in connection with numerous other asanas such as the headstand (chapter 8), so it is worthwhile to practice even if you do not intend to use it for meditation.

To come into the posture place the lateral surfaces of the ankles against the opposite thighs as close to the torso as possible. The feet should be upturned and the toes should rest against the lateral sides of the thighs. Then straighten the spine and place the hands on the knees, generating the right tetrahedron (fig. 10.15). As with other cross-legged sitting postures, the positions of the feet should be switched on alternate sittings for the sake of balance, placing the left foot first and following with the right, then placing the right foot first and following with the left.

Because the knee joints are hinges, the legs force the thighs into extreme lateral rotation when the feet are lifted onto the thighs, and when coupled with an initial flexion and abduction of the thighs, the extraordinary lateral rotation places the hip joint in a stressed and unusual position. This plus the stress on the knee is what makes this posture so difficult. Years of consistent effort may be needed to alter the anatomy of the hip joints and supporting ligaments enough to make the pose feasible. And even after this, one knee is likely to resist resting squarely against the floor unless you are sitting up on a cushion or other support.

MULA BANDHA

Yogis tell us it is important to apply mula bandha (the root lock, chapter 3) in all the sitting meditative postures. Only by doing this, they say, can we have a sound approach to meditation and govern the energy and vitality of the base of the body. It is the position of the lower extremities that determines the experience of the root lock, and this depends on four things: hip flexibility and the angle between the thighs and the pelvis, the amount of abduction of the thighs, the position of the feet and ankles, and the angle at which the perineum faces the floor. Because these differ from posture to posture, and because the experience of the root lock is fundamental to the experience of each, we shall have to consider the six standard poses individually.

In the friendship pose the base of the body feels open, and because of this, concentration is needed to hold the root lock continuously. But if a soft, padded surface is placed against the front of the perineum, this prop mildly stimulates the muscles of the pelvic floor and of the urogenital triangle (chapter 3). You can use a folded-up washcloth for this purpose, or purchase a little wedge-shaped "mula bandha cushion." In either case the prop will allow you to feel the essence of the root lock and leave your mind free for meditation.

In the adamantine pose the problem is similar. If you sit between the feet directly on the floor, you can hold the root lock only with constant attention, especially if the thighs are together, because the perineum is isolated and pulled open. On the other hand, if the knees are spread apart

Figure 10.15. The lotus posture is one of the most important symbols of yoga, floating in the water and yet anchored firmly to the earth below by a single strand. As a meditation pose the lotus pose is beyond the reach of most westerners, but even if it is not used for that purpose, it is important to so many other postures in hatha yoga that it should be practiced regularly. As for the other cross-legged sitting postures, the foot positions should be alternated daily.

the buttocks come together and the lock is easier to maintain. Sitting on a bench, you can tilt the pelvis forward, pressing the urogenital triangle against the supporting plank and thus making it simple to hold the lock.

In the easy posture the acute angle of the thighs with the trunk makes it difficult to hold the root lock for more than a few seconds. It's almost as difficult as trying to hold the lock while squatting. It is easier to hold if the thighs are more horizontal, as when the easy posture is lifted up by a cushion and the front of the perineum (the urogenital triangle) is facing a folded-up washcloth or mula bandha cushion.

Of the three classic sitting postures, the root lock is most difficult to hold in the lotus posture: the extreme position of the lower extremities tends to draw the anus open and to stretch the entire breadth of the perineum. At the other end of the spectrum is the accomplished posture (without a supporting cushion), in which the underneath heel places pressure against the central tendon of the perineum and thereby stimulates the muscles of the urogenital and pelvic diaphragms. This makes it possible to hold the lock spontaneously in that posture with little additional attention. In order of difficulty, holding the lock in the auspicious pose is somewhere between these two. As with the friendship, adamantine, and easy postures, a supporting prop at the front of the perineum makes it possible to hold the root lock with only the tiniest thread of attention.

MASTERING THE SITUATION

Yoga teachers never tire of saying that sitting postures require you to remain straight, still, and comfortable. But what do you do if you cannot follow all three requirements at the same time? Where do you compromise? Do you relax if relaxing droops your posture? Do you tense the body to maintain stillness? If so, where and how much? And how do you sit comfortably if all the classic postures are uncomfortable? Every teacher will have a different answer, and every teacher will answer differently to students of differing constitutions.

Many teachers feel that sitting straight is most important. Only by sitting with the head, neck, and trunk in proper alignment is it possible to keep a clear mind. In the zen tradition a hall monitor whacks sagging meditators with a *keisaku*, a three-foot "encouraging stick," in order to rouse their postures and energies during long stretches of sitting.

[Technical note: Like all aspects of zen, there is a lot more to doing this job than just walloping someone on the back. It's an art form that involves the precise administration of the requisite "sensory input" to the muscular region of the shoulder just medial to the spine of the scapula. There is little margin for error. It informs you—the recipient—that sitting straight will alert the mind wonderfully and that if you put your life energy into what you have accepted in the moment, the job of the hall monitor—next time—may just be to pass you by.]

Sitting still is the next priority. When you attempt to improve the sitting postures, it is always a temptation to keep adjusting them: leaning forward, arching the lumbar region, pulling the shoulders back, adjusting the position of the head, and correcting a sideways tilt. You can make all of these adjustments while you sit, but you should make them so slowly that the movements cannot be detected by an outside observer. Obvious movements will disturb your concentration, but if you slowly tighten the pertinent muscles and feel the desired shift take place over a period of 30–60 seconds, your posture and concentration will become firm without diverting your attention from meditation. Your mind may be a bundle of nerves and random thoughts anyway, but you will have no chance of centering it if you scratch, twitch, and weave around.

Comfort is third. One meditator has said that the posture should be as easy as a coat hanging on a coat rack. And certainly you will be endlessly distracted if you are uncomfortable. Pain warns of danger, and not honoring that signal will place you at hazard. And pain is a common problem: every one of the classic cross-legged sitting postures will become painful after a 20–60 minute period of sitting unless you have been practicing them and acclimating for a long time. So if you want to extend your sitting time it is legitimate to push up to the point of pain, but then stop. (It might be noted that zen traditions are generally not in agreement with this advice. Ignore pain, they say: it will pass.) In yoga, the customary attitude is not to force yourself, but to choose a posture in which you can sit straight and remain still for 10–30 minutes and yet be reasonably comfortable. The point is not to set records, but to avoid disturbing your concentration.

A BALANCED SET OF POSTURES

If you want to master the classic sitting postures it is best to use a sequential, systematic approach. If you are a beginner, limit yourself to a balanced set of asanas for several months, including standing postures of all kinds, as well as forward bending, back bending, twisting, the shoulderstand, and the headstand. The standing postures are important because they tone and balance the muscles and joints of the pelvis. Twisted standing forward bends should be performed with the feet wide apart for stretching the adductors. Backbending postures such as the cobra and locust are helpful both for increasing the lumbar lordosis and for building strength in the back muscles.

Of the inverted postures, the many variations of the shoulderstand strengthen, limber, and lengthen many of the key muscles of the body that are important for correct sitting postures. The headstand has marvelous effects in preparing you mentally for sitting. It alerts the mind and body, and it wards off sleep in the early morning. After being in the headstand, the entire body is more responsive to your efforts to sit correctly.

Many other postures and exercises are helpful in obvious ways. Cradling each leg either with your back upright or lying in the supine position stretches the piriformis and obturator internus on each side. Placing the soles of the feet against one another as close as possible to the groin in a sitting position, and following this by pressing down on the thighs stretches the adductors. The lotus posture and the preparatory half lotus are invaluable for opening the hip joints and for toughening the knees.

Exercises to improve flexibility should not be practiced in isolation. If you merely stretch the connective tissue of a joint capsule without at the same time building strength in the associated muscles, the joint will become susceptible to injury. For sitting postures the knees and ankles are the hot spots, and if you do not keep them strong with standing postures, sooner or later they are likely to be injured from the chronic strain of sitting.

THE DEFINITIVE TEST

One test of whether or not you are sitting straight is to adjust your posture near a wall. If two points on the back of the hips, two points on the upper back, and one point on the back of the head barely touch the wall, you are straight (fig. 10.16). This is the ideal; you will probably be surprised to find that you have a tendency to pitch forward.

If you wish to try to meet the ideal, you will first have to tighten the erector spinae and quadratus lumborum muscles in order to pull the body backward into a perpendicular position. It won't take much effort but the resulting posture will feel insecure, as though you would tip over if it were not for the wall. To compensate for this you will have to lift the posture more insistently with the iliopsoas muscles, but just enough to balance and not enough to pitch you forward again.

At this stage, even if the four contact points for the hips and the shoulders are barely touching the wall, most people will have to drop the head back so much to contact the wall at the fifth point that the chin comes up and they are looking up 30–45°. The most obvious remedy for this is to tuck the chin in and pull the head back enough to make contact. This does not work. It feels awkward and looks ridiculous, like plebes braced at attention in military academies. The solution: either use a cushion or increase your capacity for sacroiliac nutation so you can increase the depth of the lumbar lordosis. Either of these options or both of them in combination lift the chest and pull the back of the head to the rear without extending the neck. All of the classic sitting poses become balancing postures at this stage.

Trying to sit straight while you are just barely touching a wall at five points may be difficult, but it also underscores the three most important requirements for sitting with an erect spine without a supporting cushion: hip and sacroiliac flexibility, hip and sacroiliac flexibility, hip and sacroiliac flexibility.

MASTERING THE POSTURES

You will probably have to work patiently and for a long time to master the sitting postures. The most important thing is to sit in them with intent and purpose. Do the postures. Do not look for shortcuts. Do not yield to the temptation of placing the first foot correctly and then plopping the other one down in front of the opposite leg instead of locking it into place. If you can sit comfortably with the second foot in front of the first, you can learn to sit correctly in the classic pose. Work and play with your personal posture—just one, not all three—systematically until you are sitting close to the floor. Always sit with full concentration. There is little to be gained if you sit haphazardly. In the beginning at least five minutes is needed for settling into the posture. The body yields to your intention only gradually.

RELAXATION AND BREATHING

That you should relax completely in sitting postures is a common misconception. What actually happens is that the body and mind should be brought to a heightened state of attention and alertness, an awareness that takes its origin from your core and permeates the head, neck, trunk, and extremities. The only skeletal muscles that are entirely inactive are the muscles of facial expression. The muscles of the upper extremities should be generally relaxed, especially the shoulders, but they should be in a state of readiness, expressing just enough tone to stabilize the posture and remind you of your geometry. The lower extremities are a different story. They form your base, and even though there should be an overall feeling of ease, they will have to remain in a mild state of isometric contraction to keep the posture looking respectable.

Figure 10.16. The definitive test of sitting straight requires that the back of the head, chest, and sacrum barely touch a perpendicular surface, preferably using an unpropped cross-legged sitting posture. This is a standard that few can attain. Lack of hip flexibility is a common problem, and most people droop their heads forward. Try it anyway; it will give you feedback about your posture.

After you have settled on a meditation posture and have had some experience with it, you can make improvements with specific breathing techniques. Start with diaphragmatic breathing and notice that inhalation deepens the lumbar lordosis, pulls the chest back, and lifts the head and neck. Ordinary exhalations reverse all of these effects, permitting your head and chest to come forward and allowing the lumbar lordosis to flatten. But if you emphasize exhalation with the abdominal muscles by pressing inward gently from below, you will quickly notice that this prevents the posture from deteriorating. Make the movements as subtle as you can so that someone watching you from the side would see the posture improve only over a period of several minutes. As soon as you are straight, still, and relaxed, do your meditation.

At the end of a period of sitting you should feel alert, centered, and wide awake. But you will probably not feel like jumping up immediately to resume your daily activities. Now you can relax. Simply "let go" and allow gravity to bring you forward until the forehead is resting on the floor. You can clasp your hands behind your back, or you can make them into fists and place one on each side of the groin. Rest there for as long as you like (fig. 10.17). This is yoga mudra—the symbol of yoga.

KNOWER OF THE VEIL

Two veils are said to stand between the student and reality, between the student and what the yogis call enlightenment—the veil of body and the veil of mind. Patanjali's *Yoga-Sutras* begins with three terse aphorisms: "Now an exposition of yoga;" then, "Yoga is the inhibition of the modifications of mind;" and last, "Then the seer rests in his own true nature." Few students are prepared to make use of such stark and mysterious statements, but all accomplished yogis honor them. They know that Patanjali is not referring to hatha yoga but to meditation, and they know that to shred the veil of mind the aspirant must sit steadily and comfortably for long periods of time. To prepare for this the practices of hatha yoga are supreme. The expert in hatha yoga is even called a "knower of the veil." And within that realm alone there is much to know.

Figure 10.17. Yoga mudra, the symbol of yoga, and a fitting end to a practice of meditation.

"We are taught how to move and behave in the external world, but we are never taught how to be still and examine what is within ourselves. At the same time, learning to be still and calm should not be made a ceremony or a part of any religion; it is a universal requirement of the human body. When one learns to sit still, he or she attains a kind of joy that is inexplicable. The highest of all joys that can ever be attained or experienced by a human being can be attained through meditation. All the other joys in the world are transient and momentary, but the joy of meditation is immense and everlasting."

— Swami Rama, in *Meditation and Its Practice*, p. 9.

GLOSSARY

Abdomen The front and lateral surface of the torso between the rib cage and the pelvis.

Abdominal Has to do with the region below the rib cage and above the pelvis, to the cavity below the respiratory diaphragm and above the pelvic cavity, to the organs within that cavity, and to the sheets of superficial muscles that surround the cavity.

Abduction The swinging of the arm or thigh out from the midsagittal plane of the body; also applies to the lateral movement of the scapula; opposite of adduction.

Abductor Any muscle that swings an extremity out from the midsagittal plane of the body.

Acetabulum The socket of the hip joint; forms a deep concavity that accommodates the head of the femur.

Acute angles Those between 0° and 90°.

Adduction The drawing together of the thighs, the dropping down of the arm from an outstretched position, and the medial movement of the scapula; opposite of abduction.

Adductor Any muscle that pulls certain segments of an extremity in from a relatively outstretched position.

Aerobic Literally, the presence of oxygen; during aerobic exercise, the heart and lungs are mildly stressed but are capable of supplying the muscles with enough oxygen to keep up with their requirements; contrast with anerobic.

Agonist A muscle that assists another one in some action; opposite of antagonist.

Anal triangle The anatomical region defined by three lines, one between the two ischial tuberosities (this boundary is shared with the urogenital triangle), and two between each of those bumps and the tip of the sacrum and coccyx; see also urogenital triangle.

Anatomical position Standing upright with the forearms supinated (palms facing the front).

Anerobic (or anaerobic) Literally, an absence of oxygen; an anerobic muscular effort requires more oxygen than can be supplied (over a short period of time) to a muscle or group of muscles by the cardio-respiratory system, leading to an oxygen debt in the muscle that is paid after the activity ceases; contrast with aerobic.

Ankylosis Complete or partial fusion of a joint which had once been a movable synovial joint.

Annulus fibrosus The circular connective tissue sheath that makes up most of an intervertebral disk.

Antagonist A muscle that opposes the function of another muscle on the opposite side of a bone, often by restraining movement at a joint; opposite of agonist.

Anterior A directional term meaning toward the front of the body while standing in an upright posture with the palms of the hands facing forward.

Appendicular skeleton The bones of the upper and lower extremities; see upper and lower extremities for the specific bones; contrast with axial skeleton.

Arm The segment of the upper extremity between the shoulder and the elbow.

Articulate To form a joint.

Articular processes Little protrusions of bone that form joints between adjacent vertebral arches.

Articulation Any joint, either a movable synovial joint, a fibrous joint, or a suture.

Atrium, or atria (plural) The chambers of the heart on the right and on the left that receive blood from the vena cava (right side) and the pulmonary vein (left side).

Autonomic nervous system, or ANS The part the nervous system that controls internal organs and that operates more or less autonomously; see also sympathetic and parasympathetic nervous systems.

Axial Has to do with a central axis, either the axial skeleton of the body or an axial line through some segment of an extremity.

Axial skeleton The skeletal support for the body as a whole and for the head, neck, and torso in particular; includes the skull (cranium and mandible), seven cervical vertebrae, twelve thoracic vertebrae, twenty-four ribs (twelve on each side), the sternum, five lumbar vertebrae, the sacrum, and the coccyx; contrast with appendicular skeleton.

Axilla The region of the armpit and the deep structures within.

Axon The cellular process of a neuron that transmits nerve impulses from the cell body to the axon terminals.

Bone One of the connective tissues; contains extracellular bone salts (hydroxyapatite) that make it hard, extracellular connective tissue fibers that make it strong, and living cells that support the extracellular components.

Brachial plexus A plexus of peripheral nerves that supplies the upper extremity; derived from spinal nerves between C4 and T1; see also plexus.

Brain The brain stem, the cerebellum, and the cerebrum; see the separate listings.

Brain stem The continuation of the spinal cord into the brain.

Cartilage One of the connective tissues; contains a specialized extracellular matrix that gives it its characteristic strong but rubbery nature.

Cartilagenous joint A joint without a cavity or synovial fluid; held together by fibrous cartilage.

Cell body The part of the neuron that contains the nucleus; supports the rest of the cell metabolically.

Central nervous system, or CNS The brain and spinal cord; see the separate listings; contrast with peripheral nervous system.

Cerebellum A large segment of the brain located posteriorly to the brain stem and inferiorly to the cerebrum; affects both unconscious reflex activity and willed motor activity.

Cerebrum The largest segment of the brain, composed of the right and left cerebral hemispheres (the right brain and left brain); necessary for willed activity, for conscious appreciation of sensations such as pain, temperature, touch, pressure, vision, audition, taste, and smell, and for the higher functions of the intellect.

Cervical Has to do with the neck, its seven vertebrae and eight bilateral cervical nerves, and its anteriorly convex cervical lordosis.

Circumduction The movement of the arm or thigh in its most extreme circular course.

Coccyx The lowermost segment (the tail bone) of the axial skeleton; composed of a few tiny bones at the bottom of the sacrum.

Concentric Has to do with an ordinary muscle contraction in which the belly of the muscle as a whole shortens; opposite of eccentric.

Connective tissue One of the four primary tissues; includes loose connective tissue, fasciae, tendons, ligaments, cartilage, bone, and blood; see also epithelial, muscle, and nervous tissues.

Contraction Shortening of the individual muscle fibers in a muscle, or activity of the entire muscle; includes eccentric lengthening, concentric shortening, and isometric activity.

Coronal or frontal plane Applies to a vertical, side to side plane, or to some structure that runs in that plane; contrast with sagittal and transverse planes.

Costal Having to do with the ribs.

Counternutation Movement at the sacroiliac joints; moderate slippage in which the top of the sacrum rotates to the rear in relation to the ilia, the coccyx rotates forward, the ilia spread laterally, and the ischial tuberosities move medially; opposite of nutation.

Cross-sectional or transverse plane A plane perpendicular to the vertical axis of the body or of an extremity.

Crural Having to do with the crus, or the crura.

Crus, or crura (plural) Literally "leg"; the portion of the diaphragm (on both the right and left sides) that attaches between the central tendon of the diaphragm and the lumbar spine.

Deep A directional term meaning beneath the surface of the body; opposite of superficial.

Dendrites The cellular processes of neurons that are specialized to receive information from other neurons (from interneurons and motor neurons) or from the environment (from sensory neurons).

Denervate (or more rarely, enervate) To isolate from the nerve supply; usually applied to cutting a motor nerve to a muscle.

Dens See odontoid process.

Diaphragm The respiratory diaphragm and the pelvic diaphragm; see those listings.

Distal A directional term meaning toward the fingers or toes; opposite of proximal.

Diastolic Has to do with the time between ventricular contractions when the right and left ventricles are filling with blood; if the blood pressure is 120/80, the diastolic pressure is 80 mm Hg; opposite of systolic.

Eccentric lengthening A type of muscle contraction in which the belly of the muscle as a whole lengthens but at the same time resists lengthening, always against the force of gravity; opposite of concentric.

Enteric nervous system The component of the autonomic nervous system that lies embedded in the wall of the gut and that allows the gut to function independently.

Epithelial tissue One of the four primary tissues; forms the boundary between the external world and the internal environment of muscle and connective tissue; see also connective tissue, muscle tissue, and nervous tissue.

Extension The unfolding of a limb or straightening of the trunk from a flexed position; opposite of flexion.

Extensor muscles Act generally to extend, or unfold, the joints of the upper and lower extremities; antagonists to the flexors.

Extra- Outside of, especially outside of cells; includes everything non-cellular deep to the outermost surface of the skin; contrast with inter- and intra-.

Extremity, or appendage Either the upper or lower extremity; contains the appendicular skeleton, as opposed to the axial skeleton; see also upper extremity, lower extremity, appendicular skeleton, and axial skeleton.

Facet A small smooth region of a bone covered with articulating cartilage for forming a joint (always a gliding joint) with another bone.

Facilitation Input from axon terminals of other neurons that increases activity in the neurons on which the axon terminals impinge; produces an increase in the number of nerve impulses per second that travel down the axon of the affected neuron; opposite of inhibition.

Facilitatory Tending to induce facilitation.

Fascia Sheets of connective tissue (fibers and cells) that support and give form to organs and muscles throughout the body.

Fibrocartilage A specialized connective tissue containing a combination of connective tissue fibers and cartilage.

Fibrous Has to do with joints of that type, to dense connective tissue that makes up the capsule of joints, and to the substance of fasciae that surrounds muscle.

Fibrous joint One in which no joint cavity is present and where heavy concentrations of fibrous connective tissue hold the joint together.

Flexion The folding in of a limb; opposite of extension.

Flexor muscles Act generally to fold the joints; antagonists to the extensors.

Forearm The segment of the upper extremity between the elbow and the wrist.

Frontal plane See coronal plane.

Hip bone See pelvic bone.

Hyaline cartilage Found on the surface of long bones of the extremities; facilitates movement of synovial joints.

Hyperextension Extension beyond the norm.

Hyperventilation Overbreathing to the extent of creating a subjective level of discomfort; results in elevated levels of blood oxygen and lowered levels of blood carbon dioxide.

Hypoventilation Underbreathing; results in lowered levels of blood oxygen and abnormally high levels of carbon dioxide.

Ilium One of three segments of the hip bone; the large wing-shaped portion that articulates posteriorly with the sacrum to complete the pelvic bowl.

Inferior A directional term meaning toward the feet or below the head; opposite of superior.

Inferior articulating processes Small bony processes that extend inferiorly from the junction of the pedicle and lamina on both sides of a vertebral arch; in combination with the superior articulating processes of the next lower vertebra, the superior articulating processes form small, gliding-type facet joints.

Inferior pubic rami Wing-like extensions of the pubic bones that run inferiorly, posteriorly, and laterally from the region of the pubic symphysis to the ischial tuberosities.

Inhibition Input to neurons from axon terminals that decreases activity in the neurons on which the axon terminals impinge; produces a decrease in the number of nerve impulses per second that travel down the axon of the affected neuron; opposite of facilitation.

Inhibitory Tending to induce inhibition.

Innervate The nerve supply to a structure, as when a motor nerve innervates a muscle.

Inter- Between, especially between cells, as in intercellular substances or spaces; contrast with extra- and intra-.

Intercostal nerves Branches of spinal nerves T1-12 that supply intercostal (between the ribs) muscles and abdominal muscles.

Interneuron or association neuron Any neuron that is interposed between a motor neuron and a sensory neuron.

Intervertebral disks Cylindrically-shaped segments of fibrocartilage that link adjacent vertebral bodies; they contain a central liquid core (the nucleus pulposus) and a superficial annulus fibrosis.

Intervertebral foramina Bilateral gaps in the spine through which spinal nerves emerge carrying mixed spinal (motor and sensory) nerves; motor fibers pass peripherally to make synaptic contact with skeletal muscle, smooth muscle, and glands; sensory fibers pass centrally to bring sensory information into the central nervous system.

Intra- Inside of; refers to nuclei and other organelles within the cell; contrast with extra- and inter-.

Ischial tuberosities The sitting bones; see also ischium.

Ischium One of the three fused-together components of the pelvic bone, the others being the ilium and the pubis; its most inferior surface is the ischial tuberosity.

Isometric contraction A type of muscle contraction in which tension in the muscle increases (as a result of shortening of the individual muscle fibers) but the length of the muscle stays the same, as would happen if you were to try to lift a locomotive.

Isotonic contraction A type of muscle contraction in which individual muscle fibers produce movement about a joint, as in walking, running, and nearly all athletic endeavors.

Joint The region of apposition of two or more bones; same as articulation; includes cartilaginous, fibrous, and synovial joints; see individual listings for details.

Kyphosis A posteriorly convex curvature of the spinal column; the clinical meaning of kyphosis is an excessive thoracic curvature (humpback) that develops in the upper back, often in association with osteoporosis.

Laminae The flattened segments of the vertebral arch between the transverse processes and the spinous process.

Lateral A directional term meaning to the side, using a midsagittal plane as a point of reference.

Leg The segment of the lower extremity between the knee and the ankle.

Lordosis An anteriorly convex curvature of the spinal column; the clinical meaning is an excessive lumbar curvature (swayback) that develops in the lower back.

Lower extremity Includes the pelvic bone and hip joint, the thigh (with femur), knee joint, leg (with tibia and fibula), ankle (with tarsal bones), and feet (with metatarsals and phalanges), as well as all associated muscles, nerves, blood vessels, and skin.

Lumbar Having to do with the lower back, its five vertebrae and five bilateral lumbar nerves, and its anteriorly convex lumbar lordosis.

Lumbosacral plexus A plexus of nerves from L1 to S5 that supplies the lower abdomen, perineum, and lower exremities; see also plexus.

Matrix Substance, material, or content; usually applied here to some extracellular component of connective tissues.

Medial A directional term meaning toward the mid-line, or closer to the midsagittal plane than some other structure.

Meniscus, or semilunar cartilage The medial and lateral menisci are incomplete, donut-shaped wafers of fibrocartilage that cushion the knee joint.

Motor Has to do with output from the motor neurons (located within the central nervous system) to skeletal muscles, smooth muscle, cardiac muscle, and glands throughout the body; contrast with sensory.

Motor neuron One of three classes of neurons (the others are interneurons and sensory neurons); innervates skeletal muscle fibers; counterpart to sensory neurons.

Motor unit A motor neuron (including its dendrites and axon) plus all of the muscle fibers that it supplies.

Muscle fiber, or muscle cell The individual cells in muscle tissue.

Muscle tissue One of the four primary tissues; includes cardiac muscle in the heart, smooth muscle in the walls of internal organs, and skeletal muscle; see also epithelium, connective tissue, and nervous tissue.

Myotatic stretch reflex A reflex contraction of a muscle that occurs as a result of dynamic stretch.

Nervous tissue One of the four primary tissues; specialized for communication; see also epithelium, connective tissue and muscle tissue.

Neuron The genetic, anatomical, and functional unit of the nervous system; a cell that is specialized to receive information from the environment or other cells and to transmit information to other sites, frequently long distances; see motor neuron, interneuron, and sensory neuron.

Nucleus pulposus The liquid core of intervertebral disks; moves posteriorly within the disk during a forward bend, anteriorly during a backbend, to the right when bending left, to the left when bending right, and is compressed during a twist.

Nutation Movement at the sacroiliac joints; a moderate slippage in which the top of the sacrum rotates forward in relation to the ilia, the coccyx rotates to the rear, the ilia move medially, and the ischial tuberosities move laterally; opposite of counternutation.

Obtuse angles Those between 90° and 180°.

Occipital The cranial bone situated behind the parietal bones; located just outside the occipital lobe of the brain.

Odontoid process, or dens The tooth-like process of the axis (C2) around which the atlas (C1) rotates.

Parasympathetic, or vegetative nervous system Supports the day-to-day functioning of internal organs on an individual basis; contrast with sympathetic nervous system.

Pedicles Segments of vertebral arches; the short columns of bone that extend posteriorly from the vertebral bodies and that continue into the flatter vertebral laminae (which complete the vertebral arch posteriorly).

Pelvic Refers to the two hip bones, to the cavity that is continuous with and below the abdominal cavity, to the diaphragm that defines the base of the torso, or simply to the region of the body vaguely below the abdomen and above the thighs.

Pelvic, or hip bone A single bone (one on each side of the body) formed from three separate bones in the embryo (the ilium, the ischium, and the pubis); the two pelvic bones together with the sacrum form the pelvic bowl.

Pelvic diaphragm A combination of fasciae and muscle that closes off the base of the pelvic bowl and supports the abdominopelvic viscera; the deepest layer of the perineum, on which the genitals are superimposed externally.

Pelvis Includes both pelvic bones; forms the base of the torso, and articulates with the axial skeleton at the two sacroiliac joints.

Pericardial Has to do with the cavity (a potential space only) around the heart whose outer boundary is the fibrous pericardium and which contains a small amount of slippery fluid that allows the heart to expand and contract without trauma; pericardial cavity is comparable to the peritoneal cavity in the abdominopelvic cavity and the pleural cavity that surrounds the lungs.

Pericardium The thin layers of tissue that line the outer surface of the heart (visceral pericardium) and the inner surface of the pericardial cavity (parietal pericardium); also the fibrous pericardium, a heavy connective tissue sack that surrounds the heart and pericardial cavity collectively; see also peritoneum and pleura.

Perineal Having to do with the perineum.

Perineum A diamond-shaped region whose borders are the bottom of the pubic symphysis, the inner borders of the inferior pubic rami and ischial tuberosities, and the sacrotuberous ligaments, which extend between the ischial tuberosities and the inferior tip of the sacrum and coccyx; contains the pelvic diaphragm and all the urogenital structures located within these boundaries; the more common definition is the small region between the anus and the genitals.

Peripheral nervous system, or PNS Includes motor and sensory roots of spinal nerves, spinal nerves, and autonomic plexuses and ganglia, that is, all parts of the nervous system except the brain and spinal cord; contrast with central nervous system.

Peritoneal Has to do with the cavity (a potential space only) between the internal organs of the abdomen and pelvis.

Peritoneum The lining of the peritoneal cavity; includes visceral peritoneum lining the internal organs and parietal peritoneum lining the inner aspect of the body wall; see also pericardium and pleura.

Peroneal nerve (common peroneal nerve) Arises (with the tibial nerve) from the sciatic nerve; has two main branches, the deep peroneal nerve and the superficial peroneal nerve; the latter can be traumatized by sitting intemperately on a hard surface in the adamantine pose.

Phrenic nerve Originates from C3-5; supplies the respiratory diaphragm.

Pleura The lining of the pleural cavity; visceral pleura lines the surfaces of the lungs and parietal pleura lines the inner aspect of the body wall and the upper surface of the respiratory diaphragm; see also peritoneum and pericardium.

Pleural Has to do with the cavity (a potential space only) between the lateral surfaces of the lungs and the inner wall of the chest, and between the base of the lungs and the upper surface of the diaphragm.

Plexus An intertwining and mixing of nerves; the brachial plexus supplies the upper extremity, and the lumbosacral plexus supplies the lower extremity.

Posterior A directional term meaning toward the back of the body while standing in an upright posture with the palms of the hands facing forward; opposite of anterior.

Process A small extension of bone, not as long and prominent as a ramus and not as robust as a trochanter; in the vertebral column, the superior articulating process of one vertebra articulates with the inferior articulating process of the next higher vertebra, forming a small facet joint.

Pronation Rotation of the wrist and hand with reference to the elbow; if you stand and face the palms to the rear, the forearms are pronated; the opposite of supination.

Proximal A directional term referring to portions of the extremities relatively nearer the pelvis and chest than the fingertips and toetips; opposite of distal.

Pubic symphysis A fibrocartilaginous joint that is the site of union of the two pelvic bones.

Pubis, or pubic bone One of the three components of the pelvic bone; also, the superficial region in front of the pubic symphysis, as in the mons pubis.

Pulmonary Having to do with the lungs. The pulmonary circulation is the circuit of blood from the heart (right ventricle) to the lungs and back to the heart (left atrium); counterpart to systemic circulation.

Radius One of the two bones of the forearm; located laterally (on the thumb side) in the anatomical position; see also ulna.

Ramus (plural rami) Literally means "branch."

Rectus Straight; applied here to the side-by-side, straight up-and-down abdominal muscles, or to the rectus femoris (the straight head of the quadriceps femoris).

Respiratory diaphragm The dome-like sheet of muscle that spans the torso between the chest and the abdomen; its costal portion attaches to the base of the rib cage, and its crural portion (or crus) attaches to the lumbar spine.

Right angle A 90° angle.

Sacral Has to do with the sacrum, as in the posteriorly convex sacral kyphosis and the five bilateral sacral nerves.

Sacrum, or the sacred bone The lowermost major segment of the spine; articulates with the ilium on each side at the sacroiliac joints; five pairs of spinal nerve exit the sacrum.

Sagittal plane A plane that runs through the body in an up-and-down and front-to back orientation; a midsagittal plane bisects the body down the middle, and a parasagittal plane is parallel to the midsagittal plane but to one side; contrast with coronal and transverse planes.

Sciatic nerve The largest nerve of the lumbosacral plexus; its two large branches are the tibial nerve, which supplies the calf muscles on the posterior side of the leg, and the common peroneal nerve, which supplies muscles on the anterior side of the leg.

Semilunar cartilage See meniscus.

Sensory Has to do with conscious and unconscious input from the body to the central nervous system; conscious input includes pain, temperature, touch, pressure, vision, audition, taste, smell; unconscious input includes information for equilibrium and sensory aspects of the autonomic nervous system; also includes associated systems within the central nervous system, especially those that carry the sensory information to consciousness; contrast with motor.

Sensory neuron One of three classes of neurons (the others are interneurons and motor neurons); synapses with interneurons that carry incoming sensory information to consciousness in the cerebral cortex or that take part in unconscious reflexes; counterpart to motor neurons.

Sesamoid bone One that is within a tendon; the largest sesamoid bone in the body is the patella.

Sitting bones See ischial tuberosities.

Solar (coeliac) plexus A major nerve plexus of the autonomic nervous system that lies against the aorta between the diaphragm and the bifurcation of the aorta (illustrated on the cover of this book).

Somatic nervous system Innervates skeletal muscle; receives sensory information from somatic senses such as pain, touch, vision, and audition.

Spinal cord The part of the central nervous system that extends from the brain stem into the vertebral canal; includes twelve thoracic segments, five lumbar segments, five sacral segments, and one coccygeal segment.

Spinal nerve A mixed (motor and sensory) nerve that exits from the spinal cord segmentally in relation to each vertebra; contains both somatic and autonomic components.

Spinous processes Extend posteriorly from each vertebral arch.

Suboccipital The region just inferior to the base of the skull posteriorly.

Superficial A directional term meaning toward the surface; opposite of deep.

Superior A directional term meaning above the feet or toward the head;

opposite of inferior.

Superior articulating processes Small bony processes that extend superiorly from the junction of the pedicle and lamina on both sides of a vertebral arch; they form facet joints with the next lower vertebra; see also processes and inferior articulating processes.

Superior pubic rami Wing-like extensions of the pubic bones that run superiorly, posteriorly, and laterally from the region of the pubic symphysis to the ilia.

Supination Rotation of the wrist and hand with reference to the elbow; if you stand and face the palms to the front, the forearms are supinated; opposite of pronation.

Suture A fibrous joint that unites the flat bones of the cranium.

Sympathetic nervous system Supports the whole-body "fight-or-flight" function; gears up certain internal organs and the musculoskeletal system for emergencies; contrast with parasympathetic nervous system.

Symphysis A cartilaginous joint; see also pubic symphysis and intervertebral disk.

Synapse The point of junction between the axon terminal of one neuron and its target, usually either a muscle cell or another neuron.

Synovial fluid A slippery fluid which lubricates synovial joints, facilitating smooth and easy movement of the articulating hyaline cartilage.

Synovial joint A slippery movable joint that contains synovial fluid, a synovial membrane that secretes synovial fluid, slippery articular cartilage on the ends of long bones that contact one another in the joint, and a fibrous joint capsule that protects the surfaces and retains the synovial fluid in the vicinity of the articulating surfaces.

Systemic Refers to the body as a whole; the systemic circulation begins at the left ventricle, leaves the heart in the aorta, flows to the capillaries of the body, and returns to the heart (right atrium) by way of the vena cava; counterpart to pulmonary circulation.

Systolic Has to do with the time during which the ventricles are contracting; if the blood pressure is 120/80, the systolic pressure is 120 mm Hg; opposite of diastolic.

Thigh The segment of the lower extremity between the hip and the knee.

Thoracic Has to do with the thorax, or chest; includes twelve vertebrae and twelve bilateral thoracic nerves, the thoracic cage with twelve ribs on each side and the sternum in front, the posteriorly convex thoracic kyphosis, and the thoracic cavity; this cavity in turn contains the pericardium, heart, and pericardial cavity, as well as the lungs and pleural cavities.

Transverse, or cross-sectional plane A plane through the body that extends both from front to back and from side to side; contrast with coronal and sagittal planes.

Transverse process The bony protuberance (one on each side of each vertebra) that extends laterally from the vertebral arch at the junction of the lamina and the pedicle; see also process.

Trochanter   A large bony prominence that represents the sites of muscle attachments.

Tuberosity   A bump on a bone that represents the sites of muscle attachments; see also ischial tuberosity.

Ulna   One of the two bones of the forearm, located medially (on the little finger side) in the anatomical position; see also radius.

Upper extremity   Includes the clavicle, scapula, arm (with humerus), elbow joint, forearm (with radius and ulna), wrist (with carpal bones), hand (with metacarpals and phalanges), as well as all associated muscles, nerves, blood vessels, and skin.

Urogenital triangle   The anatomical region defined by three lines, one between the ischial tuberosities (this boundary is shared by the anal triangle), and two between each of those bumps and the inferior border of the pubic symphysis; see also anal triangle.

Vasoconstriction   Constriction of blood vessels (usually small arteries and arterioles) caused by sympathetic nervous system input to smooth muscle surrounding the vessel; contrast with vasodilation.

Vasodilation   Dilation of blood vessels (usually small arteries and arterioles) caused by biochemical factors and diminished sympathetic nervous system input to the smooth muscle surrounding the vessels; contrast with vasoconstriction.

Ventricles   The chambers from which blood is pumped out of the heart; blood is pumped into the pulmonary circuit on the right side and into the systemic circuit on the left side.

Vertebra   The bony unit of the vertebral column, or the spine; we have seven cervical vertebrae (C1-7), twelve thoracic vertebrae (T1-12), five lumbar vertebrae (L1-5), a single fused sacrum with five segments (S1-5), and a rudimentary coccyx; each vertebra contains a vertebral body, a vertebral arch, transverse processes, a spinous process, and superior and inferior articulating processes; see the individual listings.

Vertebral arch   Made of up two pedicles and two laminae which meet posteriorly to complete the arch; see also vertebral body, pedicles, laminae, and the inferior and superior articulating processes.

Vertebral body   The cylindrically-shaped portion of the vertebra which is separated from its neighbors (one below and one above) by intervertebral disks; see also vertebral arch.

Vertebral canal   The tubular portion of the vertebral column (just behind the vertebral bodies) that houses the spinal cord.

Visceral   Having to do with the viscera (internal organs) and other related structures such as smooth muscle, cardiac muscle, and glands.

ADDITIONAL SOURCES

Alter, Michael J. *Science of Flexibility*. 2nd ed. Champaign, IL: Human Kinetics, 1996.

Anderson, Sandra, and Rolf Sovik. *Yoga, Mastering the Basics*. Honesdale, PA: Himalayan Institute Press, 2000.

Basmajian, J. V. *Muscles Alive: Their Functions Revealed by Electromyography*. 5th ed. Baltimore: Williams & Wilkins, 1985.

Bengali Baba. *The Yogasutra of Patanjali with The Commentary of Vyasa*. Delhi: Motilal Banarsidass Publishers, 1976.

Bernard, Theos. *Hatha Yoga*. London: Rider, 1982.

Berne, Robert M., and Matthew N. Levy. *Cardiovascular Physiology*. 8th ed. Philadelphia: Mosby, 2001.

Brodal, A. *Neurological Anatomy in Relation to Clinical Medicine*. 3rd ed. New York: Oxford University Press, 1981.

Burke, David. Spasticity as an Adaptation to Pyramidal Tract Injury. *Adv. Neurol.* 47:401-422, 1988.

Calais-Germain. *Anatomy of Movement*. Seattle: Eastland Press, 1993.

DeTroyer, Andre, and Stephen H. Loring. Action of the respiratory muscles, Ch. 26, 443–461. In: Handbook of Physiology, Section 3, Vol. III, *Mechanics of Breathing, Part two*. Bethesda, MD: American Physiological Society, 1986.

DeTroyer, Andre. Mechanics of the Chest Wall Muscles, Ch. 6, 59–73. In: Miller, Alan D., Armand L. Bianchi, and Beverly P. Bishop, eds. *Neural Control of the Respiratory Muscles*. New York: CRC Press, 1997.

Dillard, Annie. *The Writing Life*. New York: Harper Perennial, 1989.

Gershon, Michael D. *The Second Brain: The Scientific Basis of Gut Instinct and a Groundbreaking New Understanding of Nervous Disorders of the Stomach and Intestine*. New York: HarperCollins, 1998.

Hale, Robert Beverly, and Terence Coyle. *Albinus on Anatomy*. New York: Dover Publications, 1988 (originally published by Bernard Siegfried Albinus in the mid-1700s).

Heck, J. G., ed. *Heck's Pictorial Archive of Nature and Science*. New York: Dover Publications, 1994 (originally published by Rudolph Garrigue, Publisher, New York, in 1851).

Hollingshead, W. Henry. *Functional Anatomy of the Limbs and Back*. 6th ed. Philadelphia: W. B. Saunders, 1991.

Iyengar, B. K. S. *Light on Yoga*. New York: Schocken, 1977.

Jackson, C. M., ed. *Morris's Human Anatomy: A Complete Systematic Treatise by English and American Authors*. 5th ed. Philadelphia: P. Blakiston's Son and Co., 1914.

Jami, Lena. Golgi Tendon Organs in Mammalian Skeletal Muscle: Functional Properties and Central Actions. *Physiol. Rev.* 72:623–666, 1992.

Kapandji, I. A. *The Physiology of the Joints.* 3 vols. Edinburgh: Churchill Livingstone, 1970.

Kuvalayananda. *Popular Yoga Asanas.* Rutland, VT: Charles E. Tuttle, 1971.

Lumb, Andrew B. *Nunn's Applied Respiratory Physiology.* 5rd ed. London and Boston: Butterworth Heinemann, 2000.

MacConaill, M. A., and J. V. Basmajian. *Muscles and Movements: A Basis for Human Kinesiology.* 2nd ed. Huntington, NY: Robert E. Krieger, 1977.

Morgan, Elaine, *The Scars of Evolution: What Our Bodies Tell Us About Human Origins.* New York and Oxford: Oxford University Press, 1994.

Norkin, Cynthia C., and Pamela K. Levangie. *Joint Structure and Function: A Comprehensive Analysis.* 2nd ed. Philadelphia: F. A. Davis, 1992.

Otis, Arthur B. History of respiratory mechanics, Ch. 1, 1–12. In: In: Handbook of Physiology, Section 3, Vol. III, *Mechanics of Breathing, Part one.* Bethesda, MD: American Physiological Society, 1986.

Pert, Candace B. *Molecules of Emotion: Why You Feel the Way You Feel.* New York: Scribner, 1997.

Proctor, Donald F., et al. *A History of Breathing Physiology.* New York: Marcel Dekker, 1995.

Rama, Sri Swami, *Exercise Without Movement.* Honesdale, PA: Himalayan International Institute of Yoga Science and Philosophy of the U.S.A., 1984.

Rama, Swami, *Meditation and Its Practice.* Honesdale, PA: Himalayan International Institute of Yoga Science and Philosophy of the U.S.A., 1992.

Sappey, Philip C. *Traite D'Anatomie Descriptive.* Quatrieme ed. Paris: Lecossier et Babe, Libraires-Editeurs, 1889.

Saraswati, Swami Karmananda, ed. *Hatha Yoga Pradipika.* Bihar, India: Bihar School of Yoga, 1985.

Sivananda. *Yoga Asanas.* 19th ed. Tehri-Garhwal: The Divine Life Society, 1993.

Smith, Laura K., Elizabeth L. Weiss, and L. Don Lehmkuhl. *Brunnstrom's Clinical Kinesiology.* 5th ed. Philadelphia: F. A. Davis Company, 1996.

Thomson, Allen, Edward Albert Schafer, and George Dancer Thane. *Quain's Elements of Anatomy.* 9th ed. New York: William Wood and Co., 1882.

Tigunait, Pandit Rajmani. *Inner Quest: The Path of Spiritual Unfoldment.* Honesdale, PA: Yoga International Books, 1995.

Walford, Roy, and Lisa Walford. *The Anti-Aging Plan: Strategies and Recipes for Extending Your Healthy Years.* New York and London: Four Walls Eight Windows, 1994.

Williams, Peter L., et al., eds. *Gray's Anatomy: The Anatomical Basis of Medicine and Surgery.* 38th British ed. New York and London: Churchill Livingstone, 1995.

ACKNOWLEDGEMENTS

Some of the material in this book first appeared in *Yoga International* magazine in the following articles, authored by David Coulter: "Building a Foundation" issue #1, "Postures and Breathing" #2, Relaxation and Energy" #3, "For Clarity of Mind" #4, "Anatomy of a Headstand" #5, "The Value of Being Upside Down" #6, "Siddhasana" #7, "Straight from the Hip" #9, "Letting Go" #10, "Back Repair" #12, "Moving Gracefully" #19, "Moving Slowly" #20, "Self-Preservation" #21, "Physiology of Bhastrika" #22, "All Knotted Up" #23, "Give Yourself a Lift" #26, ands "The Master Lock" #27. I also wrote a first draft of one more article ("Don't Pause" #28), but this one was completed by Mick Grady, who in the end was rightfully listed as the sole author. Some of my original commentary that survived in that piece, however, was re-worked for chapter 2 of this book. I thank *Yoga International* for permission to use materials from all of these articles in the present work.

Most of the anatomical illustrations were derived from works now in the public domain, namely Philip Sappey's 1888-1889 edition of *Traite D'Anatomie Descriptive*, one plate (fig. 2.8) from *Albinus on Anatomy* (Dover Publications), seven plates (figs. 8.8–14) from *Heck's Pictorial Archive of Nature and Science* (Dover Publications), a few plates from the ninth edition of *Quain's Elements of Anatomy* (1882), and several more from the sixth edition of *Morris's Human Anatomy* (1914). See the individual illustrations for specific credits that honor the moral rights of the author according to the doctrine of *droit moral* in Berne convention rules.

I thank many friends and colleagues who contributed their time and talents to this work. Thad Pawlikowski generated the charts for figures 2.13–14, 2.18, 2.21, 2.27, 3.18, 3.32, 8.21, and 9.15. Figures 2.1, 2.29a–e, 4.2, 8.1, 8.3, 10.1, 10.4a, and 10.13 were redrawn from numerous sources by Eben Dodd. Figure 4.8 was traced by the author from roentgenograms provided by Dr. Donald O. Broughton. Figure 6.2a was reprinted by permission from E. I. Kapandji's well-known *Physiology of the Joints*. The design and layout of the text, illustrations, and cover were done by the author, with encouragement, hints, and cautions from Lyle Olsen, Kamala Gerhardt, Jeanette Robertson, and especially Joyce Baronio, who also contributed the author's photograph.

Models for the postures, in descending order of the number of their poses, were: my son Jamie Coulter (chapters 2–10), Bill Boos (chapters 2 and 4–10), Mike Kerr (chapters 5 and 7–9), Brian Wind (chapters 1–3, 7–8, and 10), Joanne Wahrer (chapters 1, 4, and 6–9), Charles Crenshaw (chapters 3, 6, and 10), John Hanagan (chapter 6), Josh Hajicek (chapter 1), and Lyle Olson (chapter 10). I thank all of you for bringing this work to life, as well as down to earth. I also wish to thank the hundreds of hatha yoga teachers who have put up with my technical questions over the past three decades. Many of these teachers have had long association with the Himalayan Institute, and many others have been guest instructors from all over the world. Without their input this book would have had little or no legitimacy. I also thank Swami Bua, a centenarian who still teaches in Manhattan, for introducing me to the series of forty-eight twists and bends that are illustrated here—I think for the first time—in chapter 7.

The unfailing support of Swami Rama, the founder of the Himalayan Institute, and of Pandit Rajmani Tigunait, its spiritual head, created the atmosphere that made it possible to complete this work, along with a generous allowance of time for teaching, writing, and work on illustrations courtesy of the Institute directors. In addition, Deborah Willoughby, the President of the Institute and editor of Yoga International, initiated me in the early 1990s in the strictures of writing magazine articles on technical subjects for a general audience. Then at a critical moment in the mid-1990s, Madalasa Baum's enthusiasm and persistence in supporting the work generated the energy needed for lifting the project off the ground and for enlisting the editorial support of Anne Craig (who, as it happened, had already been put on notice by Swamiji in 1981 that she would do the editing).

Additional help on the manuscript came from many sources. Dr. Dale Buegel, an accomplished yogi and expert in manual medicine, read an early draft. Dr. Steve Rogers, who by good fortune is both a physician and a journalist, read the entire manuscript critically. Mick Grady, who is both a journalist and hatha yogi, commented on many individual sections, especially those related to breathing. Rolf Sovik and Sandy Anderson, who completed their own book on hatha yoga *(Yoga, Mastering the Basics)* a year ago, read the manuscript critically for content. Dr. Dalia Zwick, a professor of physical therapy, contributed her expertise on matters related to clinical kinesiology, Michael Alter, the author of *Science of Flexibility*, read and commented both on an early and a late draft, and Dr. Timothy McCall read the final draft in preparation for contributing the foreword. Thanks to all of them. And finally, I wish to thank my editor and friend, Anne Craig. The author, of course, assumes full responsibility for all errors of omission and commission that still remain.

While the text of *Anatomy of Hatha Yoga* remains the same for this second edition, a lot of work and thought has gone into colorizing the original images in order to make a more clear, informative, and aesthetic book. A year ago, when I decided to release a second edition, I did not realize how ambitious a goal I was undertaking or how much help I would need along the way. I was diagnosed with Parkinson's disease in 2001. As the disease has progressed, I've called upon my children, Jamie and Elina Coulter, to manage my affairs related to this endeavor. In addition to the continual support and guidance of my children, there are several others who have offered their expertise to help me complete the second edition. I would like to thank Andy Perez and Alexandra Keep for colorizing anatomical drawings as well as Timothy Thrasher for colorizing half-tone photos and redesigning the cover. I also wish to thank Alexandra for the challenging cover image, repeated on page 559. I appreciate Jeff Miller's efforts in formatting and finalizing this publication. Yoga for almost all of us is a work in progress, and I've continued to receive the critique and encouragement of many people. Thank you. Thank you. Thank you.

INDEX OF ANATOMICAL TERMS

Text citations are indicated in Century Schoolbook plain, and page references for anatomical drawings, diagrams, and photographs are indicated in italics. In both cases, boldfacing highlights the locations of especially useful references.

Breathing methods

INDEX OF PRACTICES

BIOGRAPHICAL SKETCH

David Coulter received a Ph. D. in anatomy from the University of Tennessee Center for the Health Sciences in 1968. From 1968 to 1986 he taught various microscopic, neuroscience, and elementary gross anatomy courses in the Department of Anatomy of the University of Minnesota (Medical School) in Minneapolis, MN. During that period he also served as a principal investigator for neuroscience research funded by the National Institutes of Health and the National Science Foundation. He next taught in the Department of Anatomy and Cell Biology at Columbia University College of Physicians & Surgeons (1986 to 1988), and since then has practiced and taught a style of bodywork called Ohashiatsu® in New York City and elsewhere. Dr. Coulter has been practicing yoga since 1974. He was initiated by Swami Veda (formerly Dr. Usharbudh Arya of Minneapolis, MN), trained under Swami Rama from 1975 to 1996, and studied under Pandit Rajmani Tigunait at the Himalayan Institute since 1988. From the inception of his interest in yoga, Dr. Coulter has been committed to correlating his understanding of the practices of that discipline with accepted principles of biomedical science. (Photo by Joyce Baronio.)